## PROGRESS IN FOOD SCIENCE AND TECHNOLOGY

Additional books in this series can be found on Nova's website under the Series tab.

Additional e-books in this series can be found on Nova's website under the e-book tab.

## BOTANICAL RESEARCH AND PRACTICES

Additional books in this series can be found on Nova's website under the Series tab.

Additional e-books in this series can be found on Nova's website under the e-book tab.

PROGRESS IN FOOD SCIENCE AND TECHNOLOGY

# PHYTOCHEMICALS

## OCCURRENCE IN NATURE, HEALTH EFFECTS AND ANTIOXIDANT PROPERTIE

PROGRESS IN FOOD SCIENCE AND TECHNOLOGY

# PHYTOCHEMICALS

## OCCURRENCE IN NATURE, HEALTH EFFECTS AND ANTIOXIDANT PROPERTIES

HAI-XUE KUANG
EDITOR

*New York*

For permission to use material from this book please contact us:
Telephone 631-231-7269; Fax 631-231-8175
Web Site: http://www.novapublishers.com



**Library of Congress Cataloging-in-Publication Data**

Phytochemicals : occurrence in nature, health effects and antioxidant properties / editor: Hai-Xue Kuang.
p. cm.
Includes index.
ISBN 978-1-62417-354-7 (hardcover)
1. Phytochemicals. 2. Phytochemicals--Health aspects. 3. Antioxidants. I. Kuang, Haixue.
QK861.P447 2012
572'.2--dc23 2012043092

*Published by Nova Science Publishers, Inc. † New York*

# Contents


**Preface** **vii**

**Chapter I** Phytochemicals as Potential Pharmacophores to Target Diabetes **1**
*Mahomoodally Mohamad Fawzi*

**Chapter II** Phytochemicals and their Potential Effects on Inflammatory Bowel Disease **25**
*Susana Gorzalczany*

**Chapter III** Bioactive Compounds in Whole Grain Cereals **39**
*Maria Teresa Pedrosa Silva Clerici and Elizabeth Harumi Nabeshima*

**Chapter IV** Antioxidant and Antimicrobial Activities of Berries Extracts of Five *Juniperus* Species under Juniperus Section from Turkey **55**
*Maria Fernanda Taviano, Andreana Marino, Ada Trovato, Rita De Pasquale and Natalizia Miceli*

**Chapter V** Phenolic Compounds, Health Effects and Antioxidant Properties of Berries **69**
*Aleksandra Konić-Ristić, Katarina Šavikin, Gordana Zdunić, Dejan Gođevac, Teodora Janković and Nebojša Menković*

**Chapter VI** Overview of the Chemical, Natural Occurrence and Antioxidant Activity of Terpenes **87**
*E. González-Burgos and M. P. Gómez-Serranillos*

**Chapter VII** Supercritical Extraction of Phloroglucinols from St. John's Wort (Hypericum Perforatum L.) – Process Optimization, Mathematical Modeling and Antimicrobial Activity **121**
*Sandra B. Glisic and Aleksandar M. Orlovic*

**Chapter VIII** Natural Antioxidants in Food **147**
*Gui-Fang Deng, Xiang-Rong Xu, Sha Li, Fang Li, Shan Wu and Hua-Bin Li*

**Chapter IX** Regulation of Inflammation by Dietary Phytochemicals: Current Perspectives **175**
*Marta González Castejón and Arantxa Rodriguez Casado*

**Chapter X** Antioxidative Phytochemicals and Skin Care **189**
*Frédéric Bonté and Gallic Beauchef*

**Chapter XI** Mineral, Flavonoid, and Fatty Acid Concentrations in Ten Diverse Lablab Purpureus (L.) Sweet Accessions **219**
*J. Bradley Morris, Michael A. Grusak, Ming Li Wang and Brandon David Tonnis*

**Chapter XII** Research Progress on Medicinal Plants of Urtica Linn. Genus **225**
*Zhang Yanxiu, Wang Ying, Ju Zhihe and Guan Feng*

**Chapter XIII** Research and Development on Chemical Constituents and Pharmacology of Acanthopanax Genus **247**
*Gaofeng Liu, Chao Zeng and Zhiping Lin*

**Chapter XIV** Simultaneous Determination of Five Triterpenoid Saponins in the Root of Caulophyllum Robustum Maxim by HPLC-ELSD **283**
*Guo-Yu Li, Yong-Gang Xia, Jun Liang, Yan-Hai Zhang and Hai-Xue Kuang*

**Chapter XV** Chemical Study and Medical Application of Saponins from the Genus Aralia **295**
*Hai-Xue Kuang, Yong-Gang Xia and Guang-Rui Mu*

**Chapter XVI** Chemical Constituents and Bioactivities of Plants of the Genus Hydrangea **317**
*Yang Chunjuan and Xiao Yang*

**Chapter XVII** Phytochemicals in Cancer Research: A Review of Their Biological Activities and Anticancer Potentials in Preclinical and Clinical Studies **357**
*Daniel L. Pouliquen and Marc Grégoire*

**Index** **395**

# Preface

The term "phytochemicals" refers to a wide variety of compounds with biological significance produced by plants. They are found in fruits, vegetables, beans, grains, and other plants. Scientists have identified thousands of phytochemicals, although only a small fraction have been studied closely, which have the potential to affect diseases such as cancer, stroke or metabolic syndrome. In recent years, plant medicine with important bioactivity has attracted the increasing attention of the researchers in the world. The book 《*Phytochemicals: Ocurrence in Nature, Health Effects and Antioxidant Properties*》 is devoted to the publication of original articles concerning phytochemicals and their biological and pharmacological effects. The spectrum of coverage is broad, encompassing extraction, separation, purification, structure identification, bioactivity and quantification of compounds in plant, the food sciences and agriculture. Early people confronted with illness and disease, discovered a wealth of useful therapeutic agents in the plant and animal kingdoms. In order to ensure a better understanding of the pharmacodynamic material base off these medical materials, we wrote this book. The authors consist of an international group of recognized researchers who possess abundant academic knowledge and research backgrounds due to their years of practicing phytochemistry. During this compilation process, we have consulted with a variety of specialists to collect advice. We have also referenced a large number of national and international documents that analyze and explore the subject based on current research results. I recommend the book for either those who have recently embarked on plant medicine research work or for those who are considering it.

In: Phytochemicals
Editor: Hai-Xue Kuang

ISBN: 978-1-62417-354-7


*Chapter I*

# Phytochemicals as Potential Pharmacophores to Target Diabetes

***Mahomoodally Mohamad Fawzi****

Department of Health Sciences, Faculty of Science,
University of Mauritius, Reduit, Mauritius

## Abstract

The sensible and liberal use of phytotherapy in the treatment and/or management of panoply of human diseases have always been supported by the ancient medicinal literature worldwide. Indeed, prior to the introduction of insulin therapy, the management of diabetes mellitus (DM) relied closely on dietary measures, which included the use of traditional plant therapies. The Papyrus Ebers of 1550 B.C. recommended a high-fiber diet of wheat grains for the management of diabetes. Currently more than 1200 species of organisms have been used ethnopharmacologically or clinically to treat/manage symptoms and complications associated to DM and panoply of reviews/reports of plants with known antidiabetic activity or traditional use as antidiabetic remedies have been published. In general, discussion pertaining to medicinal agents from plant tends to center around plant secondary metabolites or phytochemicals, i.e., non-ubiquitous constituents with no known essential role in the plant's metabolism. However, it has been postulated that bioactive plant secondary metabolites may play a pivotal role in chemical defense mechanisms and are likely molecules for the hypoglycemic agents in these plants. Recently, the growing interest of researchers in traditional medicine has led to discovery and establishment of biological activity of many potent phytochemicals possessing hypoglycemic activity (e.g. glycosides, lavonoides, terpenes, steroidal saponins, alkaloids, galactomannan gum, pepdidoglycans, polysaccharides, hydroglycans, guanidines, glycopeptides and flavonoids amongst others). A plethora of plant's derived bioactive compounds representing numerous phytochemicals has demonstrated activity consistent with their use in the traditional treatment/management of DM. Even the discovery of the widely used hypoglycemic drug, metformin has been developed from the

* E-mail address: f.mahomoodally@uom.ac.mu; mmfawzi@gmail.com.

traditional approach of using *Galega officinalis*, and which subsequently led to the synthesis of biguanides. Thus, phytochemical could be a potential source of anti-diabetic pharmacophores and template molecules geared towards alleviating chronic complications of diabetes. This present book chapter endeavors to present an overview on how phytochemicals, particularly from common traditionally used medicinal plants, may be helpful in glucose homeostasis thereby reducing the risk of DM and related complications.

**Keywords:** Phytochemicals, diabetes mellitus, medicinal plants

# Introduction

The increase in the prevalence of Diabetes mellitus (DM) across the world has become an important public health problem given that this disease ranks among the leading causes of blindness, renal failure and lower limb amputation, besides being a significant risk factor for coronary heart disease and stroke. From recent data on the frequency of diabetes in different countries, a worldwide prevalence among adults aged 20-79 years increasing from 285 million in 2010 (6.4%) to 439 million in 2030 (7.7%) has been estimated. It is now clearly established that DM is the world's largest endocrine disease involving metabolic disorders of carbohydrate metabolism characterized by fasting elevation of blood glucose level. While the fundamental mechanism underlying abnormality is a net (relative or absent) deficiency of the hormone insulin, the complications of chronically high serum glucose are devastating to the individual. If not managed properly, DM leads to severe complications. Patients with diabetes experience significant morbidity and mortality from microvascular (retinopathy, nephropathy, and neuropathy) and macrovascular (heart attacks, stroke, and peripheral vascular disease) complications [1-2]. In addition, post-prandial hyperglycemia or hyperinsulinemia are independent risk factors for the development of macrovascular complications of DM [3-4]. Diabetes has significant impact on the health, quality of life and life expectancy of patients. The current number of diabetes patients is 143 million worldwide, and 200 million people are estimated to have type 2 diabetes by 2030 [3,5]. Age - adjusted prevalence is set to rise from 5.9% to 7.1% (246-380 million) worldwide in the 20-79 yr age group, a 55% increase. In America, DM is the 7$^{th}$ leading cause of death and the number of cases of this disease is growing at an alarming rate [6]. Indeed, reports from the WHO tend to indicate that DM is one of the major killers of our time, with people in South-East Asia and western Pacific being more at risk [5, 7].

Diabetes can be classified as type I and II. Type I, often called insulin dependent diabetes mellitus (IDDM), is caused by autoimmune destruction of pancreatic β- cells, leading to a near total deficiency in insulin secretion. IDDM patients require lifelong insulin for the control of blood sugar levels. About 90% of all diabetics are considered to have non-insulin dependent diabetes mellitus (NIDDM) or type II DM. NIDDM is usually caused by a combination of pancreatic β-cell failure and insulin resistance in target tissues like liver, muscle and fat. NIDDM is associated with well-documented micro- and macrovascular complications, which increase the risk of premature mortality and result in considerable morbidity which diminishes quality of life, impacts on employment and life opportunities, and has wide-ranging economic implications for the individual, his or her family and society

[8]. The relative proportions of IDDM to NIDDM vary from 15: 85 for western populations to 5:95 in developing countries [5, 8].

The occurrence of hypoglycaemia in the treatment of NIDDM is well recognized, but is more protean in nature, having different risk factors and clinical features according to the nature of the hypoglycaemic therapy, the extent of the insulin secretory deficit and the duration of diabetes. Indeed, a large number of studies have focused on possible factors involved in the pathogenesis of NIDDM complications; most seeking effective therapies, but the exact cellular or molecular basis of these complications has not yet been fully elucidated. Nonetheless, it is generally agreed that the main aim of management is to prevent the development of diabetes related complications or delay their progression if they are present. This requires a multi-factorial approach, which includes lifestyle changes including diet, physical activity, weight control and avoidance of smoking, education and pharmacotherapy to control blood glucose, blood pressure and lipids [8]. It is still generally considered that hyperglycemia is the principal cause of diabetes complications [5, 8-9]. Nonetheless, it has been observed that oxidative damage due to hyperglycemia contributes to the microvascular pathology of diabetes that occurs particularly in the retina, renal glomerulus, and peripheral nerves, causing retinopathy, nephropathy, and neuropathy respectively.

In patients with diabetes, especially in those with poor glycemic control, oxidative stress is increased. In addition, it was reported that in aortic endothelial cells incubated with high glucose, reactive oxygen species (ROS) formation was increased by 250% within 24 h, which resulted in lipid peroxidation. Recently, numerous studies tend to suggest that oxidative damage from ROS produced during inflammation is a major predisposing factor for diabetes. Indeed, increased oxidative stress in DM may underlie the development of endothelial cell dysfunction by decreasing the availability of nitric oxide as well as by activating pro-inflammatory pathways [7]. Chronic blood sugar imbalances, or dysglycemia, may both fuel and be fueled by oxidative stress. In excess, free radicals can also deform the normal structure of fat molecules. When this happens, fats become oxidized, creating a more reactive, destructive form of cholesterol that accelerates the build-up of plaque in the arteries. Such a process can accelerate the hardening and narrowing of blood vessels, a key pathological development in the most common type of heart disease, atherosclerosis. The source of oxidative stress is a cascade of ROS leaking from the mitochondria, and this process has been associated with the onset of IDDM via the apoptosis of pancreatic β-cells, and the onset of NIDDM via insulin resistance. The underlying mechanisms in the onset of DM are complex because hyperglycemia may be both cause and effect of increased oxidative stress [10]. Recently, the NADPH oxidase activity of phagocytes has been shown to be significantly higher in vascular tissue (saphenous vein and internal mammary artery) obtained from diabetic patients [5,-6, 11-12]. The phagocyte NADPH oxidase and endothelial NADPH oxidase are sources of ROS in humans [9]. Following exposure to appropriate stimuli, both polymorphonuclear neutrophils and monocytes-macrophages undergo a so-called respiratory burst that leads to the production of ROS. The respiratory or oxidative burst is a coordinated series of metabolic events whereby activated phagocytes, such as monocytes, macrophages and neutrophils, reduce molecular oxygen to a variety of toxic reactive oxygen species [13]. At the molecular level, it involves an electron transfer chain, the NADPH-oxidase complexes, which is dormant in resting phagocytes and, following activation, involves striking increase in oxygen consumption, accompanied by activation of the hexose monophosphate shunt, which generates ROS including superoxide, hydrogen peroxide, hydroxyl radicals and hypochlorous acid [14].

Another, emerging deleterious effect of hyperglycemia is the formation of sugar-derived substances called advanced glycation end products (AGEs). AGEs form at a constant but slow rate in the normal body, starting in early embryonic development, and accumulate with time. However, their formation is markedly accelerated in DM because of the increased availability of glucose. Indeed, it has been found that prolonged exposure to hyperglycaemia is now recognized as a major factor in the pathogenesis of atherosclerosis in DM [5] and many neurodegenerative diseases. The non-enzymatic condensation reaction between the reducing sugars, such as glucose and the amino side chains in proteins (also called Maillard reaction or glycation) has been shown to play an important role in the development of chronic complications of DM [15]. A large body of evidence suggests that AGEs are important pathogenetic mediators of almost all diabetes complications, conventionally grouped into micro- or macroangiopathies. For instance, AGEs are found in retinal vessels of diabetic patients, and their levels correlate with that in serum as well as with severity of retinopathy. Clinical studies have shown that AGEs levels are up to four-fold higher in diabetic patients [15]. Thus, glycated low density lipoproteins (LDL) are poorly recognized by the specific LDL receptor, and on the contrary, are preferentially recognized by a nonspecific (scavenger) receptor present on human macrophages. The recognition of glycated LDL by the scavenger receptor pathway is thought to promote intracellular accumulation of cholesteryl esters and hence atherosclerosis [16].

# 1. Phytotherapy as Promising Alternative Therapies for Diabetes

Clinicians and consumers are fortunate in having an extensive range of pharmacological agents available to treat hyperglycaemia, but access to many of these options is limited in many middle and low income countries [8]. Currently, the treatment and management of DM is based on the variable use and combination of diet, antidiabetic oral agents (metformin, sulphanylureas, glynides, acarbose and thiazolidinediones) and insulin or its analogs, depending on the type of diabetes and the needs of the patient. Various antidiabetic drugs with different mechanisms are currently available for treatment of NIDDM, including sulphonylureas, glinides, biguanides, thiadiozines and $\alpha$-glucosidase inhibitors. In addition, incretin-related drugs such as dipeptidyl peptidase-4 inhibitors and glucagon-like peptide-1 receptor agonists, which have recently been developed, have excited considerable attention as new antidiabetic drugs [8, 17]. The prevention and treatment of chronic micro- and macrovascular complications, on the other hand, is based on the achievement and maintenance of an optimal glycaemic control and requires the combined use of adjunctive therapy such as antihypertensive drugs and cholesterol-lowering medications. Furthermore, several herbal preparations and dietary supplements, such as antioxidants, essential fatty acids, lipid metabolism activators, vitamins and trace elements, are advertised and prescribed to patients as a useful adjuvant to a diabetic diet and conventional medications in order to improve glycaemic control and reduce the impact of chronic complications. In this regard, investigators have attempted to review the current concepts dealing with the usefulness of these complementary therapies in treating diabetic patients. It has been argued that the conventional drugs for DM are associated with several adverse effects and, being expensive, they are often unaffordable in developing countries, like African, Asian and Latin American countries, where more than 80% of the population depend on herbs for treating/managing their illnesses. Therefore, the continuing search for possible alternative antidiabetic medications from plants, as recommended by the WHO, seems a justified cause [15].

With the increasing drive for more strict glucose control in NIDDM and new therapies which may carry different risks for hypoglycaemia from established therapies, and as the current methods of treatment therapy fail to achieve the ideals of normoglycemia and the prevention of diabetic complications, therefore a plethora of research has been geared towards alternative therapies. Indeed, insufficiency of current therapies for the treatment of diabetes, combined with both a lack of trust in conventional medical treatment and an inability of the economy to absorb the cost of pharmaceuticals, have created a growing public interest in dietary supplements and botanical [6, 18-24]. Further problems with conventional therapy in developing countries include insulin supply, storage, and injection, dietary control and complications of malnutrition, shortage of trained health care workers, and a lack of education from parents. Despite significant achievements in treatment modalities and preventive measures, the prevalence of diabetes has risen exponentially in the last few decades. Because of these limitations there is a continued need for new and more effective therapies. An increasing number of people are using dietary and herbal supplements, even though there is a general lack of evidence for their safety and efficacy. Moreover, the 2007 WHO regional office for the Africa report stated that one of the strategic aims was to support research in community interventions, including traditional medicine. There is, therefore, a clear need for alternate sources of both oral and potential antidiabetic drugs and alternate strategies for diabetes therapy.

Interestingly, prior to the introduction of insulin in 1922, the treatment/management of DM relied heavily on dietary measures, which included the use of traditional plant therapies. The Papyrus Ebers of 1550 B.C recommended a high-fiber diet of wheat grains and Orche. Currently more than 1200 species of organisms have been used ethnopharmacologically or experimentally to treat symptoms of DM and several reviews/reports of plants with known antidiabetic activity or traditional use as antidiabetic remedies have been published [6, 18-24]. They represent more than 725 genera in 183 families, extending phylogenetically all the way from marine algae and fungi to advanced plants such as the composites [26]. Nonetheless, only 5-15% of the higher plants have been systematically investigated for the presence of bioactive compounds, nature's biodiversity remains largely unexplored [6]. Although, oral hypoglycemic agents such as metformin and tolbuamide amongst others are the mainstay of treatment of DM and are effective in controlling hyperglycemia, they have prominent side effects and fail to significantly alter the course of diabetic complications [19, 27]. As the knowledge of heterogeneity of this disorder increases, there is a need to look for more efficacious agents with lesser side effects. Patients with diabetes may try herbal remedy that has been used traditionally in their country of origin, or they may take herbs that are prescribed by a non-physician health care provider or friend [19, 27]. It would thus appear that traditional anti-diabetic plants might provide a useful source of new oral hypoglycemic compounds for development as pharmaceutical entities, or as simple dietary adjuncts to existing therapies. Although an orally active botanical substitute endogenous to insulin seems unlikely, new phytochemicals molecules to stimulate endogenous insulin biosynthesis and secretion (and to promote insulin action) are realistic possibilities [29]. To this effect, several investigators have suggested that study of such traditional medicines with active phytochemicals might offer a natural key to unlock a diabetologist's pharmacy [19-24].

## 2. Phytochemicals as Potential Hypoglycemic Pharmacophores

Pharmacological properties of medicinal agents from plants center on plant secondary metabolites, also referred to as natural products. They are the products of metabolism not essential for normal growth, development or reproduction of an organism. These compounds serve to meet the secondary requirements of the producing organisms. They empower them to survive interspecies competition, provide defensive mechanisms and facilitate reproductive processes. Well known sources of secondarymetabolites are plants, bacteria, fungi and marine organisms such as sponges, tunicates, corals and snails. Many secondary metabolites have proved invaluable as antibacterial or antifungal agents, anticancer drugs, cholesterol-lowering agents, immunosuppressants, antiparasitic agents, herbicides, diagnostics, and tools for research. Some of these have found to play a pivotal role in treatment or prevention of a multitude of biological disorders, many of which did not have any cure until these products were discovered. It is estimated that 100,000-200,000 metabolites occur in the plants kingdom.

However, they are increasingly accepted as health promoting, maintaining, and repairing agents in cells, tissues, or the whole human body. Phytochemicals are compounds obtained from plants that exert particular health effects; generally, they are not necessarily basic nutrients (minerals, vitamins, carbohydrates, proteins or lipids), medicines or toxins [25-28]. However, it has been postulated that bioactive plant secondary metabolites may play a role in chemical defense mechanisms and are likely molecules for the hypoglycemic agents in these plants. Recently, it has been suggested that the enormous diversity of chemical structures found in these plants are not waste products, but specialized secondary metabolites involved in the relationship of the organism with the environment, e.g., as attractants of pollinators, signal products, defensive substances against predators and parasites, or in resistance against pests and diseases amongst others [28-29].

Panoply of plant's derived bioactive principles representing numerous phytochemicals has demonstrated activity consistent with their use in the treatment/management of NIDDM. The phytochemicals that are frequently associated with human health are phenolics, carotenoids, organic acids, and several miscellaneous bioactive compounds such as saponin and sterols. However, recently alkaloids, saponins, glycosides, galactomannan gum, polysaccharides, pepdidoglycans, hydroglycans, guanidines, steroids, glycopeptides, terpenoids, flavonoids, amino acids and inorganic ions obtained from plant sources have been reported to be promising hypoglycemic pharmacophores. Many of them may be found in a single plant and their combined synergistic action may be giving the observed behaviors. Several of these phytochemicals are structure-similar but bioactivity-different, and their enormous heterogeneity is usually derived from differential modification of common backbone structures. Interestingly, the discovery of the widely used hypoglycemic drug, metformin has been developed from the traditional approach of using *Galega officinalis*.

Identification of the active agent guanidine from *G. officinalis* led to the synthesis of biguanides. Thus, plants could be a potential source of anti-diabetic drugs as well as alleviating chronic complications of diabetes [28].

A plethora of mechanisms have also been postulated to justify the hypoglycemic effect of phytochemicals, such as inhibition of carbohydrate metabolizing enzymes, manipulation of glucose transporters and *β*-cell regeneration and enhancing insulin-releasing activity. Additionally, other mode of action such as its effect(s) on glucose diffusion, carbohydrate hydrolyzing enzymes and glucose transport across cell membrane have been employed in order to elucidate the possible hypoglycemic mechanism of action. Most recently the hypoglycemic response of phytochemicals from medicinal plants have been focused on: inhibition of hepatic and key carbohydrate-digesting and metabolizing enzymes; inhibition of glucose uptake from the small intestine; studies using insulin-secreting cell lines and release from isolated β-pancreatic islets cells; glucose uptake and metabolism using cultured non-insulin secreting cells; modulators of peroxisome proliferator-activated receptors and peripheral insulin binding and glucose utilization; insulin-mimetic strategies and aldose reductase pathway inhibitors. Additionally, some popular indirect pharmacological actions of antidiabetic medicinal plants on chronic complications of NIDDM have been geared on their antioxidative, anti-inflammatory and antiglycation properties.

Recently it was reported that γ-linolenic acid from plants sources possessed therapeutic effects in diabetic complications in a double blinded placebo-controlled trial. In the body, linoleic acid (LA) is converted to GLA via regulation by the enzyme δ-6-desaturase (D6D). Two metabolites of Gamma LA, dihomogammalinolenic acid and arachidonic acid are prostaglandin precursors that regulate the balance of platelet aggregation and maintain blood flow in small blood vessels. In diabetes, conversion of LA to GLA is thought to be impaired secondary to problems with D6D. Subsequent problems may occur, such as difficulty maintaining nerve membrane structure, nerve blood flow, and nerve conduction. An exogenous source of GLA bypasses the need for conversion of LA to GLA [30].

Many other phytochemicals were also shown to be promising in managing diabetic complications in diabetic animal models including astilbin, astragaloside astragalus astragalus saponins I, baicalein, berberine, breviscapine, curcumin, delphinidin, docosahexanoic acid, epigallocatechin gallate, eugenol, genistein, isoliquiritigenin, α-lipoic acid, lithospermate B, lithospermic acid B, *Lycium barbarum* polysaccharide 4, magnolol, maltol, mangiferin, oleanolic acid, protocatechualdehyde, puerarin, quercetin, resveratrol, rhein, rutin, scoparone, scutellarin, α-spinasterol, tanshinone IIA, tetrandrine, troxerutin and vitamin E [31].

Cinnamaldehyde, a phytoconstituent extracts have recently been reported to exhibit significant antihyperglycemic effect resulting in the lowering of both total cholesterol and triglyceride levels and, at the same time, increasing HDL-cholesterol in STZ-induced diabetic rats. This investigation reveals the potential of cinnamaldehyde for use as a natural oral agent, with both hypoglycaemic and hypolipidemic effects. Recent report tends to indicate that *Cinnamon* extract and polyphenols with procyanidin type-A polymers exhibit the potential to increase the amount of Thrombotic Thrombocytopenic Purpura (TTP), Insulin Resistance (IR), and Glucose Transporter-4 (GLUT4) in 3T3-L1 Adipocytes. It was suggested that the mechanism of *Cinnamon*'s insulin-like activity may be in part due to increase in the amounts of TTP, IRβ, and GLUT4 and that *Cinnamon* polyphenols may have additional roles as anti-inflammatory and/or anti-angiogenesis agents [32].

Swerchirin, xanthone from *Swertia chirayita* [(Roxb ex. Flem) Karst] has recently been reported to possess hypoglycemic activity. Researchers compared the effects of mode of action of three different hypoglycemic agents; centipiperalon, tolbutamide and swerchirin in normal as well as diabetic rats. Except in rats with severe pancreatic damage, swerchirin showed better glucose lowering effect compared to tolbutamide. Karanjin, a flavone showed significant hypoglycemic activity with a single dose of 0.5 mg/kg in normal rats. However it was ineffective in alloxan- induced diabetic rats [33].

Mechanistic studies have found that specific phytochemicals regulate insulin signaling through complex nutrigenomic and intracellular signaling networks. For example, it was reported that plant polyphenols influence gene expression, transcription, and protein expression and function, preventing metabolic disturbance in cellular systems. It was reported that dietary polyphenols activate the Sirtuin 1 gene, which regulates metabolism, inflammation-immune function and endothelial functions. Also it was reported results from a screening of phytochemicals from 206 foods and spices for activity on components of insulin signaling pathways. Ethanolic extract of licorice (*Glycyrrhiza uralensis Fisher*) was reported to contain phytochemicals that bind to peroxisome proliferator-activated receptor gamma and reduce the metabolic disturbances in diabetic mice. Recently a placebo controlled study evaluating the activities of various phytochemicals from green tea (epigallocatechin gallate), grapes (resveratrol), tomato (lutein), and soy oil (tocopherol) was conducted in overweight individuals.

There is also an increasing body of literature indicating that specific phytochemicals have discrete actions on kinase-mediated intracellular signaling processes that are disrupted in patients with chronic diseases.

For instance, flavonoids have been found to bind to Akt/protein kinase B (PKB), Fyn, Janus kinase 1, mitogen-activated protein kinase and phosphoinositide-3-kinase (PI3K). Adipocytes isolated from insulin resistant patients exhibit impaired insulin signaling, reduced insulin receptor substrate-1 (IRS-1) gene and protein expression, and dysfunctional insulin-stimulated PI3K and Akt/PKB signaling [34-36].

The effects of phytochemicals on gene expression in different tissues and cells have been of intensive research still ongoing to specify the mechanisms and novel targets of therapeutic nutrients. Polyphenolic phytochemicals may also influence expression of genes relevant for the development of NIDDM, i.e. genes regulating glucose transport, insulin secretion or action, antioxidant effect, inflammation, vascular functions, lipid metabolism, thermogenic or other possible mechanisms.

Resveratrol has been found to increase the expression of GLUT-4 in muscle of STZ-induced diabetic rats via PI3K-Akt pathways. In mice fed anthocyanin rich purple corn colour to ameliorate weight gain, gene expression of enzymes involved in the fatty acid and triacylglycerol synthesis and the sterol regulatory element binding protein-1 levels in white adipose tissue were reduced. In a cell culture model of rat adipocytes the treatment of anthocyanins PPAR gamma and target adipocyte specific genes were significantly up-regulated. Leptin and adiponectin as well as their mRNA levels were also increased by anthocyanins resulting from the increased phosphorylated MAPK. Quercetin has also been found to regulate gene expression mainly via NF-kB, xenobiotic responsive elements and antioxidant responsive elements [36].

# 3. Promising Phytochemicals from Traditional Antidiabetic Medicinal Plants

## 3.1. *Momordica charantia* – Bitter Melon

Numerous active compounds have been isolated from the traditionally used antidiabetic medicinal plant, *M. charantia*, and some mechanistic studies have also been done. It contains several (Table 1) chemical constituents (from fruits, seeds and tissue culture of seedlings), including the glycosides (mormordin and charantin) and alkaloid (mormordicine). Among the secondary metabolites of *M. charantia,* cucurbitane-type triterpenoids are one of the main bioactive hypoglycemic constituents. These compounds and their aglycones showed important biological effects beneficial to diabetes and obesity. Interestingly, it also contains an insulin-like polypeptide called 'Polypeptide-P' - a 17-amino acid, 166-residue polypeptide which did not cross-react in an immunoassay for bovine insulin. Polypeptide-P has been designated as the 'plant insulin' as it lowers blood sugar levels when injected subcutaneously in diabetic patients. Clinical study revealed that the polypeptide-p-$ZnCl_2$ produced blood sugar lowering effect. Galactose binding lectin with a molecular weight of 124,000 capable of antilipolytic and lipogenic activities in isolated rat adipocytes similar to insulin have also been isolated from the seeds of bitter melon. It was hypothesized that *M. charantia* fruit contains more than one type of hypoglycaemic component. These may include an active principle called "charantin", a homogenous mixture of β-sitosterol β-D-glucoside and 2-5-stigmatadien-3-β-ol-glucoside that can produce a hypoglycaemic effect in normal rabbits. Another hypoglycemic principle charantin a steroidal saponin found in the fruit of the plant is reported to be more potent than the drug tolbutamide. It probably acts through an extra-pancreatic mechanism such as increased tissue glucose uptake, promoting peripheral glucose utilization, decrease blood glucose synthesis through depression of the enzymes glucose-6-phosphatase, fructose-1, and 6 bisphosphatase and enhance glucose oxidation by enzyme glucose-6-phosphate dehydrogenase pathway [30, 33]. It has also been demonstrated that an extract from bitter melon had agonist activity for AMP-activated protein kinase (AMPK) in 3T3-L1 adipocytes and enhanced glucose disposal in insulin-resistant mice [34-38].

**Table 1. Phytochemicals and constituents of *Momordica charantia* [38]**

| Source | Phytochemicals |
|---|---|
| Plant body | Momorcharins, momordenol, momordicilin, momordicins, momordicinin, momordin, momordolol, charantin, charine, cryptoxanthin, cucurbitins, cucurbitacins, cucurbitanes, cycloartenols, diosgenin, elaeostearic acids, erythrodiol, galacturonic acids, gentisic acid, goyaglycosides, goyasaponins, multiflorenol, glycosides, saponins, alkaloids, fixed oils, triterpenes, proteins and steroids |
| Fruit | Momordicine, charantin, polypeptide-p insulin, ascorbigen, amino acids (aspartic acid, serine, glutamic acid, threonine, alanine, g-amino butyric acid and pipecolic acid), luteolin, fatty acids (lauric, myristic, palmitic, palmitoleic, stearic, oleic, linoleic, linolenic acid) |
| Seeds | Urease, amino acids (valine, threonine, methionine, leucine, phenylalanine, glutamic acid) |

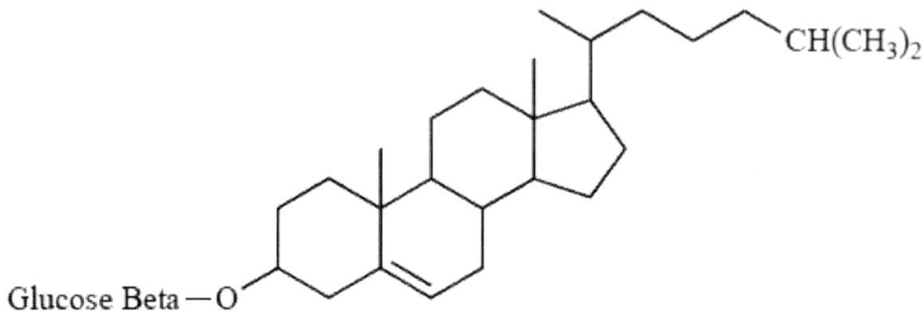


Charantin

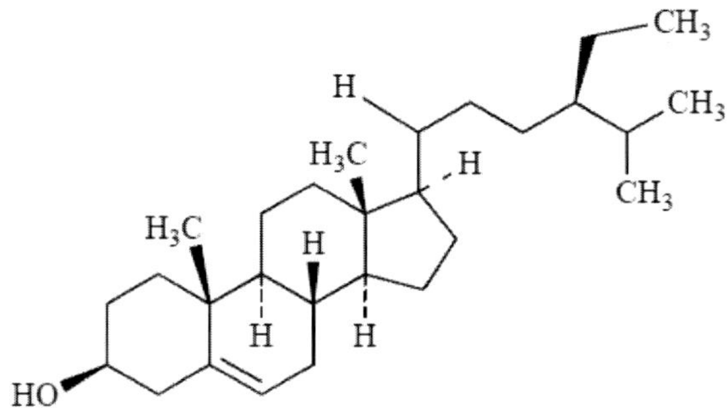


β-sitosterol

In addition, cucurbitane-type triterpenoids have been reported to exhibit various pharmacological and biological activities in addition to its antidiabetic activities [37-39]. Additionally, *M. charantia* comprises of several important nutrients: macronutrients and micronutrients, namely protein, carbohydrate, and fat, numerous minerals (calcium, iron, magnesium, phosphorus, potassium, sodium, zinc, copper, manganese selenium) and vitamins (vitamins A, D, E, K, C, B1, B2, B6, B12, folate). For instance, 100g of edible boiled bitter gourd, without salt, can provide 40% of the Recommended Daily Allowances in vitamin C, 13% folate, 8% zinc, 7% potassium and 5% vitamin K, magnesium and phosphorus [37-38].

## 3.2. *Eugenia Jambolana* Syn. *Syzygium Cumini*– Jambolan

The fruit-pulp of *E. jambolana* is rich in phytocompounds such as anthocyanins, glucoside, ellagic acid, isoquercetin, kaemferol, myrecetin and resin. The seed are claimed to contain alkaloid, jambosine, and lycoside jambolin or antimellin. Anthocyanidin backbones included cyaniding, delphinidin, peonidin, pelargonidin, petunidin, and malvidin [39]. The α-amylase inhibitory activity of *E. jambolana* may due to the presence of flavonoid diglycosides, quercetin and myricetin, hydrolysable tannins, 1-0-galloyl castalagin and casuarinin, and a triterpene- oleanolic acid. The inhibition of α-glucosidase by *E. jambolana* extracts can be attributed to the presence of tannins, rutin, quercetin, and steroids, which are reported to act as strong antioxidants and anti-inflammatory agents, with moderate free radical scavenging-linked antioxidant activity, have a high α-glucosidase inhibitory activity [34-38].

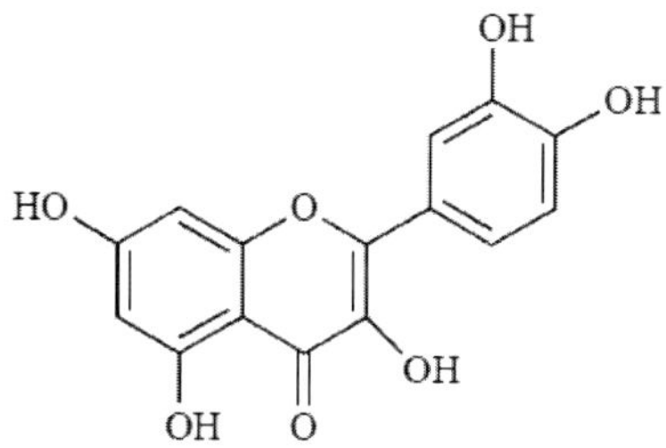


Quercetin

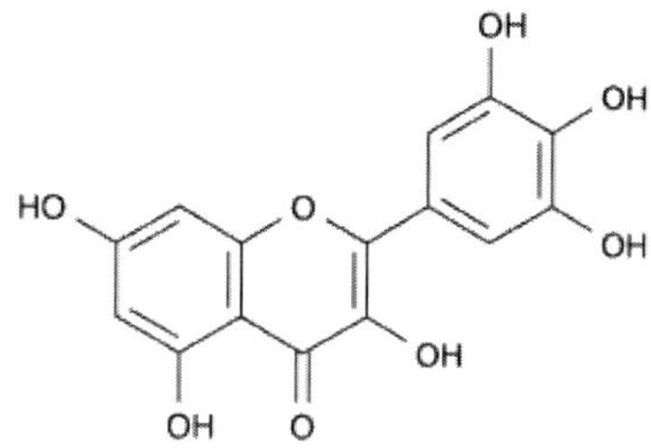


Myricetin

## 3.3. *Azadirachta indica* - Neem

The chief components present in *Azadirachta indica* oil extracted from seed are nimbin, nimbinin, and nimbidin. The seeds also contain a complex secondary metabolite azadirachtin. The leaves contain nimboin, nimbinene, 6-desacetylnimbinene, nimbandiol, nimbolide and quercetin. Leaves are also reported to contain β-sitosterol, *n*-hexacosanol and nonacosane [33]. The neem tree has almost all of its parts bitter which is due to its very high content of 'limonoids' also known as triterpenes. Phytochemical test of *A. indica* indicated highest presence of saponins, following tannins and glycosides. Lowest phytochemical content were the alkaloids, terpenes and flavonoids [40]. Although many other phytochemicals may be present in neem, clinical studies tend to show that some which are most important are azadirachtin, a tetraterpenoid molecule isolated from neem seeds and is considered to be the active ingredient of the plant [41].

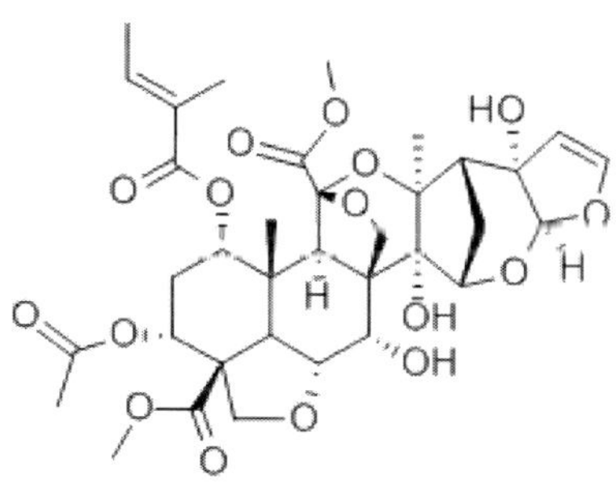

Azadirachtin

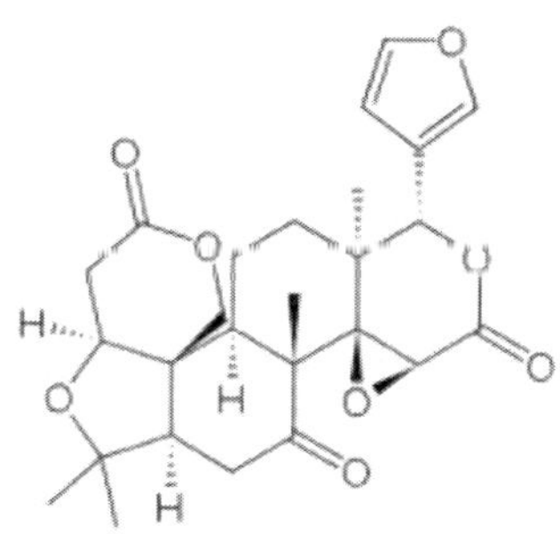

Limonoids

## 3.4. *Aloe vera* - Aloe

It is rich in anthraquinones, saccharides, enzymes, vitamins, minerals, and cholesterol amongst others. The principal constituents present in the leaves are barbaloin, chrysophanol glycoside and the aglycone, aloe-emodin. The phytosterols present in the *Aloe vera* gel have been reported to possess anti-hyperglycemic activity which may have a long term blood glucose level control effect and thus would be useful for treatment of IDDM and NIDDM. Another study reports the presence of inorganic trace elements (vanadium, zinc, copper; manganese and traces of chromium) in aloe leaf gel possessing hypoglycemic property, probably through improvement in impaired glucose tolerance thereby indirectly contributing to management of diabetes. Phytochemicals from Aloe have been found to stimulate glucose uptake and insulin release. Single as well as repeated doses of bitter principle of Aloe showed hypoglycemic effect in diabetic rats, which was through stimulation of synthesis or release of insulin from pancreatic beta cells [33].

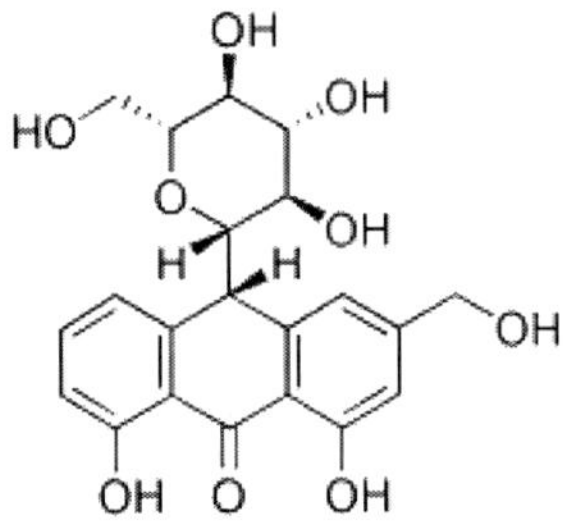

Barbaloin

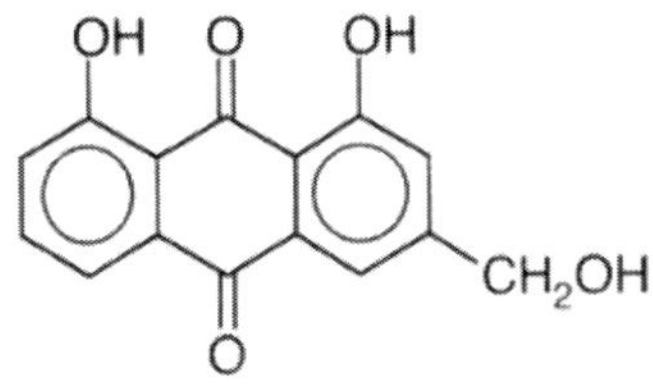

Aloe-emodin

## 3.5. *Curcuma longa* - Turmeric

Curcumin along with its mono- and bisdemethoxy derivatives, collectively called curcuminoids, constitute the major orange-yellow colouring matter and the biologically active constituents of *Curcuma longa* (turmeric). Besides curcuminoids, minor amounts of oils and resins are also present in curcumin. These include α-turmeron, β-turmeron, curlon, and zingiberen. Turmeric contains a wide variety of phytochemicals, including curcumin, demethoxycurcumin, bisdemethoxycurcumin, zingiberene, curcumenol, curcumol, eugenol, tetrahydrocurcumin, triethylcurcumin, turmerin, turmerones and turmeronols. The main curcuminoid is curcumin which is responsible for the yellow color of turmeric and the therapeutic effect of Curcuma [42]. Curcumin has antioxidant and anti-inflammatory properties. Many studies have shown that curcumin is not toxic to humans but rather it is as powerful an antioxidant as vitamins C, E and β-Carotene. Overall, turmeric consists of 2-6% of curcuminoid, 3-7% volatile (essential) oil, 2-7% fibre, 3-7% mineral matter, 6-8% protein, 5-10% fat, 6-13% moisture and 60-70% carbohydrate [43].

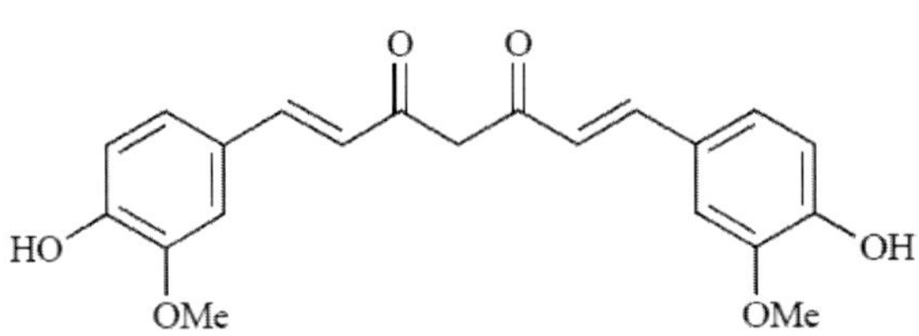

Curcumin

## 3.6. *Ocimum* spp. – Holy Basil

*Ocimum* is a genus of about 35 species in the family Lamiaceae, mostly native to the tropical and warm temperate regions of the world. The main chemical constituents in *Ocimum tenuiflorum* can be grouped as essential oils (0.516-0.596%), total carbohydrates (11.87-11.50%), total flavonoids (0.735-0.945%), proline (0.28-0.536 mmol/g leaves fresh weight) and higher amount of linalool (55.26%) and moderate amounts of methyl chavicol (6.66%), nerol (0.48-8.0%), geraniol (0.26-1.75%) and citral (4.26-6.79%) [44]. The volatile oil of Holy Basil contains 70% eugenol as well as methyl eugenol and caryophyllene. Other constituents include the triterpenoid ursolic acid, rosmarinic acid, alkaloids, saponins, flavonoids (including apigenin and luteolin and glycosides), phenylpropane glucosides and tannins. The seeds of sacred basil contain a fixed oil containing five fatty acids, including about 17% linolenic acid and about 50% linoleic acid [45]. *Ocimum sanctum* has specific aromatic odour because of the presence of essential or volatile oil, mainly concentrated in the leaf. This aromatic volatile oil mainly contains phenols, terpenes and aldehydes. The leaves of *O. sanctum* contain 0.7% volatile oil comprising about 71% eugenol and 20% methyl eugenol. The oil also contains carvacrol and sesquiterpine hydrocarbon caryophyllene [23]. *O. sanctum* also contains alkaloids, glycosides, saponines and tannins, sesquiterpenes and monoterpenes. *In vivo* studies indicated that the ethanol extract of *O. Sanctum* decreased blood glucose and increased plasma insulin in NIDDM rats [46]. Further, *O. sanctum* has aldose reductase activity, which may help in reducing the complications of diabetes such as cataract and retinopathy [46]. The constituents of *O. sanctum* leaf extracts have stimulatory effects on physiological pathways of insulin secretion, which may underlie its reported antidiabetic action. Gholap *et al.* [47] suggested that *O. sanctum* L. decreases the serum concentration of both cortisol and glucose and also exhibited antiperoxidative effect. Therefore, *O. sanctum* L. may potentially regulate corticosteroid-induced diabetic mellitus.

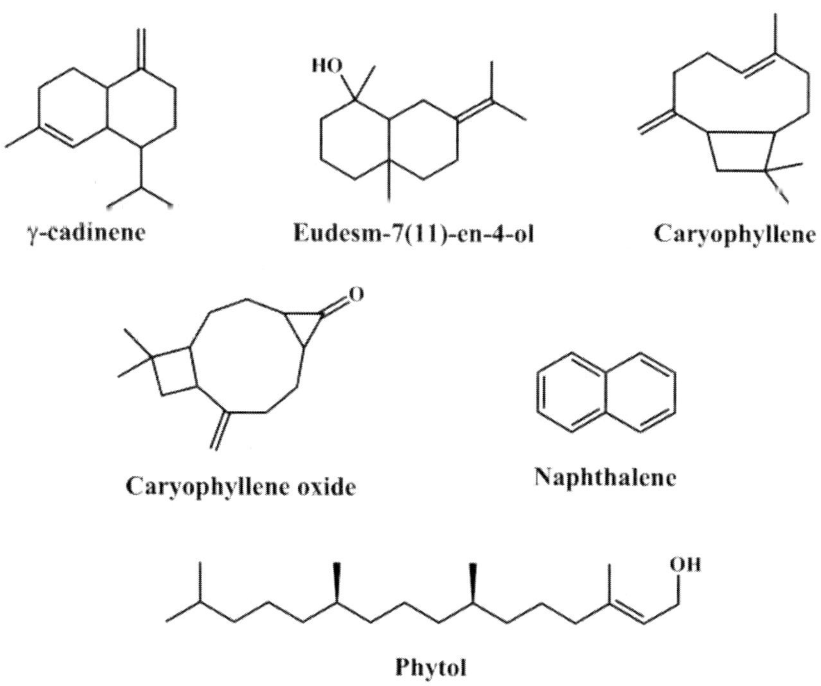


Phytochemical composition of *Ocimum* spp.

### 3.7. *Gymnema Sylvestre*

The major bioactive constituents of *G. sylvestre* are a group of oleanane type triterpenoid saponins known as gymnemic acids (gymnemosides). Other chemical constituents of the plant include saponins, stigmasterol, quercitol, and the amino acid derivatives betaine, choline, and trimethylamine. The active principles, gymnemic acids, show selective anesthetic effect on the sweet taste buds. It has been reported to be an effective anti-diabetic agent in lowering blood sugar in both IDDM and NIDDM [30, 33]. It also acts as a hepatoprotective, hypolipidemic, antiatherosclerotic, stomachic, diuretic, refrigerant, astringent and tonic. Experimental studies on rats showed comparable anti-hyperglycemic effect to that of tolbutamide. Gymnema's antidiabetic activity appears to be due to a combination of mechanisms. It increases the activity of enzymes responsible for glucose uptake and utilization, and inhibits peripheral utilization of glucose by somatotrophin and corticotrophin. Plant extracts of Gymnema have also been found to inhibit epinephrine-induced and corticosteroid-induced hyperglycemia. The drug influences the disturbed carbohydrate metabolism in hyperglycemic animals by limiting the carbohydrate turnover and thus inhibiting the vicious cycle of hyperglycemia. It may stimulate β-cell function, increase β-cell number, and/or increase insulin release by increasing cell permeability to insulin. Cytoprotection of pancreatic β-cells was demonstrated for the Gymnema extracts containing phytochemicals such as liquiritigenin, pterosupin). In pancreatectomized animals, it has no hypoglycemic effect, indicating that its effect may require some residual β-cell function. Investigations in streptozotocin treated rats indicate that this plant brings about blood glucose homeostasis through increased serum insulin levels provided by repair/regeneration of endocrine pancreas. Further clinical trials in patients with IDDM also suggest that it acts by regeneration of residual β-cells in IDDM and is insulinotropic [30, 33, 48-49].

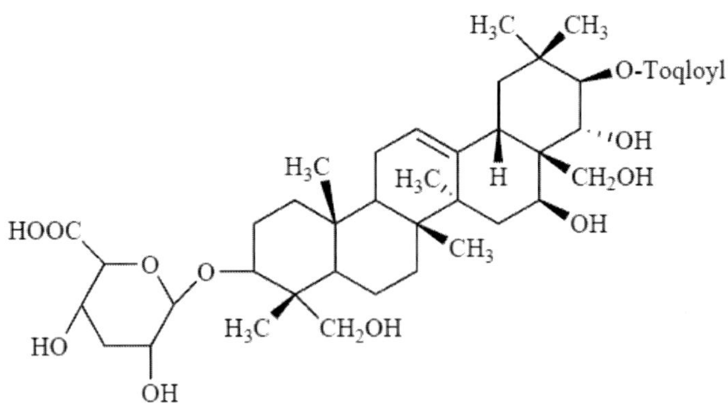


Gymnemoside

### 3.8. *Lawsonia inermis* - Henna

Much work is done in the field of phytochemical investigation of *L. inermis*. The chemical constituents isolated from Henna are divided into the following categories;

Naphthoquinone derivatives, aliphatic components, triterpenes, & sterols and lastly phenolic derivatives, coumarins, xanthones and flavonoids. Lawsone is the chief constituent responsible for the dying properties of the plant [50]. It is identified as 2-hydroxy-1, 4-naphthoquinone which is present in dried leaves in a concentration of 1.0-1.4 %. The compound was extracted from the leaves [51]. Ethanolic extract of *L. inermis* at 500 mg/kg of body weight led to significant decrease of blood sugar level to normal in streptozotocin induced diabetic rats and was found to be better than Glibenclamide [52]. The anihyperglycaemic effect of these plants are attributed to their ability to restore the function of pancreatic tissues by causing an increase in insulin output or inhibit the intestinal absorption of glucose or the facilitation of metabolites in insulin dependent processes [53].

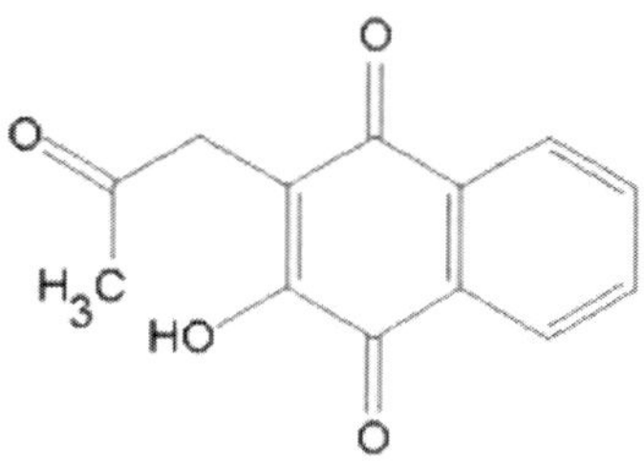

2-hydroxy-1, 4- naphthoquinone

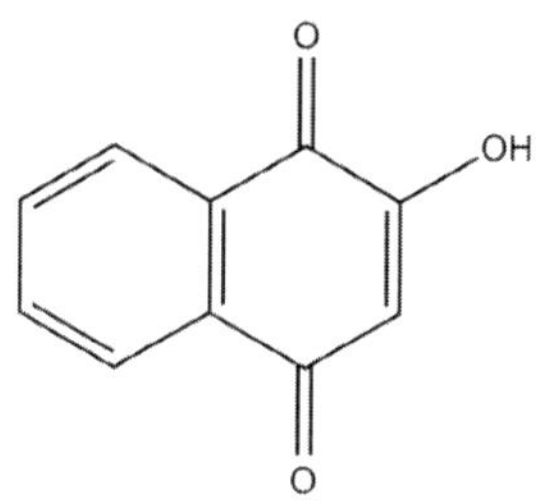

Lawsone

## 3.9. *Mangifera indica* - Mango

The different chemical constituents of *Mangifera indica* are especially the polyphenolics, flavonoids and triterpenoids. Mangiferin is a xanthone glycoside and is the major bioactive constituent, isomangiferin, tannins & gallic acid derivatives [54]. Although no change in blood glucose level was reported when the aqueous mango extract was given to streptozotocin-diabetic rats, antidiabetic activity was observed when the aqueous extract was given before/along with glucose, which could be due to decreased intestinal glucose absorption.

Analysis of the peel extract of *M. indica* in rats revealed their potential to ameliorate diet-induced change in serum lipids, thyroid dysfunctions and hyperglycemia, which could be due to polyphenols and ascorbic acid present in the test peel extract. Lupeol, a triterpene found in mango, is also acknowledged to display antioxidant, antilithiatic and antidiabetic effects and is also established to be valuable in combating oxidative stress-induced cellular injury of mouse liver by modulating cell-growth regulators.

Also, mangiferin, a xanthone glucoside, isolated from mango leaves possesses considerable antidiabetic, antihyperlipidemic and antiatherogenic properties as obvious from lowering of fasting glucose level, decrease in total cholesterol, triglycerides and LDL - cholesterol along with elevation of the HDL -cholesterol level and diminution of atherogenic index in diabetic rats [54-55].

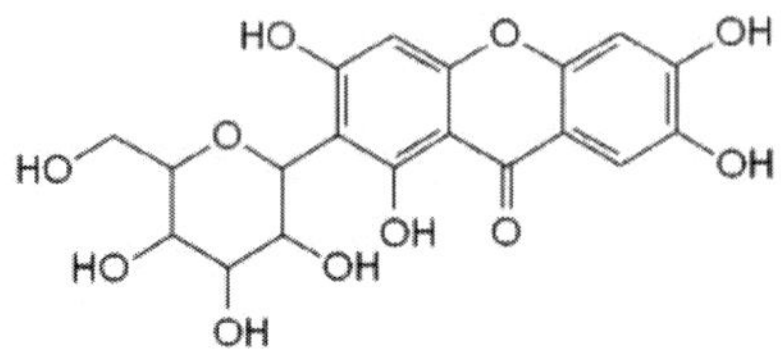

Mangiferin

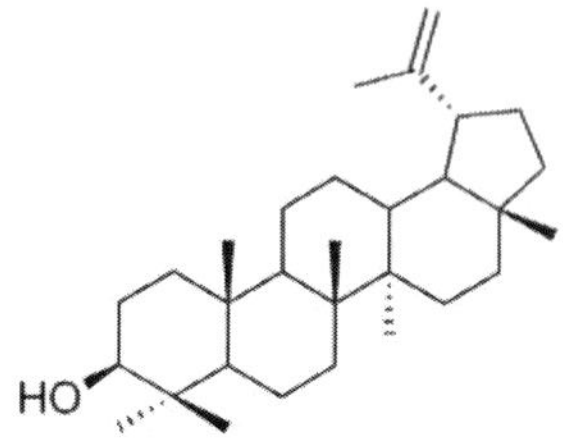

Lupeol

## 3.10. *Panax Ginseng* spp. - Ginseng

Ginseng root has been used since time immemorial for its therapeutic properties. Medicinally important ginseng species include *Panax ginseng* (*Asian ginseng*) and *Panax quinquefolius* (American ginseng). Animal research has indicated that ginseng may lower blood glucose by possibly decreasing the rate of carbohydrate absorption into the portal hepatic circulation and possibly increasing glucose transport and uptake. Another potential mechanism is modulation of insulin secretion. Some ginseng fractions have increased serum insulin levels and glucose-stimulated insulin secretion in mice. The main chemical constituents of the entire ginseng species are ginsenosides, polysaccharides, peptides, polyacetylenic alcohol and fatty acids; with ginsenosides being the most important pharmacological constituent. Ginsenosides have been the object of countless researches as they are believed to be the key principles behind the efficacy and potential effects of ginseng. Animal study data indicate that both Asian ginseng and American ginseng have a prominent hypoglycemic effect, which may be due to ginsenoside Rb-2 and, more specifically, because of panaxans I, J, K and L in IDDM models. American ginseng has also been reported to inhibit the tumor necrosis factor-alfa (TNF-α)-stimulated free fatty acid release and also attenuated the TNF-α inhibition of adiponectin secretion. Hypoglycemic activity of ginseng has also been established by clinical studies. However, a few adverse effects of ginseng have been observed, most common of which are nervousness and excitation, which are diminished with continued use or dosage reduction. Massive overdose of drug can lead to ginseng abuse syndrome. The suggested dosage is 1-3 g of the plant's crude root daily or 200- 00 mg of standardized extract [30-31].

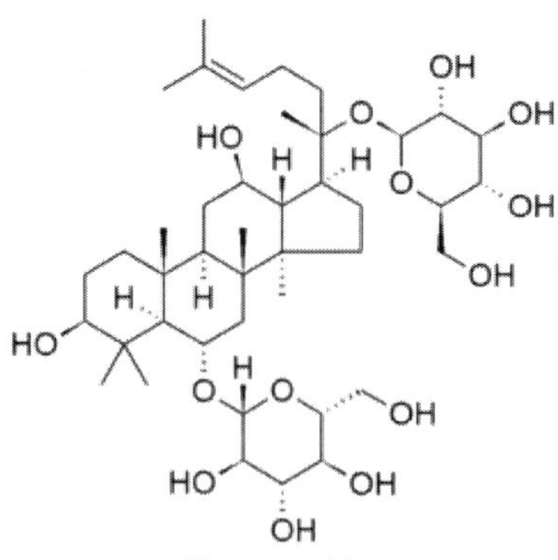

Ginsenosides

### 3.11. *Murraya koenigii* - Curry Leaf

It has been shown in recent studies that *M. koenigii* leaves have anti-diabetic properties and hence can be used as an adjunct to dietary therapy and drug treatment for controlling NIDDM. Apart from containing nutrients like proteins, carbohydrates, fibre, minerals and vitamin C, *M. Koenigii* is also a rich source of biologically active carbazole alkaloids, tocopherol, beta-carotene and lutein [56]. Some of the bioactive agents present are girinimbin, iso-mahanimbin, koenine, koenigine, koenidine, koenimbine, mahanimbicine, bicyclomahanimbicine, phebalosin, triterpenoid alkaloids like cyclomahanimbine and tetrahydromahanmbine, murrayastine, murrayaline, pypayafolinecarbazole among many others [57]. In a study, the hypoglycemic activity of *M. koenigii* was compared with the effect of a standard hypoglycemic drug, tolbutamide, and it was observed that a single oral administration of variable dose levels (200-400 mg/kg) of aqueous extract of *M. koenigii* lowered blood glucose level in both normal (fall of 14.68%) and diabetic (fall of 27.96%) rabbits after 4 hours of oral administration of 300 mg/kg [58]. The mechanism of action for this effect was due to an increase in the concentration of hepatic glycogen and glycogenesis due to a rise in glycogen synthetase enzyme involved in converting glucose to glycogen. It was also documented that there was a decrease in the activity of glycogen phosphorylase which breaks down glycogen to glucose-1-phosphate. There was also a decrease in gluconeogenic enzymes leading to glycogenolysis (conversion of glycogen to glucose monomers) and gluconeogenesis (generation of glucose from non-carbohydrate carbon substrate) [50]. Dineshkumar *et al.* [59] also investigated the effect of mahanimbine (carbazole alkaloid from *M. koenigii* leaves) on blood glucose and serum lipid profiles on streptozotocin-induced diabetic rats. Streptozotocin is toxic to insulin-producing Beta cells of the Islet of Langerhans in mammals. In research, it is used to provide an animal model for IDDM. Mahanimbine (50 and 100 mg/kg) were administrated as a single dose per week to the diabetic rats for 30 days. In the diabetic rats, the elevated fasting blood sugar, triglycerides, low density lipoprotein, very low density lipoprotein levels were reduced and high density lipoprotein level was increased by mahanimbine at a dose of 50 and 100mg/kg. It was thus showed that mahanimbine exhibited anti-diabetic and hypolipidemic effects. Most of the pharmacological properties mentioned above are due to the carbazole alkaloids present in *M. koenigii* plant indicating that these alkaloids can be used to derive synthetic drugs. In a study, the bioassay guided fractionation of the acetone extract of the fresh leaves of *M. koenigii* yielded three bioactive carbazole alkaloids mahanimbine, murrayanol, and mahanine [60]. In a recent study, three bioactive carbazole alkaloids kurryam, koenimbine and koenine were identified from the n-hexane extract of the seeds of *M. koenigii*. With more research, new alkaloids are being isolated from *M. koengii* plant as shown by a recent experiment in which a new carbazole alkaloid, murrayakoeninol, was isolated from its leaves [61-63].

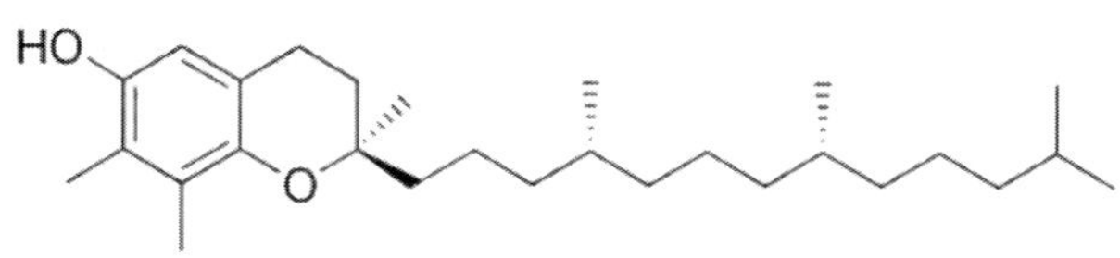


Tocopherol

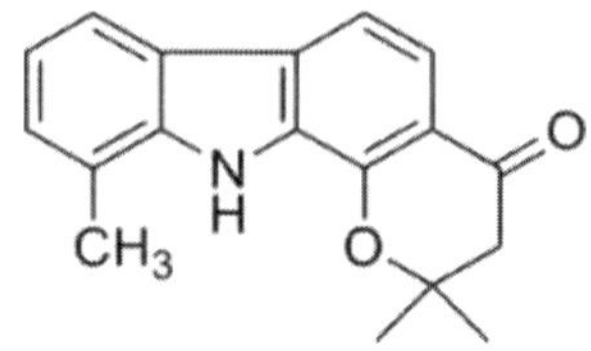


Girinimbine

## 3.12. *Coix lacryma-jobi* L - Adlay

Some of the bioactive chemical components found in *Coix lacryma-jobi* are: Fatty acids, such as hexadecanoic acid, octadecanoic acid, octadecadienoic acid, palmitic acid, stearic acid, oleic acid and linoleic acid which are present in coix seeds. Stigmasteriol, coixenolide, and coixol are also found in this herb. Some of the chemicals found in the plant like coniferyl alcohol, syringic acid, ferulic acid, syringaresinol, 4-ketopinoresinol, and a new lignan, mayuenolide has been reported to possess strong antioxidant activities. The mature dried seeds of Adlay (*Coix lacryma-jobi* L. var. *ma-yuen Stapf*) are used as a healthy food for diabetes in China. Adlay has been reported to have active constituents as the polysaccharides coixan A, B and C. Coixans isolated from the dried coix seeds decreased blood glucose in normal rats, while it increased the serum insulin level. The proposed anti-diabetic mechanism of coixans is that it may prevent alloxan induced injury to β-cells of the pancreatic islet [64-66].

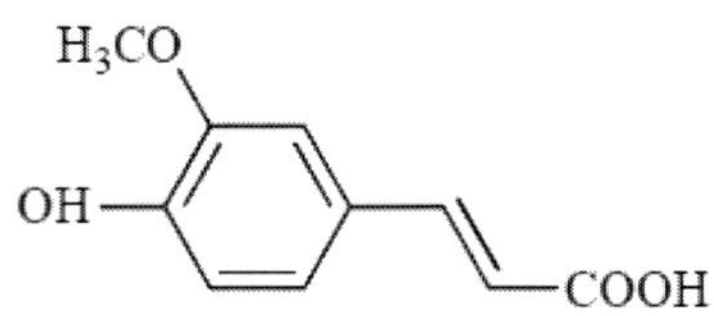


Ferulic acid

## 3.13. *Emblica officinalis* Gaertn - Amla

Fruit extract of *Emblica officinalis Gaertn.* (Indian gooseberry) has been reported to significantly decrease the blood glucose level after its intraperitoneal administration in alloxan-induced diabetic rats [67]. Similarly, the anti-diabetic activity of fresh juice and hydro-alcoholic extract of fruits of *E. officinalis* was studied in streptozotocin (STZ) induce type 1 diabetic rats. It produces significant increase in serum insulin level and AUC insulin of diabetic rats compared to that of diabetic control [68]. The fruit of Amla is known as a rich source of vitamin C, and also contains a mixture of phenolic compounds [69]. It primarily contains tannins, alkaloids, Emblicanin, phenolic compounds, amino acids and carbohydrates [70]. Compounds isolated from Amla were gallic acid, ellagic acid, 1-O-galloyl-beta-D-glucose, 3,6-di-O-galloyl-Dglucose, chebulinic acid, quercetin, flavonoids, kaempferol 3-O-alpha-L (6"methyl) rhamnopyranoside and kaempferol 3-O-alpha-L (6"ethyl) rhamnopyranoside [70]. Although the medicinal property of amla is often attributed to its high content of vitamin C, there are some *in vivo* studies indicating that antioxidant activities of amla cannot be attributed to ascorbic acid alone and that the overall effect is due to other polyphenols such as ellagic acid, gallic acid and tannins amongst others [71]. Also the tannoids of *E. officinalis* have been proved to be potent inhibitors of aldose reductase and suggest that exploring the therapeutic value of natural ingredients that people can incorporate into everyday life may be an effective approach in the management of diabetic complications [67-70].

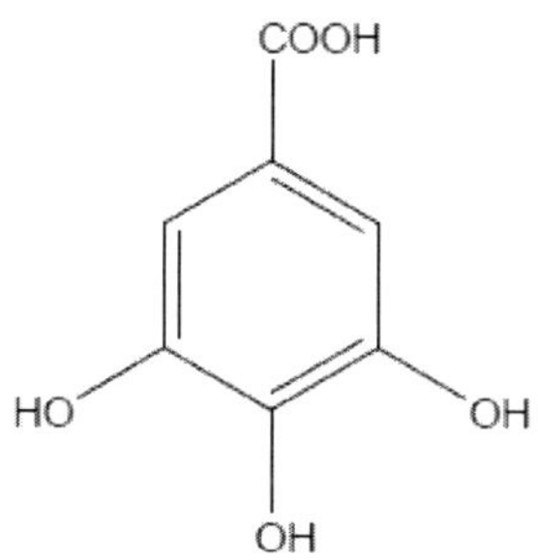


Gallic acid

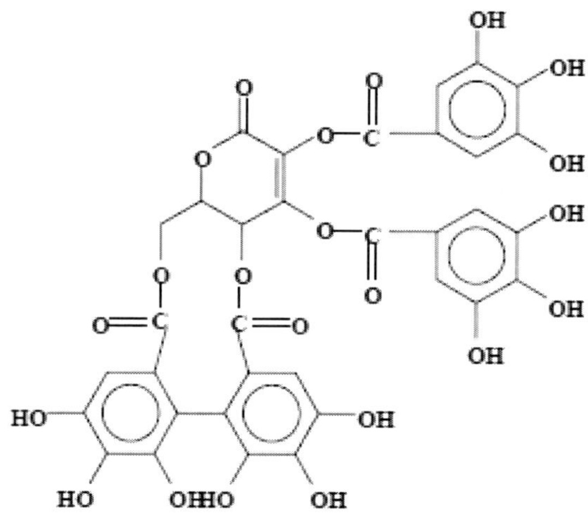


Emblicanin

## 3.14. *Trigonella Foenum Graecum* - Fenugreek

Fenugreek has been used as a medicinal agent to manage diabetes, constipation, and hyperlipidemia in many parts of the world. Chemical constituents of the plant include saponins, many of which are glycosides of diosgenin. The seeds also contain the alkaloids trigonelline, gentianine, and carpaine compounds. Other components of the seeds include several C-glycosides.

The seeds contain up to 50% mucilaginous fiber. Other seed constituents include 4-hydroxyisoleucine, an amino acid, and fenugreekine. Fenugreek is thought to delay gastric emptying, slow carbohydrate absorption, and inhibit glucose transport. It has been shown to increase erythrocyte insulin receptors and improve peripheral glucose utilization, thus showing potential pancreatic as well as extrapancreatic effects. Various components of the seeds have varying activities.

For example, the component called fenugreekine, a steroidal sapogenin peptide ester, may have hypoglycemic properties. Trigonelline, another component, may exert hypoglycemic effects in healthy patients without diabetes, but other studies have shown that fenugreek has no effect on fasting or postprandial blood glucose levels in non-diabetic subjects [30].

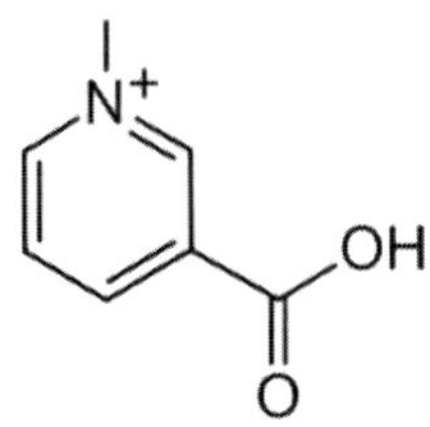


Trigonelline

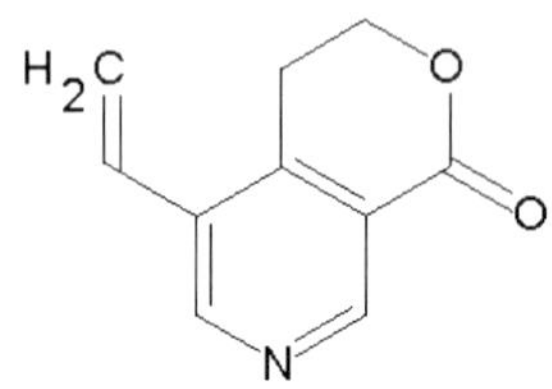


Gentianine

## Conclusion

Diabetes with its accompanying complications has become a major pandemic throughout the world and the intensity of this crisis is predicted to increase drastically even further in the next 20 years. Indeed, due to the complexity of diabetes, evaluating phytochemicals will continue to be a prolific avenue for both basic and clinical scientific investigations. Clinical studies investigating the effects of specific phytochemicals as potential pharmacophores in the management/treatment of diabetes have yielded promising results, including normalization of blood sugar level, prevention and delaying the digestion and absorption of glucose amongst others. It is generally agreed that phytochemicals such terpenoids, alkaloids, flavonoids, phenolics, and some other categories are promising lead compounds. To this effect, it would thus appear that traditional anti-diabetic plants might provide a useful source of new oral hypoglycemic compounds for development as pharmaceutical entities, or as simple dietary adjuncts to existing therapies. Although an orally active botanical substitute endogenous to insulin seems unlikely, new phytochemicals molecules to stimulate endogenous insulin biosynthesis and secretion (and to promote insulin action) are realistic possibilities.

## References

[1] P. Hamet, *Diab. Obes. Metab.*14 (S1), 20-29 (2012).
[2] R. A. Defronzo, *Ann. Intern. Med.* 131, 281-303 (1999).
[3] R. Bilous, and R. Donnelly, *4th Edn. Blackwell Pub. Ltd. UK.* (2010).
[4] A. D. Mooradian, and J. Thurman, *Drugs.* 57, 19-29. (1999).
[5] A. Dembinska-Kiec, O. Mykkanen, B. Kiec-Wilk and H. Mykkanen, *Bri. J.Nut.* 99E-S1, ES109–ES117. (2008).
[6] T. G. Guilliams, *The Standard.* 5, 1-7 (2002).
[7] S. A. Amiel, T. Dixon, R. Mann and K. Jameson, *Diab. Med.* 25, 245–254 (2008).
[8] S. Colagiuri. *Diab. Obes.Metab.* 14 (S1), 3–8 (2012).
[9] M. Peppa, J. Uribarri and H. Vlassara, *Clin. Diab.* 21, 186-187 (2003).
[10] N. F. Wiernsperger, *Diab. Metab.* 29, 579-585 2003.
[11] K. Schultz, L.M. Rasmussen, and T. Ledet, *Metab.* 54, 287-295 (2005).

[12] U. Landmesser, D.G. Harrison and H. Drexler, *Eur. J. Clin. Pharmacol.* DOI 10.1007/s00228-005-0012. (2005).

[13] K. A. Kang, S. Chae, S. Koh, J. Kim, J. H. Lee, H. J. You, and J. Hyun, *Biol. Pharm. Bull.* 28, 1154-1160 (2005).

[14] R. R. Kitagawa, M.S. Raddi, M.N. Khalil, W. Vilegas, and L.M. Fonseca, *Biol. Pharm. Bull.* 26, 905-908 (2005).

[15] D. Aronson, and E.J. Rayfield, *Cardio. Diab.* 1, 1-1 (2002).

[16] R. L. Klein, M. Laimins, and M.F. Lopes-Virella, *Diab.* 44, 1093-1098 (1995).

[17] S. Seino, H. Takahashi, T. Takahashi and T. Shibasaki, *Diab. Obes. Metab.* 14 (S1), 9–13, 2012.

[18] M. F. Mahomoodally, A. Gurib-Fakim and A. H. Subratty, *J. Cell & Mol. Biol. 3*, 39-44 (2004)

[19] M. F. Mahomoodally, A. Gurib-Fakim and A. H. Subratty, *Fund. Clin. Pharmacol.* 19, 87-92. (2004).

[20] M. F. Mahomoodally, A. Gurib-Fakim and A. H. Subratty, *Biol. Pharm. Bull.* 27, 216-218. (2004).

[21] M. F. Mahomoodally, A. Gurib-Fakim and A. H. Subratty, M. I. Choudhary, and S. Nahar-Khan. *The Sci. World J.* 285284, doi:10.1100/2012/285284 (2012).

[22] M. F. Mahomoodally, A. Gurib-Fakim and A. H. Subratty, *Bri. J. Biomed. Sci.* 63, 12-17 (2005).

[23] M. F. Mahomoodally, A. Gurib-Fakim and A. H. Subratty, *J. Ethnopharmacol.* 110, 257-263 (2007).

[24] M. F. Mahomoodally, A. Gurib-Fakim and A. H. Subratty, *J. Food Biochem.* doi:10.1111/j.1745-4514.2010.00504. (2012).

[25] R. Marles, and N. R. Farnsworth. *Phytomed.* 2, 137-189 (1995).

[26] R. Kasi, B. Ramachandran and S. Subramanian, *Biol. Pharm. Bull.* 27, 1212-1217 (2004).

[27] *V. Rao*, In Tech Publisher (2012).

[28] C. Bailey, and C. Day, *Diab. Care.* 12, 553-564. 1989.

[29] L. Pieters, and A. J. Vlietinck, *J. Ethnopharmacol.* 100, 57–60. 2005.

[30] S. McWhorter, *Diab.Spect.* 14, (2001).

[31] E. A. Omar, A. Kam, A. Alqahtani, K. M. Li, V. Razmovski-Naumovski, S. Nammi, K. Chan, B. D. Roufogalis, and G. Q. Li, *Cur. Pharm. Des.* 16, 3776-3807 (2010).

[32] V. Jakhetia, R. Patel, P. Khatri, N. Pahuja, S. Garg, A. Pandey, and S. Sharma, *J. Adv. Sci. Res,* 1(2), 19-23 (2010).

[33] N. Aggarwal and Shishu, *J Nut Dis Ther.* 1, 1(2011).

[34] J. S. Bland, *Func. Foods Heal. Dis.* 2, 1-12 (2011).

[35] J. G. Babish, L. M. Pacioretty, J. S. Bland, D.M. Minich, J. Hu, and M. L. Tripp, *J Med Food.* 13, 535-47 (2010).

[36] G. Banos, I. Perez-Torres, E.I. Hafidi, *Hematol. Agents Med. Chem.* 6, 237-52 (2008).

[37] S. Y. Lee, S. H. Eom, Y. K. Kim, N. Park and S. Park. *J. Med. Plants Res.* 3(13), 1264-1269 (2009).

[38] A. Paul, S. S. Raychaudhuri, *Elect. J. Bio.* 6(2), 43-51 (2010).

[39] E. Sousa De Brito, M. C. Pessanha De Araújo, R. Elesbão Alves, C. Carkeet, B. A. Clevidence, and J. A. Novotny, *J. Agric. Food Chem.* 55, 9389-9394 (2007).

[40] A. A. Biu, S. D. Yusufu and J. S. Rabo, *Biosci. Res. Com.* 21(6), 281-283 (2009).

[41] K. Biswas, I. Chattopadhyay, and R. K. Banerjee, *Cur. Sci.* 82(11), 1336-45 (2002).
[42] B. B. Aggarwal, C Sundaram, N. Malani, and H. Ichikawa, *The Clayton Found.Res.* 1-13 (2006).
[43] S. K. Gupta, A. Lather, V. Jaiswal, S. Garg, A. J. Kumar, *J. Pharm. Biomed. Sci.* 4, 1-7 (2010).
[44] Balaji R et al, 2011. International Journal of Pharma. Research and Development. 3(1):20-27
[45] H. Wolmuth. *Phyto Res.* 13, 275-291(1999).
[46] J. M. Hannan, L. Marenah, B. Ali, P.R. Rokeya, Y.H. Abdel-Wahab, *J. Endocrinol.* 189,127-36 (2006).
[47] S. Gholap, and A. Kar, *Pharmazie*, 59, 876-8 (2004).
[48] D. Peng Li, *Asian J. Pharmacodyn. Pharma.* 6(2), 83-102 (2006).
[49] A. Al-Romaiyan, B. Liu, H. Asare-Anane, C. R. Maity, S. K. Chatterjee, and N. Koley, *Phyto Res.* 24, 1370-6 (2010).
[50] H.S. Muhammad, and S. Muhammad, *Afr. J. Biotec.* 4 (9), 934-937 (2005).
[51] V. K. Jiny, K. S. Silvipriya, S. Resmi and C. Jolly, *Inventi Journals*, 1(1) 2010.
[52] A. Choobey, M. Ojha, A. Mishra, S. Mishra, and U. K. Patil, Intern. *J. Pharm. Sci. Res.* 1(8), (2010).
[53] I. Syamsudin, and H. Winarno, *Res. J. Pharmacol.* 2(2), 20-23 (2008).
[54] K. A. Shah, M. B. Patel R. J. Patel, and P. K. Parmar, *Phamacog. Rev.* (2010).
[55] V. Khan, N. Abul Kalam, M. Akhtar, M. Aqil, M. Mujeeb, and K. K. Pillai, *J. Pharm. Bioallied. Sci.* 4(1), 27–42 (2012).
[56] M. B. Ningappa, B. L. Dhananjaya, R. Dinesha, R. Harsha, and L. Srinivas, *Food Chem.*1-4. (2009).
[57] S. D. Bonde, L. S. Nemade, M. R. Patel, and A. A. Patel, *Intern. J. Pharm. Phytopharmacol. Res.* 1(1), 23-27, (2011).
[58] A. N. Kesari, R.K. Gupta, G. and Wata. *J. Ethnopharmacol.* 28, 247-251 (2005).
[59] B. Dineshkumar, A. Mitra, M. Mahadevappa, 22-30 (2010)
[60] R.S. Ramsewak *et al. J. Agric. Food Chem.* http://pubs.acs.org. (1999).
[61] S. Mandal, *et al. Fitoterapia.* 81(1), 72-4 (2010).
[62] M. Chakraborty, and S. Saha, S. Mukhapadhyay, *Nat. Prod.* Commun. 4(3), 355-8 (2009).
[63] B. A. Khan, A. Abraham, and S. Leelamma, *Ind.J. Biochem. Biophys.* 32(2), 106-8 (1995).
[64] S. C. Li, C. M. Chen, S. H. Lin, W. Chiang and C. K. Shih, *J. Sci. Food Agri.* 91(3), 547-52 (2011).
[65] Y. T. Yu, T. J. Lu, M. T. Chiang, and W. Chiang, *J. Food Drug Anal* 13(4), 361-367 (2005).
[66] O. Kim Sung, S Yun, and E. Lee E, *The Amer. J. Chinese Med.* 35(2), 297-308. (2007).
[67] S. A. Qureshi, W. Asad, And V. *Pak. Nut.* 8 (2), 125-128 (2009).
[68] P. R. Tirgar, K. V. Shah, V. P. Patel, T. R. Desai, R. K. Goyal, *Res. J. Pharm. Biol. Chem. Sci.* 1(4), 672-682 (2010).
[69] J. Charoenteeraboon, C. Ngamkitidechakul, N. Soonthornchareonnon, K. Jaijoya, and S. Sireeratawong, *Songklanakarin J. Sci. Tech.* 32 (6), 599-604 (2010).
[70] K. H. Khan, *Bot. Res. Intern.* 2(4), 218-228 (2009).

[71] S. M. Khopde, K. I. Priyadarsini, H. B. Mohan, V. Gawandi, J. G. Satav, J. Yakhmi, M. M. Vbanavaliker, M. K, BIYANI, and M. P. Mittal, *Cur. Prac.* 8(2). (2001).

In: Phytochemicals
Editor: Hai-Xue Kuang

ISBN: 978-1-62417-354-7


*Chapter II*

# Phytochemicals and their Potential Effects on Inflammatory Bowel Disease

***Susana Gorzalczany****

Pharmacology Department, Facultad de Farmacia y Bioquímica,
Universidad of Buenos Aires, Argentina

## Abstract

The incidence of gastrointestinal diseases has become a public health concern with high social costs. Inflammatory bowel disease (IBD), which includes ulcerative colitis and Crohn´s disease, is a chronic intestinal disorder. Their etiology is at present unknown and the therapy remains palliative rather than curative, exhibiting distressing side effects. There is a clear demand for new agents for the treatment of IBD. The interest of isolated compounds from natural sources as potential drugs for the treatment of IBD is growing in the last years. Phytochemicals include different kinds of chemical structures. Many molecules showed antioxidant activity, anti-inflammatory effect and immumodulatory properties and might be useful as adjunct therapy for IBD. The effect of phytochemicals in the complex cell-signaling pathway of gastrointestinal diseases will be discussed and their mechanism of action analyzed.

## Abbreviations

IBD: inflammatory bowel disease
CD: Crohn´sdisease
UC: ulcerative colitis
TNF: Tumor necrosis factor
TNBS: 2,3,6 trinitrobenzene sulfonic acid
DSS: dextran sodium sulfate

* E-mail address: sgorza@ffyb.uba.ar.

COX-2: cyclooxygenase-2
NFκB: nuclear factor κB
IL: interleukin
MPO: mieloperoxidase
HO-1: hemoxygenase-1
MAPK: mitogen activated protein kinase
NO: nitric oxide
PGE2: prostaglandin 2
LTB4: leukotrien B4
iNOS: inducible isoform of nitric oxide synthase
p.o.: oral administration
IFNγ: interferon gamma
Iκκ: IκB kinase
TCR: T cell receptor
CD8: cluster of differentiation
PPARγ: peroxisome proliferator activated receptor gamma

## Inflammatory Bowel Disease

Inflammatory bowel disease (IBD) comprises Crohn´s disease (CD) and ulcerative colitis (UC), both chronic inflammatory conditions of the gastrointestinal tract. CD is traditionally described as inflammation affecting any part of the gastrointestinal tract from mouth to anus. Crohn´s can affect all layers of the gastrointestinal tract and produces a characteristic lesion, the epithelioid granuloma. Abdominal pain, weight loss, and variable degrees of diarrhea are symptoms of CD. UC affects the colon only with inflammation in continuity from the rectum for a varying distance proximally. Diarrhea, hematochezia, tenesmus, and defecatory urgency are classic symptoms of active ulcerative colitis. It affects only the superficial layer of the colon and unlike CD, it does not cause strictures or fistulae [1, 2, 3].

The highest prevalence of IBD worldwide was reported in Canada and Europe, whereas Asia had a lower prevalence of IBD. Although there are a few epidemiologic data from developing countries, studies of migrant populations and populations of developing countries demonstrated a slow increase in the incidence of UC whereas the incidence of CD remained low. But the increase in the incidence of UC preceded the increase in the incidence of CD by about 10–15 years. In the last years the incidence and prevalence of IBD has been increasing in many regions of the world, indicating IBD as a global disease [4, 5, 6, 7].

IBD is a chronic condition with relatively low mortality. Healthcare utilization by patients is high and disease associated morbidity is significant for those with progressive and difficult to manage disease [8, 9].

Despite the important advances in the last years, many of the primary features of human IBD remain unexplained and its etiology remains uncertain [10]. Genetics are known to play a role in the development of CD [11], especially. But the increase in incidence of IBD during the last decades in Western countries and in countries with a westernized lifestyle and the limited familial concurrence of IBD indicate that environmental factors also take part in development of these diseases [12, 2]. Moreover, it has also been suggested that the gut microbiota participate in the progress and persistence of IBD and numerous modifications of intestinal microbiota composition have been identified [13]. To sum up, the ongoing inflammation in IBD results from persistent overly aggressive inflammatory responses to a subset of commensal microorganisms in a genetically susceptible host with exposure to environmental triggers. The diagnosis of either CD or UC is based on clinical symptoms combined with typical findings on endoscopy, radiology and pathology. Biomarkers are currently used in conjunction with routine testing in clinical care of patients with IBD. Most biomarkers used are not disease specific (C-reactive protein, Erythrocyte sedimentation rate, platelet count, fecal calprotectin, fecal lactoferrin, antineutrophy cytoplasmic antibodies, anti Saccharomyces cerevisiae antibodies, etc.). Also, an activity index, such as the Crohn'sdisease activity index and Harvey-Bradshaw index, including subjective and objective markers of disease (diarrhea, abdominal pain, general well-being, extra-intestinal manifestations, anti-diarrheal use, abdominal mass, hematocrit, and weight) are used [3, 14].

As both CD and UC remain medically incurable conditions associated with potentially significant morbidity, the focus of the treatment must be to reduce or eliminate symptoms, optimize nutritional status, promote normal growth, prevent complications and minimize the psychological effects of these chronic illnesses.

The choice of treatment is dependent on the type of disease (CD or UC) and the extent, distribution and presenting features. Nutritional modalities (enteral nutrition) have also shown promise as primary therapy for children with CD, although not for UC. Corticosteroids (usually, prednisolone) start with high dose for 2-4 weeks before gradual reduction as symptoms improve. Once remission has been obtained, immunosuppressant agents are needed to maintain this in most patients. The thiopurines, 6-mercpatopurine or its prodrug azathioprine, are the drugs most commonly used to maintain remission. Second line immunosuppressants, such as methotrexate, can be considered if there is failure or intolerance on first line treatment. Immunomodulators such as cyclosporine and tacrolimus, which block the production of interleukin (IL)-2, have been demonstrated to be effective in controlling severe-fulminant UC and fistulizing CD. But they have significant potential toxicity, especially renal and neurologic, and blood levels, so they must be followed carefully to maximize efficacy while minimizing the risk of toxicity. Sulfasalazine and the 5-aminosalicylates can also be effective in mild cases of CD. Also, biological agents are introduced in the last years in order to improve the management of IBD refractory to conventional therapies. Infliximab is the best studied chimeric monoclonal antibody used against TNF-α. It can be used in the acute setting as a rescue therapy in patients with severe exacerbation refractory to second line immunosuppressants. Although infliximab has been used successfully in severe growth failure and pubertal delay, some cases of fatal hepatosplenic T-cell lymphoma have been described. Although other agents are used too, such as adalimumab and cerolizumab, more research on the benefits of early biological treatment on disease progression is needed [15, 16, 17, 18, 19, 20].

## Animal Models of Inflammatory Bowel Disease

The availability and use of animal models led us to test different compounds or agents that could lead to a possible cure for IBD or improve the available treatment. In the last years experimental models with a variable range of clinical manifestations similar to those observed in human IBD have been developed. These models contributed greatly to the advance of our knowledge about the underlying mechanism of the disease as well as treatment. An appropriate animal model should display certain key characteristics, the gut should exhibit morphological alterations and signs of inflammation and its pathophysiology should be similar to the human IBD. There are different kinds of animal models: genetic colitis models (Knockout and transgenic models), inducible colitis models (microbials, chemical, physical), spontaneous colitis models and adoptive transfer models [21, 22].

In 1993 Sadlack et al. reported that 50% of mice deficient for interleiukin-2 died between 4 and 9 weeks after birth but the remaining mice developed an inflammatory bowel disease with a striking clinical and histological similarity to ulcerative disease in humans [23]. In the same year Kunh et al. reported that IL-10 deficient mice suffered chronic enterocolitis[24], providing evidence for a primary role of the immune system in the etiology of IBD.

Rats transgenic for human HLA-B2 and $\beta_2$- microglobulin genes spontaneously develop colitis, gastritis, providing a suitable experimental model for investigating the influence of normal luminal bacteria on the cellular immune mechanism in chronic intestinal inflammation [25].

Different chemicals have been used to induce models of IBD such as acetic acid, iodoacetamide, indomethacin, peptidoblycan-polysaccharide, 2,4,6 trinitrobenzene sulfonic acid (TNBS), oxazolone and dextran sodium sulfate (DSS) [26, 27, 28]. The development of colitis due to repeated chemical triggering reflects the chronic aspects of intestinal inflammation and they have the advantage of evaluating the efficacy of prolonged prophylactic or therapeutic treatment of chronic colitis.

Although no single animal model reflects all of the pathogenic and clinical features of human IBD, each experimental model contributed to improve our understanding of the mechanisms underlying irritation and perpetuation of chronic intestinal inflammation. In the majority of the models only the large bowel rather than the small intestine is affected. Unfortunately, the onset and severity of disease in most of the spontaneous or genetic models are highly variable and generally, depend largely on environmental factors such as the enteric flora. So, for the purpose of evaluating novel therapeutic agents experimentally, induced acute colitis models are widely used although they resemble only some aspects of human disease. Chemically induced models provide an easy means to study factors involved in pathogenesis and to test new therapeutic agents for human IBD with high reproducibility and an additional advantage of cost-efficiency [29, 27].

Therefore, in animal models it is possible to test innovative experimental interventions in order to develop therapeutic strategies for IBD. Indeed, they have identified potential candidate molecules for human therapy [30]. Therefore, careful attention may be required to translate animal studies to clinical settings, ensuring that both safety and efficacy have been displayed [31].

## Phytochemicals

Plants contain ubiquitous organic molecules termed primary metabolites that are essential to cell structure and basic metabolism. These compounds also serve as substrates for the synthesis of secondary plant metabolites, which are accumulated at lower concentrations. After their ingestion, certain secondary metabolites interact beneficially with the regulatory domains of functional proteins that are shared by plants and herbivore. For instance, vitamins act as enzyme coactivators and their insufficient intake results in some disease. Nevertheless, other kinds of bioactive compounds from plants also exist. In this case, its deficiency state has not been described, denying them the title of essential micronutrients. They appear to protect against different diseases in virtue of their biological activities. After the Second World War, phytochemicals, especially polyphenols, gradually became a topic of intensive investigation in various plant-related scientific domains, including applied research areas such as agriculture, ecology, food science and nutrition as well as medicine [32, 33].

Polyphenols constitute one of the most numerous and widely distributed groups of natural products in the plant kingdom. More than 8000 phenolic structures are currently known. They are chemically characterized as compounds with phenolic structural features. Interestingly, many epidemiological and experimental studies have demonstrated that phenolic acids, referred to as "polyphenols" in the literature although they are really monophenolics compounds, flavonoids and non-flavonoid polyphenols could be considered important to human health. Polyphenols or polyphenol rich diets provide significant protection against the development and progression of many chronic pathological conditions including cancer, diabetes, cardiovascular disease and aging [34, 35, 36].

## Phenolic Acids and IBD

Vanillic acid, an oxidized form of vanillin is a benzoic acid derivate used as a flavoring agent. Kim et al. (2010) demonstrated that vanillic acid reduced the severity of signs of dextran sodium sulfate (DSS)-induced colitis (weight loss, shortening of colon length and disease activity index), suppressed the expression of cyclooxygenase-2 (COX-2), the activation of transcription nuclear factor κB (NF-κB) p65 and reduced the increased plasma levels of IL 6 [37].

In the same sense, caffeic acid phenethyl ester caused a significant reduction in the colon injury and myeloperoxidase activity induced for 2,4,6 trinitrobenzene sulfonic acid (TNBS) in vivo [38]. Therefore, in Caco-2 cells, a human colon cancer in culture, caffeic acid inhibited the oversecretion of $H_2O_2$ and TNF-α as well, induced by the IL8, a central pro-inflammatory chemokine involved in the pathogenesis of IBD [39].

In addition, 3,5-dicaffeoyl-4malonylquinic acid isolated from *CentellaAsiatica* reduced the diarrhea and the loss of body weight, with a reduction of colonic mieloperoxidase (MPO) activity and NF-κB activation induced by intracolonic instillation of dinitrobenzene sulphonic[40]. Another prodrug of 4-geranyloxy-ferulic acid showed cancer chemopreventive ability in colitis-related colon carcinogenesis and lowered colonic inflammation induced by DSS, enhanced hemoxygenase-1 (HO-1) expression in the colon of mice. HO-1 is involved in the heme degradation process in the colonic mucosa, representing a novel target to achieve chemoprevention and chemoprotection of colonic inflammation [41]. Moreover a mixture of phytosterylferulates isolated from rice bran improving the clinical signs of DSS-induced colitis in mice, such as body weight, and stool consistency as well as decreasing intestinal bleeding and preserving colon length. The oral administration inhibited the transcripts of TNFα, IL1b and IL-6 and reduced the expression of COX-2 [42].

# Flavonoids and IBD

Flavonoids have the C6–C3–C6 general structural backbone in which the two C6 units are of phenolic nature. Due to the hydroxylation pattern and variations in the chromane ring, flavonoids can be further divided into different sub-groups such as chalcones, flavones, flavonols, flavanones, isoflavonoids.

Many pure polyphenolic compounds have been proposed as protective agents for colon inflammation and their effects were observed using in vivo or in vitro models [43, 44].

In 1976, Galsanov et al. described, for the first time, that quercitrin promoted a decline of focal lymphohistiomonocytic infiltrations, of eosinophilic in the mucosa and of scleroti changes in the large intestine in an experimental model of enterocolitis[45]. More recently, these observations were confirmed using a specific model for colonic inflammation[46, 47].

Quercitrin (3-rhamnosylquercetin) and rutin (3-O-rhamnosyl-glucosy-quercetin), lack anti-inflammatory properties in vitro but act as pro-drugs, attenuating the production of the critical proinflammatory mediators in DSS-induced colitis in mice. Nevertheless, quercetin failed to ameliorate colitis despite the efficacy of its glycoside [48, 49, 50].

Apigenin (5 mg/kg/day by oral administration), isolated from *Cordiadichotoma*, has shown promising activity in an acetic acid induced UC experimental model, reducing neutrophil infiltration as evidenced by suppression of colon MPO and improvement of histological features [51].

Luteolin, a major flavonoid of *Lonicera japonica* suppressed TNF-α induced IL-8 production in human colonic epithelial cells, inhibiting TNF-α induced phosphorylation of p38 MAPK [52]. Although luteolin showed beneficial effects on spontaneous colitis of transgenic mice, it aggravated DSS-induced experimental colitis by blocking NF-kB dependent protective molecules in enterocytes [53].

Kaempferol could be an effective anti-inflammatory agent for IBD since it inhibited allergic responses in relation to intestinal epithelium barrier function in RBL-2H3 and Caco-2 cells and decreased the plasma levels of nitric oxide (NO), prostaglandin E2 (PGE2), leukotrien B4 (LTB4) and the MPO activity in colonic mucosa in a DSS-induced UC model [54, 55].

Altough myricetin is claimed to have anti-colorectal cancer action, its relationship with IBD has not yet been studied[56].

Naringin and the biosynthesized naringenin demonstrated the ability to reduce edema of the gut, reduce levels of nitrates through inhibition of inducible isoform of nitric oxide synthase (iNOS) and reduce lipid peroxidation by oxygen radicals scavenging in DSS colitis in mice [57].

Flavanols of falvan-3 ols are often commonly called catechins. Catechin is the isomer with *trans* configuration and epicatechin is the one with *cis* configuration. Flavanols are found in many fruits, particularly in the skins of grapes, apple and blueberries. Monomeric flavanols and their derivatives (e.g. gallocatechins) are the major flavonoids in tea leaves and cacao bean. Catechin and epicatechin can form polymers, which are often referred to as proanthocyanidins [58, 59, 36, 60]. Although tea polyphenols, especially catechins of green tea, have been the primary agent responsible for the beneficial activity [61], during the production of black tea, substantial proportions of catechins are converted to theaflavins and thearubigins by a polyphenol oxidase that also possess promisory activities [62]. In 1998 it was demonstrated for the first time, that supplementation of catechins significantly decreases colonic damage induced by TNBS in rats [63]. Oral administration (p.o.) of theaflavin 3-3′-digallate (5 mg/kg/day) significantly improved TNBS-induced colitis associated with decreased mRNA and protein levels of TNFα, IL-12, interferon-gamma (IFNγ) and iNOS in colonic mucosa and markedly inhibited TNBS-induced increases in nuclear localization of NF-κB, cytosolic IκB kinase (Iκκ) activity in colon tissue [64]. In the same sense 40 mg/kg/day (p.o.) of thearubigin reduced the upregulation of proinflammatory T-helper 1 cytokine, leading to the suppression of iNOS, lipid peroxidation and tissue injury [65]. In addition, epigallocatechin-3-gallate improved the clinical course, attenuating histological damages to the colon and reducing the tissue concentrations of malondialdehyde [66, 67, 68, 69, 70]. In an attempt to improve the bioavailability of epigallocatechin-3-gallate, peracetylation of this compound was applied and its effect on a DSS model demonstrated was that it inhibited histone 3 lysine 9 (acetyl-H3K9) acetylation, decreased the proinflammatory mediator levels by down-regulating of PI3K/Akt/NFκB phosphorylation and p65 acetylation and increased heme oxygenase-1 expression via activation of extracellular signal-regulated protein kinase (ERK) [71].

In addition, apple procyanidins attenuated mortality and prevented the body weight loss associated with DSS and oxazolone-induced colitis, increasing the proportions of T cell receptors (TCR gammadelta) and clusters of differentiation 8 (CD8) in intraepithelial lymphocytes and suppressed IFNγ[72]. Furthermore large procyanidins, such as hexamericprocyanidins, although not absorbed by cells, have the capacity to modulate TNF-α- induced NF-κB activation in Caco-2 cells. The activity of these high molecular weight procyanidins could be limited to the gut lumen, representing compounds with potential anti-inflammatory actions at the gastrointestinal tract [73].

Although pelargonidin has been reported to inhibit NO production by reducing iNOS expression through inhibiting the activation of transciptor factor NF-κB, its protective effect on the IBD model still remains unknown [74].

Peptidergic capsaicin-sensitive afferent nerves are involved in the pathogenesis of IBD, accordingly the topical application of capsaicin protected against an acute colonic injury provoked by TNBS+ethanol in rats [75].

With less level of evidence genistein[76], phloretin, a dihydrochalcone isolated from apples and strawberries [77], cirsilineol, a natural flavone isolated from *Artemisia vestita* [78] showed beneficial effects in rodent models of IBD.

## Other Polyphenols and IBD

In addition several non-flavonoid polyphenols are considered important to human health. Resveratrol is a polyphenolicphytoalexin found in peanuts, grapes and red wine. It is recognized as one of the most promising natural molecules in the prevention and treatment of chronic inflammatory disease and autoimmune disorders [79, 80].

Resveratrol (5-10 mg/kg/day) reduced the degree of colonic injury, the index of neutrophil infiltration and the COX-2 expression as well as enhanced apoptosis in TNBS-treated rats [81]. Moreover, resveratrol supplementation prevented mucosal damage and attenuated the expression of iNOS in an experiment necrotizing enterocolitis in rats [82]. As well, beneficial effects are reported on DSS-induced ulcerative colitis by increasing the cell protection against reactive oxygen species (ROS)-induced damage (enhanced activity of superoxide dismutase and glutathione peroxidase in colonic tissue and reduced the subunits of NADPH oxidase $g22_{phox}$ and $gp91_{phox}$). Furthermore, resveratrol enhanced the level of the anti-inflammatory cytokine IL-10 and the expression of silent mating type information regulation-1 (SIRT1). Several investigations have demonstrated, in the same model, that resveratrol decreased the expression levels of cytokines (TNFα, IL-8, INFγ, IL-1β), prostaglandin E synthase-1, COX-2, iNOS, via downregulation of p38 (MAPK) and NF-kB activity, a major transcription factor known to upregulate proinflammatory gene expression [83, 84, 85, 86, 87].

Over the last few years, a number of in vivo and in vitro studies have provided evidence in support of important pharmacological properties of ellagic acid. It is a naturally occurring plant phenol found in certain fruits, nuts and vegetables such as strawberries and raspberries. The oral administration of ellagic acid (10-20 mg/kg) diminished the severity and extension of the intestinal injuries induced by TNBS and DSS, decreasing neutrophil infiltration, over expression of COX-2 and iNOS plus the activation of p38MAPK. p38MAPK is a key modulator of several genes that control infiltration of monocytic cells, the activity of leukocytes, acute intestinal inflammation and cytokine production [88, 89].

Curcumin is a non-flavonoid polyphenol isolated from turmeric (*Curcuma longa*) that provides its pigmentation. Curcumin could be useful to prevent or treat various pathophysiological processes, including carcinogenesis, wound healing and inflammation [90]. At a dose of 50-100 mg/kg p.o. caused a reduction of levels of cytokine TNFα, enhanced activity of MPO as well as COX-2 and iNOS expression. It was able to reduce the activation of p38 MAPK in colonic tissue of mice with colitis induced by TNBS. The peroxisome proliferator-activated receptor gamma (PPARγ) could also be involved in the improvement induced by curcumine[91, 92, 93]. It has been also reported that curcumine could change cytokine profiles, from the pro-inflammatory Th1 to the anti-inflammatory Th2 type [94].

Also PPARγ was involved in the beneficial effect of verbascoside in IBD animal models [95].

## Clinical Trials

Epidemiologic studies suggest a protective effect of vegetables and fruits against cancer, cardiovascular and neurodegenerative diseases [96, 97, 98, 99].

Although epidemiologic studies are useful for evaluation of the human health effects of long-term exposure to physiologic concentrations of polyphenols, controlled randomized clinical trials are necessary to answer the question of data from animal models and in vitro systems are relevant for human disease outcomes. An important phenomenon is that, after absorption, polyphenols are subject to phase II metabolism that may greatly influence their bioactivity, but few studies have been published about this subject.

Among the various polyphenols apigenin, epigallocatechingallate, and resveratrol are likely to have a potential value in preventive and therapeutic roles in cancerogenic conditions of the gastrointestinal tract [100, 101].

Since many polyphenols possess anti-inflammatory and anti-oxidative properties, they are strong candidates for new treatment options for patients with chronic inflammatory bowel diseases. In spite of the great amount of compounds, only very few of them have been studied in clinical trials. Epigallocatechingallate, in form of Polyphenon E®, is an example of one such compound [102]. Moreover, curcumin is also being investigated in clinical trials phase I and III, alone or associated with aminosalicylic acid, both for adults and pediatrics patients to treat mild to moderately severe ulcerative colitis [103, 104, 105, 106].

Nevertheless, in general, very few clinical trials have been performed to determine the ability of polyphenols to treat IBD.

## General Considerations

Numerous studies have been published identifying molecular mechanisms of polyphenol effects. Mostly evidence of efficacy has been derived from animal models. Taking into account the limited bioavailability of polyphenols, it will be necessary to study their efficacious in humans.

The available information lets us know the potential utility of polyphenols or new agents related to them for the possible treatment of inflammatory bowel disease, suggesting alternative treatment strategies for IBD patients that are refractory to corticosteroids or thiopurins or as adjuncts to conventional treatment. Nevertheless, its use requires extensive clinical testing before in order to optimize its potential use in patients. It is important to identify which polyphenols might be the most useful for treatment of IBD. Another challenge will be to develop specific preparations of alone or associated polyphenols in order to deliver these compounds to the target structures, enhancing its bioavailability.

Despite tremendous progress that has been made in our understanding of pathobiology of IBD through research on in vivo or in vitro models in recent years, many features of human disease remain unexplained. Consequently, it is more difficult to choose the best treatment for these patients. Therefore, there is a great interest in finding new approaches to prevent or treat IBD. The research of compounds isolated from natural sources is now part of a dynamic discovery process and they could be a novel way to improve current treatment.

## References

[1] R. Hansen, F. L. Cameron, G. L. Hold, E. M.El-Omar, R. K.Russell, *Pediatrics and Child Health.* 20, 473 (2010).

[2] V. Andersen, A. Osen, F., Carbonnel, A. Tjonneland, U. Vogel, *Digestive andLiverDisease.* 44, 185 (2012).

[3] H. N. Iskandar and M. A., Ciorba, *Translational Research.* Preprint, 2012 at http://www.sciencedirect.com/science/article/pii/S1931524412000023.
[4] M. Russel, *European Journal of Internal Medicine*11, 191 (2000).
[5] C. Abraham and J.H. Cho, *N. Engl.* J. *Med.*21, 2066 (2009).
[6] J. Cosnes, C. Gower–Rousseau, P. Seksik, A. Cortot, *Gastroenterology* 140, 1785 (2011).
[7] N. Molodecky, I. Shian Soon, D. M., Rabi, W. A. Ghali, M. Ferris, G., Chernoff, E. I. Benchimol, R. Panaccione, S. Ghosh, H.W. Barkema and G. G. Kaplan, *Gastroenterology* 142, 46 (2012).
[8] T. Longobardi, C. N. Bernstein. *Am. J. Gastroenterol.* 102, 683 (2007).
[9] J. Buchanan, S. Wordsworth, T. Ahmad, A. Perrin, S.Vermeire, M. Sans, J. Taylor, D. Jewell. *Journal of Crohn's and Colitis*5, 301 (2011).
[10] J. Colombel, A. Watson, M. Neurath, *Gut* 57, 429 (2008).
[11] P. Rosenstiel, C. Sina, A. Franke, S. Schreiber *Seminars in Immunology*21, 334 (2009).
[12] T. S. Hansen, T. Jess, I. Vind, M. Elkjaer, M. Fey Nielsen, M. Gamborg, P. Munkholm*Journal ofCrohn'sandColitis*5, 577 (2011).
[13] O. Thompson-Chagoyán, J. Maldonado, A. Gi. *Clinical Nutrition* 24, 339 (2005).
[14] J. D. Lewis *Gastroenterology* 140, 1817 (2011).
[15] J. H. Kwon, R. J. Farre. *Critical Reviews in Oncology/Hematology* 56, 169 (2005).
[16] J. Markowitz. *Digestive and Liver Disease* 40, 16 (2008).
[17] S. Saibeni, T. Virgilio, R. D'Incà, L. Spina, A. Bortoli, M. Paccagnella, M. Peli, R. Sablich, G. Meucci, E. Colombo, G. Benedetti, C.M. Girelli, G. Casella, G. Grasso, R. de Franchis, M. Vecchi. *Digestive and LiverDisease* 40, 814 (2008).
[18] A. Orlando, F. Mocciaro, G. Civitavecchia, D. Scimeca, M. Cottone*Digestive and Liver Disease* 40, S236 (2008).
[19] H. Lochs. *e-SPEN, the European e-Journal of Clinical Nutrition and Metabolism* 5, e100 (2010).
[20] F. Magro, F. Portela, *BioDrugs*24, 3 (2010).
[21] J. C., Hoffman, N. N. Pawlowsku, A. A. Kühl, W. Höhne, M. Zeits, *Pathobiology* 70, 121 (2003).
[22] A. R. Jurjus, N. N. Khoury, J. M. Reimund, *Journal of Pharmacological and Toxicological Methods* 50, 81 (2004).
[23] B. Sadlack, H. Merz, H. Schorle, A. Schimpl, A. Feller and I. Horak, *Cell* 75, 253 (1993).
[24] R. Kühn, J. Löhler, D. Rennick, K. Rajewsky, W. Müller, *Cell* 75, 263 (1993).
[25] H.C. Rath, *Pathobiology* 70, 131 (2003).
[26] I. Okayasu, S. Hatakeyama, M. Yamada, T. Ohkuse, Y. Inagaki, R. Nakaya, *Gastroenterology* 98, 694 (1990).
[27] S. Wirtz, M. F. Neurath, *Advanced Drug Delivery Reviews* 59, 1073 (2007).
[28] A. Mizoguchi, *Progress in Molecular Biology and Translational Science* 105, 263 (2012).
[29] M. E. A. Borm, G. Bouma, *Drug Discovery Today: Disease Models* 1, 437 (2004).
[30] W. Strober, I. Fuss, P. Mannon, *J. Clin. Investi.* 117, 514 (2007).
[31] A. Mizoguchi, E.Mizoguchi, *Current Opinion in Pharmacology* 10, 578 (2010).
[32] H. Shapiro, P. Singer, Z. Halpern, R. Bruck, *Gut* 56, 426 (2007).

[33] S. Quideau, D. Deffieux, C. Douat-Casassus, L. Pouységu, *AngewandteChemie International Edition*50, 586 (2011).
[34] A. Scalbert, I. Johnson, M. Saltmarsh, *Am. J. Clin. Nutr.* 81, 215S (2005).
[35] K. Pandey, S. Rizvi,*Oxid. Med. Cell Longev.* 2, 270 (2009).
[36] R. Tsao, *Nutrients*2, 1231 (2010).
[37] S. J. Kim, M. C. Kim, J. Y. Um, S. H. Hong, *Molecules*19, 7208 (2010).
[38] R. O. Ek, M. Serter, K. Ergin, Y. Yildiz, S. Cecen, T., Kavak T., C. Yenisey, *Dig. Dis. Sci.* 53, 1609 (2008).
[39] Z. Zhao, H.S. Shin, H. Satsu, M. Totsuka, M. Shimizu, *J. Agric. FoodChem.* 56, 3863 (2008)
[40] R. di Paola, E. Esposito, E. Mazzon, R. Caminiti, R.D., Toso, G. Pressi, S. Cozzocrea, *FreeRadic. Res.* 44, 74 (2010).
[41] S. Miyamoto, F. Epifano, M. Curini, S. Genovese, M. Kim, R. Ishigamori-Suzuki, Y. Yasui, S. Sugie, T. Tanaka, *NutritionandCancer*60, 675 (2008).
[42] M. S. Islam, T. Murata, M. Fujisawa, R., Nagaska, H. Ushio, A. M. Bari, M. Hori, H. Ozaki, *British JournalofPharmacology* 154, 812 (2008).
[43] B. Romier, Y. J., Schneider, Y. Larondelle, A. During, *Nutr. Rev.* 67, 363 (2009).
[44] C. Frontela, R. Canali, F. Virgili, *Gastroenterol. Hepatol.* 33, 307 (2010).
[45] S. B. Galsanov, A. D. Tourova, E. D. Klimenko, *BiullEksp. Biol. Med.* 81, 623 (1976).
[46] F. Sánchez de Medina, B. Vera, J. Gálvez, A. Zarzuelo, *Life Sci.* 70, 3097 (2002).
[47] D. Camuesco, M. Comalada, A. Concha, A. Nieto, S. Sierra, J. Xaus, A. Zarzuel, J. Gálvez, *Clin. Nutr.* 25, 466 (2006).
[48] D. Camuesco, M. Comalada, M. E. Rodrìguez-Cabezas, A. Nieto, M. D. Lorente, A. Concha, A. Zarzuelo, J. Gálvez, *Br. J. Pharmacol.* 143, 908 (2004).
[49] M. Comalada, D. Camuesco, S. Sierra, I. Ballester, J. Xaus, J. Gálvez, A. Zarzuelo, *Eur. J. Immunol.*35, 584 (2005).
[50] K. H. Kwon, A. Murakami, T. Tanaka, H. Ohigashi, *BiochemicalPharmacology* 69, 395 (2005).
[51] A. B. Ganjare, S. A. Nirmal, A. N. Patil, *Fitoterapia*82, 1052 (2011).
[52] J. Kim, D. Kim, O. Kang, Y. Choi, H. Park, S. Choi, T. Kim, K. Yun, Y. Nah, Y. Lee, *International Immunopharmacology*5, 209 (2005).
[53] T. Karrasch, J. Kim, B. Jang, C. Jobin, *Plos ONE* 7, e596 (2007).
[54] E. Lee, G. Ji, M. Sung, *Inflamm. Res.* 59, 847 (2010).
[55] M. Park, G. Ji, M. Sung, *Dig. Dis. Sci.* 57, 355 (2012).
[56] P. Nimala, M. Ramanthan, *J. Exp. Ther. Oncol.* 9, 101 (2011).
[57] M. Amaro, J. Rocha, H. Vila-Real, M. Figueira, H. Filipe, *Food Research International* 42, 1010 (2009).
[58] C. Williams, R. Grayer, *Nat. Prod. Rep.*21, 539 (2004).
[59] N. Veitch, R. Graver, *Nat. Prod. Rep.* 25, 555 (2008).
[60] S. Sang, J. Lambert, C. Ho, C. Yang, *Pharmacol. Res.* 64, 87 (2011).
[61] C. Dufresne, E. Farnworth, *J. Nutr. Biochem.* 12, 404 (2001).
[62] D. Balentine, S. Wiseman, L. Bouwens, *Crit. Rev. Food Sci. Nutr.* 37, 693 (1997).
[63] K. Sato, A. Kanazawa, N. Ota, T. Nakamura, K. Fujimoto, *J. Nutr. Sci. Vitaminol.* 44, 769 (1998).

[64] A. Ukil, S. Maity, P. Das, *British Journal of Pharmacology* 149, 121 (2006).
[65] S. Maity, A. Ukil, S. Karmakar, N. Datta, T. Chaudhuri, J. Vedasiromoni, D. Ganguly, P. Das, *European Journal of Pharmacology*, 470, 103 (2003).
[66] M. Mochizuki, N. Hasegawa, *Journal of Health Science* 51, 362 (2005).
[67] Y. Lin, Y. Chen, Y. Song, *Zhonghua Yi XueZaZhi*13, 2965 (2007).
[68] Z- Ran, C. Chen, S. Xiao, *Biomed. Pharmacother.* 62, 189 (2008).
[69] M. Mochizuki, N. Hasegawa, *Phytother. Res.*24, S120 (2010).
[70] M. Brücknere, S. Westphal, W. Domschike, T. Kucharzik, A. Lügering, *J. Crohns Colitis* 6, 226 (2012).
[71] *Y.* Chiou, N. Ma, S. Sang, C. Ho, Y. Wang, M. Pan, *J. Agric. Food Chem.* Preprint, 2012 at http://pubs.acs.org/doi/abs/10.1021/jf300441p.
[72] Y. Yoshioka, H. Akiyama, M. Nakano, T. Shoji, T. Kanda, Y. Ohtake, T. Takita, R. Matsuda, T. Maitani, *Int. Immunopharmacol.* 20, 1802 (2008).
[73] A. Erlejman, G. Jaggers, C. Fraga, P. Oteiza, *Archives of Biochemistry and Biophysics* 476, 186 (2008).
[74] M. Hämäläinen, R. Nieminen, P. Vuorela, M. Heinonen, E. Moilanen, *Mediators of Inflammation* 1 (2007).
[75] Z. Szepes, J. Kiss, T. Molnk, D. Lamarque, G. Jancsbh, F. László,*J. Physiol. Paris* 91, 123 (1997).
[76] J. Seibe, A. Molzberger, T. Hertrampf, U. Leschowski, P. Diel, *Eur. J. Nutr.* 48, 213 (2009).
[77] J. Lee, S. Regmi, J. Kim, M. Cho, H. Yun, C. Lee, J. Lee, *Infect. Immun.* 79, 4819 (2011).
[78] Y. Sun, X. Wu, Y. Yin, F. Gong, Y. Shen, T. Cai, X. Zhou, X. Wu, Q. Xu, *BiochemicalPharmacology*79, 229 (2010).
[79] L. Frémont, *Life Sci.* 66, 663 (2000).
[80] H. Kalantari, D. Das, *Biofactors*36, 401 (2010).
[81] A. Martín, I. Villegas, C. La Casa, C. de la Lastra, *Biochem. Pharmacol.* 67, 1399 (2004).
[82] O. Ergün, G. Ergün, G. Oktem, N. Selvi, H. Dogan, M. Tunçyürek, G. Saydam, A. Erdener, *J. Pediatr. Surg.* 42, 1687 (2007).
[83] K. Williams, C. Randall Fuller, L. Dieleman, C. DaCosta, K. Clark, R. Balfour Sartor, P. Lund, *Gastroenterology* 118, A556 (2000).
[84] J. Youn, J. Lee, H. Na, J. Kundu, Y. Surh*Nutr. Cancer* 61, 847 (2009).
[85] S. Sánchez-Fidalgo, A. Cárdeno, I. Villegas, E. Talero, C. de la Lastra, *European Journal of Pharmacology* 633, 78 (2010).
[86] U. Singh, N. Singh, B. Singh, L. Hofseth, R. Price, M. Nagarkatti, P. Nagarkatti*J. Pharmacol. Exp. Ther.* 332, 829 (2010).
[87] J. Yao, J. Wang, L. Liu, Y. Li, A. Xun, W. Zeng, C. Jia, X. Wei, J. Feng, L. Zhao, L. Wang, *Archives of Medical Research* 41, 288 (2010).
[88] Y. Ogawa, K. Kanatsu, T. Iino, S. Kato, Y. Jeong, N. Shibata, K., Takada, K. Takeuchi, *Life Sciences* 71, 827 (2002).
[89] M. Rosillo, M. Sánchez-Hidalgo, A. Cárdeno, C. Alarcón de la Lastra, *Biochemical Pharmacology* 82, 737 (2011).
[90] G. Irving, A. Karmokar, D. Berry, K. Brown, W. Steward, *Best Practice &Research Clinical Gastroenterology* 25, 519 (2011).

[91] M. Zhang, C. Deng, J. Zheng, J. Xia, D. Sheng, *International Immunopharmacology*6, 1233 (2006).

[92] L. Camacho-Barquero, I. Villegas, J. Sanchez-Calvo, E. Talero, *International Immunopharmacology*7, 333 (2007).

[93] S. Mouzaoui, I. Rahim, B. Djerdjour, *International Immunopharmacology*12, 302 (2012).

[94] H. Hanai, K. Sugimoto, *Curr. Pharm. Des.* 15, 2087 (2009).

[95] E. Esposito, E. Mazzon, I. Paterniti, R. Dal Toso, G. Pressi, R. Caminiti, S. Cuzzocrea, *PPAR Res.* Article ID 917312 (2010).

[96] Y. Surh, *Nat. Rev. Cancer* 3, 768 (2003).

[97] I. Arts and P. Hollman, *Am. J. Clin. Nutr.* 81, 317S (2005).

[98] S. Ramos, *J. Nutr. Biochem.*18, 427 (2007).

[99] M. Hu, *Chang Gung Med. J.* 34, 449 (2011).

[100] M. Goodman, R. Bostick, O. Kucuk, D. Jones, *Free Radical Biology & Medicine* 51, 1068 (2011).

[101] H. Hoensch, R. Oertel, *World J. Gastrointest. Oncol.* 15, 7 (2011).

[102] G. Dryden NCT number 00718094, www.clinicaltrials.gov.

[103] R. Mitchell, M. Byme, J. Baillie, *Gastroenterology* 130, 84 (2006).

[104] A. Lang, N. Salomon NCT number 01320436, www.clinicaltrials.gov.

[105] I. Dotan NCT number 00793130, www.clinicaltrials.gov.

[106] D. Suskind NCT number 00889161, www.clinicaltrials.gov.

In: Phytochemicals
Editor: Hai-Xue Kuang

ISBN: 978-1-62417-354-7


*Chapter III*

# Bioactive Compounds in Whole Grain Cereals

***Maria Teresa Pedrosa Silva Clerici*[1*]**
***and Elizabeth Harumi Nabeshima*[2]**
[1]College of Nutrition, Federal University of Alfenas – UNIFAL – MG, Brazil,
[2] Institute of Food Technology (ITAL), Brazil

## Abstract

Cereals are edible portions of the grass family, *Gramineae*, whose grains and products are an important source of energy, carbohydrate, protein and fiber, containing a range of micronutrients such as vitamin E, some of the B vitamins, magnesium, and zinc. Currently, it has been recognized as an important source of phytochemical compounds. The final nutrient content of a cereal after milling will depend on the extent to which the outer bran and aleurone layers are removed, as this is where the fiber, vitamins, minerals and bioactive compounds tend to be concentrated. Epidemiological studies have shown that regular consumption of whole grain cereal products is associated with reduced development of chronic diseases, such as cardiovascular diseases, type 2 diabetes, and some kinds of cancers. Phenolic acids and flavonoids represent the most common form of phenolic compounds found in whole grains, and they are among the major and most complex groups of phytochemicals. The increased use of whole grain flour in breads, pastas and other products can, in addition to promoting nutritional benefit, promote a better exploitation of cereal and reduce the by-product formation. This chapter aims to show studies regarding the great potential of use of whole grain cereals as a source of bioactive compounds.

[*] E-mail address: mariateresapedrosa@gmail.com.

# Introduction

Cereal grains may be separated into three main fractions [1, 2]:

- Endosperm contains 50-75% of starch and storage proteins (8-18%). In wheat, the proteins of endosperm are capable of forming a gluten network, that is important for baking and pasta industries, so the conventional milling process is focused on producing wheat flour without another parts of the grain. In corn, endosperm is used for starch extraction.
- Bran derives from the outer coverings of those grains and contains little or no endosperm or hull material. This part was used to feed animals, but the bran has been used for producing breads, snacks, and other products, because it contains fibers and phytochemical compounds.
- Germ is rich in lipid and has been used for oil extraction and its residue has been used for animal feeding. Germ is rich in vitamin E, phenolic acids, proteins and minerals, so it is important for production of functional foods. Corn oil is the most important product of germ.

Since ancient times, cereals are part of the population's diet. Until the industrialization and development of modern mills, cereals were consumed as whole grain and/or flour (bran, germ and endosperm), without the hull only. The products had dark color, characteristic strong flavor and low storage stability, because of the presence of lipids that rancidified easily.

The refined or white wheat flours, which have almost all the endosperm, are obtained from complex mills and have advantages, such as great shelf-life and provide consumers with more attractive products, since the soft flavor and white color of flours allow for obtaining a variety of products, such as breads, snacks, cakes, instant flours from cereals with different flavors. As the removal of the germ and bran from the grains affects the storage stability, the demand for refined products boosted the development of other processed and refined cereals, such as corn, which began to be degerminated with removal of its bran, and rice, which began to be modified in order to become white and polished. As a consequence, there was an increase in the production and consumption of refined cereal grains in the diet of several countries. Many studies now indicate a relationship between refined cereals and the increase in cases of obesity, coronary heart disease, cancer, and diabetes, i.e., Western diseases [3].

Since it was observed that in countries showing high intake of whole grains there is a lower incidence of Western diseases, there has been a great stimulus to the consumption of whole grain products. The beneficial effects of whole grains related to the fiber and resistant starch contents are well known [4, 5]. However, research has shown that the other bioactive compounds, such as carotenoids and phenolic compounds, in whole grain exhibit antioxidant activity and may be beneficial to health [6]. The history of the evolution of whole wheat and white flours was described by Spilher [7], in the first book devoted to the detailed study on whole-grain foods, published by Marquart and Fulcher [3].

This chapter will address the key research focusing on the role of bioactive compounds in whole grain of cereals, showing the importance of the current methods of identification and the effects of processing on these compounds.

## Whole Grain

Whole grain can be defined as cereal grains produced as intact, ground, cracked or rolled grains, whose key compounds (endosperm, germ and bran) are present at the same relative proportion as in the intact grain [8]. In food products, the whole grain can be intact, consisting of the original composition of bran, germ, and endosperm, throughout the entire lifetime of the product, or reconstituted, in which one or more of the original components of a whole grain is recombined to the relative proportion naturally occurring in the grain kernel [9]. Some countries establish regulations for including the expression "whole grain" in the label:

In the United States of America, the statement on whole grain products was approved in 1999 by the Food and Drug Administration (FDA), and in order to print this health claim on the packaging labels, the food must contain ≥ 51% of whole grain per weight, i.e., this grain must be the first listed in the ingredient list. In addition, the product must be associated to reduce total fat content. Cereal grains may include amaranth, barley, buckwheat, bulgur, corn (including popcorn), millet, quinoa, rice, rye, oats, sorghum, teff, triticale, wheat, and wild rice [10]. In this list are some pseudo grains, as per their similarities comparable in terms of nutritional composition and use [11].

The United Kingdom has similar requirements for whole grain related to foods containing 51% or more of whole grain ingredients by weight per serving. The term "whole grain" refers to the major cereal grains including wheat, rice, maize and oats [12].

In Sweden the whole grain product shall contain at least 50% whole grain. When applicable, the criteria for the "key hole" shall also be fulfilled. For some products this regulation requires 2/3 wholegrain and maximum 13% sugar for breakfast cereals, gruels and porridges; and 2/3 whole grain and maximum 10% fat for bread biscuits, rusks and pasta product [13].

## Nutritional Value

Table 1 shows the composition of macronutrients of the main whole grains consumed worldwide [14 to 19].

In addition to being sources of energy, due to its high starch content, they are rich sources of fiber (lignans, β-glucan, inulin, and resistant starch), vitamins, minerals, and phytochemicals (phenolics, carotenoids, vitamin E, sterols, and phytates). This complex mixture of bioactives in whole grain is considered unique and complementary to those in fruits and vegetables when consumed together. The additive and synergistic effects of phytochemicals on fruits, vegetables, and whole grains are responsible for their health benefits [20, 21].

The most studied bioactive compounds in grains are the following:

- Lignans are a group of dietary phytoestrogen compounds. The common plant lignans in the human diet include secoisolariciresinol, matairesinol, lariciresinol, pinoresinol, and syringaresinol [22].
- β-glucan is a group of linear polymers of glucose molecules connected by 70% of β-(1–4) and 30% of β-(1–3)–linkages. β-glucan is more flexible, soluble and viscous

than cellulose. β-glucan is commonly found in cell walls of many grains, such as oats, barley, and wheat [23].

- Phytochemicals are defined as biologically active compounds of plant origin that, when ingested, provide certain functional benefits beyond basic nutrition [20].

**Table 1. Nutritional value of whole grain (%)**

| Whole grain | Moisture | Protein | Fat | Ash | Carbohydrate | Fiber | Reference |
|---|---|---|---|---|---|---|---|
| Barley | 8.79 | 11.73 | 1.52 | 2.52 | 77.67 | - | 14 |
| Brown rice* | - | 7.1-8.3 | 1.6-2.8 | 1-1.5 | 73-87 | 0.6-1.0 (Crude) | 16 |
| Corn | 10.37 | 9.42 | 4.74 | 1.2 | 74.26 | 7.3 (Dietary) | 14 |
| Oat* | - | 13-16 | 6-7 | 2 | 54-64 | 11-13 (Dietary) | 15 |
| Rye* | Dry matter | 8.5-10 | - | 1.59-1.86 | 53- 55.7 | - | 17 |
| Sorghum* | 10.4 | 10.4-12.3 | 3.6 | 1.67 | - | - | 18 |
| Wheat * | Dry matter | 10-14 | 3 | 2 | 67-70 | 10-13 (Dietary) | 19 |

*Several cultivars or genotypes.

Phenolic acids can be subdivided into two major groups:

- Hydroxybenzoic acid derivatives, including p-hydroxybenzoic, protocatechuic, vannilic, syringic, and gallic acids.
- Hydroxycinnamic acid derivatives, including p-coumaric, caffeic, ferulic, and sinapic acids.

The phenolic compounds may be in free or bound form in cereal grains and, as opposed to what occurs in fruits, according to Liu [24] the bound phenolics contributed to 85% of the total amount in corn, 76% in wheat, 75% in oats, and 62% in rice, while free phenolics contributed to 15% of the total amount in corn, 24% in wheat, 25% in oats, and 38% in rice.

Adom and Liu [25] observed that the quantification of phenolics in grains must have a more appropriate methodology, since total phenolic contents of whole grains may be underestimated when the same methodology is used for fruits, which have higher amount of phenolic compounds in the free form. As an example, these authors showed that the ratio of free, soluble-conjugated and bound forms of ferulic acid in corn and wheat was 0.1:1:100.

The concentration of phenolic compounds and antioxidant activity may vary according to the type of cereal, the methodology used, and also the varieties of cereals, which can be observed in Table 2 [26 to 32]. For example, Adom et al. [33] observed that ferulic acid content of whole grains differs among 11 wheat varieties. They observed a significant difference (up to two-fold) among wheat varieties, as well as in ferulic acid, mostly in the bound form in all varieties.

**Table 2. Antioxidant activity of grain cereals and products**

| Products | Total Phenolics Content [4] | DPPH scavenging activity[5] | TEAC[6] | Reference |
|---|---|---|---|---|
| Bread wheat[1] | 1-1.3 mg of CE/g d.m. | 8.82-10.62 ($IC_{50}$. mg d.m.) | | 26 |
| Durum wheat[1] | 1.5- 1.6 mg of CE/g d.m. | 8.88-10.35 ($IC_{50}$. mg d.m.) | | 26 |
| Rye[1] | 1.8- 2.3 mg of CE/g d.m. | 6.85-7.50 ($IC_{50}$. mg d.m.) | | 26 |
| Rye [1] | 2.2-2.4 mg/g d.m. | | 14.3-15.2 PBS extract (μmol trolox/g d.m.) | 27 |
| Hull-less barley[1] | 2.8-3.8 mg of CE/g d.m. | 1.38-1.81 ($IC_{50}$. mg d.m.) | | 26 |
| Hull less oat [1] | 2.2-3.0 mg of CE/g d.m. | 6.57-8.66 ($IC_{50}$. mg d.m.) | | 26 |
| Whole-wheat spaghetti[1] | 773- 1423 (μg equivalent of ferulic acid/g) | 1.08-1.87 μmol equivalent of trolox/g | | 28 |
| Sorghum grain[2] | 0.5- 0.8 mg GAE/g flour | 17.11- 18.32 mg/mL | | 29 |
| Whole flour wheat[1] | 613-1006.5 μmol GAE/100g | | | 30 |
| Oat [3] | | | 28.5 mmol TEAC/kg | 31 |
| Rice [3] | | | 20 mmol TEAC/kg | 31 |
| Barley [3] | | | 24 mmol TEAC/kg | 31 |
| Corn [3] | | | 33 mmol TEAC/kg | 31 |
| Wheat flour | 6.96 mg/g d.m. | 3.95 mmol Trolox/kg d.m. | | 32 |
| Buckwheat flour | 7.25 mg/g d.m. | 3.8 mmol Trolox/kg d.m. | | 32 |
| Amaranth flour | 2.71 mg/g d.m. | 3.6 mmol Trolox/kg d.m. | | 32 |
| Quinoa | 2.8 mg/g d.m. | 6.2 mmol Trolox/kg d.m. | | 32 |

[1] Several genotypes or cultivars.
[2] Several extracts.
[3] Extraction procedure: Water+ethanol+bound (%).
[4] Where: DPPH, 2,2-diphenyl-1-picrylhydrazyl; CE, catechin equivalents; GAE, gallic acid equivalent; d.m.= dry matter.
[5] Where: IC50 is expressed as the weight of the flour from which the extract was prepared to quench 50% DPPH radicals under experimental conditions.

According Liu [21], bound phytochemicals cannot be digested by human enzymes, could survive stomach and small intestine digestion, and therefore may possibly reach the colon. The colonic microbiota may release the bound phytochemicals through fermentation, and thus provide site-specific health benefits in colon or other tissues after absorption.

Carotenoids are classified into hydrocarbons (carotenes) and their oxygenated derivatives (xanthophylls). In whole grains it may be commonly found lutein, zeaxanthin, β-cryptoxanthin, β-carotene, and α-caroten [34, 35].

Vitamin E has two types of structures, tocopherols and tocotrienols, which are considered lipid-soluble antioxidants.

Tocopherol and tocotrienol have similar structures, but tocopherols contain saturated phytol side chains, while tocotrienols have three carbon-carbon double bonds in the phytol side chain. Vitamin E compounds are found in many foods including whole grains, where they are mostly present in the germ fraction [26].

Panfili et al. [36] analyzed the concentration of vitamin E compounds in whole grains and found 75 mg/kg dry weight (DW) total tocopherol in soft wheat and barley; 33–43 mg/kg DW β-tocotrienol in hulled and dehulled wheat; 45 mg/kg DW δ-tocopherol in corn; 56 and 40 mg/kg DW a-tocotrienol in oats and barley, respectively.

Phytosterols are a collective term for plant sterols and stanols, which are similar in structures to cholesterol, differing only in the side chain groups. Plant sterols have a double bond in the sterol ring, and plant stanols, the hydrogenated counterparts of corresponding sterols, lack a double bond in the sterol ring. The most common plant sterols are sitosterol, campesterol, and stigmasterol, and the most common plant stanols are sitostanol, campestanol, and stigmastanol [37].

## Other Nutrients

Cellulose, hemicellulose, lignin, inulin, and resistant starch are distributed in the bran and starchy endosperm parts of whole grains [1, 2].

Whole grain dietary fiber may have these effects through multiple physiological mechanisms that include binding and eliminating cholesterol, binding bile acids, modulation of hormonal activity, stimulation of the immune system, facilitating toxicant transit through the digestive tract, production of short chain fatty acids in the colon, dilution of gut substances, lowering caloric content and glycemic index of foods, improving insulin response, providing bulk in foods, and scavenging free radicals [5].

The consumption of whole grain foods has been recommended in many international dietary guidelines, due the health benefits associated with reduced risks of various types of non-transmissible chronic diseases [38]. According to Topping [39], the effects of bioactive compounds of whole grain are abundantly clear on health and lower disease risks.

Nabeshima et al. [40] summarized the physiological effects of bioactives in whole grain, attributing to the compound phenolics, carotenoids, vitamin E and lignans the antioxidant effects that reduce the risk of chronic diseases associated with oxidative problems, such as cardiovascular and cancers. The lignans also have potential estrogenic activity. Phytosterols promote the lower serum total and LDL cholesterol concentration. Finally, the dietary fiber and non-amylaceous polysaccharide have effects on improvement in bowel function and decreased risk of colon cancer and LDL cholesterol levels.

In the last years many researchers have been concentrating their studies about phytochemicals in whole grain on proving their antiradical activities and/or epidemiological effects by using in vitro [41 to 43] and in vivo tests [44 to 49], respectively.

In the face of scientific evidences, a number of health claims for whole grain have been approved in various countries. The first health claim was allowed in 1999 by the U.S. Food and Drug Administration [10], which permitted the following claims "Diets rich in whole grain foods and other plant foods and with low total fat, saturated fat, and cholesterol may reduce the risk of heart disease and some cancers." In the United of Kingdom [12] and in Sweden [13], health claims for whole grain and heart health were also approved under their voluntary codes. The European Union regulation is still limited on health claim, having none approved for whole grain. However, the nutritional claims allow the use of "rich in fiber" labels no products, therefore the food must contain at least 6 g or 3 g of fiber/100 g/100 kcal of the food. Claims about nutrients or substances such as phytochemicals, polyphenols, or ferulic acid may be made under the statement "contains [name of the nutrient or other similar substance]", but they are limited to "contain" claims, "rich in" claims would not be allowed, since there are no established Dietary Reference Values (DRV) for these substances [50].

According to Nabeshima et al. [40], the Brazilian regulation is similar to that of the European Union; there are also no approved regulations for the statement on whole grain products, and only the statement on the content of fibers is permitted.

## Antioxidant Activity of Cereals

Whole grains, especially corn, wheat, and oats, have been shown to have high antioxidant activity. The total antioxidant activity of corn was 181,470.86 mmol of vitamin C equiv/g of grain, and was the highest ($p>0.01$) compared to wheat, oats, and rice. The total antioxidant activities of wheat (76.70 71.38 mmol/g of grain) and oats (74.67-71.49 mmol/g of grain) were similar ($p<0.05$), but higher than the total antioxidant activity in rice (55.77-71.62 mmol/g of grain). Similarly, when expressed on a per serving basis, the total antioxidant activity of whole grain was comparable to that of some fruits and vegetables [25].

As opposed to phytochemicals contained in fruits and vegetables, which are mainly in free or soluble-conjugate (glycosides) forms, the phytochemicals from cereals are localized in the cell walls of grains, are difficult to digest, and may reach the colon and be digested and released by microflora and esterases, showing therefore a beneficial local action. Colonic digestion of such materials by microflora may release the bulk of the bound phytochemicals to exert their health benefits locally and beyond absorption. For instance, gastrointestinal esterase (from intestinal mucosa and microflora) can release ferulic acid and diferulic acids from cereal bran [51].

# Whole Grain Products and Bioactive Compounds

Whole grain products have become available on the market; however, most of the papers presented in the literature deal with studies on technological quality [52 to 54], and there is little or no literature on the antioxidant profile of products derived from whole wheat compared with the traditional products, as well as the effects of food formulation, food processing and cooking on the availability of these beneficial components [18, 28, 32, 55, 56].

The main differences between the consumption of bioactive compounds from fruits and cereals are related to the fact that fruits can be consumed raw, in its natural or processed form, having high concentration of free phenolic compounds, which may experience changes according to the way it is consumed. However, in whole grain cereal they will always undergo cooking processes, aiming to gelatinize the starch; this process may vary according to the product, but usually, high temperatures and long periods are used, which can cause many reactions that may affect the composition of bioactive compounds in the final product. In the following, some studies showing these effects on different cereal products will be presented.

## Bread

Bread is one the most important sources of carbohydrates, in the form of starch, and is consumed as part of the usual diet in many countries. Although wheat, as refined flour, is crucial in bakery products, for many reasons other grains are replaced in the formula of bread. In breads containing bioactive compounds it is very important to know the availability and level of these compounds, because during the bread making process they can increase or decrease [57].

Technologically, the use of whole grain flour promotes deleterious effects on the formation of gluten network, which affects the quality of the dough, such as water absorption and fermentation time, and the final product, such as decreased specific volume, increased firmness of crumb, darker appearance, and change in flavor and texture. This fact has provided many developments to improve the quality of whole bread.

Due to the conditions of baking at high temperatures, research to determine the quality and quantity of bioactive compounds in bread is beginning to be carried out.

Chlopicka et al. [32] investigated the effect of adding, in two different doses (15% and 30%), pseudocereal flour (buckwheat, amaranth, and quinoa) on the antioxidant properties of breads. Buckwheat flour had the highest phenolic content (7.25 ± 0.23 mg/g d.w.). The addition of buckwheat flour to wheat bread, particularly in higher doses, was more effective in enhancing antioxidant activity, as evaluated by means of FRAP (Ferric Reducing Ability of Plasma) and DPPH (2,2-diphenyl-1-picrylhydrazyl), which increased by 2.36 fold, and 3.64 fold, respectively, in comparison with other pseudocereal flours (amaranth, quinoa), which at higher doses caused changes in the above parameters within the ranges of 1.20 to 1.79 fold, and 0.60 to 1.71 fold. Buckwheat bread had the highest content of phenolic compounds, although the bread making process influenced the content of total flavonoids in bread causing the reduction of about 2 to 4 fold when compared to flours.

## Pasta

Pasta is the non-fermented product obtained by mixing wheat flour or semolina with cold water, kneaded cold or hot, with or without addition of other ingredients. The most suitable type of wheat flour or semolina is the "Durum" wheat, which has yellow color, high protein content of 12-14%, high elasticity and low extensibility of gluten [19, 58, 59].

The use of whole grain flour promotes the dilution of gluten and negatively affects the sensory and technological quality of pasta, which will present a darker color, change in flavor and texture and greater tendency to break during cooking. However, because of the high consumption, the whole grain is presented as a good option for diet due to its health benefits. In addition, pasta has the advantage of having lower glycemic index [60, 61].

Hirawan et al. [28] conducted a comparative study on the antioxidant properties among brands of regular and whole wheat spaghetti, and verified the effect of cooking on the total phenolic content (TPC). It was evaluated the antioxidant properties of 10 samples of regular and whole wheat spaghetti that are commercially available. Whole wheat spaghetti exhibited significantly higher levels of TPC (1389 μg/g) than regular wheat spaghetti (865 μg/g); however, TPC in both regular and whole wheat spaghetti was 48–78% of the original content after cooking.

Whole wheat spaghetti (234 μg/g) had significantly higher content of ferulic acid than regular spaghetti ($p < 0.05$). However, there were no significant differences in mean DPPH and peroxyl radical scavenging activities between the two categories. The discrepancies between the trends of the total phenolic content and the antioxidant capacity were probably caused by the presence of other antioxidant components, such as the Maillard reaction products formed during pasta drying.

In other study, Fares et al. [55] verified the effect of processing and cooking on phenolic acids profile and antioxidant properties of durum wheat (*Triticum turgidum* L. var. durum) pasta samples enriched with debranning fractions of an Italian wheat (*Triticum aestivum* L.) cultivar.

The cleaned grain was debranned by a friction debranning machine, resulting in the main fractions: the fraction 1 was composed of outer layers, testa and greater amounts of hyaline and aleurone layers, and the fractions 2 and 3 were composed of minor portions of hyaline and aleurone layers and slight amounts of endosperm. Five kinds of spaghetti were produced, a control made with 100% durum wheat semolina, and four enriched pasta samples: EP1 made with 6% of fraction 1, EP2 made with 10% of fraction 1, EP3 made with 6% of fractions 2 and 3, EP4 made with 0% of fractions 2 and 3.

When semolina was processed into spaghetti, the main effect of processing was a reduction of the free phenolic acids fraction both in control and enriched samples (30.8%, 39%, 9.3%, 22.6% and 30.6% for control, EP1, EP2, EP3, and EP4, respectively). In contrast, an increase in bound phenolics fraction was observed. The boiling water could have enhanced the extraction of bound phenolics from the food matrix, primarily ferulic acid ester linked to cell walls.

Cooking, however, affected almost of all the bound fraction of phenolic acids, which showed a consistent increase in all samples (46%, 34%, 43%, 37% and 22% for control EP1, EP2, EP3, and EP4, respectively).

### Extruder

Extrusion processing has become an important food process in the manufacture of ready-to-eat, cereals, snacks, pasta, pet foods, and textured vegetable protein. In many of these applications, the extruder is a high temperature, short time process which minimizes losses in vitamins and amino acids. Product color, flavor, shape and texture are also affected by the extrusion process. Extrusion has been widely applied in the production of nutritious foods. Emphasis is placed on the use of extrusion to denature antinutritional factors and the improvement of protein quality and digestibility [62]. Thus, the study of application of this technology can be interesting to obtain whole grain products.

Dlamini et al. [18] evaluated the effect of traditional African processing and extrusion cooking, using a twin-screw cooker extruder (Clextral BC92, FIRMINY Cedex, France), on total phenols, tannin content and antioxidant activities of products from different sorghum types. The products evaluated included porridges and extrudates made from whole and decorticated tannin and non-tannin sorghums. The decortication reduced the antioxidant activity of both tannin and non-tannin sorghum by 82–83% due to the removal of the pericarp and the testa, which decreased phenols. The extrusion cooking significantly reduced antioxidant activity, by up to 86%, compared to that of the unprocessed grain. In this work it was used a steam heated, co-rotating twin-screw cooker extruder where the feed was tempered to 18% moisture prior to extrusion. The higher moisture content probably promoted phenolic and tannin polymerization, which affected extractability of phenols and tannins, and reduced antioxidant activity. In contrast, the low moisture (<15%), high shear and high temperature extrusion conditions of the single screw extruder may cause depolymerization of the condensed tannins and convert them into low molecular weight oligomers that are more extractable. The higher antioxidant activity of extruded NK 283 (18 μmol TE/g) observed, compared to raw decorticated grain (7 μmol TE/g), may result from increased extractability of phenols from the grain cell walls and proteins. The significant reduction in antioxidant activity in cooked high tannin sorghum products can be largely attributed to the interaction of tannins with prolamins. The retention of antioxidant activity, particularly in fermented and unfermented porridges, means that whole tannin sorghum can be processed into foods with potential health benefits.

Other study was performed by Zielinski et al. [27], that evaluated the effect of extrusion cooking of whole grain of cereals cultivated in Poland on the inositol phosphates, tocopherols, tocotrienols, reduced glutathione, melatonin, phenolic acids, as well as the content of microelements (Cu, Zn, Mn and Se). Wheat cv. Almari, barley cv. Gregor, rye cv. Dankowskie Złote and oat cv. Sławko were used in this study. An experiment was carried out on a counter-rotating twin-screw extruder (type 2S-9/5 Metalchem, Poland), with a die bore diameter of 6 mm, a constant screw configuration and process temperatures of 120-160-200°C. The extrusion caused a significant decrease in all compounds tested, except for the phenolic ones. According to the species of cereals, with losses of 4 to 50% for inositol hexaphosphate, 63 to 94% for tocopherols and tocotrienols, 20 to 50% for reduced glutathione, and 17 to 63% for melatonin. For phenolic compounds (free and released from ester bonds) there was an increase of 200 to 300% in the initial value.

Mora-Rochin et al. [63] evaluated the effect of traditional nixtamalization and extrusion cooking processes on total phenolics, ferulic acid, and anthocyanins, as well as on antioxidant activity of tortillas produced from regular (white and yellow) and pigmented (blue and red)

Mexican maize processed into tortillas. In this study, it was considered that during the nixtamalization process, the maize grains were limecooked, stored overnight and the residue resulting from washing of nixtamal was dried and ground. The cook and wash wastewaters are rich in solids that contain important phytochemicals, and the extrusion cooking is a continuous process that offers the main advantage of the null generation of wastewater. The authors observed that the tortillas prepared from extruded flours retained between 76.2-93.9% and 58-96.7% of total phenolics and total ferulic acid (TFA), respectively, compared to 50.5-75.7% and 19.6-55.8% assayed in traditional tortillas. Approximately 97-99% of TFA in raw kernels and their tortillas were in their bound form. The retention of TFA in traditional tortillas was significantly lower compared to tortillas from extruded flours. Traditional tortillas contained freer ferulic acid compared to tortillas produced from extruded flours, indicating that the first process liberated bound ferulic acid with cell walls more efficiently. Blue maize lost more than 55% of the anthocyanins when processed into extruded or traditional tortillas. Approximately 68-92% of the ORAC associated with raw kernels or their tortillas were due to bound compounds. Traditional and extruded tortillas lost 16.4-52.4% and 6.8-24.8%, respectively, of the total ORAC associated with raw grains. The results indicated that the proposed extrusion cooking strategy was important in retaining higher levels of phytochemicals, such as ferulic acid, and antioxidant activity in tortilla.

## Miscellaneous

Lozano-Alejo et al. [64] studied the losses of carotenoids and tryptophan following nixtamalization and subsequent frying, common processes used in Mexico to prepare maize foods including tortillas. It was measured carotenoid and tryptophan contents for grain, nixtamalized (lime-cooked) maize, and processed snacks of 13 maize genotypes including landraces, quality protein maize (QPM) and non-QPM hybrids. An average 36% loss of provitamin A and an 8% increase in tryptophan were observed following nixtamalization and subsequent snack preparation by deep-frying.

Another research about the effect of antioxidants on the processing was conducted by Inglett et al. [56]. The authors investigated the effectiveness of extracting phenolic compounds and antioxidant activity from buckwheat with water, 50% aqueous ethanol, and 100% ethanol using microwave irradiation or a water bath for 15 min at various temperatures (23-150°C). This work is based on the fact that many of the health benefits of buckwheat have been attributed to its high levels of phenolic compounds, which exhibit antioxidant activity. The results showed that phenolic contents in microwave irradiated extracts were higher than those heated with a water bath, but not necessarily showing a higher antioxidant activity. The highest phenolic content, 18.5 ± 0.2 mg/g of buckwheat, was observed in the extract that was microwave irradiated in 50% aqueous ethanol at 150 °C. This occurred probably because the microwave irradiation disrupted the cellular structure of buckwheat more effectively than heating in a water bath, thus releasing more phenolic compounds. The highest antioxidant activities, 5.61 ± 0.04–5.73 ± 0.00 μmol Trolox equivalents/g of buckwheat, were found in the 100% ethanol extracts obtained at 100 and 150 °C, regardless of the heat source.

Other processing studied was the sprouting, because it promotes increase of nutrients during the germination of the seed. Van Hung, Hatcher and Barker [65] investigated the phenolic composition and corresponding antioxidant activity of the sprouted wheats representing two different classes of wheat, CWRS, hexaploid wheat, and CWAD, tetraploid wheat. Seven phenolic acids (4-hydroxybenzoic, vanillic, caffeic, syringic, p-coumaric, ferulic and sinapic) are detected in the CWRS and CWAD alcoholic and alkaline extracts using the ultra-performance liquid chromatography (UPLC) system. Syringic acid was the main compound in the free phenolic alcoholic extracts of the wheat meal representing 77.0% and 75.3% of the total amount of detected free phenolic compounds for CWRS and CWAD, respectively. However, the major released phenolic compound detected in the alkaline hydrolysed extracts was ferulic acid accounting for 72.3% and 71.0% for CWRS and CWAD total bound phenolics, respectively. During germination, syringic acid levels rose as the length of germination time increased, resulting in the increase in total phenolic compound and antioxidant activity of the sprouted wheat flours. There was an increase in total phenolic compounds and the antioxidant activity of the alcoholic extracts from the CWRS and CWAD wheat flours as the germination time was extended. As a result, the sprouted wheats exhibit better nutritional properties than ungerminated wheats and could be used to improve the nutritional value in food products.

## Future Challenges

The biggest challenges in the future for the research on bioactive compounds lie in the fact that the research should be directed to the effects of processing, from the fermentation processes to cooking, on the quality and quantity of these compounds in the final product.

The studies have shown a decrease in bioactive compounds during the rigorous heat treatments, which should lead to research on the preservation of these compounds and their antioxidant activity.

## Conclusion

This chapter showed that whole grains of cereals and pseudocereals have higher amount of phenolic conjugates which may be beneficial for a protective effect on the colon, and many studies point to the use associated with fruits, which have free phenolics, as a way of promoting a better beneficial effect of the product.

Finally, as per this chapter, the effect of heat treatments, used for cooking of cereal products, presents a series of reactions that decrease the concentration of bioactive compounds and the antioxidant activity of the processed product, which points to a future in developing technologies for the preservation of these compounds during processing.

# References


[1] Y. Pomeranz, *Wheat: Chemistry and Technology*, 3.Ed., St. Paul: American Association of Cereal Chemists 1, 514 (1988).

[2] K. Kulp, *Handbook of cereal science and technology*, 2.Ed., New York: CRC, 808 (2000).

[3] L. Marquart, J. M. Jowes, E. A. Cohen and K. Poutanen, The future of whole grains, Marquart and Fulcher, *Whole Grain and Health*, Blackwell Publishing, 3 (2007).

[4] J. F. Adams and A. Engstrom, *Cereal Foods World* 45, 75 (2000).

[5] B. P. Lamsal and J. M. Faubion, *Food Reviews International* 25, 2, 103 (2009).

[6] H.-Y. Kim, O.-H.Kim and M.-K. Sung, *Journal of the American College of Nutrition* 22, 217 (2003).

[7] G. A. Spiller, *Dietary Fiber in Health and Nutrition*. CRC Press: Boca Raton (1994).

[8] AACC (American Association of Cereal Chemists) (2005) Definition/ Reports. At *http://www.aaccnet.org/definitions/wholegrain.asp*

[9] N. Okarter, C.-S. Liu, M. E. Sorrells and R. H. Liu, *Food Chemistry* 119, 249 (2010).

[10] FDA. Food and Drug Administration (1999). FDA Modernization Act (FDAMA) Claims, At *<http://www.fda.gov/Food/LabelingNutrition/LabelClaims/ FDAModernizationActFDAMAClaims/default.htm>*.

[11] L. Marquart, K. L. Wiemer, J. M. Jones and B. Jacob, *Proceedings of the Nutrition Society* 62, 151 (2003).

[12] JHCI, Joint Health Claims Initiative, Approved generic health claim (2002), At *<http://www.jhci.org.uk/approv/wgrainh.htm>*.

[13] SNF, Swedish Nutrition Foundation, New generic health claim (coronary) heart disease, whole grain (2003), At. *<http://www.snf. ideon.se/snf/en/rh/Wholegrain.htm>*.

[14] United States Department of Agriculture (USDA), World Agricultural Production, At. *<http://www.fas.usda.gov/wap/circular/2010/10-08/productionfull08-10.pdf>*.

[15] L. C. Gutkoski and I. Pedó, *Aveia: composição química, valor nutricional e processamento*, São Paulo: Varela, 2000.

[16] B. O. Juliano, Rice in human nutrition (1993), At. http://www.fao.org/docrep/ t0567e/T0567E08.htm#Gross nutrient composition

[17] H. Zielinski, A. Ceglinska and A. Michalska, *Food Chemistry* 104, 980 (2007).

[18] N. R. Dlamini, J. R. N. Taylor and L. W. Rooney, *Food Chemistry* 105, 1412 (2007).

[19] D. A. V. Dendy and B. J. Dobraszczyk, *Cereals and Cereal Products: Chemistry and Technology,* Gaithersburg: Aspen, 429 (2001).

[20] R. H. Liu, *Journal of Nutrition 134, 3479 (2004).*

[21] R. H. Liu, *Journal of Cereal Science* 46, 207 (2007).

[22] L. U. Thompson, *Critical Review in Food Science and Nutrition* 34, 473 (1994).

[23] A. A. M. Andersson, N. Ruegg and P. A. Man, *Journal of Cereal Science* 47, 399 (2008).

[24] R. H. Liu, *American Journal of Clinical Nutrition* 78, 517S (2003).

[25] K. K. Adom and R. H. Liu, *Journal of Agricultural and Food Chemistry* 50, 6182 (2002).

[26] S. Zilic, V. H.T. Sukalovi, D. Dodig, V. Maksimovi, M. Maksimovi and Z. Basic, *Journal of Cereal Science* 54, 417 (2011).

[27] H. Zielinski, H. Kozłowska and B. Lewczuk, *Innovative Food Science & Emerging Technologies* 2, 3, 159 (2001).
[28] R. Hirawan, W. Y. Ser, S. D. Arntfield and T. Beta, *Food Chemistry* 119, 258 (2010).
[29] V. G. Kamath, A. Chandrashekar and P. S. Rajini, *Journal of Cereal Science* 40, 283 (2004).
[30] G. Dinelli, A. S. Carretero, R. Silvestro, I. Marotti, S. Fub, S. Benedettelli, L. Ghiselli and A. F. Gutierrez, *Journal of Chromatography A*, 1216, 43, 7229 (2009).
[31] A. Serpen, V. Gokmen, N. Pellegrini and V. Fogliano, *Journal of Cereal Science* 48, 816 (2008).
[32] J. Chlopicka, P. Pasko, S. Gorinstein, A. Jedryas and P. Zagrodzki, *LWT - Food Science and Technology* 46, 2, 548 (2012).
[33] K. K., Adom, M. E.Sorrells and R. H. Liu, *Journal of Agricultural and Food Chemistry* 51, 7825 (2003).
[34] E. S. M. Abdel-Aal, J. C. Young, I.Rabalski, P. Hucl and J. Fregeau-Reid, *Journal of Agricultural and Food Chemistry* 55, 787 (2007).
[35] K. K. Adom, M. E. Sorrells and R. H. Liu, *Journal of Agricultural and Food Chemistry* 53, 2297 (2005).
[36] G. Panfili, A. Fratianni and M. Irano, *Public Health Nutrition* 52, 6373 (2004).
[37] F. Ntanios, *European Journal of Lipid Science* 103, 102 (2001).
[38] Q. Liu and H. Y. Yao, *Food Chemistry* 102, 732 (2007).
[39] D. L.Topping and P. M. Clifton, *Physiological Reviews* 81, 1031 (2001).
[40] E. H. Nabeshima, M. T. P. S. Clerici, J. M. Hashimoto and Y. K. Chang, In. Clerici, M.T.P.S. *Bread Consumption and Health.* Nova Publishing Inc 1, 1-22 (2012).
[41] C. M. Liyana-Pathirana and F. Shahidi, *Journal of Cereal Science* 45, 238–247 (2007).
[42] B. Verma, P. Hucl and R. N.Chibbar, *Cereal Chemistry* 85, 4, 544-549 (2008), At <http://www.wholegrainscouncil.org/whole-grains-101/definition-of-whole-grains>.
[43] J. Chlopicka, P. Pasko, S. Gorinstein, A. Jedryas and P. Zagrodzki, *Food Chemistry* 105, 1412 (2007).
[44] J. Slavin, *Proceedings of the Nutrition Society* 62, 1, 129 (2003).
[45] A. B. Ross, A. Kamal-Eldin, E. A.Lundin, J.-X. Zhang, G. Hallmans and P. Åman, *Journal of Nutrition* 133, 7, 2222 (2003).
[46] M. K. Jensen, P. Koh-Banerjee, F. B. Hu, M. Franz, L. Sampson, M. Grønbœk and E. B. Rimm, *American Journal of Clinical Nutrition* 80, 6, 1492 (2004).
[47] A. Fardet, *Nutrition Research Reviews* 23, 1, 65 (2010).
[48] L. A. H. Rosén, E. M. Östman, P. R. Shewry, J. L. Ward, A. A. M. Andersson, V. Piironen and A.-M. Lampi and I. M. E. Björck, *Journal of Agricultural and Food Chemistry* 59, 22, 12139 (2011).
[49] H. Kim, M. Turowski, W. H. K. Anderson, S. A. Young, Y. Kim and W. Yokoyama, *Journal of Agricultural and Food Chemistry* 59, 14, 7672 (2011).
[50] N. M. Binns, *Cereal Chemistry* 87, 2, 162 (2010).
[51] M. F. Andreasen, P. A. Kroon, G. Williamson and M. T. Garcia-Conesa, *Free Radical Biology and Medicine* 31, 304 (2001).
[52] C. B. Silva and Y. K. Chang, *Instituto Nacional de Propriedade Industrial*, PI0801816-2 A2, apr. 24 (2008).
[53] A. D. C. Brito, E. H. Nabeshima, C. R. Gomes-Ruffi, C. L. C. V. Cruz, V. S. N. Silva and F. M. Montenegro, Estudo do melhoramento da farinha de trigo integral para

produção de pães de forma (Study of improvement in whole wheat flour for bread production). *4º Congresso Interinstitucional Iniciação Científica*, p.1-6. At *www.cnpm.embrapa.br/5ciic/4ciic/Artigos/RE10243.pdf* (2010).

[54] E. L. Almeida, *Estudo da qualidade pão francês pré-assado congelado elaborado com farinha do trigo integral: influência da formulação, processo e estocagem congelada. Thesis.* Food Engineering Faculty – State University of Campinas (Unicamp) (2011).

[55] C. Fares, C. Platani, A. Baiano and V. Menga, *Food Chemistry* 119, 1023 (2010).

[56] G E. Inglett, D. J. Rose, D. Chen, D. G. Stevenson and A. Biswas, *Food Chemistry* 119, 1216 (2010).

[57] K. Dewettinck, F. Van Bockstaele, B. Kühne, D. Van De Walle, T. M. Courtens and X. Gellynck, *Journal of Cereal Science* 48, 243 (2008).

[58] E. J. Pyler, *Baking Science and Technology*, Merrian, Kansas, Sosland Publishing Company 1, 588 (1988).

[59] C. F. Ciacco and Y. K. Chang, *Tecnologia de Massas Alimentícias.* São Paulo: Secretaria da Indústria, Comércio e Tecnologia, pp.182 (1982).

[60] D. J. A. Jenkins, T. M. S. Wolever and A. L. Jenkins, *Diabetes Care* 11, 2, 149 (1988).

[61] D. J. A. Jenkins, C. W. C. Kendall, L. S. A. Augustin, S. Franceschi, M. Hamidi, A. Marchie, A. L. Jenkins and M. Axelsen, *Am J Clin Nutr* 76(suppl), 266S (2002).

[62] J. M. Harper and J. P. Clark, C R C *Critical Reviews in Food Science and Nutrition* 11, 2, 155 (1979).

[63] S. Mora-Rochin, J. A. Gutiérrez-Uribe, S. O. Serna-Saldivar, P. Sánchez-Peña, C. Reyes-Moreno and J. Milán-Carrillo, *Journal of Cereal Science* 52, 502 and 508 (2010).

[64] N. Lozano-Alejo, G. V. Carrillo, K. Pixley and N. Palacios-Rojas, *Innovative Food Science and Emerging Technologies* 8, 385–389 (2007).

[65] P. Van Hung, D. W. Hatcher and W. Barker, *Food Chemistry* 126, 1896 (2011).

In: Phytochemicals
Editor: Hai-Xue Kuang

ISBN: 978-1-62417-354-7


*Chapter IV*

# Antioxidant and Antimicrobial Activities of Berries Extracts of Five *Juniperus* Species under *Juniperus* Section from Turkey

***Maria Fernanda Taviano**, *Andreana Marino*, *Ada Trovato*, *Rita De Pasquale* and *Natalizia Miceli***

Pharmaco-Biological Department, University of Messina,Messina, Italy

## Abstract

In Turkish folk medicine numerous *Juniperus* species, particularly those under the *Juniperus* section, are frequently employed to treat several diseases.

This work was designed to evaluate the antioxidant and antimicrobial potential of methanol berries extracts of *Juniperus*species under*Juniperus* section from Turkey: *Juniperuscommunis* L. var. *communis*, *Juniperuscommunis* L. var. *saxatili s*Pall., *Juniperusdrupacea* Labill., *Juniperusoxycedrus*L. subsp. o*xycedrus*, *Juniperusoxycedru s*L. subsp. *macrocarpa* (Sibth. & Sm.) Ball.

The total phenolic content of *Juniperus* spp. extracts, evaluated by Folin-Ciocalteau method, ranged from 5.14 ± 0.06 mg GAE/g extract (*Joo*) to 59.17 ± 1.65 mg GAE/g extract (*Jcc*), and it decreased in the following order: *Jcc*>*Jd*>*Jom* = *Jcs*>*Joo*.

The antioxidant properties of the extracts were examined in different *in vitro* systems. *Jd* showed the best activity in the DPPH (1,1-diphenyl-2-picrylhydrazyl) test ($IC_{50}$=0.38 ± 0.02 mg/mL), in the $Fe^{2+}$ chelating assay ($IC_{50}$=2.26 ± 0.06 mg/mL), and in the TBA (ThiobarbituricAcid)test ($IC_{50}$=2.47 ± 1.13 µg/mL); *Joo*resulted in the most active extract (4.05 ± 0.09 ASE/mL) in the reducing power assay. In addition, the antioxidant activity of the extracts was correlated to their phenolic content. The obtained results indicated that *Juniperus* spp. extracts possess both primary and secondary

* E-mail address: mtaviano@unime.it.

antioxidant activity; moreover, phenolic compounds appear to be responsible for the primary antioxidant effects.

Since *J. communis* berries are traditionally used for their antiseptic action, it seemed interesting to evaluate theantimicrobial potential of all *Juniperus* spp. included in *Juniperus* section. The antimicrobial activity of *Juniperus* spp. extracts against bacteria and yeasts was evaluated by standard methods. The efficacy was appreciable on gram-positive only. Among the extracts *Jd* showed the higher bacteriostatic activity (*MIC*: 78.12-312.50 μg/mL) than *Jcs* and *Jcc* (*MIC*: 156.25-1250.00 μg/mL) followed by *Joo* and *Jom*(*MIC*: 625.00-1250.00 μg/mL). The *Staphylococcus aureus* wasthemost susceptible strain tested (*MIC*: 78.12-625.00 μg/mL).

The results give a scientific basis to the traditional utilization of these *Juniperus* species, also demonstrating their potential as sources of natural antioxidant and antimicrobial compounds.

# Introduction

Antioxidants are substances that delay or prevent the oxidation of cellular oxidizable substrates [1]. They exert their effects by scavenging reactive oxygen species (ROS), activating a series of detoxifying proteins, or preventing the generation of ROS [2].

Consumption of antioxidants is thought to provide protection against oxidative damage and contribute positive health benefits; epidemiological studies have shown a clear inverse correlation between the consumption of antioxidant-rich foods and the onset of several diseases related to oxidative stress including cardiovascular disease, cancer, atherosclerosis, diabetes, immune deficiency diseases and ageing [3]. Antioxidants can also be used in food industry to suppress lipid peroxidation and consequently to improve food quality and stability.

Among the isolated compounds, polyphenols represent the main class of natural antioxidants. Phenolic compounds, including phenols, phenolic acids, flavonoids, tannins and lignans, act as antioxidants in several ways; they are radical scavengers because phenolic groups are excellent nucleophiles and are also able to inhibit lipid peroxidation, acting as breakers of oxidation reaction by binding with free radicals generated through lipid peroxidation. In addition, phenolic compounds could act as chelators of metal ions that induce oxidation [3].

The use of traditional medicinal plants for primary healthcare have steadily increased worldwide in recent years. Currently, multiple drug resistances in human pathogenic microorganisms have developed because of the indiscriminate employment of commercial antimicrobial drugs commonly used in the treatment of infectious diseases; for this reason, several studies were carried out in order to identify new antimicrobial substances from various sources, like plants, utilized in traditional medicine for several thousand years. Plants are richand wide in variety of secondary metabolites with antimicrobialproperties; polyphenols are well documented to have microbicide activity against a huge number of pathogenic bacteria [4].

Based on these statements, several research studies are focused on the antimicrobial and antioxidant properties of plant-derived natural products.

*Juniperus* genus (Cupressaceae) comprises about 60 monoecious or dioecious evergreen species, trees or shrubs distributed throughout the Northern hemisphere [5]. The genus is usually divided into three distinct sections or subgenera: *Juniperus* (syn: *Oxycedrus*, 9 or 10 species), *Caryocedrus* (one species, *J. drupacea*Labill.), and *Sabina* (approximately 50 species) [6]. However, concerning this genus, Flora of Turkey reports only two sections, *Juniperus* and *Sabina*, and *J. drupacea* is included in *Juniperus* section [7].

Numerous *Juniperus* species are frequently employed in Turkish folk medicine, particularly those under *Juniperus* section. *J. communis* L., common juniper, is an aromatic shrub, whose berries are known for their biological properties [8]. According to the Flora of Turkey, *J.communis* L. includes two varieties,var. *communis* and var. *saxatilis* Pall. In folk and official medicine, *J. communis* represents a drug with diuretic, antiseptic, stomachic and carminative properties [9]. Decoction of branch with berries is used to relieve kidney infections [10]. *J. oxycedrus*L., according to the Flora of Turkey,comprises two subspecies: subsp. *oxycedrus*and subsp. *macrocarpa*[11]. *J. oxycedrus* is used to prepare empyreumatic oil by destructive distillation of the branches and wood, the so-called oil of Cade, which is widely employed in human and veterinary dermatology [12]. Moreover, *J. oxycedrus* is used in folk medicine for the treatment of various ailments, such as hyperglycaemia, obesity, tuberculosis, bronchitis and pneumonia; leaves and fruits are applied externally for parasitic disease [13, 14]. Leaves, resins, barks and fruits extracts of *J. oxycedrus* were found to inhibit the growth of numerous microorganisms [12]. *J. drupacea*Labill has a limited distribution in the Eastern Mediterranean region. In Turkish folk medicine, *J. drupacea* leaves and berries decoctions are used internally as anthelminthic [15].

This work was designed to evaluate some biological properties of berries methanol extracts of five *Juniperus*species belonging to *Juniperus* section collected from Turkey. Total polyphenols content was determined spectrophotometrically. The antioxidant properties of *Juniperus* spp. extracts were examined in different *in vitro* systems and correlated to the phenolic content. The antimicrobial potential of *Juniperus* spp. extracts against bacteria and yeasts was evaluated.

## Materials and Methods

### Chemicals

Tryptone Soya Broth, Sabouraud Dextrose Agar, Müeller Hinton Broth, and Müeller Hinton Agar were supplied from Oxoid (Basingstoke, UK). NaOH, HCl, and *n*-butanol were purchased from Merck Chemicals (Milan, Italy). $FeCl_2$ was obtained from Carlo Erba (Milan, Italy). Unless indicated otherwise, all chemicals were purchased from Sigma-Aldrich (Milan, Italy).

### Plant Material

The ripe berries of the selected plants, *Juniperuscommunis*L. var. *communis* (*Jcc*), *Juniperuscommunis*L. var. *saxatilis*Pall. (*Jcs*), *Juniperusdrupacea*Labill. (*Jd*), *Juniperusoxycedrus*L. subsp. *oxycedrus*(*Joo*), *Juniperusoxycedrus*L. subsp. *macrocarpa*(Sibth. & Sm.) Ball. (*Jom*), were collected from different regions of Turkey. Prof.

Ayşegül Güvenç at the Department of Pharmaceutical Botany, Ankara University, Turkey confirmed the taxonomic indentification of plant materials. Voucher specimens are deposited in the Herbarium (AEF) of the Faculty of Pharmacy, Ankara University. Collection sites and herbarium number of the samples are listed in Table 1.

## Extraction Procedure

The berries without seeds were dried at room temperature and powdered. The methanol extract was obtained by macerating the powdered berries (20 g) in methanol (200 mL) at 50 °C for 8 h, under continuous shaking (700 rpm/min) and in the dark, twice. After filtration, the extractive solution was evaporated to dryness by rotary evaporator. The yields of the extracts referred to 100 g of dried berries, are reported in Table 1.

**Table 1. Collection sites and herbarium numbers of the plant samples, and percentage yields of *Juniperus* spp. berries methanol extracts**

| Juniperus species | Collection sites | Herbarium numbers | Yields (w/w %) |
|---|---|---|---|
| J. communisL. var. communis | Ağrı: Doğubeyazıt, Korhan Y. | AEF 23854 | 22.59 |
| J. communisL. var. saxatilisPall. | Ankara: Kızılcahamam, Işık Dağı | AEF 23801 | 25.83 |
| J. drupaceaLabill. | Antalya: Antalya-Konya road | AEF 23610 | 21.31 |
| J. oxycedrusL. subsp. oxycedrus | Ankara: Kızılcahamam, Işık Dağı | AEF 23799 | 15.50 |
| J. oxycedrusL. subsp. macrocarpa (Sibth. & Sm.) Ball. | İzmir: Çeşme, around Çiftlikköy | AEF 23855 | 20.96 |

The yields are referenced to 100 g of dried berries.

## Total Phenolic Content

The total phenolic content of *Jcc*, *Jcs*, *Jd*, *Joo*, and *Jom* extracts was measured by the Folin-Ciocalteau reagent [16]. One hundred μL of each sample solution were mixed with 0.2 mL Folin-Ciocalteu reagent, 2 mL of distilled water, and 1 mL of 15% $Na_2CO_3$, and the absorbance was measured at 765 nm, after 2 h incubation at room temperature, with a model UV-1601 spectrophotometer (Shimadzu, Milan, Italy). Gallic acid was used as a standard and the total phenolics were expressed as mg gallic acid equivalents (GAE)/g extract (dw) ± standard deviation (S.D.). The mean value of total phenolic content was obtained from triplicate experiments.

## Antioxidant Activity

### *Free Radical Scavenging Activity*

The free radical scavenging activity of *Jcc*, *Jcs*, *Jd*, *Joo*, and *Jom* extracts was determined using the DPPH (1,1-diphenyl-2-picrylhydrazyl) method [17]. The extracts were tested at different concentrations (0.07-0.7 mg/mL). An aliquot (0.5 mL) of each sample solution was added to 3 mL of methanol DPPH solution (0.1 mM) prepared daily. The optical density change at 517 nm was measured 20 min after the initial mixing with a model UV-1601 spectrophotometer (Shimadzu). Butylatedhydroxytoluene (BHT) was used as reference standard. The scavenging activity was measured as the decrease in absorbance of the samples *versus* DPPH control solution. The radical scavenging activity percentage (%) was calculated by the formula $[(A_o - A_c)/A_o]x100$, where $A_o$ is the absorbance of the control and $A_c$ is the absorbance in the presence of the sample or standard. The assays were carried out in triplicate, and the results are expressed as mean 50% inhibitory concentration ($IC_{50}$) ± standard deviation (S.D.).

### *Measurement of Reducing Power*

The reducing power of *Jcc*, *Jcs*, *Jd*, *Joo*, and *Jom* extracts was determined according to the method of Oyaizu [18]. Different amounts of each extract (0.07-0.7 mg/mL) in 1 mL of solvent were mixed with 2.5 mL of phosphate buffer (0.2 M, pH 6.6) and 2.5 mL of 1% potassium ferricyanide. The mixture was incubated at 50°C for 20 min. The resulting solution was cooled rapidly, mixed with 2.5 mL of 10% trichloroacetic acid, and centrifuged at 3000 rpm for 10 min. The upper layer of the solution (2.5 mL) was mixed with 2.5 mL of distilled water and 0.5 mL of 0.1% fresh $FeCl_3$, and the absorbance was measured, after 10 min, at 700 nm; increased absorbance of the reaction mixture indicated increased reducing power. As blank, an equal volume (1 mL) of solvent was mixed with a solution prepared as described above. Ascorbic acid and BHT were used as reference standards. The assays were carried out in triplicate, and the results are expressed as ascorbic acid equivalent (ASE/mL) ± standard deviation (S.D.). When the reducing power is 1 ASE/mL, the reducing power of 1 mL extract is equivalent to 1 µmol ascorbic acid [19].

## Ferrous ions ($Fe^{2+}$) Chelating Activity

The $Fe^{2+}$chelating activity of *Jcc*, *Jcs*, *Jd*, *Joo*, and *Jom* extracts was estimated by the method of Dinis et al. [20]. Briefly, different amounts of each extract (0.07-0.7 mg/mL) in 0.4 mL solvent were added to a solution of 2 mM $FeCl_2$ (0.05 mL). The reaction was initiated by the addition of 5 mMferrozine (0.2 mL) and the total volume was adjusted to 4 mL with methanol. The mixture was shaken vigorously and left standing at room temperature for 10 min; the absorbance was then measured spectrophotometrically at 562 nm. The control contains $FeCl_2$ and ferrozine, complex formation molecules. Ethylenediaminetetraacetic acid (EDTA) was used as reference standard. The percentage of inhibition of the ferrozine–($Fe^{2+}$) complex formation was calculated by the formula $[(A_o - A_c)/A_o]x100$, where $A_o$ is the absorbance of the control and $A_c$ is the absorbance in the presence of the sample or standard.

The assays were carried out in triplicate, and the results are expressed as mean 50% inhibitory concentration ($IC_{50}$) ± standard deviation (S.D.).

## Thiobarbituric Acid (TBA) Test

Inordertoassesstheefficacy of using*Jcc*, *Jcs*, *Jd*, *Joo*, and *Jom* extracts to protect liposomes from lipid peroxidationa TBA test wasused[21]. The extracts were tested at different concentrations (0.016-1 mg/mL). Liposomes were prepared from bovinebrain extract in a phosphate buffered saline (5 mg/mL). The reaction mixture (ET) consisted of 0.2 mL of liposomes, 0.1 mL of aqueous $FeCl_3$ (1 mM), 0.1 mL of aqueous ascorbic acid (1 mM), 0.5 mL of PBS, and 0.1 mL of the sample solution. After incubation at 37°C for 20 min, the TBA test was performed by adding 0.1 mL of 2% butylatedhydroxytoluene (BHT) in EtOH followed by 0.5 mL of 1% w/v TBA in 50 mMNaOH and 0.5 mL of 25% HCl. The tubes were heated at 90°C for 30 min. After cooling, chromogens were extracted with 2.5 mL of *n*-butanol.

The mixture was vortexed to ensure complete extraction of the chromogen and then centrifuged at 3500 rpm for 15 min. at room temperature. The absorbance of the MDA-TBA complex in the upper layer was determined spectrophotometrically at 532 nm. The percentage of lipid peroxidation inhibition was assessed by using the following formula: [(FRM - B) - (ET – B - EA) / (FRM - B)] x 100, where FRM is theabsorbance of thecontrolreactionand ET is absorbance in the presence of thesample. Theabsorbance of liposomesalone (B) andextractalone (EA) wasalsotakenintoaccount. Propyl gallate was used as a reference compound. The assays were carried out in triplicate, and the results are expressed as mean 50% inhibitory concentration ($IC_{50}$) ± standard deviation (S.D.).

## Antimicrobial Activity

The following strains were used as indicators for antimicrobial testing and were obtained from the in-house culture collection of Pharmaco-Biological Department (University of Messina, Italy): *Staphylococcus aureus* ATCC 6538P, *Staphylococcus epidermidis* ATCC 12228, *Enterococcus hirae* V3, *Bacillus subtilis* P3, *Escherichia coli* ATCC 25922, *Pseudomonas aeruginosa* ATCC 9021, *Proteus mirabilis* G4 were grown at 37°C in Tryptone Soya Broth; *Candida albicans* ATCC 10231 and *Candida parapsilosis* P7 were grown at 35 °C on Sabouraud Dextrose Agar (SDA).

The minimum inhibitory concentrations (MICs) of*Jcc*, *Jcs*, *Jd*, *Joo*, and *Jom*extracts were determined in Müeller Hinton Broth (MHB) using a broth dilution micromethod in polystyrene microtitre plates according to the Clinical Laboratory Standards Institute (CLSI) guidelines [22]. The micro*plates were maintained* under *shaking* conditions to prevent sedimentation and to increase aeration. The final concentrations of the extracts adopted to evaluate the antibacterial activity were 1.22 to 2500 μg/mL. Ofloxacin (range from 0.2 to 0.3 μg/mL) and tetracycline (range from 2 to 8 μg/mL) were used as positive reference standard drugs. Two growth controls consisting of MHB medium and MHBMeOH were included for each strain.

Microbial growth was determined by visual readings and spectrophotometric values (550 nm) using the Microplate Reader, Model 550 (BIO-RAD Laboratories Milano, Italy). The MIC was considered to be the lowest concentration of each extract at which there was no microbial growth after 24 h of incubation.

As an indicator of bacterial growth, 20 μL of 2,3,5-Triphenyl-2H-tetrazolium chloride (TTC) dissolved in water was added to the wells and incubated at 37°C for 30 min. Where bacterial growth was inhibited, the solution in the well remained clear after incubation with TTC. The minimum bactericidal concentrations (MBCs) were determined by seeding 20 μL from all clear MIC wells onto Müeller Hinton Agar (MHA) plates. The MBC was defined as the lowest extract concentration that killed 99.9% of the final inoculums after 24 h of incubation. The data from at least three replicates were evaluated and model results were calculated.

The minimum inhibitory concentrations (MICs) of *Juniperus* spp. extracts on yeasts were determined in RPMI1640 buffered to a pH of 7.0 with MOPS according to the National Committee for Clinical Laboratory Standards [23].

The micro*plates were maintained* under *shaking* conditions to prevent sedimentation and to increase aeration. The final concentrations of the extracts adopted to evaluate the activity were 1.22 to 2500 μg/mL. Amphoterycin B (0.15 μg/mL) was used as a positive reference standard drug.

Two growth controls consisting of RPMI 1640 medium and RPMI1640MeOH were included for each yeast. Microbial growth was determined by visual readings and spectrophotometric values (550 nm) using the microplate reader (BIO-RAD). The MIC was considered to be the lowest concentration of each extract at which there was no microbial growth after 24 h-48 h of incubation.

The minimum fungicidal concentrations (MFCs) were determined by seeding 20 μL from all clear MIC wells onto SDA plates [23]. The MFC was defined as the lowest extract concentration that killed 99.9% of the final inoculums after 24 h-48 h of incubation. The data from at least three replicates were evaluated and model results were calculated.

# Results and Discussion

## Total Phenolic Content

Phenolic compounds are called high-level antioxidants because of their ability to scavenge free radicals and activate oxygen species such as singlet oxygen, superoxide free radicals and hydroxyl radicals [24]. The antioxidant activities of plant extracts are often explained with their total phenolic content with good correlation; therefore, it is worthwhile to determine their total amount in the extracts utilized for the study. The total phenolic content of *Juniperus* spp. extracts ranged from 5.14 ± 0.06 mg GAE/g extract (*Joo*) to 59.17 ± 1.65 mg GAE/g extract (*Jcc*), and it decreased in the following order: *Jcc*>*Jd*>*Jom* = *Jcs*>*Joo* (Figure 1).

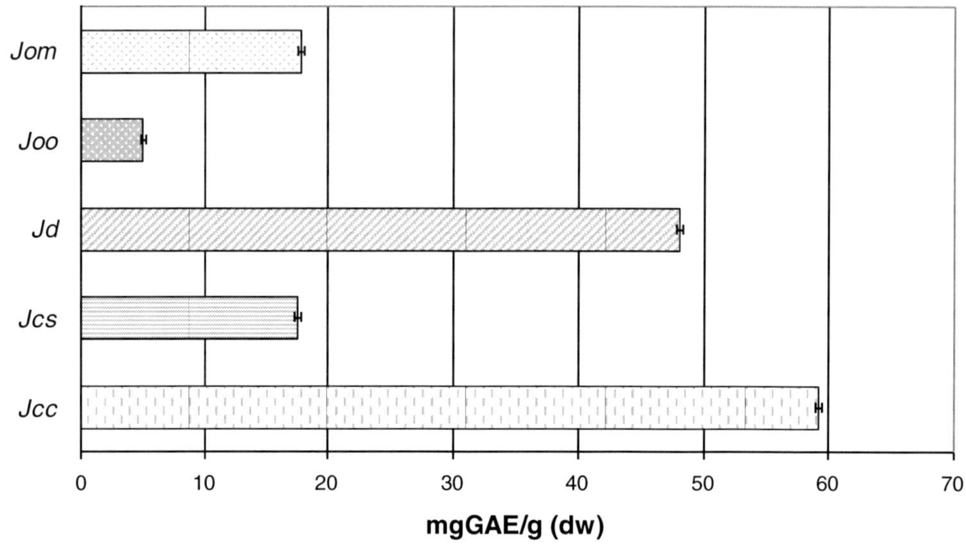


Figure 1. Determination of Total phenolic content (calculated as gallic acid equivalents) of *Juniperus*spp. berries methanol extracts. Values are expressed in mgGAE/g extract ± S.D.

## Antioxidant Activity

Different antioxidant compounds may act through different mechanisms, such as prevention of chain initiation, decomposition of peroxides, prevention of continued hydrogen abstraction, free radical scavenging, reducing capacity, and binding of transition metal ion catalysts [25]. Consequently, four antioxidant capacity tests with various approaches and mechanisms have been carried out: the DPPH test, which measures the ability to donate hydrogen to the radical; reducing power assay, which measures the ability to donate electrons to Fe (III); chelating activity, which determines the ability of substances to chelate Ferrous ions ($Fe^{2+}$);the TBA test was carried out in order to evaluate the anti-lipid peroxidation activity of the extracts.

Antioxidant properties, particularly radical scavenging activities, are very important due to the deleterious role of free radicals in foods and in biological systems. In this study, the free radical scavenging activity of *Juniperus*spp. berries extracts was determined using the DPPH test.

DPPH (2,2-Diphenyl-1-picrylhydrazyl) is a stable radical in methanol with a violet color because of the delocalization of the spare ***electron*** throughout the molecule. Accepting a proton in the reaction with the oxygen atom of a radical scavenger's OH group, the reduced DPPH-H (2,2-Diphenyl-1-picrylhydrazine) is formed, which is yellow [26]. The degree of discoloration indicates the amount of DPPH scavenged; the greater the bleaching action, the higher the antioxidant activity, and this is reflected in a lower $IC_{50}$ value. The method has been used extensively to predict antioxidant activities of single compounds, as well as of different plant extracts because of the relatively short time required for the analysis; in addition, it is reproducible and strongly correlated with phenolic compounds [27].

*Juniperus* spp. extracts have a noticeable effect on scavenging free radicals; the activity of the extracts varied from 0.38 ± 0.02 mg/mL (*Jd*) to 5.92 ± 0.04 mg/mL (*Joo*). Based on $IC_{50}$ values, the activity of the extracts and the standard decreased in the order: BHT >*Jd*>*Jcc*>*Jcs*>*Jom*>*Joo* (Table 2). Finally, a good correlation between total phenolic content and DPPH radical scavenging activity was found ($r^2$=0.7379).

Different studies have indicated that the electron donation capacity, reflecting the reducing power, of bioactive compounds is associated with antioxidant activity [28]. In the assay for the determination of reducing power the presence of antioxidants compounds in the sample determines the reduction of $Fe[(CN)_6]_3$ to $Fe[(CN)_6]_2$; this reduction is highlighted by spectrophotometric measurement of the intensity of Perl's Prussian blue complex, $Fe4[Fe(CN\text{-})_6]_3$, which has a strong absorbance at 700 nm. In this assay, the yellow color of the test solution changes to various shades of green and blue depending on the reducing power of antioxidant samples. An increase in absorbance of the reaction mixture would indicate an increase in the reducing capacity, and this is reflected in a lower ASE/mL value [26].

*Juniperus* spp. extracts exhibited reducing power; the activity varied from 4.05 ± 0.09 ASE/mL (*Joo*) to 64.14 ± 1.20 ASE/mL (*Jcs*). The reducing power of methanol extracts and standard decreased in the following order: BHT >*Joo*>*Jd*≥*Jcc*>*Jom*>*Jcs*(Table 2). No correlation was found between phenolic content and reducing power ($r^2$=0.1135).

Among the transition metals, iron is known as the most important lipid oxidation pro-oxidant due to its high reactivity [29]. An effective ferrous ion chelator affords protection against oxidative damage by removing iron that may otherwise participate in HO· generation via the Fenton type reactions. Ferric ions ($Fe^{3+}$) also produce radicals from peroxides although the rate is ten-fold less than that of ferrous ion. Ferrous ions ($Fe^{2+}$) are the most powerful pro-oxidant among the various species of metal ions [30, 31]. Minimizing ferrous ions may afford protection against oxidative damage by inhibiting production of ROS and molecular damage. In this method, Ferrozine can quantitatively form complexes with $Fe^{2+}$, producing a red chromophore with absorbance that can be measured at 562 nm. In the presence of chelating agents, complex formation is disrupted, resulting in a reduction in the red color of the complex. Measurement of color reduction therefore allows estimation of the metal chelating activity of the coexisting chelator. Lower absorbance indicates higher metal chelating activity. Metal chelation is an important antioxidant property [32] and hence *Juniperus* spp. berries methanol extracts were assessed for their ability to compete with ferrozine for ferrous ion in the solution.

The obtained results show that *Juniperus* spp. extracts interfered with the formation of theferrous-ferrozine complex. Among extracts*Jd*exhibited the highest activity ($IC_{50}$=2.26 ± 0.06 mg/mL), whereas *Jcc*is the less active ($IC_{50}$=7.51 ± 0.62 mg/mL). The metal chelating effect of methanol extracts and EDTA decreased in the order: EDTA >*Jd*>*Joo*>*Jcs*>*Jom*>*Jcc*. (Table 2). A low correlation between metal chelating activity and total phenolic content was found ($r^2$=0.3489).

Lipid peroxidation can inactivate cellular components that play a major role in oxidative stress in biological systems. Further, several toxic byproducts of the peroxidation can damage other biomolecules including DNA away from the site of their generation. Therefore, compounds possessing anti-lipid peroxidation activity are really important for health benefits [33]. The TBA reaction is based on the fact that peroxidation of most membrane systems leads to formation of smallamounts of freemalonaldehyde (MDA). One molecule of MDA reacts with two molecules of TBA toyield a coloredproduct, which in an acid environment absorbs light at 532 nm and it is readily extractable in to organic solvents. Theintensity of coloris proportional to the concentration of MDA; higher MDA levels indicate an increased lipid peroxidation. The incorporation of any antioxidant compound in the lipid peroxidation assay reaction mixture will lead to a reduction of the extent of peroxidation [21].

The results of the determination of anti-lipid peroxidation activity show that *Juniperus* spp. extracts have anti-lipid peroxidation ability. Based on $IC_{50}$ values, the activity of the extracts ranged from 2.47 ± 1.13 μg/mL (*Jd*) to 1112.25 ± 9.83 mg/mL (*Joo*); the activity of extracts and standard decreased in the order: propyl gallate>*Jd*>*Jcc*>*Jcs*>*Jom*>*Joo*(Table 2). Inaddition, a positive correlation was found between total phenolic content and anti-lipid peroxidation activity ($r^2$=0.5592).

In this study, a positive correlation between antiradical activity and total phenolic content was found. Extensive investigations on antiradical and antioxidant activities of small phenolics, including flavonoids and phenolic acids, have been reported; moreover high molecular weight phenolics (tannins) have shown ability to quench free radicals and their effectiveness depends more on the molecular weight, the number of aromatic rings, and the nature of the hydroxyl groups' substitution than the specific functional groups [34].

**Table 2. Free radical scavenging activity (DPPH test), Reducing power, Ferrous ions ($Fe^{2+}$) chelating activity, and Anti-lipid peroxidation activity (TBA assay) of *Juniperus*spp. berries methanol extracts**

| ***Juniperus*species** | **DPPH test** $IC_{50}$ (mg/mL) | **Reducing power assay** ASE/mL | **$Fe^{2+}$ chelating activity** $IC_{50}$ (mg/mL) | **TBA test** $IC_{50}$ (μg/mL) |
|---|---|---|---|---|
| *J. communis*L. var. *communis* | 0.63 ± 0.09 | 12.82 ± 0.10 | 7.51 ± 0.62 | 4.44 ± 0.70 |
| *J. communis*L. var. *saxatilis*Pall. | 1.84 ± 0.10 | 64.14 ± 1.20 | 3.04 ± 0.07 | 120.07 ± 3.60 |
| *J. drupacea*Labill. | 0.38 ± 0.02 | 12.63 ± 0.14 | 2.26 ± 0.06 | 2.47 ± 1.13 |
| *J. oxycedrus*L. subsp. *oxycedrus* | 5.92 ± 0.04 | 4.05 ± 0.09 | 2.61± 0.12 | 1112.25 ± 9.83 |
| *J. oxycedrus*L. subsp. *macrocarpa* (Sibth. & Sm.) Ball. | 3.15 ± 0.05 | 47.11 ± 3.81 | 4.16 ± 0.61 | 293.91 ± 7.07 |
| Standard | BHT 0.0677 ± 0.0002 | BHT 1.75 ± 0.05 | EDTA 0.0065 ± 0.0001 | Propyl gallate 0.21 ± 0.40 |

Values are expressed as the mean ± S.D. (n = 3).

As reported in Figure 1, the greater concentration of phenolic compounds was found in*Jcc* and *Jd*, resulting more elevated in *Jcc* (59.17 ± 1.65 mg GAE/g) than *Jd*(48,06 ± 1.65 mg GAE/g); nonetheless, *Jd* showed greater free radical scavenging than *Jcc*. This could depend on the different quali-quantitative phenolic profile of the extracts. As we reported in previous works, both the extracts contain flavonoids, which are present in higher amounts in *Jcc*, whereas phenolic acids are the most abundant class of phenolic compounds revealed in *Jd*; the occurrence of phenolic acids in *Jd* represents a major difference with *Jcc* berries extract, in which phenolic acids were not detected [35, 36]. Based on these findings, it can be hypothesized that the greater free radical scavenging activity of *Jd* is related to the synergic effect of both flavonoids and phenolic acids.

With regard to chelating activity, our results show that *Juniperus*spp. extracts may offer protection against oxidative damage through their chelating properties; however, low correlation between metal chelating activity and total phenolic content was found. It is known that the extracts are very complex mixtures of many different compounds and that non-phenolic antioxidants can contribute to their antioxidant activity; therefore, it can be speculated that the observed activity could depend on other phytochemicals that are extracted together with phenolics.

The results of the determination of anti-lipid peroxidation activity are positively correlated to the phenolic content of *Juniperus*spp. extracts. Moreover, like the radical scavenging activity, *Jd* showed the highest anti-lipid peroxidation activity, followed by *Jcc*. Most lipid oxidation inhibition assaysmeasure a combination of the transition metal (usually iron) chelation and radical scavenging properties [24]. Hence, on the basis of the obtained results, it can be hypothesized that the anti-lipid peroxidation activity of *Juniperus* spp. extracts is related mainly to the radical scavenging properties of phenolic compounds.

## Antimicrobial Activity

The extracts showed efficacy *versus* the Gram-positive bacteria. No activity was detected against Gram-negative bacteria and yeasts tested. Comparing the results of antibacterial activity, it can be seen that *Jd*extract showed the higher bacteriostatic activity (*MIC*: 78.12-312.50 μg/mL), then *Jcs* and *Jcc* (156.25-1250.00 μg/mL), followed by *Joo* and *Jom*(625.00-1250.00 μg/mL) (Table 3).

Among the tested extracts*Jd, Jcc* and *Jcs*showed good bacteriostatic properties. The activity of *Jd* and *Jcc* could be related to their phenolic content, and the greater effect of *Jd* could depend on the different quali-quantitative phenolic profile of the extracts. For a phenolic compound to function effectively as an antimicrobial it should be able to function at the lipid–water interface and therefore has to be partially hydrophobic. Many of the phenolic compounds have activity against Gram-positive bacteria and have no activity or negligible activity against Gram-negative bacteria [37]. The partial hydrophobicity of some phenolic compounds allows them to act efficiently at the membrane–interface of Gram-positive bacteria; this causes the loss of membrane integrity and the dissipation of the proton motive force [38]. Gram-negative bacteria, which have an external lipo-polysaccharide layer and additional minor membranecomponents besides an intact plasma membrane around its cell, have potentially more buffering capacity and hydrophobicity [39]. In addition, the bacteriostatic activity detected on Gram-positive bacteria may be due to the formation of pseudomulticellular aggregates.

In spite of its lower phenolic content,*Jcs* possesses good bacteriostatic properties, resulting even more elevated than *Jcc*. Thus, the noticeable antimicrobial activity of *Jcs* depends not only on phenolic compounds but also on other different secondary metabolites such as terpenoids, whose antimicrobial activity has been widely confirmed [40, 41].

**Table 3. The MIC/MBC values (μg/mL) of *Juniperus* spp. berries methanol extracts**

| Micro-organisms | *Jcc* | | *Jcs* | | *Jd* | | *Joo* | | *Jom* | | Oflo-xacin |
|---|---|---|---|---|---|---|---|---|---|---|---|
| | MIC | MBC | MIC | MBC | MIC | MBC | MIC | MBC | MIC | MBC | MIC |
| Gram-positive | | | | | | | | | | | |
| *S. aureus*ATCC 6538P | 156.25 | 2500.00 | 156.25 | > 2500 | 78.12 | > 2500 | 625.00 | > 2500 | 625.00 | > 2500 | 0.3[a] |
| *S. epider-midis* ATCC 12228 | 1250.00 | > 2500 | 1250.00 | > 2500 | 312.50 | > 2500 | 1250.00 | > 2500 | 1250.00 | > 2500 | 0.3[a] |
| *E. hirae* V3 | 625.00 | > 2500 | 156.25 | > 2500 | 312.50 | > 2500 | 625.00 | > 2500 | 1250.00 | > 2500 | 0.2[a] |
| *B. subtilis* P3 | 312.50 | > 2500 | 156.25 | > 2500 | 312.50 | > 2500 | 312.50 | > 2500 | 1250.00 | > 2500 | 0.2[a] |

The extracts were not active against any of the Gram-negative bacteria or the yeasts tested (data not shown).

## Conclusion

The obtained results give a scientific basis to the ethnopharmacological use of these Turkish *Juniperus*species under *Juniperus* section in the treatment of infections and skin diseases, also demonstrating the potential of these plants as sources of natural antioxidants and antimicrobial compounds. Bioassay-guided fractionationprocedures are necessary to characterize and isolate the active constituents.

## References


[1] B. Halliwell and J. M. Gutteridge, *Method. Enzymol.* 186, 1 (1990).

[2] J. M. Matés, *Toxicology* 153, 83, (2000).

[3] H. Han and B-K. Baik, *Int. J. Food Sci. Technol.* 43, 1971 (2008).

[4] M. M. Cowan, *Clin. Microbiol. Rev.* 12, 564 (1999).

[5] S. M. Walters, A. Brady, C. D. Brickell, J. Cullen, P. S. Green, J. Lewis, V. A. Matthews, D. A. Webb, P. F. Yeo, J. C. M. Alexander, *The EuropeanGarden Flora*, Cambridge University Press, (1986).

[6] R. P. Adams, *Biochem. Syst. Ecol.*26, 637 (1998).

[7] M. J. E. Coode and J. Cullen, *Flora of Turkey and the East Aegean Islands*. Vol 1. Edinburgh at the University press, Edinburgh, UK (1965).

[8] S. Vichi, M. Riu-Aumatell, M. Mora-Pons, J. M. Guadayol, S. Buxaderas and E. Lopez-Tamames,*Food Chem.* 105, 1748 (2007).

[9] G. Stanić, I. Samaržija and N. Blažević, *Phytother. Res.* 12, 494 (1998).

[10] E. M. Ritch-Krc, S. Thomas, N. J. Turner and G. H. N. Towers, *J. Ethnopharmacol.* 52, 85 (1996).

[11] A. Farjon, *Flora of Turkey and the East Aegean Islands*. Vol. 11., Edinburgh at the University Press, Edinburgh, UK 2000.

[12] I. Karaman, F. Sahin, M. Güllüce, H. Ogütçü, M. Sengül and A. Adigüzel, *J. Ethnopharmacol.*85, 231 (2003).

[13] M. R. Loizzo, R. Tundis, F. Conforti, A. M. Saab, G. A. Statti and F. Manichini, *FoodChem.* 105, 572 (2007).

[14] E. Sezik, E. Yesilada, M. Tabata, G. Honda, Y. Takahishi, T. Fujita, T. Tanaka and Y. Takeda, *Econ. Bot.*51, 195 (1997).

[15] E. Yesilada, G. Honda, E. Sezik, M. Tabata, K. Goto and Y. Ikeshiro, *J. Ethnopharmacol.*39, 31 (1993).

[16] X. Gao, M. Ohlander, N. Jeppsson, L. Bjork, and V. Trajkovski, *J. Agric. Food Chem.* 48, 1485 (2000).

[17] M. Ohnishi, H. Morishita, H. Iwahashi, T. Shitzuo, S. Yoshiaki, M. Kimura and R. Kido, *Phytochemistry*36, 579 (1994).

[18] M. Oyaizu, *Jpn. J. Nutr.* 44, 307 (1986).

[19] P. Apáti, K. Szentmihályi, S. T. Kristó, I. Papp, P. Vinkler, E.Szoke and A. Kéry, *J. Pharm. Biomed. Anal.* 32, 1045 (2003).

[20] T. C. P. Dinis, V. M. C. Madeira and L. M. Almeida, *Arch. Biochem. Biophys.*315, 161 (1994).

[21] F. Conforti, G. A. Statti, R. Tundis, F. Menichiniand P. J. Houghton, *Fitoterapia*73, 479 (2002).

[22] *Clinical and Laboratory Standards Institute*, Wayne, PA (2008).

[23] *National Committee for Clinical Laboratory Standards*, Wayne, Pa. (2002).

[24] R. Jayabalan, P. Subathradevi, S. Marimuthu, M. Sathishkumar and K. Swaminathan, *Food Chem.*109, 227 (2008).

[25] S. Bounatirou, S. Smiti, M. G. Miguel, L. Faleiro, M. N. Rejeb, M. Neffati,M. M. Costa, A. C. Figueiredo, J. G. Barroso and L. G. Pedro, *FoodChem.*105, 146 (2007).

[26] İ. Gülçin, M. Elmastaş and H. Y. Aboul-Enein, *Phytother. Res.*21, 354 (2007).

[27] P. Maisuthisakul, S. Pasuk and P. Ritthiruangdej, *J. Food Comp. Anal.*21, 229 (2008).

[28] P. Siddhuraju, P. S. Mohan and K. Becker, *Food Chem.* 79, 61 (2002).

[29] İ.Gülçin, F. Topal, S. B. O. Sarıkaya, E. Bursal, G. Bilsel and A. C. Gören, *Rec. Nat. Prod.* 5, 158(2011).

[30] İ. Gülçin, V. Mshvildadze, A. Gepdiremen and R. Elias, *Phytother. Res.* 20, 130 (2006).

[31] B. Halliwell and J. M. C. Gutteridge, *Biochem. J.* 219, 1 (1984).

[32] İ. Gülçin, *Int. J. Food Sci. Nutr.* 56, 491 (2005).

[33] J. S. Rathee, B. S. Patro, S. Mula, S. Gamre and S. Chattopadhyay, *J. Agric. FoodChem.*54, 9046 (2006).

[34] P. Siddhuraju and S. Manian, *FoodChem.*105, 950 (2007).

[35] N. Miceli, A. Trovato, P. Dugo, F. Cacciola, P. Donato, A. Marino, V. Bellinghieri, T. M. La Barbera, A. Güvenç and M. F. Taviano, *J. Agric. FoodChem.*57, 6570 (2009).

[36] N. Miceli, A. Trovato, A. Marino, V. Bellinghieri, A. Melchini, P. Dugo, F. Cacciola, P. Donato, L. Mondello, A. Güvenç, R. De Pasquale, M. F. Traviano, *FoodChem. Toxicol.* 49, 2600 (2011).

[37] H. Nikaido and M. Vaara, *Microbiol. Rev.* 1, 1 (1985).

[38] P. M. Davidson, *FoodMicrobiology. Fundamentals and Frontiers.* ASM Press, Washington DC, USA (2001).

[39] D. A. Vattem, Y.-T Lin, R. G. Labbe and K. Shetty, *Innov. Food Sci. Emerg. Technol.*5, 81 (2004).

[40] N. Filipowicz, A. Piotrowski, J. R. Ochocka and M. Asztemborska, *Planta Med.* 72, 850, (2006).

[41] T. Fujita, E. Sezik, M. Tabata, E. Yeşilada, G. Honda, Y. Takeda, T. Tanaka and Y. Takaishi, *Econ. Bot.* 49, 406 (1995).

In: Phytochemicals
Editor: Hai-Xue Kuang

ISBN: 978-1-62417-354-7


Chapter V

# Phenolic Compounds, Health Effects and Antioxidant Properties of Berries

*Aleksandra Konić-Ristić*[1], *Katarina Šavikin*[2,*], *Gordana Zdunić*[2], *Dejan Gođevac*[3], *Teodora Janković*[2] *and Nebojša Menković*[2]

[1]Institute for Medical Research, University of Belgrade, Belgrade, Serbia
[2]Institute for Medicinal Plant Research "Dr. Josif Pančić", Belgrade, Serbia
[3]Institute for Chemistry, Technology and Metallurgy, University of Belgrade, Belgrade, Serbia

## Abstract

The biological value of berries relies on their nutritive and non-nutritive constituents. They are good sources of vitamins (vitamin A, C, E, folic acid), pro-vitamins and related compounds (carotene and lutein), minerals (calcium and selenium) and phytosterols, but the proposed health benefit is usually connected with phenolics as the main biologically active compound of berries. Berry phenolics include anthocyanins, flavonols, flavanols, condensed tannins (proanthocyanidins), hidrolyzable tannins (ellagitannins and gallotannins) and phenolic acids. Berries are recognized as the best dietary sources of anthocyanins. Different berries are distinguished by the presence of different classes of phenolics or compounds within the same class, their content and chemical structure (substituents, glycosylation, conjugation), potentially influencing their physiological properties and health effects. Analytical assessment of the phytochemical content of berries, the effects of storage and processing and putative interaction with other dietary compounds is essential for evaluation of their health benefit. It is of great importance to emphasize the biodiversity in the context of different phytochemical content in wild versus cultivated species, different varieties of the same species, or influenced by different cultivating conditions. Beneficial effects of berries consumption in health promotion and prevention of chronic diseases, such as cancer or cardiovascular diseases are shown in numerous epidemiological studies. In *in vitro* and animals studies berry fruits, their extracts and isolated compounds showed antioxidant,

[*]Email address : ksavikin@mocbilja.rs.

anticancer, antidiabetic, hypoholesterolemic, anti-inflammatory, vasoprotective, cardioprotecive, antiplatelet, antitrombotic or antimicrobial effects. Some of these effects are also shown in clinical trials in humans. Phenolic compounds of berries and their extracts have strong antioxidant potential that is often used to explain their health-promoting properties. Recent studies postulated that their antioxidant potential, as a basic mechanism of biological activity, is compromised by their low bioavailability. Consequently further research in this area is targeted towards other mechanisms of action, the evaluation of main sites of action and further elucidation of potential of berry phenolic compounds as antioxidants, based on their structure or metabolism, both *in vitro* and within the human body.

## Introduction

It is the intent of this paper to summarize data regarding the phytochemical composition of berries, influences of processing and storage, bioavailability, biological activity and health effects in clinical trials.

Berry fruits are widely consumed throughout the world in fresh or frozen form, within the different dishes or as a constituent of functional foods. Their extracts or isolated compounds are also used as bioactives in different dietary supplements. The biological value of berries relies on their nutritive and non-nutritive constituents. They are good sources of vitamins (vitamin A, C, E, folic acid), pro-vitamins and related compounds (carotene and lutein), minerals (calcium and selenium) and phytosterols, but the proposed health benefit is usually connected with phenolics as the main biologically active compounds of berries [1, 2, 3, 4]. Berries phenolics include anthocyanins, flavonols, flavanols, condensed tannins (proanthocyanidins), hidrolyzable tannins (ellagitannins and gallotannins) and phenolic acids. The total phenolic content in berries may vary significantly. The highest levels are found in black elderberry (1950 mg/100 g) and black chokeberry (1752 mg/100 g) followed by black raspberry (980 mg/100 g), and blackcurrant (821 mg/100 g), [5].

Scientists are interested in the potential health benefits of flavonoids associated with fruit and vegetable-rich diets. No adverse effects have been associated with high dietary intakes of flavonoids from plant-based foods. This lack of adverse effects may be explained by the relatively low bioavailability and rapid metabolism and elimination of most flavonoids [6]. A number of factors may affect the flavonoid content of foods, including agricultural practices, environmental factors, ripening, processing, storage, and cooking. Although individual flavonoid intakes may vary, a total flavonoid intake in Western populations is estimated to be about 150-200 mg/day [7, 8].

## Phenolic Compounds

Different berries are distinguished by the presence of different classes of phenolics or compounds within the same class, their content and chemical structure (substituents, glycosylation, conjugation), potentially influencing their physiological properties and health effects.

The most common phenolic acids in berries are the hydroxybenzoic and hydroxycinnamic acids. They are rarely occur as free acids but are found in conjugated forms as esters and glycosides. Among hydroxycinnamic acids, p-coumaric, caffeic, chlorogenic (5–

caffeoylquinic acid), neochlorogenic (3–caffeoylquinic acid), ferulic acids are usually detected. Cholorogenic acid is the predomonant hydroxycinamic acid ester and also one of the most abundant polyphenolic compounds in berries [9]. The most frequently reported hydroxybenzoic acids in berries are p-hydroxybenzoic, protocatehuic and gallic acid. The important hydroxybenzoic acids, ellagic acid, has been detected in berries of the genera *Rubus* and *Fragaria*, and could be present as free ellagic acid but is usually found as ellagic acid glycosides or as hydrolysable tannins, ellagitannins [2, 10, 11]. Ellagic acid derivatives and ellagitannins are dominantly localized in seed while fewer amounts are detected in fruits.

Berries are recognized as the best dietary sources of anthocyanins. Anthocyanins are the most important pigments of the vascular plants responsible of the shiny orange, pink, red, violet and blue colours in the flowers and fruits of some plants [12]. There is a huge variety of anthocyanins spread in nature and the anthocyanidins (aglycons) are the basic structures of the anthocyanins (Figure 1). The anthocyanidins consist of an aromatic ring (A) bonded to an heterocyclic ring (C) that contains oxygen, which is also bonded by a carbon–carbon bond to a third aromatic ring (B). Anthocyanidins rarely occur in nature because of their instability so, the anthocyanidins are dominantly found glycosylated with glucose, galactose, arabinose or rutinose, and to a lesser extent with xylose as in chokeberry, sambubiose as in elderberry, or sophorose as in raspberry and is known as anthocyanins [2, 13]. Among anthocyanins, glycoside derivatives of cyanidin, delphinidin and pelargonidin are the most common in nature, being found in about 80% of pigmented leaves, 69% in fruits and 50% in flowers [14]. Cyanidin derivates dominate in chokeberry, blackberry, black and red raspberry; delphinidin derivates are dominant in black currants while strawberries are the most abundant with pelargonidin derivates. Bilberries and blueberries are characterised with diverse derivates of all six most common aglycones [2, 10]. The amount of anthocyanins in berries is affected by a series of factors such as cultivar, applied agricultural techniques, environmental factors, ripening stage. In different studies, the amount of total anthocyanins varied significantly and the highes amounts were detected in blackberry, chokeberry and bilberry [10, 15, 16, 17,].

Proanthocyanidins (condensed tannins) are widely present in berries and are comprised of oligomeric and polymeric flavan-3-ols. The most common constitutive units are (epi)catechins and (epi)gallocatechins. The physical, chemical, and biological properties of proanthocyanidins depend on their structure and especially on their degree of polymerization [18]. Highly polymerized proanthocyanidin forms dominated in chokeberry, black currant, blueberry, red currant and bilberry [19]. The highest levels of proanthoc-yanidins are found in black chokeberry (659 mg/100 g) among all berries [5] with the most abundant fraction of polymers (542.6 mg/100g fw, mean value), [20] (USDA Database for proanthocyanidin content of selected foods). Jarkko et al. [18] analysed proanthocyanidin content in bilberry, European cranberry, lingonberry, black currant, chokeberry, and strawberry and it varied significantly, especially in the bilberry and lingonberry samples, between 133 and 192 mg/100 g (mean = 148 mg/100 g, CV 30%, $n$ = 6) and 298 and 548 mg/100 g (mean = 390 mg/100 g, CV 36%, $n$ = 6), respectively.

Flavonols are one of the most widespread groups of flavonoids in food. Among flavonol aglycones, quercetin, kaempferol and myricetin are the predominant compounds in berries (Figure 2) and to a lesser exent isorhamnetin [21, 22]. They differ in the number and position of OH groups in the B ring of the flavonoid skeleton and are usually occur as $O$ – glycosides with sugars attached at the C3 position. The highest flavonol quantities are found in chokeberry (88 mg/100 g) and black elderberry (42 mg/100 g), [5]. Häkkinen et al. [1]

analysed the amounts of quercetin, myricetin, and kaempferol in 25 edible berries using optimized RP-HPLC method. Quercetin was found in all berries and the contents being highest in bog whortleberry (158 mg/kg, fresh weight), lingonberry (74 and 146 mg/kg), cranberry (83 and 121 mg/kg) and chokeberry (89 mg/kg). Myricetin were detected in cranberry, black currant, crowberry, bog whortleberry, blueberries, and bilberry while kaempferol was detected only in gooseberries (16 and 19 mg/kg) and strawberries (5 and 8 mg/kg). Also, Häkkinen et al. [9] observed marked differences in the phenolic profiles among the berries, with certain similarities within families and genera. The major phenolic compound analyzed in the genus *Vaccinium* was quercetin in lingonberry and cranberry, and its level was high also in blueberries and bilberry. In the genus *Ribes*, quercetin was the main compound in gooseberry, red currant and black currant.

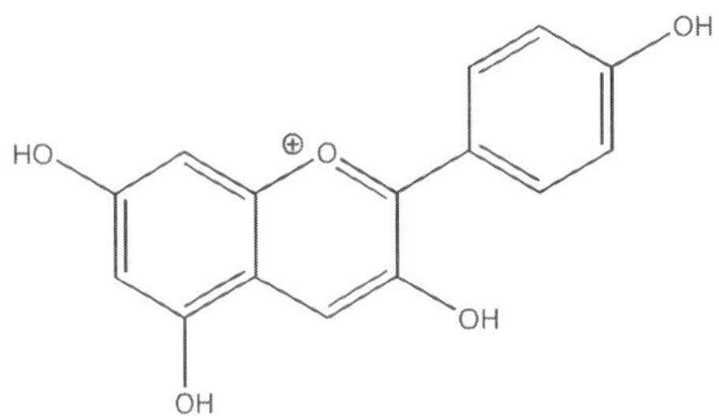

Pelargonidin

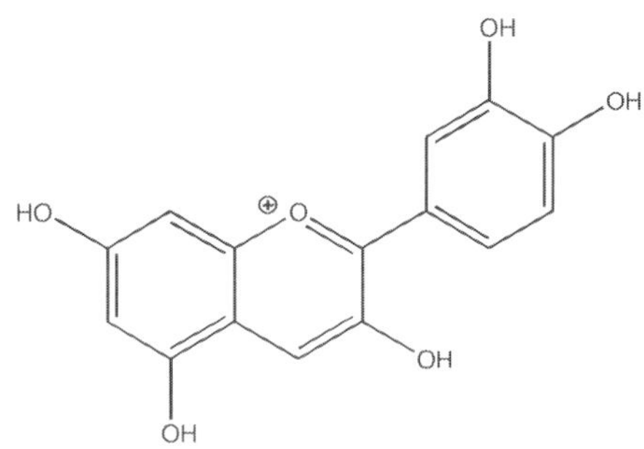

Cyanidin

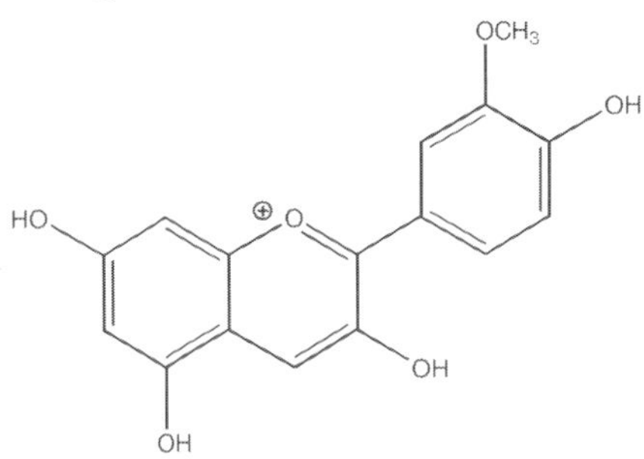

Peonidin

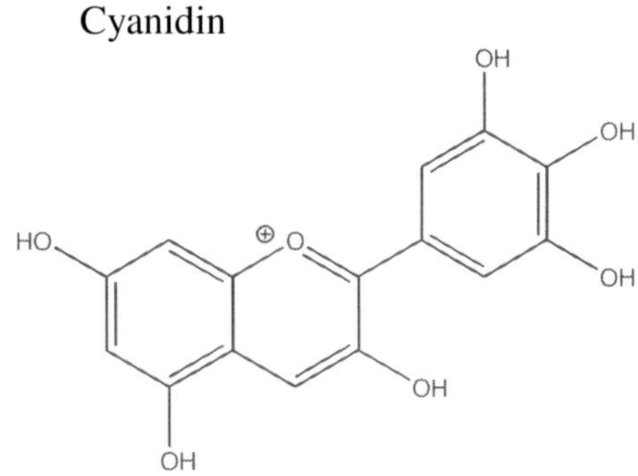

Delphinidin

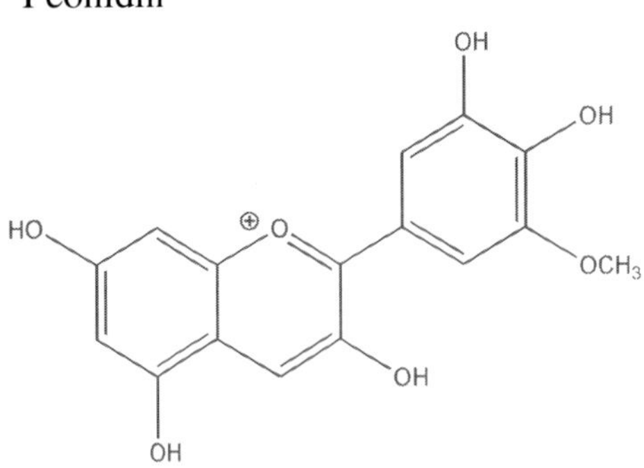

Petunidin

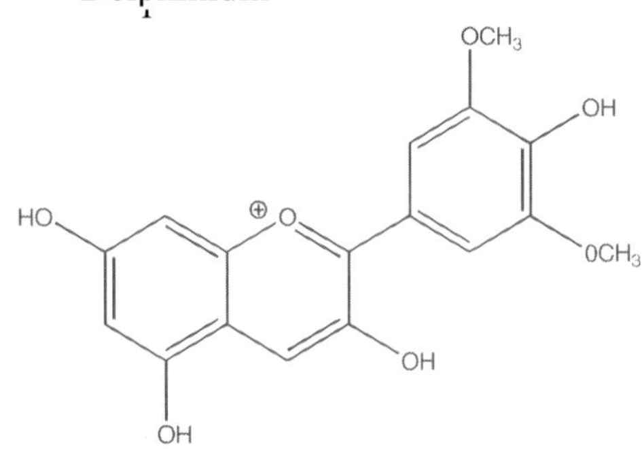

Malvidin

Figure 1. The most common anthocyanidins in berries.

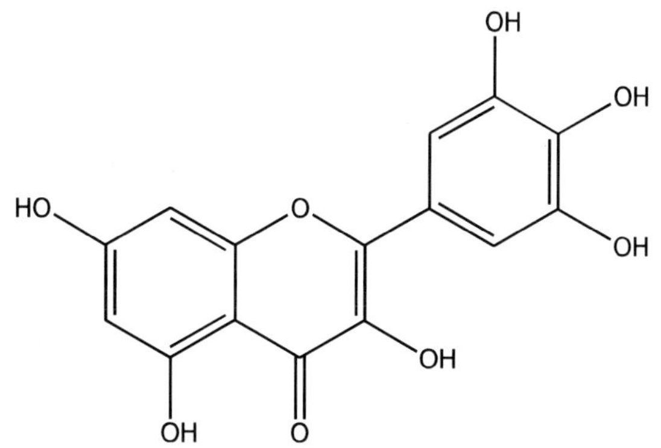

Myricetin

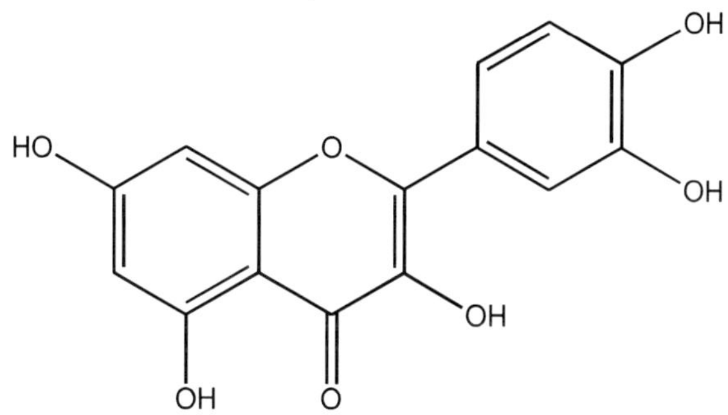

Quercetin

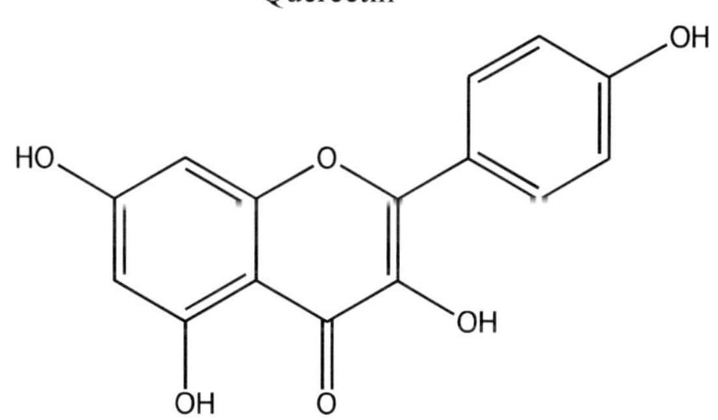

Kaempferol

Figure 2. The most common flavonol aglycones in berries.

# Effects of Storage and Processing

The most commonly consumed berries belong to the Ericaceae and Rosaceae families, and are strawberries (*Fragaria x ananassa*), raspberries (*Rubus idaeus*), blackberries (*Rubus spp.*), bilberries (*Vaccinium myrtillus*), black currants (*Ribes nigrum*) and red currants (*Ribes rubrum*). Berries are consumed mainly as fresh fruit. However, because of the seasonal character, berries are not only available fresh but also distributed as frozen and thermally processed products such as jams, juices, purees, jellies. Various factors during the processing of berries may influence the contents of anthocyanins, phenolic compounds and other antioxidants. Therefore, it is interesting to assess whether frozen berries or thermally processed products could also represent a good source of bioactive compounds.

Freezing is the most common processing and preservation technology applied to berries because of its power to reduce chemical and biological reactions in foods, preservation of nutrients and important quality attributes, and relatively low operating costs [23]. When berries are frozen, chemical changes can occur, including concentration of solutes and chemicals, enzyme activity and oxidative reactions. The effects of cold storage at -20°C on the contents of vitamin C, flavonoids and anthocyanins are different in different berries. The amount of total phenolics decreased by 10-56% in frozen red currant, bilberry and black raspberry [3, 4], while in black currant increased by even 46-171%. Increases in anthocyanin content have been reported for strawberry and red currant (15-85%), whereas in black currant and bilberry reduction for 5-52% was noticed. Vitamin C is well preserved in frozen black currants, red currants and red raspberries, showing minor changes during storage [4, 24].

Berries are often processed into juice or juice concentrate or canned for subsequent use in beverages, syrups or other products [25]. The effectiveness of extraction of polyphenols into juice is influenced by factors such as the extraction solvent, temperature, and processing time. The amounts of total phenolics and anthocyanins in juices are lower compared to the values for corresponding berries. For example, after pressing, only 20% of total anthocyanins from black currant berry fruits were present in the juice [26]. Anthocyanins and flavonoids are located mainly in the skins of fruits which were removed by filtering during processing and remain as residue. Generally, total phenolics and anthocyanins were higher in juices obtained from dark blue berries such as black currant, black raspberry and bilberry than in red currant or raspberry juices, supporting the relation between phenolic content and the colour of juices [27, 28]. Freshly pressed berry juices are rich in polyphenols, while much lower levels have been found in commercial juices because they are diluted. Also, various treatments during processing, such as crushing, heating, enzyme treatment, pressing, pasteurization, can reduce polyphenol content [29]. Furthermore, significant losses of polyphenols in juices occur during storage. Fruit juices should be consumed within 48 h after opening because juices near to their sell-by dates had much lower polyphenol contents than had fresher ones.

Another berry product commonly consumed in human diet is jam. The effects of jam preparation are different on various classes of compounds. Cooking strawberries, raspberries, black currants, red currants or blackberries with sugar caused small losses of total phenolic content [24, 30]. On the other hand, jam processing caused a decrease of total anthocyanins in bilberry, black currant, red currant, blackberry and raspberry [3, 30]. The greatest loss was noticed in bilberry (85%), while in black currant cultivars the reduction was between 30 and 60%. This might be due to the fact that anthocyanins are very sensitive to temperature, and a combined time/temperature process can greatly reduce the level of pigments in the final product [31]. Temperatures > 70°C and long time of extraction have been shown to cause rapid anthocyanin degradation.

However, regardless of the losses in the amount of total phenolics and total anthocyanins, frozen berries and related processed products are still excellent sources of nutritional substances. The consumption of berries juices or jams made in the households or in food industry is becoming a part of a healthy lifestyle.

# Health Effects

## Berries as Direct Antioxidants

A great amount of data showed that berry fruits, juices and extracts are powerful antioxidants *in vitro*. It is also shown that the major contributors to the antioxidant potential of berries are phenolic compounds and vitamin C. Antioxidant activity differs between the different groups of phenolic compounds (anthocyanins, ellagitannins, proanthocyanidins) but also between specific phenolics within the same group. As a result, antioxidant effect of berries is strongly dependent on the choice of berry raw material or extraction (production) procedure.

Borges et al. [32] evaluated the total antioxidant capacity (AOC) of black currants, blueberries, raspberries, red currants, and cranberries using FRAP assay. They reported that the total AOCs were high for all five berries. Black currant had the highest AOC (51.6 μmol $Fe^{2+}$/g) while AOC values for blueberries, raspberries, red currants and cranberries were 30.0, 27.7, 24.6, 18.6 μmol $Fe^{2+}$/g respectively. The extent to which vitamin C, phenolic compounds, and flavonoids contribute to the AOC of the different fruits was measured using HPLC-PDA-MS and HPLC-PDA with an on line antioxidant detection system, based on TEAC assay. Results showed that in major antioxidants of black currant were anthocyanins (5521 nmol/g) and vitamin C (2238 nmol/g), with contribution to the AOC of 73 and 18% respectively. In blueberries anthocyanins were the major bioactives (4810 nmol/g) with 84% of AOC. Raspberries were characterized with high ellagitannins (1352 nmol/g), vitamin C (1014 nmol/g) and anthocyanins (885 nmol/g) content, but ellagitannins were the most potent antioxidants, with 58% of AOC, compared with contribution of other two bioactives (11 and 16% respectively). Red currants had similar content of anthocyanins and vitamin C (328 and 318 nmol/g), but vitamin C showed stronger antioxidant capacity (47% of AOC vs. 21% for anthocyanins). In cranberries, although vitamin C content was higher (1107 nmol/g) then content of anthocyanins (725 nmol/g), the contribution to AOC was lower (23% and 39% respectively). These data showed that although AOC of berries depend on numerous factors including variety, degree of ripeness, season, and storage conditions, it is also influenced by specific bioactive profile and the content of specific compounds within the different groups of phenolic compounds. The type of bioactive that is the major contributor to the antioxidant activity of whole berries could be the reason for observed differences between antioxidant activity of raw berries compared to their products (juices). Jakobek et al. [33] evaluated the antioxidant activity of different berry fruits using DPPH assay, showing highest values for chokeberry and elderberry (72,2 and 62,14 μmol Trolox equivalents (TE)/ml, respectively), high value for black currant (30.15 μmolTE/ml) and lower values for red raspberry (8,2 μmolTE/ml) and strawberry (4,39 μmolTE/ml). Radical scavenging activity (RSA) measured by DPPH assay was also evaluated for pure juices of berries and results showed similar distribution among species [27]. Black currant juice had the strongest activity (0.91 ± 0.07 IC50 mg/ml), followed by black raspberry juice, blueberry juice, raspberry juice and red currant juice and RSA values were 1.28 ± 0.09, 1.68 ± 0.13, 2.40 ± 0.21, 3.14 ± 0.23 IC50 mg/ml, respectively. Besides marked AOC and RSA berry fruits and present phenolic compounds are also strong inhibitors of lipid oxidation *in vitro*, using different experimental models (bulk oils, emulsions, liposomes, LDL), [34].

Chemical antioxidant capacity assays on food or isolated compounds, based either on hydrogen atom transfer (total radical trapping antioxidant potential-TRAP and oxygen radical absorbance capacity-ORAC) or single electron transfer (ferric ion-reducing antioxidant power-FRAP, DPPH assay and Trolox equivalent antioxidant capacity-TEAC) are widely used in measurement of direct action of compounds present in food with radicals and transition metals [35].

Correlation of antioxidant capacity of food and dietary compounds measured using chemical assays and their antioxidant potential *in vivo* is compromised by the fact that *in vivo* antioxidant capacity of dietary compounds is influenced by important physiological processes such as bioavailability, metabolism and tissue targeting. For that reason chemistry antioxidant measures are mostly used for evaluation the most promising dietary antioxidants, effects of processing and comparison of various dietary sources of bioactives or bioactives themselves [36].

A step-forward in *in vitro* evaluation of antioxidant potential of dietary compounds is measurement of cellular antioxidant activity (CAA), using primary cultures of different blood cells [37] or malignant cells in culture [38], taking in account some aspects of cell uptake, distribution, and metabolism of antioxidant compounds. Wolfe et al. [39] evaluated the cellular antioxidant activity of 25 most commonly consumed fruits in USA, in HepG2 model. Wild blueberry had the highest CAA value followed by pomegranate and blackberry, with similar CAA values, and strawberry, blueberry, and raspberry, representing next cluster. Berry fruits evaluated also had the highest phenolic content and the highest ORAC values. Correlation analyses showed that total phenolics were significantly correlated to ORAC and CAA values. ORAC values for fruits were also significantly positively related to CAA values, but with lower correlation coefficients. It is noteworthy that in similar study evaluating CAA of 20 different flavonoids [40] there was no significant association between flavonoid ORAC and CAA values. Moreover, many of the flavonoids with the highest ORAC values had no CAA or low values.

The cytokinesis-block micronucleus (CBMN) assay is an *in vitro* study that has been used to evaluate antioxidant and protective effects of phenolic compounds on induced chromosome aberration in human lymphocytes.

Two ellagitannins, lambertianin C and sanguiniin H-6, and an ellagic acid glycoside, 4-α-L-arabinofuranosylellagic acid, isolated from blackberry seed extract were tested for protective effect on human lymphocytes using CBMN assay. The results of lambertianin C and sanguiniin H-6 show that their antioxidant potential, at the concentrations of 1.0 μg/ml, was similar to that of amifostine, a synthetic antioxidant and cytoprotective agent which was used as a positive control. Better antioxidant potential exhibited 4-α-L-arabinofuranosylellagic acid, with the optimal dose of 1.0 μg/ml. Higher doses of these compounds (2.0 μg/ml) showed lower decrease in micronucleus frequency. Blackberry seed extracts also decreased the frequency of micronuclei. However, antioxidant potential of the main constituents of blackberry seed extracts (ellagitannins) cannot explain the strong effect of extract itself. Synergistic effects of the individual compounds and/or strong activity of minor components were assumed [41]. Raspberry and currant seed extracts, rich in poliphenolics, also exhibited beneficial effects on human lymphocytes DNA, as measured by CBMN [42, 43].

## Anticancer Potential of Berries – *In Vitro* Studies

Numerous *in vitro* studies have shown that berry extracts and their bioactive compounds inhibit cell proliferation, modulate cell cycle arrest, and induce apoptosis in cancer cells with marked selectivity towards normal cells.

Study of Seeram [44] showed that blackberry, black raspberry, blueberry, cranberry, red raspberry, and strawberry extracts inhibit the growth of human oral, breast, colon, and prostate cancer cell lines in a dose-dependent manner. Black currant, red currant, red raspberry, black raspberry and blueberry juices were evaluated for antiproliferative activities against cervix cell line (HeLa), melanoma cell line (Fem X), colon cell line (LS 174), breast cell line (MCF-7) and prostate cell line (PC-3) [27]. All examined juices showed antiproliferative activity in dose-dependent manner and strongest inhibition of cell growth was observed for the black currant, red currant and raspberry juices. Regression analysis showed that there was no statistically significant correlation between radical scavenging activity and antiproliferative activity regardless cell line, but there was significant correlation between antocyanin content and antiproliferative activity against HeLa, LS 174 and MCF-7 cell lines.

Strawberry, raspberry, both highbush and lowbush blueberries, blackberry, and cranberry juices inhibited growth of human stomach, prostate, intestine, and breast cancer cell lines, and also induce apoptosis and cell cycle against used cell lines [45]. Anticancer activity of berry fruits also includes inhibition of matrix metalloproteinases [46] and down-regulation of VEGF [47], important factors in the process of primary tumor metastasis and tumor progression. The correlation between levels of antioxidants or antioxidant activity of fruits and berry extracts and their antiproliferative activity is not consistent and often shows cell-type specificity, implicating more specific mechanisms than direct antioxidant action [27, 45, 48]. At the same time concentrations of berry extracts used in cell culture experiments are much higher than concentrations that could be achieved by dietary intake, suggesting that data obtained so far should be used as milestone for further research of more sensitive mechanisms and biomarkers of adaptogenic action of berries bioactives towards cancer prevention.

## Effects of Berries on Antioxidant Status – Human Intervention Studies

As discussed before, high AOC of foods or bioactive, as well as their CAA should not be considered as an indicator of the potential for altering *in vivo* antioxidant status. The major factor precluding extrapolation between *in vitro* and *in vivo* systems, regarding effects of berry fruits, extracts and phenolics on antioxidant status, is their extensive metabolism and "poor" bioavailability. Bioactive compounds of berries are metabolized in upper gastrointestinal tract, by colonic microflora and in different tissues of human body, undergoing excessive transformations and providing different simple phenolics, aromatic acids and other products as metabolites, and many of them are still unidentified [44]. Levels of specific metabolites, produced mainly by the action of detoxifying enzymes, in circulation are very low, especially compared with constitutive antioxidants of human body [36]. For that reason effects of berries consumption on antioxidant status, observed in human intervention studies, could be a result of direct antioxidant effects of bioactive, but more likely their indirect effect, mediated through various signaling pathways, enzymes, cytokines, and receptors. It should be noted that berry phytochemicals may also impact cellular processes that are completely independent of antioxidant mechanisms [49].

*In vivo* antioxidative effects of berry fruits, extracts or bioactive compounds intake has been evaluated in several studies by measuring total antioxidant capacity, ie. direct action of antioxidants present in plasma, serum or tissues with radicals, by measuring markers of oxidative damage in body fluids and tissues, or their influence on activity of enzymatic systems, considered as a major part of antioxidant defense.

Korean raspberry induces glutathion peroxidase activity, without effects on lipid peroxidation, after 4 weeks of consumption of 30g freeze-dried fruits daily in healthy men [50]. In subjects with metabolic syndrome 50g of blueberries daily, during 8 weeks, induced decrease in oxLDL cholesterol levels compared with the control [51]. Similar results were obtained in the same population group, in the study investigating the effects of 50 gr freeze-dried fruits daily, during 8 weeks [52]. At the same time cranberry juice (750ml/day during 2 weeks), did not show any effects on glutathion- peroxidase,catalase and superoxe-dismutase activities, levels of malonyl aldehyde or H2O2-induced DNA damage in lymphocytes in healthy probands [53]. In hemodialysis patients, consuming 200ml/day of red berries juice during 4 weeks significantly decreased DNA-damage in lymphocytes and nuclear factor kappaB activity and increase in glutathione levels and status [54].

Weisel et al. [55] completed a 4 week intervention with a red berry juice, containing red raspberry juice, which improved levels of glutathione and reduced DNA oxidative damage in healthy adult males. However, when a dessert made from a similar juice was given to elderly subjects, no effect was found on oxidative stress status after a lesser duration of 2 weeks [56].

Indirect antioxidant activity could also be based on their interaction with other nutrients before the consumption or within the gastrointestinal tract. Prevention of oxidation of food lipids and other nutrients by phenolic compounds could reduce the burden of pro-oxidant intake, influencing *in vivo* antioxidant or redox status. At the same time, induction of oxidative stress, by consumption of a meal or exercise or in an unhealthy population with high risk of disease improves the ability to measure significant antioxidant effects of dietary antioxidants [57]. Taking in account postprandial state is an important contributing factor to chronic disease [58], several human studies evaluated influence of berry consumption within or after meals, and their possibility to counterbalance putative negative effects.

In a study of Kay and Holub [59], a single-blinded crossover study was performed in a group of eight middle-aged male subjects which consumed a high-fat meal and a control supplement or the same high-fat meal supplemented with 100 g freeze-dried wild blueberry powder 1 week later. Significant increases in serum antioxidant status compared to the controls were observed. Blueberries consumed with high fat meal also effectively induced total antioxidant capacity (measured by ORAC assay) in healthy subjects compared with controls [60].

Using exercise as a source of oxidative burden, Morillas-Ruiz et al. [61] have evaluated effects of a beverage containing raspberry, black grape, and red currant concentrates and found significantly decreased levels of protein and DNA oxidation in the treatment group receiving the antioxidant-rich beverage after exercise, compared to pre-exercise and control levels.

## Effects of Berries on Cardiovascular Health - Human Intervention Studies

Numerous epidemiological studies showed positive correlation between berry fruits consumption and cardiovascular (CV) health [62, 63, 64, 65]. One of the first study initiating further investigation of berries consumption benefits for CV health was a study by Knekt et al. [66] showing that high intake of berries, alongside other factors, was associated with a 60% decline in heart disease and stroke. Taking epidemiological studies as a good start point, and based on numerous *in vitro* data, beneficial effects of berries on CV health are investigated in numerous human intervention studies using "traditional" (dyslipidemia, hyperglycemia, obesity, blood pressure) and "non-traditional" (platelet function, chronic inflammation, endothelial function, oxidative stress) risk factors as biomarkers.

Consumption of berry fruits, juices or corresponding extracts could be effective in optimization of lipid profile based on results of numerous randomized, placebo-controlled clinical trials, in healthy subjects or subjects with cardiovascular risk factors.

Long- term (8 weeks) moderate consumption of different berries or berry juices (bilberries, lingonberries, black currant, strawberries, chokeberries and raspberries) in a single-blind, randomized, placebo-controlled intervention trial with patients with cardiovascular risk factors exert beneficial effects on HDL-cholesterol levels, but without statistically significant effects on LDL- levels and triglycerides [67].

Abidov et al. [68] showed that 4 week consumption of dietary supplement that contains 900 mg of standardized extracts and concentrates of prune, pomegranate, apple, grape, raspberry, blueberry, white cherry and strawberry, resulted in statistically significant reduction of total plasma cholesterol, LDL cholesterol and elevation of plasma HDL levels compared to the placebo group, in none-obese, non-smoking, non-diabetic hypercholesteremic male volunteers.

Similarly, in controlled randomized trial, 8 week of consumption of 50g freeze-dried strawberries per day, equivalent to 350 g fresh fruits, significantly decreased total and low-density lipoprotein cholesterol and small low-density lipoprotein particles in obese men and women with metabolic syndrome. Serum triglycerides and high-density lipoprotein cholesterol, as well glucose, blood pressure, and waist circumference were not affected [51]. These data confirmed previously published results of non-controlled dietary intervention study with daily consumption of juice containing 25 g freeze-dried strawberry powder strawberry, for four weeks [69]. Obtained results showed that short-term supplementation of freeze-dried strawberries appeared to exert hypocholesterolemic effects by causing significant decrease of total cholesterol and LDL-cholesterol plasma levels and non-significant decreasing trend for ox-LDL in women with metabolic syndrome.

Opposite to this finding, in a study with similar design (50g freeze-dried fruits per day for 8 weeks) blueberries did not alter lipid status of subjects with metabolic syndrome [52].

Results obtained from the intervention studies with cranberry fruits or juices regarding their potential to regulate dyslipidemia are not consistent. In healthy female subjects consuming 750 ml of cranberry juice per day or control drink, for 2 weeks, total cholesterol, triglycerides (TG), HDL and LDL levels were unchanged both in intervention and control group [53]. Dietary intervention with 480 ml/day of cranberry juice but during longer period (8 weeks) in subjects with metabolic syndrome showed similar results, there were no significant improvements in lipid profile, glucose levels, blood pressure and markers of inflammation [70].

Cranberry juice concentrate powder, used as intervention in randomized controlled study of Chambers and Camire [71] during 12 weeks, was not effective in regulation of lipid status, fasting glucose or HbA1C in patients with diabetes type 2.

Inconsistently with the studies reported above, two studies investigating effects of cranberries and their bioactives on lipid status reported positive results. Ruel et al. [72] showed dose-dependent increase in HDL cholesterol with 125, and 250 ml per day of cranberry juice vs. control, in healthy subjects, during 4 weeks in every consecutive phase. HDL level was not further increased with 500 ml of the same juice. After 12 weeks there was significant reduction in body weight, BMI and waist circumference also. Favorable effects of cranberry extracts powder vs. control during 12 weeks were also obtained in a study of Lee et al. [73] in subjects with diabetes type 2, showing decrease in total and LDL-cholesterol and total:HDL-cholesterol ratio.

Several studies investigated effects of chokeberry on lipid status, although many of them were un-controlled. Sikora et al. [74]showed that after 2 months of supplementation with 300 mg of chokeberry extract plasma levels of triglycerides, total cholesterol and LDL-cholesterol decreased, while HDL levels remained unchanged, both in healthy and subjects with metabolic syndrome. The same results were obtained by other authors, in patients with metabolic syndrome consuming 300 mg of chokeberry extract during 8 weeks [75, 76] or 250 ml of chokeberry juice during 6 weeks [77], and in subjects with mild hypercholesterolemia consuming 250 ml of chokeberry juice during 6 weeks [78]. In randomized controlled study, Naruszewicz et al. [79]reported that in patients with myocardial infarction on statin therapy, consumption of 255 mg chokeberry extract daily during 6 weeks, did not changed lipid status compared with the control.

In an un-controlled study Simeonov et al. [80] showed that long-term consumption of 200 ml of pure chokeberry juice decreased total cholesterol and lipid levels in patients with non-insulin dependent diabetes mellitus.

In randomized, double-blind, placebo-controlled trial consumption of 320 mg/day of supplement containing 17 different natural purified anthocyanins from bilberry and black currant, during 12 weeks by subjects with dyslipidemia resulted in increased HDL-cholesterol and decreased LDL-cholesterol levels [81].

Several studies investigated effects of berry consumption on lipid status in postprandial state. This approach was rationalized by potential of berry phenolics to inhibit oxidation of lipids, shown in numerous *in vitro* studies, thus acting as potential antioxidants of food lipids and beneficial agents in attenuating acute dietary lipid load.

Consumption of blueberries (50g freeze-dried fruits per day for 8 weeks) resulted in decrease in systolic and diastolic blood pressure of subjects with metabolic syndrome [52]. Similarly, consumption of different berries or berry juices (bilberries, lingonberries, black currant, strawberries, chokeberries and raspberries) during 8 weeks reduced systolic blood pressure compared to the control in patients with cardiovascular risk factors [67].

All five studies evaluating blood pressure as biomarker showed decrease in systolic and diastolic blood pressure after 6 weeks of regular consumption of chokeberry extract (255g/day) in patients with myocardial infarction [79] and chokeberry juice (250ml/day) in subjects with metabolic syndrome [77], and also after 12 weeks of consumption of chokeberry juice (400ml/day) in patients with non-insulin dependent diabetes mellitus [80]. In subjects with metabolic syndrome 300 mg of chokeberry extract during 8 weeks [75, 76] also decreased systolic and diastolic blood pressure.

In patients with myocardial infarction [79] chokeberry consumption statistically significant decreased serum levels of IL-6, C-reactive protein, ICAM, VCAM, monocyte chemotactic protein-1 (MCP-1) and adiponectin, compared to the control group. In subjects with mild hypercholesterolemia consumtion of 250 ml of chokeberry juice during 6 weeks [78] improved serum nitric oxide level, flow-mediated dilatation and brachial artery diameter.

Consumption of chokeberry reduced fasting glucose levels in subjects with metabolic syndrome compared to the control [77], and fasting glucose levels and HbA1C in patients with non-insulin dependent diabetes mellitus, compared to the baseline values [80].

In a study of Ruel et al. [72] consumption of cranberry juice during 12 weeks resulted in 12 significant reduction in body weight, BMI and waist circumference in healthy subjects.

Sikora et al. [74] showed that regular supplementation with 300 mg of chokeberry extract in patients with metabolic syndrome had beneficial effects on platelet aggregation, after one month. Platelet function of patients with cardiovascular risk factors, measured with Platelet Function Analyzer (PFA-100), was reduced compared to the control after 8 weeks of consumption of different berries or berry juices [67].

## Conclusion

Beneficial effects of regular consumption of different berries in health promotion and prevention of chronic diseases are shown in numerous human intervention studies. As shown from the review berry fruits, juices or extracts are potent modulators of different parameters, recognized as risk factors or biomarkers of chronic diseases such as cancer or cardiovascular diseases, investigated in healthy subjects or in risk population. Phenolic compounds of berries and their extracts have strong antioxidant potential that is often used to explain their health-promoting properties.

Recent studies postulated that their antioxidant potential, as a basic mechanism of biological activity, is compromised by their low bioavailability. It is still unclear if direct antioxidant capacity of berries and their bioactive agents is responsible for these beneficial effects or they are mediated by other antioxidant-independent mechanisms. It is also very important to determine bioactive compound(s) responsible for observed beneficial effects and to evaluate dose-response correlation.

Consequently further research in this area is targeted towards other mechanisms of action, the evaluation of main sites of action and further elucidation of potential of berry phenolic compounds as antioxidants, based on their structure or metabolism, both *in vitro* and within the human body.

## Acknowledgments

The authors acknowledge their gratitude to the Ministry of Education and Science of Serbia for financial support, project number 46013.

## References


[1] S.H. Häkkinen, S.O. Kärenlampi, I.M. Heinonen, H.M. Mykkänen, and A.R. Törrönen, *J. Agric. Food Chem.* 47, 2274 (1999).

[2] J.M. Koponen, A.M. Happonen, P.H. Mattila, and A.R. Törrönen, *J. Agric. Food Chem.* 55, 1612 (2007).

[3] K. Šavikin, G. Zdunić, T. Janković, S. Tasić, N. Menković, T. Stević, B. Đorđević, *Plant Food Hum. Nutr.* 64, 212 (2009).

[4] B. Đorđević, K. Šavikin, G. Zdunić, T. Janković, T. Vulić, C. Oparnica, and D. Radivojević, *Plant Food Hum. Nutr.* 65, 326 (2010).

[5] Phenol-explorer Data base on polyphenol content in foods http://www. phenol-explorer.eu.

[6] W. Wiczkowski, E. Romaszko and M. K. Piskula, *J. Agric. Food Chem.* 58, 12130 (2010).

[7] C. Manach, A. Scalbert, C. Morand, C. Remesy and L. Jimenez, *Am. J. Clin. Nutr.* 79 727 (2004).

[8] L. Gu, M.A. Kelm, J.F. Hammerstone, G. Beecher, J. Holden, D. Haytowitz, S. Gebhart, and R.L. Prior, *J. Nutr.* 134, 613 (2004).

[9] S. Häkkinen, M. Heinonen, S. Kärenlampi, H. Mykkänen, J. Ruuskanen, and R.Törrönen, *Food Res. Int.* 32, 345 (1999).

[10] M.P. Kahkonen, A.I. Hopia, and M. Heinonen, *J. Agric. Food Chem.* 49, 4076 (2001).

[11] S. Häkkinen, S. Kärenlampi, H. Mykkänen, M. Heinonen, and R. Törrönen, *Eur. Food Res. Technol.* 212, 75 (2000).

[12] A. E. Pazmiño-Durán, M. M. Giusti, R. E. Wrolstad and B. A. Glória, *Food Chem.* 75, 211 (2001).

[13] I. Konczak, and W. Zhang, *J. Biomed. Biotechn.* 5, 239 (2004).

[14] P.M. Dey, and J.B. Harborne, *Plant phenolics methods in plant biochemistry*, Academic Press Limited, London, (1993) p326.

[15] S.Y. Wang and H. S. Lin, *J. Agric. Food Chem.* 48, 140 (2000).

[16] X. Wu, L. Gu, R.L. Prior and S. McKay, *J. Agric. Food Chem.* 52, 7846 (2004).

[17] S. Benvenuti, F. Pellati, M. Melegari, and D. Bertelli, *J. Food Sci.* 69, 164 (2004).

[18] K.H. Jarkko, A.R. Törrönen, and H.M. Pirjo, *J. Agric. Food Chem.* 57, 7899 (2009).

[19] J.K. Hellstrom, A. Torronen, and P.H. Mattila, *J. Agric. Food Chem.* 57, 7899 (2009).

[20] USDA Database for the proanthocyanidin content of selected foods, http://www.nal.usda.gov/fnic/foodcomp

[21] USDA Database for flavonoid content of selected foods, http://www.nal.usda.gov/ fnic/foodcomp

[22] L.R. Howard, and T.J. Hager, *Berry fruit, value-added products for health promotion*, CRC Press: Taylor and Francis Group, USA (2007) p73.

[23] Y. Zhao, *Berry fruit, value-added products for health promotion*, CRC Press: Taylor and Francis Group, USA (2007) p291.

[24] S. Häkkinen, S. Kärenlampi, H. Mykkänen, and R. Törrönen, *J. Agric. Food Chem.* 48, 2960 (2000).

[25] G. Skrede, R. E. Wrolstad and R. W Durs, *J. Food Sci.* 65, 357 (2000).

[26] M. Sandell, O. Laaksonen, R. Jarvinen, N. Rostiala, T. Pohjanheimo, K. Tiitinen and H. Kallio, *J. Agric. Food Chem.* 57, 3718 (2009).

[27] A. Konić-Ristić, K. Šavikin, G. Zdunić, T. Janković, Z. Juranić, N. Menković, and I. Stanković, *Food Chem.* 125, 1412 (2011).

[28] J. Piljac-Žegarac, L. Valek, S. Martinez and A. Belšcak, *Food Chem.* 113, 394 (2009).

[29] P. Mattila, J. Hellström, G. McDougall, G. Dobson, J-M. Pihlava, T. Tiirikka, D. Stewart and R. Karjalainen, *Food Chem.* 127, 1216 (2011).

[30] M. Plessi, D. Bertelli and A. Albasini, *Food Chem.* 100, 419 (2007).

[31] I. Zabetakis, D. Leclerc and P. Kajda, *J. Agric. Food Chem.* 48, 2749 (2000).

[32] G. Borges, A. Degeneve, W. Mullen, and A. Crozier, *J. Agric. Food Chem.* 58, 3901 (2010).

[33] L. Jakobek, M. Seruga, I. Novak, M. Medvidovic-Kosanovic, *Deut. Lebensm.-Rundsc.* 103, 58 (2007).

[34] M. Heinonen, *Mol. Nutr. Food Res.* 51, 684 (2007).

[35] D. Huang, B. Ou, and R.L. Prior, *J. Agric. Food Chem.* 53, 1841 (2005).

[36] P.C. Hollman, A. Cassidy, B. Comte, M. Heinonen, M. Richelle, E. Richling, M. Serafini, A. Scalbert, H. Sies, and S. Vidry, *J. Nutr.* 141, 989S (2011).

[37] D. Honzel, S.G. Carter, K.A. Redman, A.G. Schauss, J.R. Endres, and G.S. Jensen, *J. Agric. Food Chem.* 56, 8319 (2008).

[38] K. L. Wolfe and R. H Liu, *J. Agric. Food Chem.* 55, 8896 (2007).

[39] K. L. Wolfe, X. Kang, X. He, M. Dong, Q. Zhang and R. H. Liu, *J. Agric. Food Chem.* 56, 8418 (2008).

[40] K. L. Wolfe and R. H Liu, *J. Agric. Food Chem.* 56, 8404 (2008).

[41] D. Gođevac, V. Tešević, V. Vajs, S. Milosavljević, and M. Stanković, *J. Food Sci.* 76, C1039 (2011).

[42] D. Gođevac, V. Tesević, V. Vajs, S. Milosavljević, and M. Stanković, *Food Chem. Toxicol.* 47, 2853 (2009).

[43] D. Gođevac, V. Tešević, V. Vajs, S. Milosavljević, G. Zdunić, and M. Stanković, "Currant seed extracts showing protective effect on human lymphocytes DNA", The International PSE Symposium of Phytochemicals in Nutrition and Health, Sept. 2011 (Bari, Italy) p69.

[44] N. P. Seeram, *J. Agric. Food Chem.* 56, 630 (2008).

[45] M. E. Olsson, K. E. Gustavsson, S. Andersson, A. Nilsson and R. D. Duan, *J. Agric. Food Chem.* 52, 7264 (2004).

[46] M. D. Matchett, S. L. MacKinnon, M. I. Sweeney, K. T. Gottschall-Pass and R. A. Hurta. *J. Nutr. Biochem.* 17, 117 (2006).

[47] T. Chen, M.E. Rose, H. Hwang, R.G. Nines, and G.D. Stoner, *Carcinogenesis* 27, 2301 (2006).

[48] J. God, P. L. Tate and L. L. Larcom, *Nutr. Res.* 30, 777 (2010).

[49] R. L. Prior, L. Gu, X. Wu, R. A. Jacob, G. Sotoudeh, A. A. Kader and R. A. Cook, *J. Am. Coll. Nutr.* 26, 170 (2007).

[50] J.E. Lee, E. Park, J.E. Lee, J.H. Auh, H.K. Choi, J. Lee, S. Cho, and J.H. Kim, *Nutr. Res. Pract.* 5, 429 (2011).

[51] A. Basu, M. Du, M.J. Leyva, K. Sanchez, N.M. Betts, M. Wu, C.E. Aston, and T.J. Lyons, *J. Nutr.* 140, 1582 (2010).

[52] A. Basu, D.X. Fu, M. Wilkinson, B. Simmons, M. Wu, N.M. Betts, M. Du, and T.J. Lyons, Strawberries decrease atherosclerotic markers in subjects with metabolic syndrome. *Nutr. Res.* 30, 462 (2010).

[53] S.J. Duthie, A.M. Jenkinson, A. Crozier, W. Mullen, L. Pirie, J. Kyle, L.S. Yap, P. Christen, and G.G. Duthie, *Eur. J. Nutr.* 45, 113 (2006).

[54] T. M. Spormann, F. W. Albert, T. Rath, H. Dietrich, F. Will, J. P. Stockis, G. Eisenbrand and C. Janzowski, *Cancer Epidemiol. Biomarkers Prev.* 17, 3372 (2008).

[55] T. Weisel, M. Baum, G. Eisenbrand, H. Dietrich, F. Will, J. P. Stockis, S. Kulling, C. Rüfer, C. Johannes and C. Janzowski, *Biotechnol. J.* 1, 388 (2006).

[56] J. García-Alonso, M.J. Periago, M.L. Vidal-Guevara, M.C. Ramírez-Tortosa, A. Gil, and G. Ros, *Arch. Latinoam. Nutr. Jun.* 54, 180 (2004).

[57] A. V. Rao and D. M. Snyder, *J. Agric. Food Chem.* 58, 3871 (2010).

[58] B. Burton-Freeman, *Br. J. Nutr.* 104, S1 (2010).

[59] C.D. Kay, and B.J. Holub, *Br. J. Nutr.* 88, 389 (2002).

[60] G. Mazza, C. D. Kay, T. Cottrell and B. J. Holub, *J. Agric. Food Chem.* 50, 7731 (2002).

[61] J. Morillas-Ruiz, P. Zafrilla, M. Almar, M. J. Cuevas, F. J. López, P. Abellán, J. A. Villegas and J. González-Gallego, *Eur. J. Appl. Physiol.* 95, 543 (2005).

[62] P. M. Ridker, C. H. Hennekens, J. E. Buring and N. Rifai, *N. Engl. J. Med.* 342, 836 (2000).

[63] T. H. Rissanen, S. Voutilainen, J. K. Virtanen, B. Venho, M. Vanharanta, J. Mursu and J. T. Salonen. *J. Nutr.* 133, 199 (2003).

[64] N. P. Seeram, S. L. Adams, Y. Zhang, R. Lee, D. Sand, S. H. Scheuller and D. Heber, *J. Agric. Food Chem.* 54, 9329 (2006).

[65] H. D. Sesso, J. M. Gaziano, D. J. Jenkins and J. E. Buring, *J. Am. Coll. Nutr.* 26, 303 (2007).

[66] P. Knegt, R. Jarvinen, A. Reunanen, and J. Maatela, *Br. Med. J.* 312, 478 (1996).

[67] I. Erlund, R. Koli, G. Alfthan, J. Marniemi, P. Puukka, P. Mustonen, P. Mattila, and A. Jula, *Am. J. Clin. Nutr.* 87, 323 (2008).

[68] M. Abidov, M. Jimenez Del Rio, A. Ramazanov, O. Kalyuzhin, and I. Chkhikvishvili, *Georgian Med. News* 140, 78 (2006).

[69] A. Basu, M. Wilkinson, K. Penugonda, B. Simmons, N.M. Betts, and T.J. Lyons, *Nutr. J.* 8, 43 (2009).

[70] A. Basu, N.M. Betts, J. Ortiz, B. Simmons, M. Wu, and T.J. Lyons, *Nutr. Res.* 31, 190 (2011).

[71] B.K. Chambers, and M.E. Camire, *Diabetes Care* 26, 2695 (2003).

[72] G. Ruel, S. Pomerleau, P. Couture, S. Lemieux, B. Lamarche and C. Couillard, *Br. J. Nutr.* 96, 357 (2006).

[73] I.T. Lee, Y.C. Chan, C.W. Lin, W.J. Lee, and W.H. Sheu, *Diabet Med.* 25, 1473 (2008).

[74] J. Sikora, M. Broncel, M. Markowicz, M. Chałubiński, K. Wojdan and E. Mikiciuk-Olasik, *Eur. J. Nutr.* Aug 18 (2011).

[75] M. Broncel, M. Koziróg-Kołacińska, G. Andryskowski, P. Duchnowicz, M. Koter-Michalak, A. Owczarczyk, and J. Chojnowska-Jezierska, *Pol. Merk. Lek.* 23, 116 (2007).

[76] M. Broncel, M. Kozirog, P. Duchnowicz, M. Koter-Michalak, J. Sikora, and J. Chojnowska-Jezierska, *Med. Sci. Monit.* 16, CR28 (2010).

[77] A. Skoczyńska, I. Jędrychowska, R. Poręba, A. Affelska, B. Turczyn, A. Wojakowska and R. Andrzejak, *Pharm. Rep.* 59, 177 (2007).

[78] R. Poreba, A. Skoczynska, P. Gac, M. Poreba, I. Jedrychowska, A. Affelska-Jercha, B. Turczyn, A. Wojakowska, J. Oszmianski and R. Andrzejak, *Ann. Agric. Environ. Med.* 16, 305 (2009).

[79] M. Naruszewicz, I. Laniewska, B. Millo and M. Dłuzniewski, *Atherosclerosis* 194, e179 (2007).

[80] S. B. Simeonov, N. P. Botushanov, E. B. Karahanian, M. B. Pavlova, H. K. Husianitis and D. M. Troev, *Folia Med.* (Plovdiv) 44, 20 (2002).

[81] Y. Qin, M. Xia, J. Ma, Y. Hao, J. Liu, H. Mou, L. Cao and W. Ling, *Am. J. Clin. Nutr.* 90, 485 (2009).

In: Phytochemicals
Editor: Hai-Xue Kuang

ISBN: 978-1-62417-354-7


*Chapter VI*

# Overview of the Chemical, Natural Occurrence and Antioxidant Activity of Terpenes

***E. González-Burgos and M. P. Gómez-Serranillos****
Department of Pharmacology, Faculty of Pharmacy.
Universidad Complutense de Madrid (UCM), Madrid, Spain

## Abstract

Terpenes are one of the most abundant groups of natural products that occur mostly in plants, being identified more than 30,000 different compounds. Chemically, terpene group is one of the most diverse in structure, sharing isoprene units as a common structural motif. The number of isoprene units that defined each terpene compound classifies them into several groups including hemiterpenes, monoterpenes, sesquiterpenes, diterpenes, sesterpenes, triterpenes, tetraterpenes and politerpenes.

Terpenes, as a natural product group, have sparked a great interest in research because of their relevant and broad spectrum of health-promoting effects such as anti-inflammatory, antimicrobial, antitumor and antioxidant, among others. Nowadays, new findings in pharmacological area are in constant growing.

Since oxidative stress has been involved in the pathophysiological development of approximately one hundred of different diseases including cancer, autoimmune diseases, heart attack and neurodegenerative diseases, there have been numerous attempts to cope with this redox imbalance. Exogenous compounds with antioxidant capacity provide a protective effect against free radical overproduction and its consequence cellular damage, being this one the most promising therapeutic strategy. Over the last recent years, there is an increasing interest in the search of antioxidant compounds among terpenes, existing emerging number of reports focused on this issue.

This chapter will cover the chemistry, the natural occurrence and the most recent findings in antioxidant activity of terpenes.

* E-mail address: pserra@farm.ucm.es Dra. M. Pilar Gómez-Serranillos. Full Professor, Department of Pharmacology, Faculty of Pharmacy. Universidad Complutense de Madrid (UCM), Madrid Spain.

# Introduction

Terpenes are an extensive and varied class of more than 30,000 different secondary metabolites derived biosynthetically from isoprene. The name "terpene" comes from "terpentin" in German (meaning "turpentine" in English) in reference to the fact that the first identified compounds of this class were found in this fluid from pine tree resin [1].

Terpenes are widespread among living organisms, including bacteria, fungi, algae, plants and some animals (mainly insects and marine organisms). These compounds are the major ones of essential oils, resins waxes and latex [2]. In plants, the concentration of terpenes is higher in the young organs than in the old ones [3]. They are found particularly abundantly in reproductive structures and foliage during and immediately following flowering [4].

Terpenes have a relevance biological activity since they actively participate in multiple physiological, structural, metabolic, defensive and communication roles such as electron carriers (i.e. ubiquinone, plastoquinone), hormones for the regulation of growth (i.e. abcisic acid, gibberellins), photosynthetic pigments (i.e. carotenoids, phytol), membrane structural components (phytosterols), attractants to seed dispersers and pollinators and protectants (toxins, repellants, competitive phytoxins antibiotics, cited as example the secretions of iridoid monoterpenes of leaf beatles and the sticky mixture of mono-, sesqui- and diterpene of termites ) [5, 6]. At industry scale, terpenes have multiple interesting applications, providing flavors and fragrances in food, chemical and pharmaceutical industries. Moreover, they function as active insect antifeedants such as pyrethrins and azadirachtin in agrochemical industry and they offer useful polymers such as rubber and chicle with major industrial development. Furthermore, they are also use as drug-production bases such as the diterpene taxol, obtained from the bark of *Taxus brevifolia* L., with an extensive anticancer use against a broad range of tumors and the sesquiterpene lactone artemisin, used for the treatment of malaria caused by *Plasmodium falciparum* parasite [7, 8].

The great multitude and variability of terpene compounds have given rise to a major investigation of possible pharmacological uses, obtaining promising results for their anti-inflammatory, antifungical, antiviral, antimicrobial, antioxidant as well as above stated antiparasitic and anticancer properties [2]. In recent years, the researchers of terpenes as antioxidants compounds have grown considerably. The antioxidant defense mechanisms are considered as one of the most promising therapeutic approaches for the prevention and treatment of those diseases associated with oxidative stress. Therefore, in this chapter, we will focus on the antioxidant activities of terpenes.

# Chemical and Biosynthetic Pathways

Terpenes are acyclic, monocyclic and polyciclic compounds resulted from the combination of several isoprene units rearrangementing in a regular "head to tail" or irregular "head to head" and "head to middle" manner. Isopentenyl pyrophosphate (IPP) and its isomer dimethylallyl diphosphate (DMAPP) are both the functional isoprene units in living organisms [9].

According to the number of isoprene units, the terpenes contain, they are classified in several groups including hemiterpenes (1 isoprene unit; $C_5$); monoterpenes (2 isoprene units;

$C_{10}$); sesquiterpenes (3 isoprene units; $C_{15}$); diterpenes (4 isoprene units; $C_{20}$); triterpenes (6 isoprene units; $C_{30}$); tetraterpenes (8 isoprene units; $C_{40}$) and polyterpenes (8 isoprene units < n > 30,000; $C_{5n}$). The general formula for terpenes is $(C_5H_8)_n$, where "$C_5H_8$" refers to the molecular formula of isoprene and "n" to the number of times that this motif is repeated in the terpene structure [9].

Since IPP and its isomer DMAPP are both the structural basis for terpene synthesis, the formation of these motifs is one of the major processes in biosynthetic route.

Two different routes have been described for IPP synthesis: mevalonic acid pathway and deoxyxylulose phosphate / methylerytritol (DOXP / MEP) pathway. The mevalonic acid pathway, which was discovered by Flesh & Rohmer in 1988 during their studies on the biosynthesis of "hopanoids", has been considered until the mid-1990s as the only one for the biosynthesis of isopenthenyl diphosphate (IPP) [10]. Experiments with isotopically labeled compounds, began by the scientific research team led by Rohmer, demonstrated in 1993 the possible existence of an alternative route to mevalonic acid via for the synthesis of this intermediate compound for terpenoids in *Escherichia coli* for the first time[11]. Subsequent studies conducted in other living organisms confirmed the alternative mevalonic acid independent route [12].

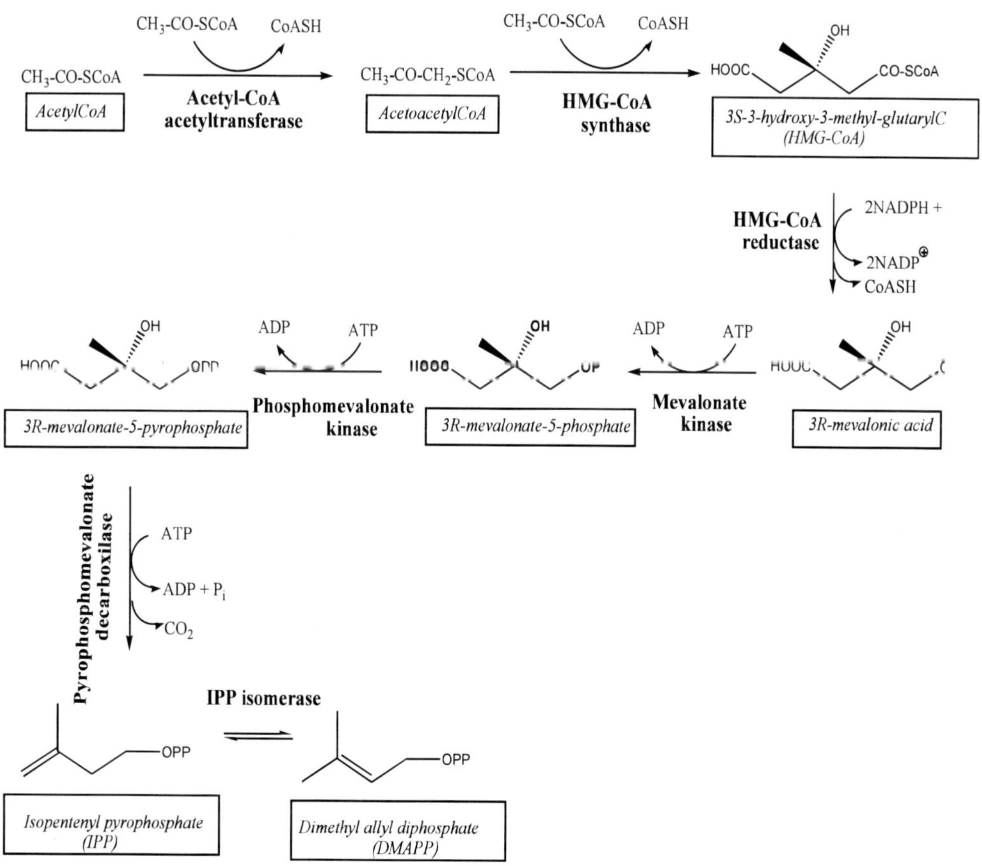


Figure 1. Mevalonic acid pathway.

### Mevalonic Acid Pathway

This route of IPP biosynthesis is common for bacteria, fungi, mammals and plants. In higher plants, this classic pathway occurs in cytosol, where it takes place the production of the precursors of the sesquiterpenes ($C_{15}$) and its multiples such as the triterpenes ($C_{30}$).

The classical IPP biosynthesis pathway, showed in Figure 1, starts with the condensation of three molecules of acetyl-CoA through Claisen-type reactions, catalyzed by thiolase and hydroxymethyl-glutaryl-CoA synthase enzymes, yielding hydroxymethyl glutarylCoA (HMGCoA). Following, this reaction product is reduced to mevalonic acid (MVA) by the action of HMG-CoA reductase using two equivalents of NADPH. Three consecutive phosphorylations and one decarboxylation, catalyzed by mevalonate kinase, phosphomevalonate kinase and pyrophosphomevalonate decarboxylase, produce finally the isopentenyl diphosphate (IPP) [13, 14].

### *1-deoxy-D-xylulose-5-phosphate / 2-C-methyl-D-erythritol-4-phosphate (DOXP/MEP) pathway*

This alternative route to mevalonate pathway has been identified in several green algae, in some parasites, in the plasmids of higher plants and in bacteria, both Gram-negative (i.e. *Escherichia coli*, *Salmonella typhimurium*, *Pseudomonas aeruginosa*) and Gram positive (i.e. *Mycobacterium phlei*, *Mycobacterium smegmatis*, *Streptomyces exfoliatus*). In higher plants, this IPP biosynthetic route, operated in the plastids, is implicated in the synthesis of the precursors of the monoterpenes ($C_{10}$), diterpenes ($C_{20}$) and other multiples.

The first reaction in DOXP/MEP pathway (Figure 2), catalyzed by DXP synthase in presence of divalent cations ($Mg^{2+}$, $Mn^{2+}$) and thiamine, is the condensation of pyruvate with glyceraldehyde-3-phosphate generating 1-deoxy-D-xylulose-5-phosphate (DXP) as product. Then the formed 1-deoxy-D-xylulose-5-phosphate is converted into 2-C-methyl-D-erythritol-4-phosphate by the action of 1-deoxy-D-xylulose-5-phosphate reductoisomerase that requires of $Mn^{2+}$ and NADPH for its activity. The next enzymatic steps involve the conversion of 2-C-methyl-D-erythritol-4-phosphate to 2-C-methyl-D-erythritol-2,4-cyclodiphosphate through different reactions in which the following enzymes MEP cytidyltransferase, CDP-ME kinase and ME CDP synthase participate. The next reaction steps, related to IPP and DMAPP formation, have not yet been completely defined. It is believed that 2-C-methyl-D-erythritol-2,4-cyclodiphosphate, through reduction and dehydration reactions, yielding 4-hydroxy-2-methyl-butenyl-4-diphosphate (HMBPP) [12, 15].

The first common step of mevalonic acid and 1-deoxy-D-xylulose-5-phosphate/2-C-methyl-D-erythritol-4-phosphate (DOXP/MEP) pathways is the isomerization of dimethylallyl diphosphate (DMAPP) from isopentenyl diphosphate (IPP) catalyzed by IPP isomerase. Two different types of isopentenyl pyrophosphate isomerase (IPP isomerase), type-1 and type-2, have been identified. Any correlation between the type of isomerase and the biosynthetic route for IPP formation has been identified [8, 12].

- *Type-1 IPP isomerase* requires both zinc and magnesium ions for interconverting IPP and DMAPP using a protonation-deprotonation mechanism. This metalloprotein has been detected in eukaryotes, in plant plastids and in some eubacteria including *Mycobacterium tuberculosis* and *Escherichia coli*.

- *Type-2 IPP isomerase,* a FMN- and NADPH- dependent enzyme which operates through a hydrogen atom addition and abstraction reaction, has been identified in *Archaebacteria* and in few Gram-positive coccoci such as *Staphylococcus aureus, Enterococcus faecalis* and *Streptococcus pyogenes.*

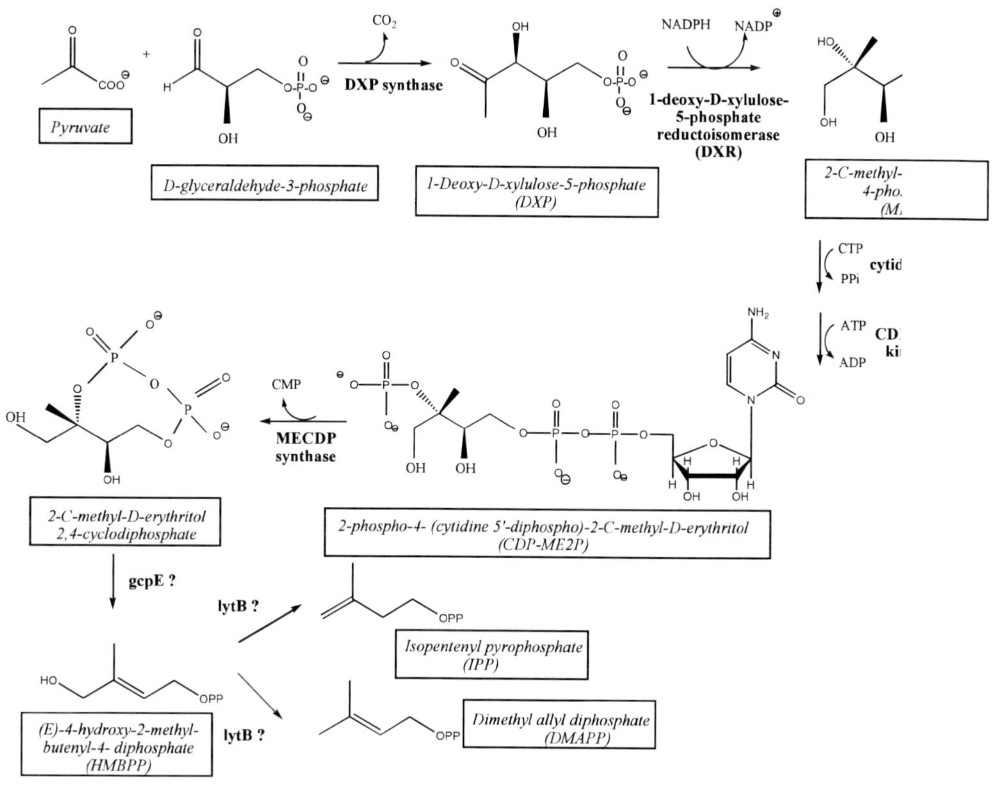


Figure 2. 1-deoxy-D-xylulose-5-phosphate / 2-C-methyl-D-erythritol-4-phosphate (DOXP/MEP) pathway.

The next step in biosynthetic route (Figure 3) involves the ionization of DMAPP to the allylic cation, which reacts then with the double bond of IPP, generating the first series of prenyl diphosphate homologs, geranyl pyrophosphate (GPP, $C_{10}$). The repetition of this process, adding one, two or more IPP units yields to the formation of pharnesyl pyrophosphate (FPP, $C_{15}$), geranyl geranyl pyrophosphate (GGPP, $C_{20}$) and polyprenyl pyrophosphate ($C_{5n}$), respectively. All these consecutive condensations of the IPP and its isomer, using electrophilic reaction mechanisms, for generating the before cited terpenoid carbon skeletons are catalyzed by prenyltransferases. Subsequent, the action of terpene synthases or terpene cyclases produce the skeletons of terpene families such as monoterpenes, diterpenes, triterpenes or polyterpenes [16].

Although a compartmental separation involving both IPP biosynthetic routes has been widely reported in higher plants, recent studies have shown possible exchanges of intermediates ($C_5$ and $C_{15}$) between the cytosol and the plastids [8].

Finally, secondary modifications including oxidation, reduction, conjugation, isomerization, among others, lead to the great diversity of known terpene structures.

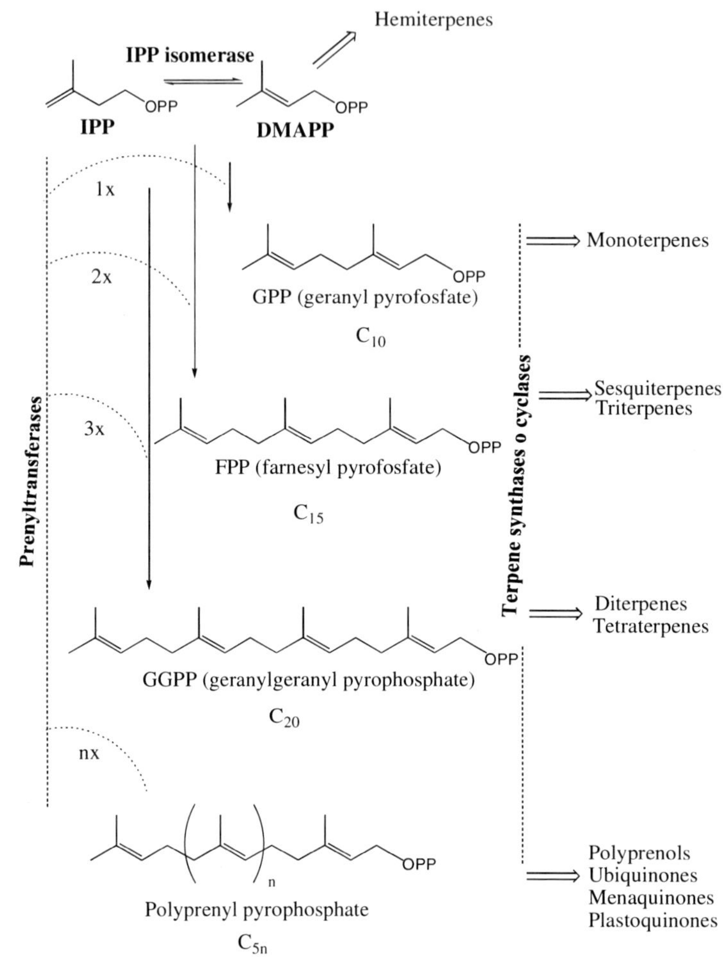


Figure 3. osynthesis of IPP, DMAPP, geranyl pyrophosphate (GPP), pharnesyl pyrophosphate (FPP), geranyl geranyl pyrophosphate (GGPP) and terpene skeletons.

# Antioxidant Role of Terpenes

The concept of "oxidative stress" defines a pathological situation resulting from the production of excessive amounts of reactive oxygen species (ROS) that cannot be compensated by the antioxidant system (enzymatic and non-enzymatic). Excessive ROS production causes lipid peroxidation, protein oxidation and mutagenesis at DNA level, which can lead to irreversible cell damage and even cell death mainly by apoptosis [17, 18]

Over the last decades, the "oxidative stress hypothesis' has gained importance as one of the causes or consequences of the pathogenesis of many diseases including cardiovascular diseases, neurodegenerative diseases, diabetes, cancer, among others. The most promising strategy for the prevention and treatment of oxidative stress-related events appears to be the use of antioxidants [17, 18].

## I. Monoterpenes

The essential oils are a complex mixture of different components constituted mainly by both hydrocarbons and oxygenated monoterpenes and sesquiterpenes. The antioxidant activity of many essential oils obtained from a great variability of plant species such as *Eucalyptus tereticornis, Origanum vulgare, Thymus vulgaris, Rosmarinus officinalis, Salvia officinalis and Syzygium aromaticum*, among others, have been widely reported [19].

It is generally assumed that the total antioxidant capacity of essential oils is the result of a combined action (antagonism, additivity, synergism) of containing components. Over the last years, more and more *in vitro* and *in vivo* studies have attempted to establish which of these compounds possess antioxidant properties and which ones pro-oxidant properties and how important is the contribution of individual components to the biological action.

The chemical structures of monoterpene with antioxidant activity as well as summary table of their activities are shown in Figure 4 and Table 1.

Roberto and Baratta (2000) [20] studied the antioxidant effectiveness of approximately 100 individual components found abundantly in essential oils composition, using two lipid model systems; one for measuring the formation of the secondary lipid peroxidation thiobarbituric acid reactive species (TBARS), utilizing egg yolk as lipid oxydable substrate and the other one for evaluating the peroxides that are formed during linoleic acid oxidation in a micellar system. In this antioxidant screening study, of over thirteen monoterpene hydrocarbons evaluated, *terpinolene*, *α-terpinene*, *γ-terpinene* and *sabinene* at the higher tested concentrations showed significant antioxidant activity in both lipid methods. This activity is attributed to the presence of strongly activated methylene groups in their structure. The antioxidant activity of oxygenated monoterpenes was evaluated in 34 components with a significant varied structural. Among all these, the two components with phenolic nature, *thymol* and *carvacrol*, were the most active ones, showing antioxidant values similar to α-tocoferol at the higher concentrations assayed. Allylic alcoholes such as *nerol*, *perillyl alcohol*, *geraniol* and *cis-verbenol* were demonstrated to possess moderate antioxidant activity, whereas this activity resulted scarce for monoterpenes aldehydes and ketones type.

The monoterpene *terpinolene* from the oil of *Pinus mugo* L. protected of oxidation of LDL (low-density lipoprotein), being this effect beneficial for cardiovascular diseases [21]. The monoterpene hydrocarbons *γ-terpinene* slowed down the linoleic acid peroxidation through a mechanism of action that involves a chain-terminating combination reaction of hydroperoxyl radicals (HOO$^{\cdot}$) and linoleylperoxyl radicals (LOO$^{\cdot}$). The antioxidant potency of γ-terpinene was similar to that of α-tocopherol. However, α-tocopherol was a prooxidant compound at high concentrations and γ-terpinene did not, at least, at the concentrations used in this assay [22]. Studies employing *in vitro* oxidation systems induced by the free radical initiator 2,2'-azobis(2-amidinopropane hydrochloride) (AAPH) demonstrated that both *α-terpinene* and *γ-terpinene* acted as scavenger agents protecting DNA, methyl linoleate and erythrocytes from the oxidant damage originated. γ-terpinene showed higher antioxidant compared to α-terpinene, attributed interestingly this unusual behavior to the fact that γ-terpinene has a methylene group at the bisallyl position in a nonconjugated diene in its structure instead of a conjugated diene as in α-terpinene [23].

**Table 1. Summary table of the antioxidant activities of monoterpenes**

| MONOTERPENES | | | |
|---|---|---|---|
| **Monoterpene type** | **Compound** | **Type of assay** | **References** |
| Monoterpene hydrocarbons | terpinolene | *In vitro*; cell-free assays<br>*In vitro*; culture system | [20,21] |
| | α-terpinene | *In vitro*; cell-free assays<br>*In vitro*; culture system | [20,23] |
| | γ-terpinene | *In vitro*; cell-free assays<br>*In vitro*; culture system | [20,22] |
| Oxygenated monoterpenes | borneol | *In vivo* | [33] |
| | carvacrol | *In vitro*; cell-free assays<br>*In vitro*; culture system<br>*In vivo* | [24-26, 29,31] |
| | eugenol | *In vitro*; cell-free assays<br>*In vitro*; culture system<br>*In vivo* | [27,28,32] |
| | thymol | *In vitro*; cell-free assays<br>*In vitro*; culture system | [24-26,30] |

The phenol structures of *thymol*, *carvacrol* and *eugenol* have been related to their antioxidant properties. *In vitro*, both cell-free assays and cultures system, have revealed that these monoterpene compounds possess scavenging actions against a variety of reactive oxygen species (ROS) and reactive nitrogen species (RNS) [24-28]. Moreover, *thymol* together with *carvacrol*, have been proposed as good natural antioxidants to preserve food since they inhibited Fe(III)-ascorbate induced lipid peroxidation in liposomes, scavenged peroxyl radicals and protected to DNA from bleomycin-Fe(III) complex damage [29]. In addition, the antioxidant properties of these monoterpene compounds have also been demonstrated in several *in vivo* studies. *Thymol* has shown a hepatoprotective role on mice. The oral pretreatment with thymol, before the injection of $CCl_4$, protected liver from the lipid peroxidation and from the changes in antioxidant enzymes activities, recovering this organ its normal oxidant/antioxidant status [30]. *Carvacrol*, through free radical scavenging and lipid peroxidation activity, relieved the pain associated to oxidative stress chemical and physical-stimulated (acetic acid, capsaicin, glutamate, formalin and thermal) in mice [31]. The administration of *eugenol* via oral at a hepatoprotective reported dosage did not cause changes neither on antioxidant enzymes activities nor on lipid peroxidation levels however increased GSH levels and glutathione-S-transferases (GSTs) in the intestine of rats, suggesting that this monoterpene compound could help with the elimination of toxic substances from this part of the gastrointestinal tract [32].

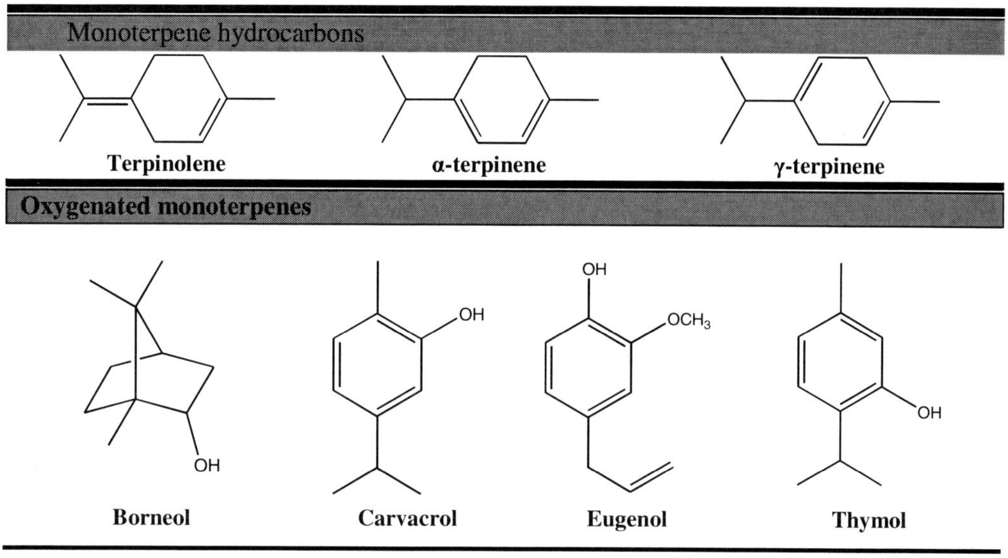


Figure 4. Chemical structure of monoterpenes with antioxidant properties.

In addition to all these compounds, others such as *borneol* have also showed antioxidant activity. The oral administration of monoterpene borneol ameliorated the alterations on lipid peroxidation levels and on activity of antioxidant defense system in NG-nitro-L-arginine methyl ester (L-NAME) induced hypertensive rats [33].

## II. Sesquiterpenes

Sesquiterpenes are also one constituent of the essential oils. The chemical structures of sesquiterpenes with antioxidant activity as well as summary table of their activities are shown in Figure 5 and Table 2.

Roberto and Baratta (2000) [20] observed that the antioxidant activity of sesquiterpene hydrocarbons was much lower and even insignificant than that of monoterpene hydrocarbons. On the other hand, sesquiterpenes oxygenated derivatives had an antioxidant effectiveness similar to that of the monoterpenes oxygenated derivatives, being again the allylic alcohols (i.e. *farnesol, guaiol, (+)-8-(15)-cedren-9-ol*) the most active ones.

Among the different sesquiterpene compounds found in the aerial parts of *Artemisia iwayomogi* Kitamura, those guaianolide-type compound with a double bond in position C2-C9 were the most active ones. Particularly, *1α,4α-dihydroxy-guaia-3(4), 9(10), 11(13)-trien-6α,12-olide and 4α-hydroxy-8α-acetoxy-1α-methyl-guaia-3 (4), 9(10), 11(13)-trien-6α,12-olide* were effective as peroxyl radicals scavengers and as ferric reduction compounds. Besides, the compound 4α-dihydroxy-guaia-3 (4), 9(10), 11(13)-trien-6α,12-olide had the ability of metal chelating [34].

Sesquiterpene lactone with α,ß-unsaturated carbonyl groups have shown to activate the Nrf2/ARE pathway and subsequently, inducing antioxidant enzymes to combat oxidative stress. That is the case of the sesquiterpene lactones *arucanolide*, *calealactone A*, *2,3-epoxy calealactone A*, *calealactone B* and *2,3-epoxyjuanislamin* from *Calea urticifolia* (Mill.) DC., *parthenolide* from *Tanacetum parthenium* (L.) Sch.Bip. and *zerumbone* from *Zingiber zerumbet* Smith [35, 36].

**Table 2. Summary table of the antioxidant activities of sesquiterpenes**

| SESQUITERPENES | | | |
|---|---|---|---|
| **Sesquiterpene type** | **Compound** | **Type of assay** | **References** |
| Oxygenated sesquiterpenes | (+)-8-(15)-cedren-9-ol | *In vitro*; cell-free assays | [20] |
| | 1α,4α-dihydroxy-guaia-3 (4), 9(10), 11(13)-trien-6α,12-olide | *In vitro*; cell-free assays | [34] |
| | 2,3-epoxyjuanislamin | *In vitro*; culture system | [35] |
| | 4α-hydroxy-8α-acetoxy-1α -methyl-guaia-3 (4), 9(10), 11(13)-trien-6α,12-olide | *In vitro*; cell-free assays | [34] |
| | arucanolide | *In vitro*; culture system | [35] |
| | calealactone A | *In vitro*; culture system | [35] |
| | calealactone B | *In vitro*; culture system | [35] |
| | farnesol | *In vitro*; cell-free assays | [20] |
| | guaiol | *In vitro*; cell-free assays | [20] |
| | zerumbone | *In vitro*; culture system | [36] |

**Oxygenated sesquiterpenes**

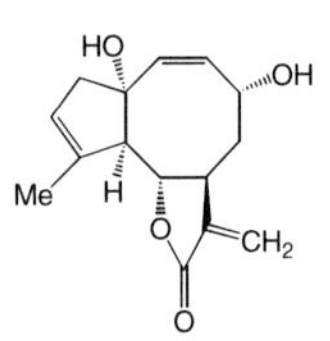

**1α,4α-dihydroxy-guaia-3 (4), 9(10), 11(13)-trien-6α,12-olide**

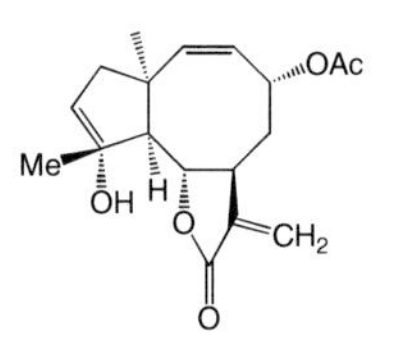

**4α-hydroxy-8α-acetoxy-1α -methyl-guaia-3 (4), 9(10), 11(13)-trien-6α,12-olide**

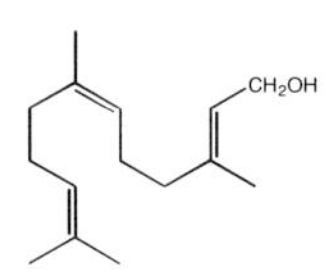

**Farnesol**

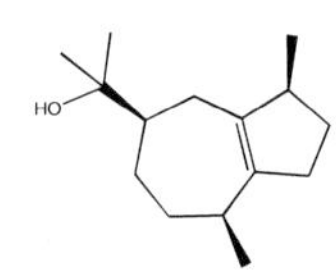

**Guaiol**

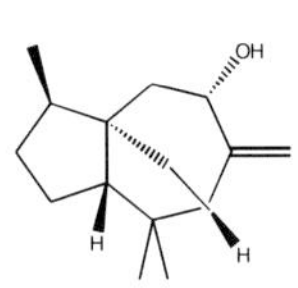

**(+)-8-(15)-cedren-9-ol**

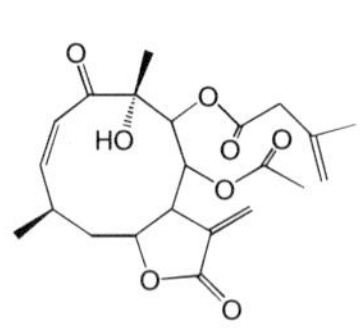

**Arucanolide**

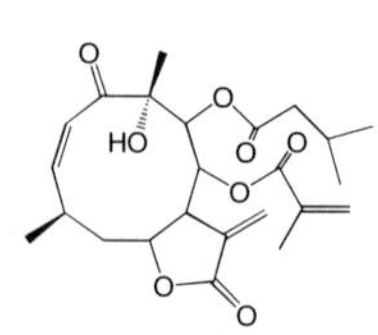

**Calealactone A**

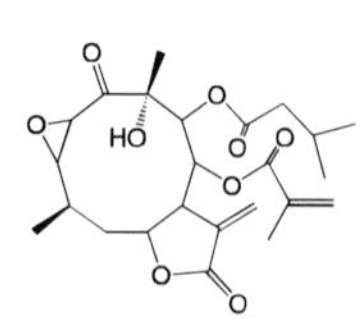

**2,3-epoxy calealactone A**

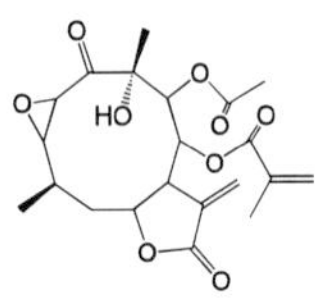

**Calealactone B**

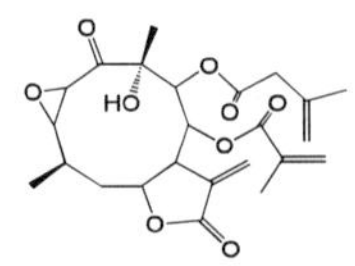

**2,3-epoxyjuanislamin**

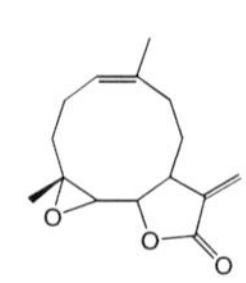

**Parthenolide**

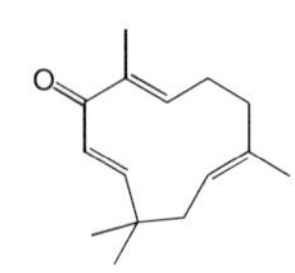

**Zerumbone**

Figure 5. Chemical structure of sesquiterpenes with antioxidant properties

## III. Diterpenes

Diterpenes contain four isoprene units ($C_{20}H_{32}$). Over 5000 diterpenes, with about 200 different skeletons such as pimarane, labdane, abietane and kaurane have been identified in living organisms, mainly in plants [37]. The botanical families Pinaceae, Leguminosae, Cistaceae, Labiateae and Euphorbiaceae, among others, are rich in diterpenes. In addition to plants, some animals area good source of diterpenes such as soft corals and sea fans. Diterpenes possess a wide variety of biological activities including anti-inflammatory, anti-apoptotic, anti-ulcerogenic, antimicrobial and antioxidant [2].

The number of studies focused on the antioxidant role of diterpenes has grown considerably in recent years, constituting a great potential of molecules with antioxidant interesting properties. Several of the most notable findings in this field are summarized below. The chemical structures of diterpenes with antioxidant activity as well as summary table of their activities are shown in Figure 6 and Table 3.

Among the different *in vitro* cell-free assays to estimate the antioxidant capacity of a compound, the DPPH method is the most extensive used for the characterization of antioxidant properties of diterpenes. For instance, the antioxidant activity of several abietane-type diterpenes isolated from the heartwood of the native taiwanese tree *Taiwania cryptomerioides* Hayata has been reported by using DPPH method. Results revealed that the most active scavenger diterpene was *ferruginol*, followed by *hinokiol* > *secoabietane dialdehyde* > *6β-acetoxy-7α-hydroxyroyleanone* = *isopimarinol* > *sugiol*. Comparing structural characteristics of these six abietane diterpenes, the most antioxidant activity observed corresponds to those compounds with one phenol group and without an electron-withdrawing group next to the aromatic ring [38]. Likewise, three years later Chao et al. [39] demonstrated that *sugiol* isolated from the bark of *Calecedrus formosana* Florin possessed a low free radical scavenging (DPPH) activity. Interestingly, in this same study, sugiol inhibited significantly ROS production in LPS-stimulated in the J7747 macrophage cell line as shown carboxy-2',7'-dichlorofluorescein diacetate (CM-DCFH) assay. Result of the different antioxidant behavior of sugiol in cell free system and culture system *in vitro* systems, the researchers of this previous work emphasize on the importance of experimental research using cells and animal models to avoid overlook of the antioxidant effectiveness of pure compounds in *in vitro* noncell-based assays.

The DPPH method has been also employed for quantifying and comparing the antioxidant activity of several compounds, both polyphenols and diterpenes, from *Rosmarinus officinalis* L., used in extensive in traditional medicine for its antioxidant properties. The major phenolic abietane diterpene derived from this medicinal plant, *carnosic acid*, showed significant DPPH scavenging ($IC_{50}$ values of 32 μM), being this activity greater than that for several polyphenolic compounds such as ascorbic acid ($IC_{50}$ values of 47 μM), vanillic acid ($IC_{50}$ values of over 200 μM) and naringin ($IC_{50}$ values of over 200 μM) with manifested potent antioxidant capacity in other reports. Researchers suggest that carnosic acid, together with the polyphenol rosmarinic acid, are the best scavengers in *Rosmarinus officinalis*, contributing greatly to the antioxidant activity attributed to this plant [40]. In another study, the abietane diterpene *inuroyleanol*, isolated from the roots of *Salvia barrelieri* Etlinger, showed to have greater antioxidant efficacy than reference substances such as butylated hydroxytoluene (BHT) and α-tocoferol as well as than other abietane diterpenes found in this plant species

(7α-acetoxyroyleanone-12-methyl ether, royleanone, horminone, 7-acetylhorminone and cryptojaponol). The presence of two hydroxyl groups and one methoxyl on the aromatic ring of inuroyleanol explains the major antioxidant potency of this compound comparing structurally with the other tested abietane diterpenes [41].

**Table 3. Summary table of the antioxidant activities of diterpenes**

| DITERPENES | | | |
|---|---|---|---|
| **Diterpene type** | **Compound** | **Type of assay** | **References** |
| Abietane | 6β-acetoxy-7α-hydroxyroyleanone | *In vitro*; cell-free assays | [38] |
| | 7α-acetoxy-6β,12-dihydroxy-8,12-abietadiene-11,14-dione | *In vitro*; cell-free assays | [43] |
| | carnosic acid | *In vitro*; cell-free assays<br>*In vitro*; culture system | [40,49,50] |
| | carnosol | *In vitro*; culture system | [49-51] |
| | epirosmanol | *In vitro*; culture system | [51] |
| | ferruginol | *In vitro*; cell-free assays | [38] |
| | hinokiol | *In vitro*; cell-free assays | [38] |
| | inuroyleanol | *In vitro*; cell-free assays | [41] |
| | isopimarinol | *In vitro*; cell-free assays | [38] |
| | parvifloron D | *In vitro*; cell-free assays | [43] |
| | parvifloron E | *In vitro*; cell-free assays | [42] |
| | parvifloron F | *In vitro*; cell-free assays | [42] |
| | plectranthol A | *In vitro*; cell-free assays | [42] |
| | plectranthol B | *In vitro*; cell-free assays | [42] |
| | rosmanol | *In vitro*; culture system | [49,51] |
| | secoabietane dialdehyde | *In vitro*; cell-free assays | [38] |
| | serofendic acid | *In vitro*; culture system | [56,57] |
| | sugiol | *In vitro*; cell-free assays<br>*In vitro*; culture system | [38,39] |
| Cassane | benthaminin 2 | *In vitro*; cell-free assays | [45] |
| | deoxycaesaldekarin C | *In vitro*; cell-free assays | [45] |
| Dolastane | amijiol acetate | *In vitro*; cell-free assays | [46] |
| | amijiol-7,10-diacetate | *In vitro*; cell-free assays | [46] |
| Icetexane | 8,11,13-icetexatriene-10-hydroxy, 11,12,16-triacetoxyl | *In vitro*; cell-free assays | [44] |
| | 8,11,13-icetexatriene-7,10,11-dihydroxy-12,13-dihydrofuran | *In vitro*; cell-free assays | [44] |
| Kaurane | cafestol | *In vitro*; culture system<br>*In vivo* | [58,59] |
| | ent-3ß-acetoxykaur-15-en-17-ol | *In vitro*; culture system | [52] |
| | ent-3ß-butanoyloxykaur-15-en-17-ol | *In vitro*; culture system | [52] |
| | ent-kaur-15-en-17-ol | *In vitro*; culture system | [52] |
| | ent-kaur-15-en-3ß,17-diol | *In vitro*; culture system | [52] |
| | ent-kauran-16ß,17-diol | *In vitro*; culture system | [52] |
| | kahweol | *In vitro*; culture system<br>*In vivo* | [58,59] |
| Labdane | 13-epi-3ß,19-dihydroxylabda- | *In vitro*; culture system | [55] |

| DITERPENES | | | |
|---|---|---|---|
| **Diterpene type** | **Compound** | **Type of assay** | **References** |
| | 8(17),11E-dien-16,15-olide | | |
| | 14-deoxy-11,12-didehydroandrographolide | *In vitro*; culture system | [55] |
| | 19-hydroxylabda-8(17),13-dien-16,15-olide | *In vitro*; culture system | [55] |
| | 19-hydroxylabda-8(17)-en-16,15-olide | *In vitro*; culture system | [55] |
| Labdane | 3ß,19-dihydroxylabda-8(17),11E-dien-16,15-olide | *In vitro*; culture system | [55] |
| | andrographiside | *In vivo* | [60] |
| | andrographolide | *In vivo* | [61] |
| | neoandrographolide | *In vivo* | [60] |
| Myrsinane | 14-deoxo-3-O-propionyl-5,15-di-O-acetyl-7-O-benzoylmyrsinol-14ß-nicotinoate | *In vitro*; culture system | [53] |
| | euphorbiaproliferin B | *In vitro*; culture system | [53] |
| | euphorbiaproliferin H | *In vitro*; culture system | [53] |
| | euphorbiaproliferin I | *In vitro*; culture system | [53] |
| | euphorbiaproliferin J | *In vitro*; culture system | [53] |
| | euphorprolitherin C | *In vitro*; culture system | [53] |
| | proliferin A | *In vitro*; culture system | [53] |
| | proliferin B | *In vitro*; culture system | [53] |
| | proliferin C | *In vitro*; culture system | [53] |
| Neo-clerodane | A2 chlorohydrin | *In vitro*; culture system | [54] |
| | ajugaciliatin B | *In vitro*; culture system | [54] |
| | ajugaciliatin G | *In vitro*; culture system | [54] |
| | ajugaciliatin H | *In vitro*; culture system | [54] |
| | ajugamarin A1 chlorohydrin | *In vitro*; culture system | [54] |
| | ajugamarin H1 | *In vitro*; culture system | [54] |
| | ajuganipponin A | *In vitro*; culture system | [54] |
| | ajugatakasin A | *In vitro*; culture system | [54] |
| | ajugatakasin A1 | *In vitro*; culture system | [54] |
| | ajugatakasin B | *In vitro*; culture system | [54] |
| Pimarane | acanthoic acid | *In vivo* | [62] |
| Totarane | 19-hydroxytotarol | *In vitro*; culture system | [47,48] |
| | 4ß-carboxy-19-nortotarol | *In vitro*; culture system | [47,48] |
| | sugiol | *In vitro*; culture system | [47,48] |
| | totaradiol | *In vitro*; culture system | [47,48] |
| | totaral | *In vitro*; culture system | [47,48] |
| | totarol | *In vitro*; culture system | [47,48] |

The genus *Plectranthus* (Labiatae family) comprises over 350 different species containing abietane diterpenes. The diterpenes *plectranthol A*, *plectranthol B*, *parvifloron E* and *parvifloron F*, isolated from the leaves of *Plectranthus nummularius* Briq., have shown significant radical scavenging activity against DPPH. The antioxidant efficacy of all of these abietane diterpenes (plectranthol A - $EC_{50}$ 0.073 mM; plectranthol B - $EC_{50}$ 0.099 mM;

parvifloron E - $EC_{50}$ 0.086 mM and parvifloron F - $EC_{50}$ 0.131 mM) was higher than that of the positive reference compound α-tocoferol ($EC_{50}$ 0.134 mM) [42]. The DPPH scavenging activity has been also demonstrated for other oxidized abietanes from *Plectranthus* spp., particularly for *parvifloron D* and for *7α-acetoxy-6β,12-dihydroxy-8,12-abietadiene-11,14-dione* isolated from *Plectranthus ecklonii* Benth. and *Plectranthus grandidentatus*, respectively. The antioxidant effectiveness of parvifloron D ($IC_{50}$ 0.11 mM) was much higher than that of 7α-acetoxy-6β,12-dihydroxy-8,12-abietadiene-11,14-dione ($IC_{50}$ 1.85 mM), showing the first compound comparable activity to the antioxidant reference substance butylated hydroxytoluene (BHT) ($IC_{50}$ 0.10 mM) [43].

In addition to the abietane diterpenes whose antioxidant activity has been widely reported, other type of diterpenes has also demonstrated remarkably DPPH radical-scavenging activities such as the icetexane diterpenes *8,11,13-icetexatriene-10-hydroxy, 11,12,16-triacetoxyl* ($SC_{50}$ 24.8 μg/mL) and *8,11,13-icetexatriene-7,10,11-dihydroxy-12,13-dihydrofuran* ($SC_{50}$ 7.01 μg/mL) from the roots of *Premna tomentosa* L. [44] and the cassane diterpenes *benthaminin 2* ($IC_{50}$ 42.7 μM) and *deoxycaesaldekarin C* ($IC_{50}$ 97.4 μM) from the root bark of *Caesalpinia benthamiana* [45].

Apart from DPPH method, the antioxidant activity of diterpenes has been also estimated with other *in vitro* cell-free assays, particularly ABTS assay. So, *amijiol acetate and amijiol-7,10-diacetate*, both dolastane type diterpenes from the brown alga *Dictyota dichotoma* var. *implexa* occurring in the Red Sea, inhibited the oxidation of ABTS by 83.35% and 80.24%, respectively. These percentages of inhibition were similar than that of the reference substance L-ascorbic acid (88.61%), suggesting that both tested diterpenes are potent antioxidant compounds [46].

In addition to noncell-based assays, the numbers of cell-based assays and *in vivo* studies on the role of diterpenes as protective agents, based on their antioxidants properties, have grown over the last recent years.

Lipid peroxidation, an early event under oxidative stress situation, causes structural and functional alterations of the membrane and changes in membrane fluidity. Totarol diterpenes from the conifer specie *Podocarpus nagi* (Podocarpaceae family) are effective antioxidant compounds in preventing lipid peroxidation oxidative stress-mediated induced by different agents. Particularly, *totarol, totaradiol, 19-hydroxytotarol, totaral, 4β-carboxy-19-nortotarol* and *sugiol* inhibited $Fe^{3+}$-ADP/NADPH and $Fe^{3+}$-ADP/NADH stimulated lipid peroxidation in microsomes and in mitochondria as well as they inhibited linoleic acid oxidation. Moreover, the compound totarol prevented changes in mitochondrial respiratory enzymes causing by ROS/RNS stimulated NADPH oxidase and protected human blood cells against oxidative hemolysis [47, 48].

For phenolic abietane diterpenes containing in *Rosmarinus officinalis*, several cellular models have confirmed the potential antioxidant activity of a few of these compounds. In an *in vitro* study using human skin fibroblasts NB1RGB, Kunio and Chekio demonstrated that the abietane diterpenes *carnosol, rosmanol* and *carnosic acid* from *Rosmarinus officinalis* L. had both direct and indirect antioxidant effects. On the one hand, these phenolic diterpenes inhibited ROS production hydrogen peroxide-induced by scavenging mechanisms. On the other hand, they induced, via Nrf2, the gene expression levels of the antioxidant enzymes superoxide dismutase (SOD) and catalase (CAT) as well as of the gamma-glutamylcysteine synthetase (GCS), which is the rate limiting enzyme of the synthesis of the endogenous antioxidant glutathione [49]. Based on these antioxidant properties of the abietane diterpenes derived from rosemary, Wijeratne and Cuppett showed that both *carnosic acid* and *carnosol* inhibited lipid peroxidation and avoided DNA damage in Caco-2 cells under oxidative stress situation induced by oleic acid hydroperoxide (OAHPx). The carnosic acid was more potent antioxidant than carnosol [50]. Moreover, in another study, *carnosol*, *epirosmanol* and *rosmanol* strongly inhibited the lipid peroxidation in human LDL and cell membranes through scavenging mechanisms, being higher the activity of these compounds for the higher in OH group number [51].

The neuroprotective activity of a series of kaurane, myrsinol and neo-clerodane type diterpene compounds isolated from the air-dried bulbs of *Fritillaria ebeiensis* G.D. Yu and G.Q.Ji [52], from the roots of *Euphorbia prolifera* Buch-Ham [53] and from the whole plant *Ajuga ciliate* Bunge [54], respectively, has been recently reported using $MPP^+$ as an oxidant inductor and the human dopaminergic neuroblastoma SH-SY5Y cells as cell model. Of the five kaurane-type diterpenes from *Fritillaria ebeiensis*, *ent-kaur-15-en-3β,17-diol* at the concentration of 30 μM was the most active protective compound. The other kaurane diterpenes, *ent-3β-butanoyloxykaur-15-en-17-ol*, *ent-kaur-15-en-17-ol*, *ent-3β-acetoxykaur-15-en-17-ol* and *ent-kauran-16β,17-diol* showed a moderate but significant protection. Concerning the myrsinol-type diterpenes, the total 19 compounds isolated from *Euphorbia prolifera* exerted a protective effect against $MPP^+$, being noteworthy for *euphorbiaproliferin B*, *euphorbiaproliferin H*, *euphorbiaproliferin I*, *euphorbiaproliferin J*, *euphorprolitherin C*, *14-deoxo-3-O-propionyl-5,15-di-O-acetyl-7-O-benzoylmyrsinol-14β-nicotinoate*, *proliferin C*, *proliferin A* and *proliferin B* at a concentration of 30 μM. On the other hand, the *ajugaciliatin B*, *ajugaciliatin G*, *ajugaciliatin H*, *A2 chlorohydrin*, *ajugamarin A1 chlorohydrin*, *ajugatakasin B*, *ajugamarin H1*, *ajugatakasin A*, *ajugatakasin A1* and *ajuganipponin A*, which belong to neo-clerodane diterpene-type showed a moderate protective activity.

In addition to the previously mentioned diterpenes isolated from plants, other compounds of this class from fungus and animals have also shown to exert neuroprotective activity. For instance, *19-hydroxylabda-8(17)-en-16,15-olide*; *3β,19 dihydroxylabda-8(17),11E-dien-16,15-olide*; *13-epi-3β,19-dihydroxylabda-8(17),11E-dien-16,15-olide*; *19-hydroxylabda-8(17),13-dien-16,15-olide* and *14-deoxy-11,12-didehydroandrographolide* isolated from the fruiting body of the parasitic fungus *Antrodia camphorata* Wu, Ryvarden & Chang have demonstrated to protect against Abeta(25-35) apoptosis-induced in primary cultures of cortical neurons from rat in a range of concentrations of 5 to 20 μM [55]. As example of animal diterpene-derived, highlight the atisane-type diterpenoid *serofendic acid* (15-hydroxy-17-methylsulfinylatisan-19-oic acid), isolated from a lipophilic fraction of fetal calf serum, which protect primary neuronal cultures of striatal and cortical rats against ROS generation induced by paraquat, hydrogen peroxide and nitric oxide through antioxidant mechanisms. The serofendic acid antioxidant activity involves ROS scavenging activity, attributed to methylsulfoxide group-containing in its structure, and up-regulation of enzymatic and non-enzymatic endogenous antioxidant [56, 57].

In the liver, different toxic substances such as carbon tetrachloride ($CCl_4$), hexachlorocyclohexane (BHC), butyl hydroperoxide (t-BHP) and acetaminophen (APAP) are metabolized to highly reactive free radicals by the action of the enzyme cytochrome $P_{450}$. The formation of these free radicals may initiate lipid peroxidation leading to liver injury. Several different diterpenes have been reported in the literature as hepatoprotective agents against free radical metabolites, involving antioxidant mechanisms in this protective activity. The intragastrical administration of the diterpenes presented in the coffee, *kahweol* and *cafestol*, protected mice from liver injury $CCl_4$-induced, being this action related, among other mechanisms, to their ability to scavenge free radicals, to inhibit lipid peroxidation and avoid GSH depletion [58]. These two diterpenes have also shown an hepatoprotective role against oxidative stress butyl hydroperoxide (t-BHP)-induced in both *in vitro* and *in vivo* assays carried out on isolated cultured rat hepatocytes and male mice [59].

**Abietane-type diterpenes**

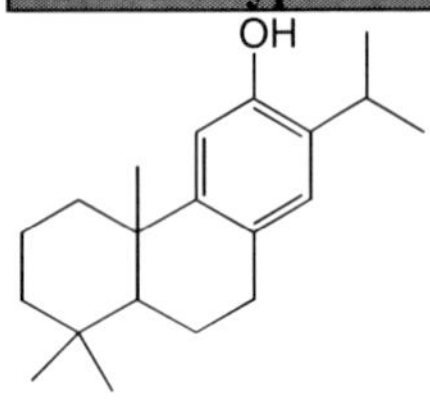

**Ferruginol**

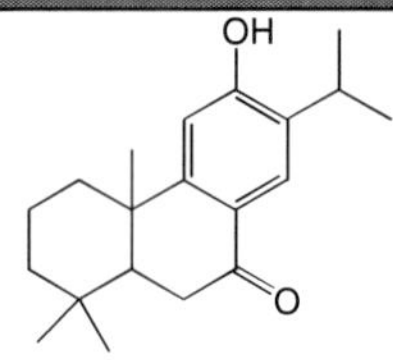

**Sugiol**

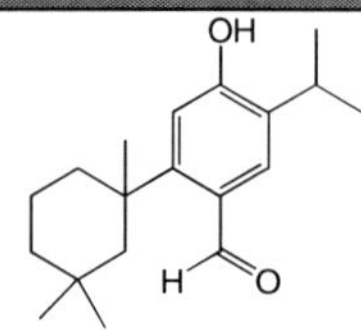

**Secoabietane dialdehyde**

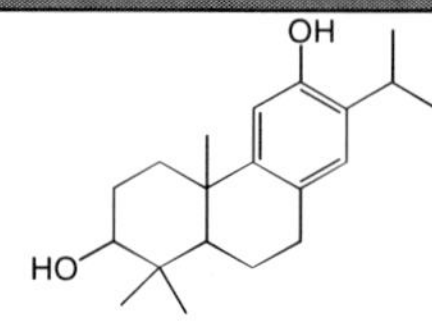

**Hinokiol**

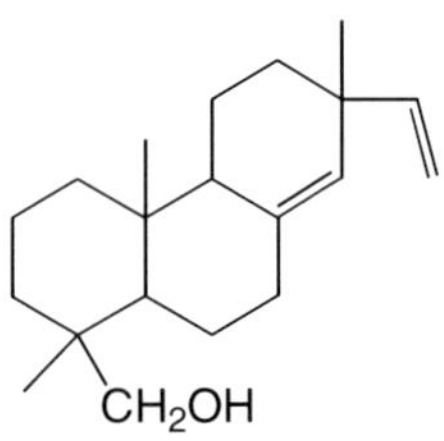

**Isopimarinol**

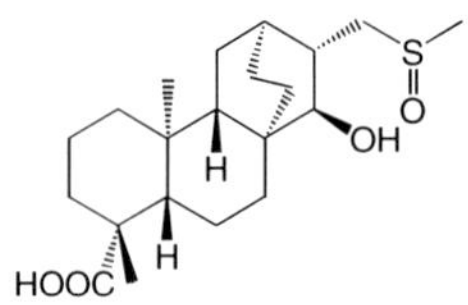

**Serofendic acid**

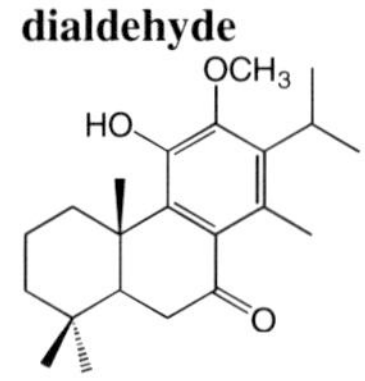

**Inuroyleanol**

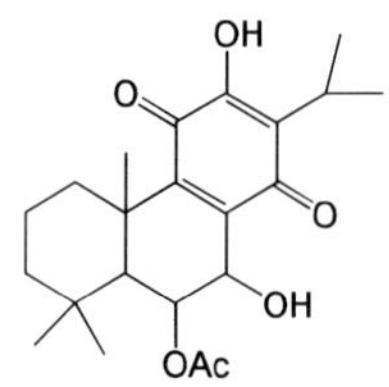

**6ß-acetoxy-7α-hydroxy-royleanone**

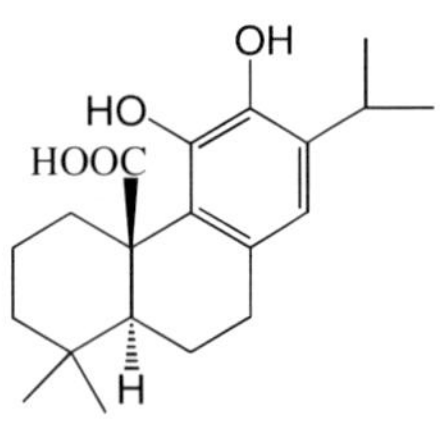

**Carnosic acid**

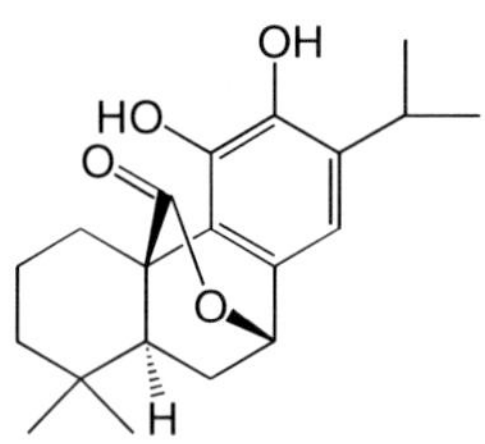

**Carnosol**

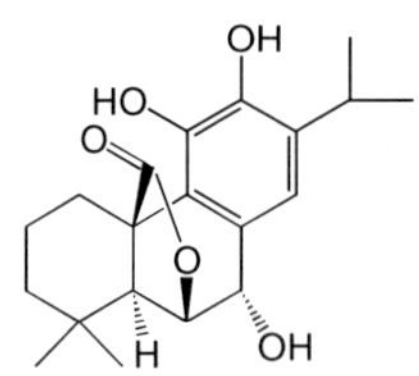

**Rosmanol**

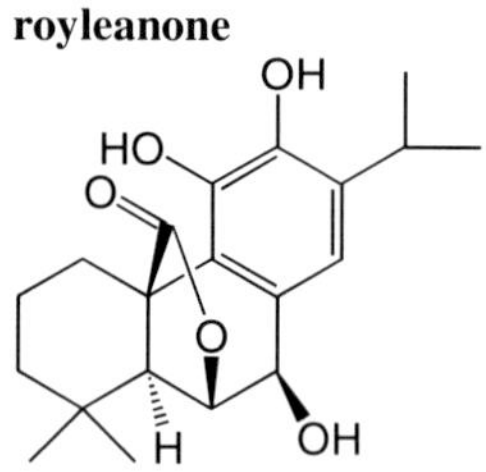

**Epirosmanol**

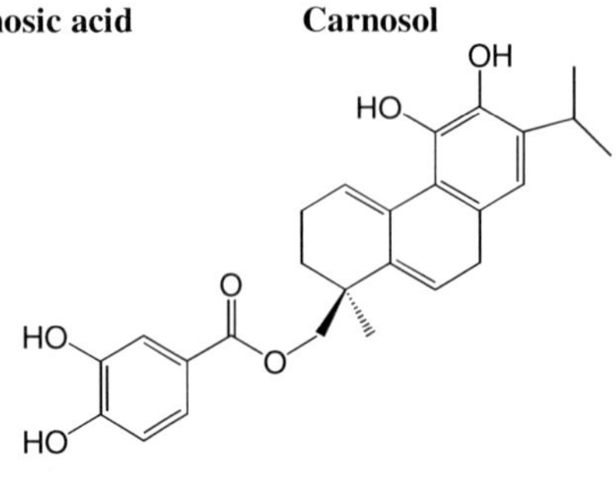

**Plectranthol A**

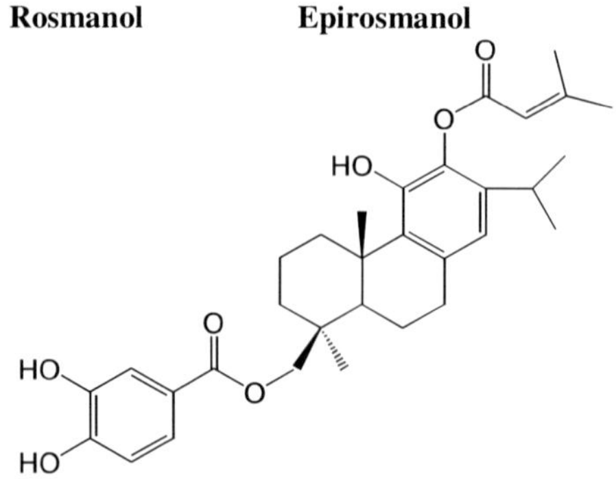

**Plectranthol B**

**Abietane-type diterpenes**

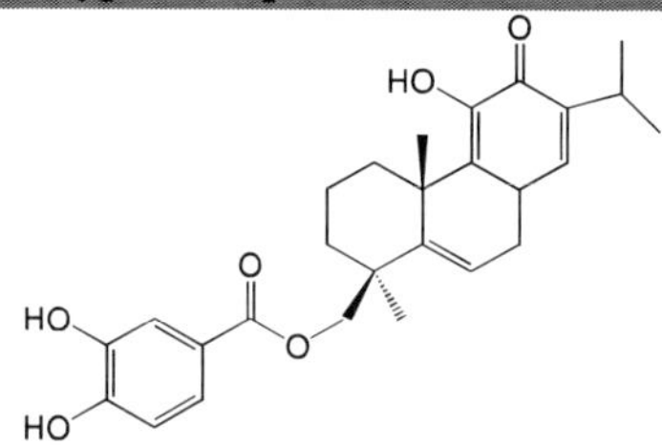

**Parvifloron E**

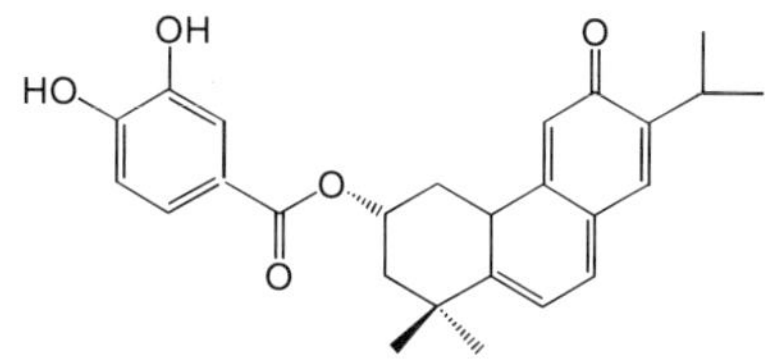

**Parvifloron F**

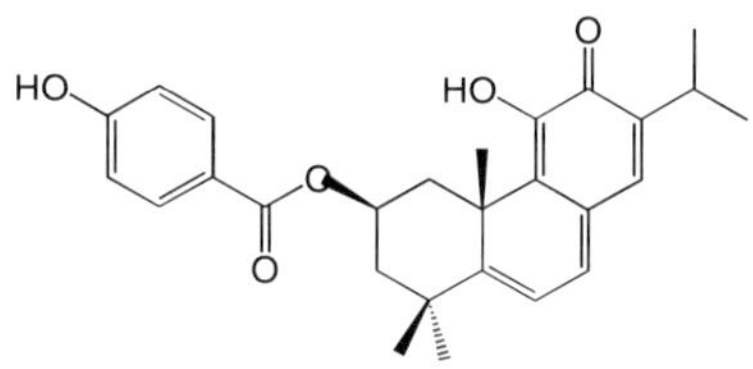

**Parvifloron D**

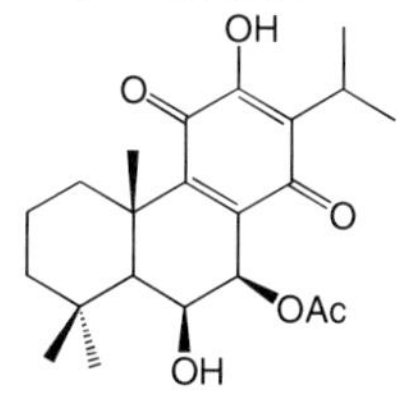

**7α-acetoxy-6β,12-dihydroxy-8,12-abietadiene-11,14-dione**

**Icetexane-type diterpenes**

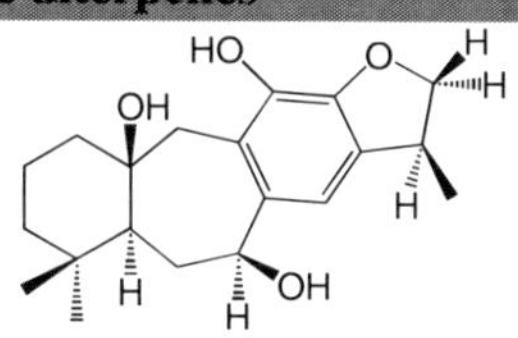

**8, 11,13-icetexatriene-7,10,11-dihydroxy-12,13-dihydrofuran**

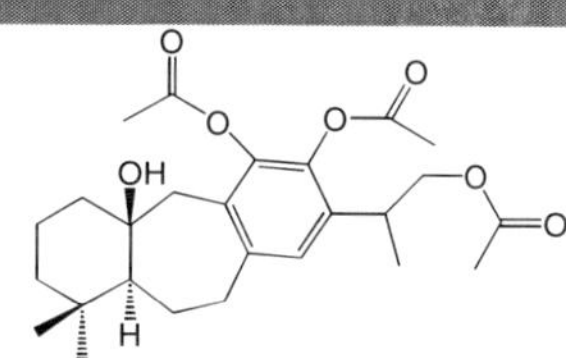

**8,11,13-icetexatriene-10-hydroxy-11,12,16-triacetoxyl**

**Casane-type diterpenes**

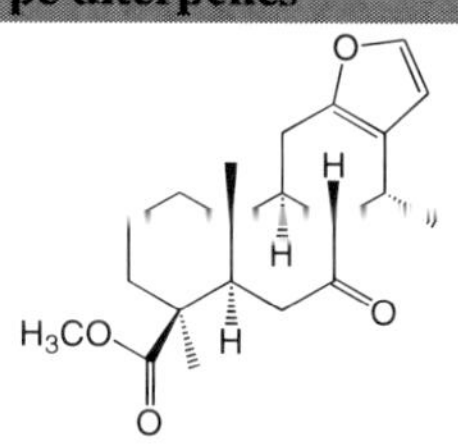

**Deoxycaesaldekarin C**

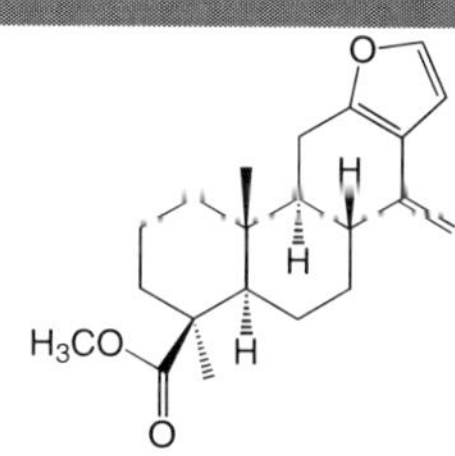

**Benthaminin 2**

**Dolastane-type diterpenes**

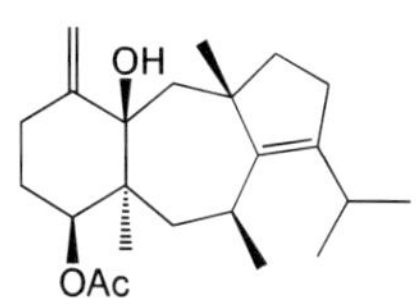

**Amijiol acetate**

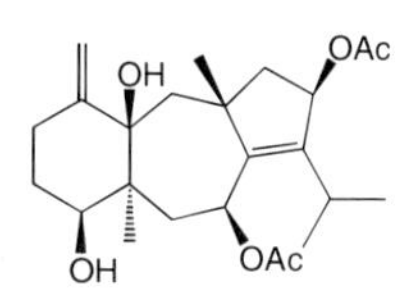

**Amijiol-7,10-diacetate**

Figure 6. (Continued)

## Totarane-type diterpenes

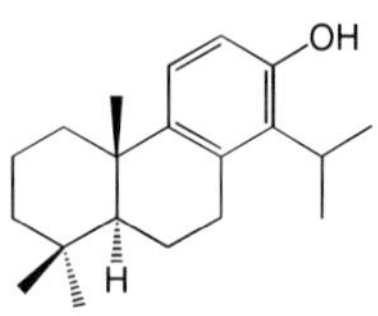

**Totarol**

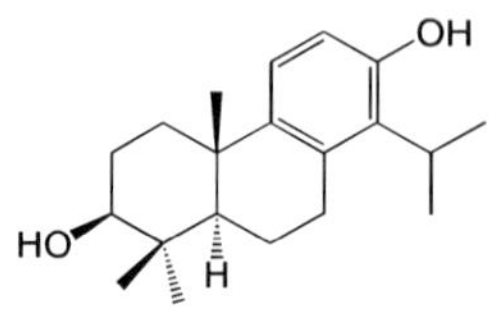

**Totaradiol**

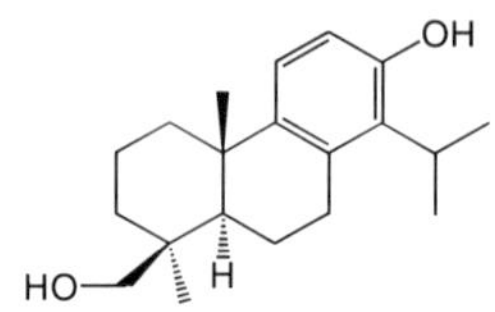

**19-hydroxitotarol**

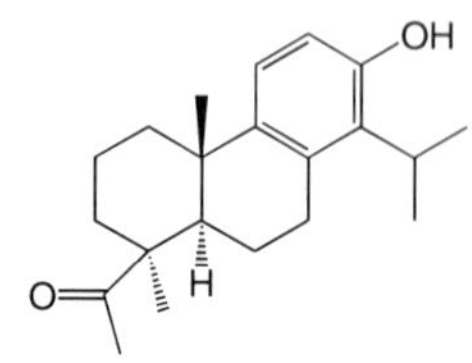

**Totaral**

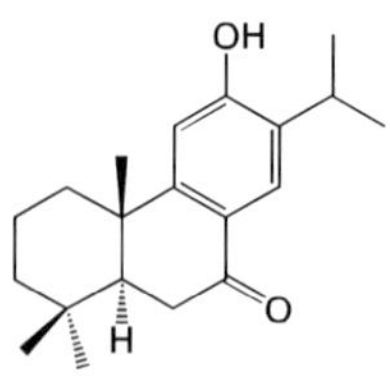

**Sugiol**

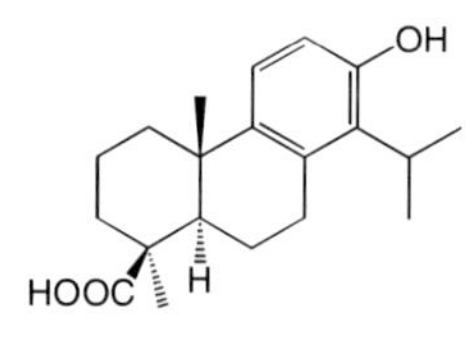

**4ß-carboxy-19-nortotarol**

## Pimarane-type diterpenes

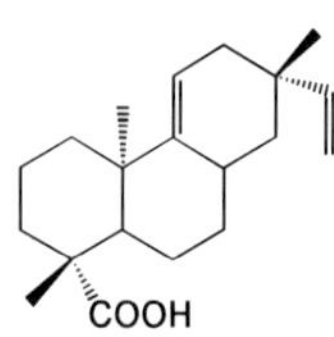

**Acanthoic acid**

## Kaurane-type diterpenes

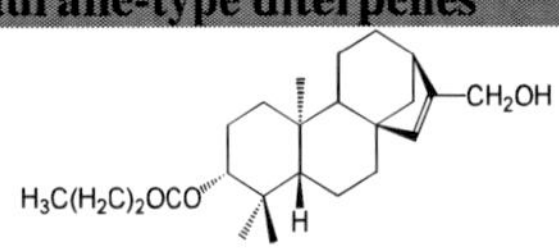

**ent-3ß-butanoyloxykaur-15-en-17-ol**

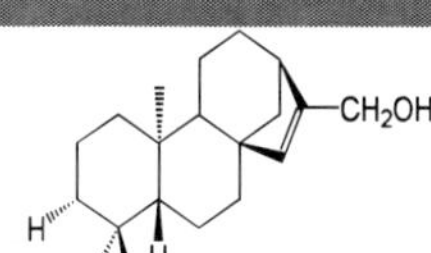

**ent-3ß-butanoyloxykaur-15-en-17-ol**

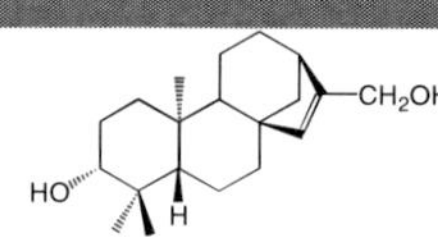

**ent-kaur-15-en-3ß,17-diol**

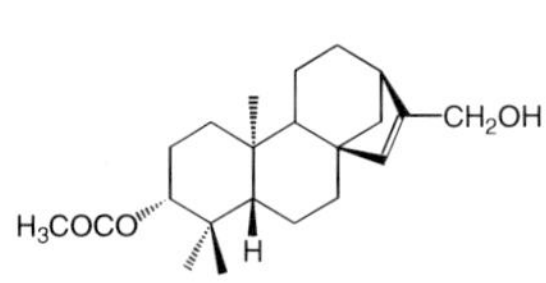

**ent-3ß-butanoyloxykaur-15-en-17-ol**

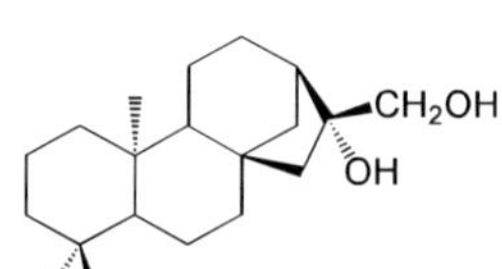

**ent-kauran-16ß,17-diol**

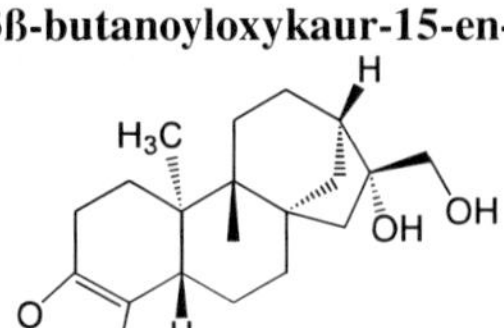

**Cafestol**

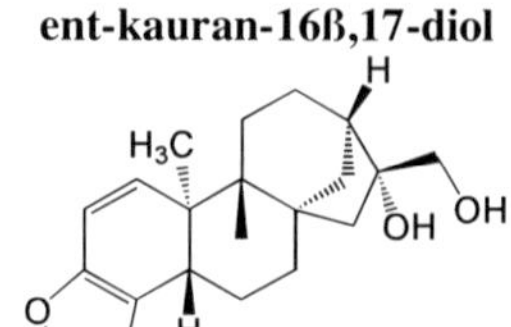

**Kahweol**

**Neo-clerodane -type diterpenes**

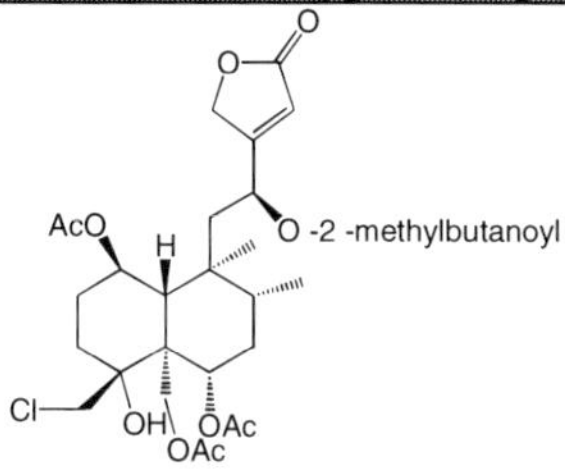

**Ajugaciliatin B**

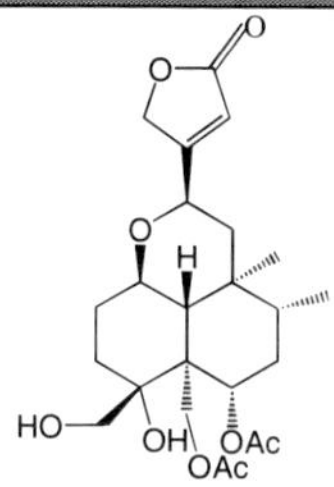

**Ajugaciliatin G**

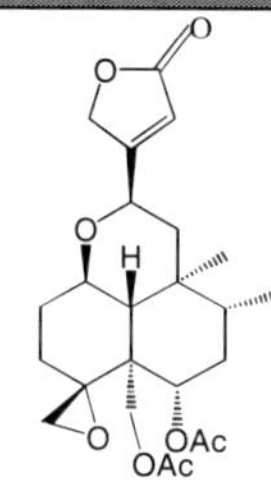

**Ajugaciliatin H**

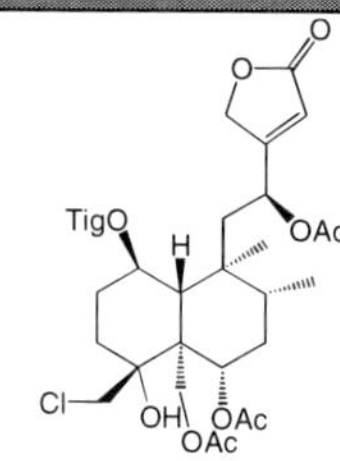

**A2 chlorohydrin**

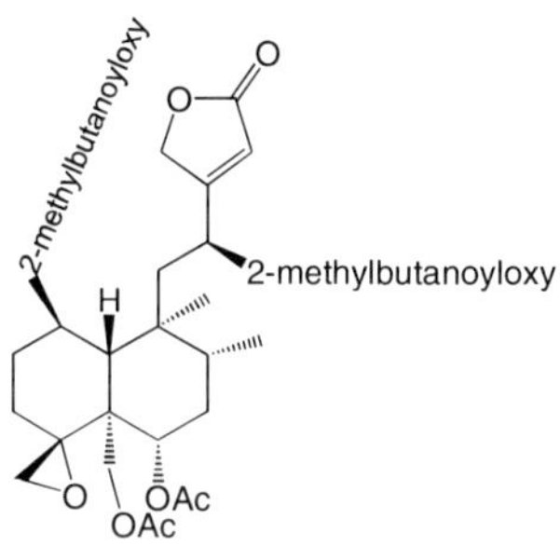

**Ajugatakasin B**

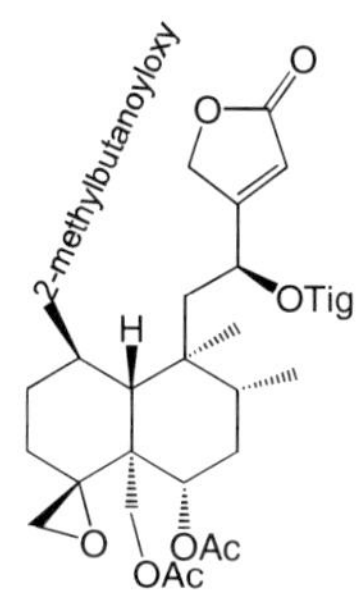

**Ajugamarin H1**

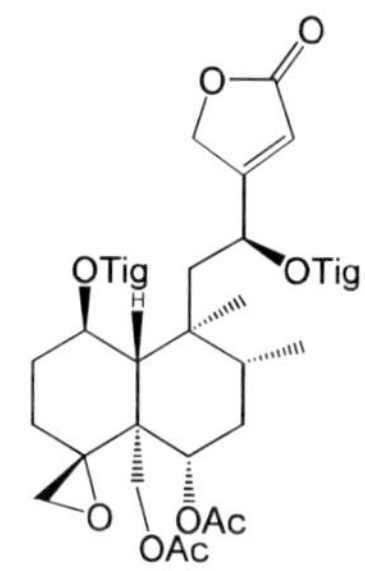

**Ajugatakasin A**

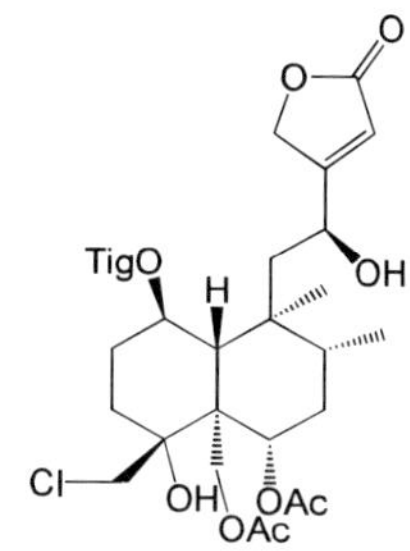

**Ajugamarin A1 chlorohydrin**

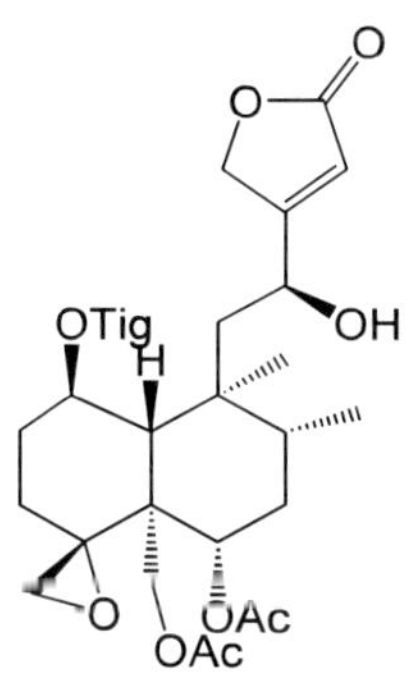

**Ajugatakasin A1**

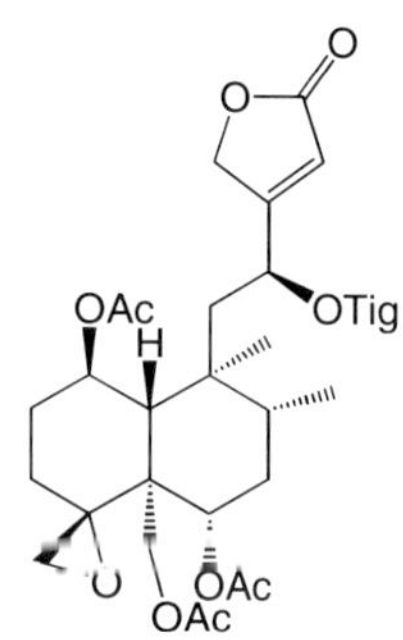

**Ajuganipponin A**

**Labdane-type diterpenes**

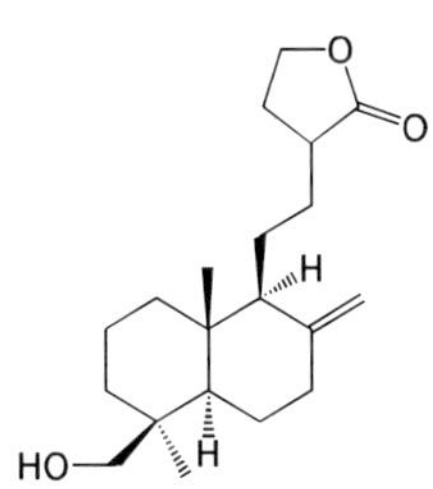

**19-hydroxylabda-8(17)-en-16,15-olide**

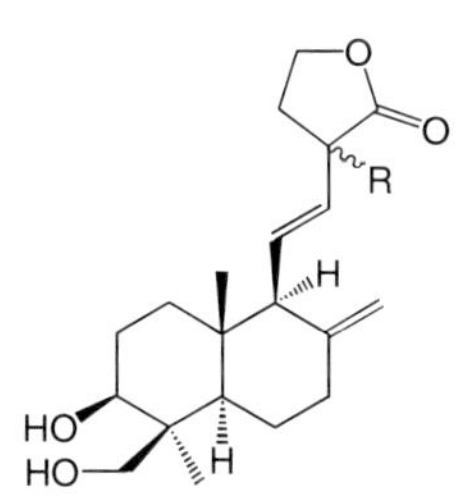

R = H (α or β)

**3ß,19-dihydroxylabda-8(17),11E-dien-16,15-olide**

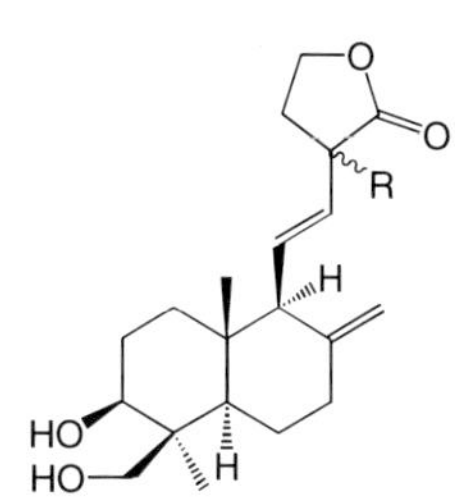

R = H (β or α)

**13-*epi*-3ß,19-dihydroxylabda -8(17),11E-dien-16,15-olide**

Figure 6. (Continued)

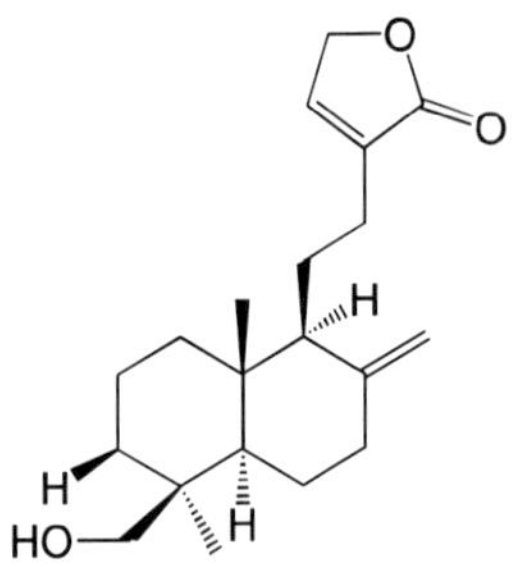

**19-hydroxylabda-8(17),13-dien-16,15-olide**

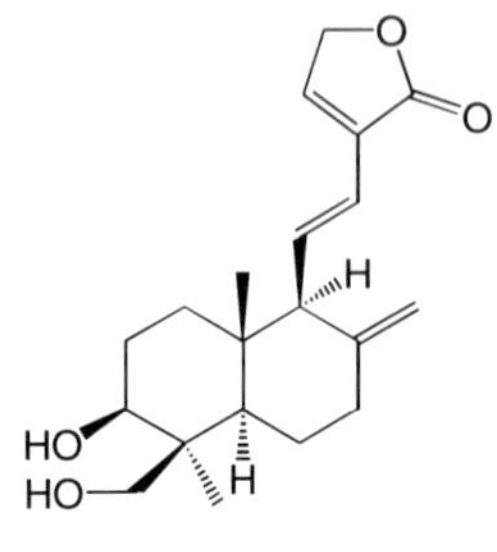

**14-deoxy-11,12-didehydroandrographolide**

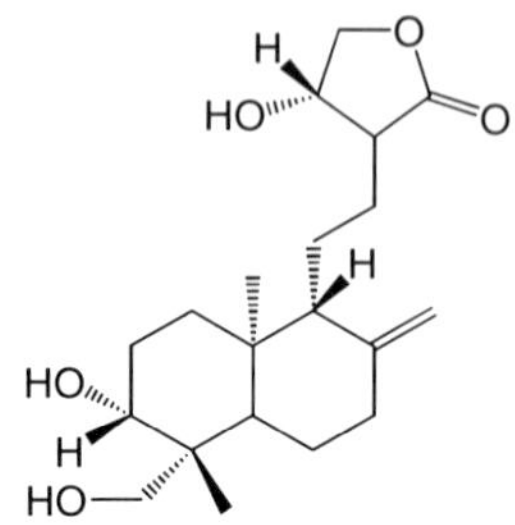

**Andrographolide**

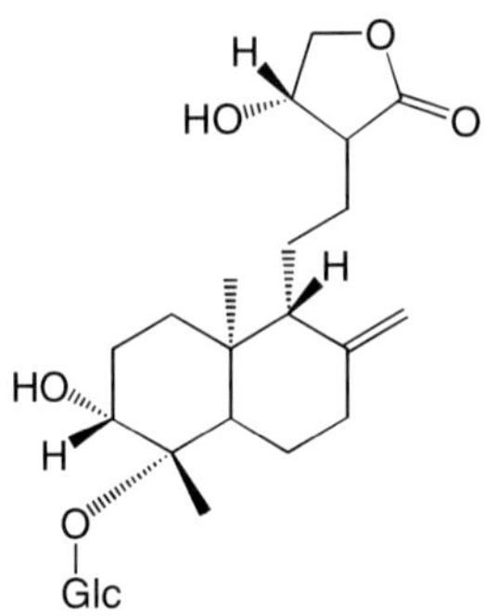

**Andrographiside**

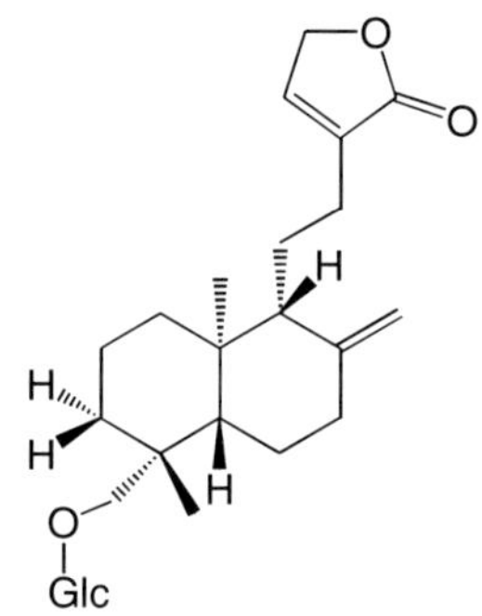

**Neoandrographolide**

## Myrsinane-type diterpenes

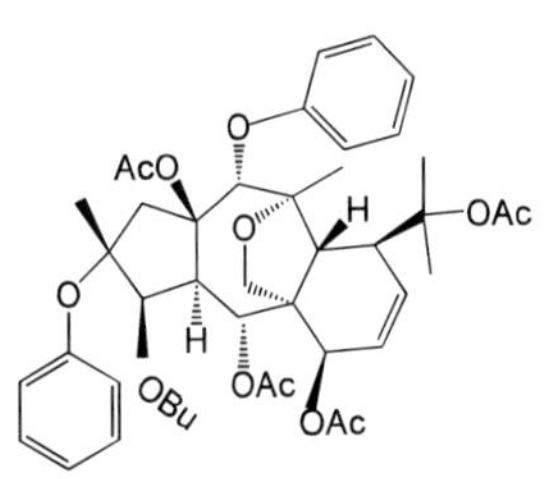

**Euphorbiaproliferin B**

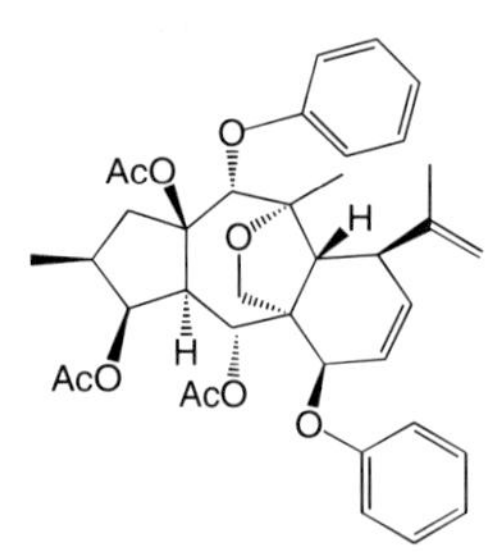

**Euphorbiaproliferin H**

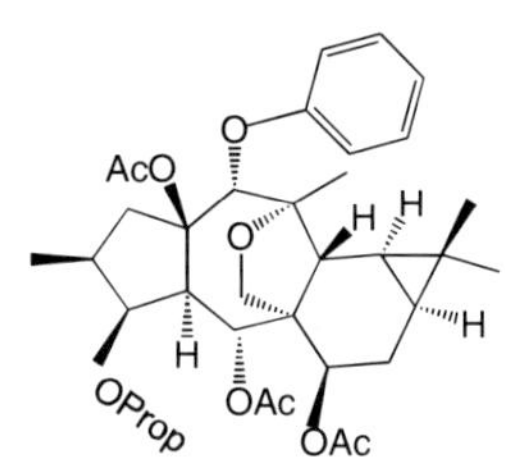

**Euphorbiaproliferin I**

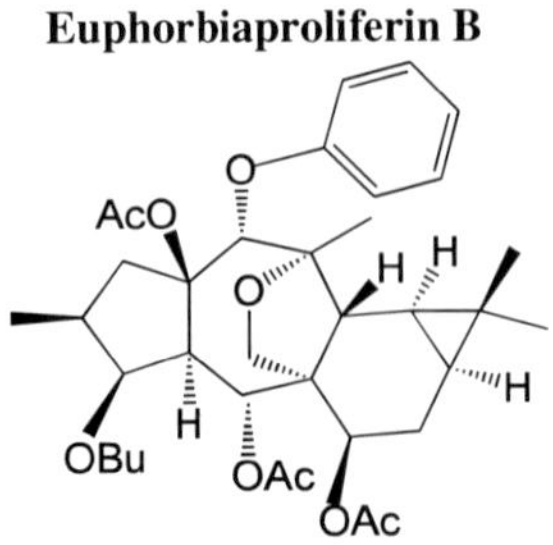

**Euphorbiaproliferin J**

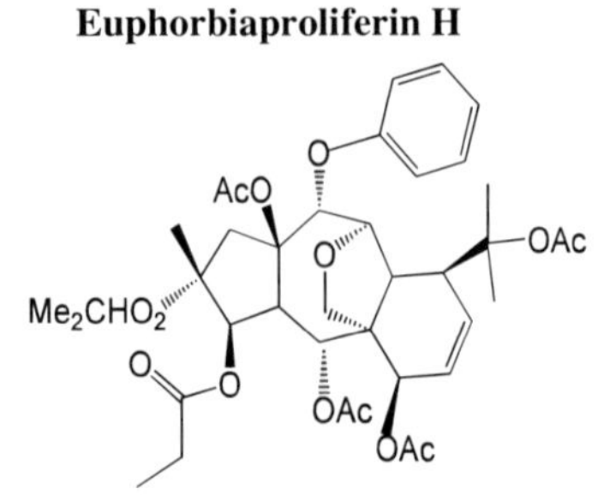

**Proliferin A**

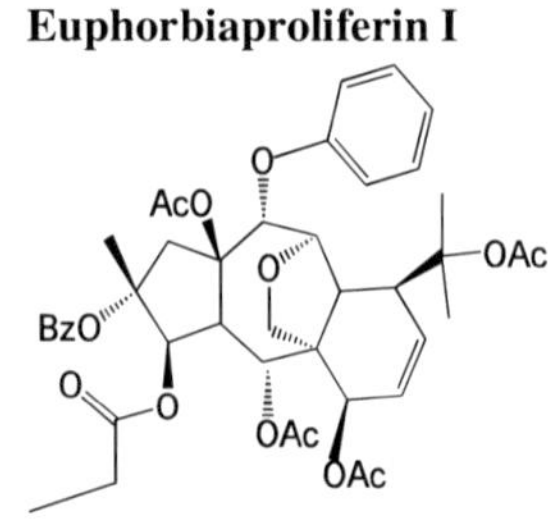

**Proliferin B**

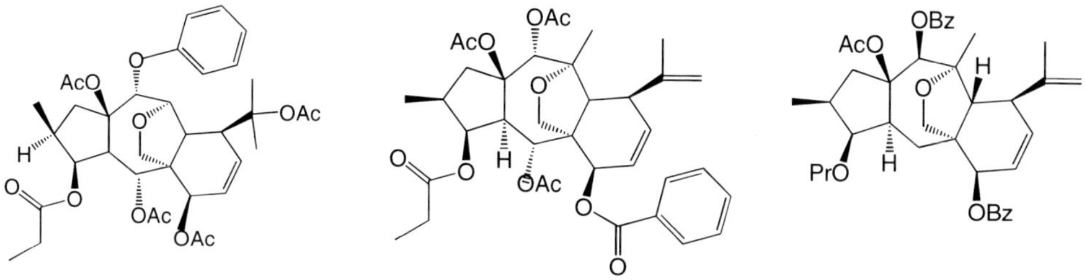

**Proliferin C** | **14-deoxo-3-O-propionyl-5,15-di-O-acetyl-7-O-benzoylmyrsinol-14ß-nicotinoate** | **Euphorprolitherin C**

Figure 6. Chemical structure of diterpenes with antioxidant properties.

As another example of hepatoprotection, the diterpenes *neoandrographolide* and *andrographiside* from the medicinal plant *Andrographis paniculata*, administered intraperitoneally in mice, act as hepatoprotective agents against carbon tetrachloride ($CCl_4$) by increasing cellular antioxidant defense and decreasing lipid peroxidation, respectively [60]. Furthermore, other component from *Andrographis paniculata* named as *andrographolide* attenuated hexachlorocyclohexane (BHC)-induced liver lesions in mice by activation of the antioxidant system [61].

The pimarane diterpene *acanthoic acid* isolated from the roots of *Acanthopanax koreanum* Nakai exerted also a hepatoprotective action against acetaminophen (APAP) in mice. The toxic metabolite N-acetyl-p-benzo-quinoneimine (NAPQI) derived from APAP metabolization caused a depletion of intracellular glutathione (GSH), increased lipid peroxidation and decreased the activity of the antioxidant enzymes catalase (CAT), superoxide dismutase (SOD) and glutathione peroxidase (GPx), being this oxidative stress situation mediated by the transcription factor hypoxia-inducible factor-1 (HIF-1α). The hepatoprotective effect of acanthoic acid in view of this oxidant condition is related to its ability to inhibit the expression of HIF-1α and thus, prevented oxidative stress-related events [62].

Even though the diterpenes are a large group of natural products, vey few studies have evaluated until the moment the antioxidant potential of these compounds. Moreover, any correlation between the structure and the activity has been found in those available research works, except for the diterpenes that contain a phenol group. In view the experimental results of the previous mentioned *in vitro* and *in vivo* studies and in view for future research, diterpenes offer a wide range of possible potential antioxidant compounds.

## IV. Triterpenes

The group of triterpenes (30 carbons atoms; six isoprene units) comprises at least 4,000 different compounds subdivided into 20 classes according to its structure (i.e. lanostane, oleanane, lupane, hopane). The chemical structures of triterpenes with antioxidant activity as well as summary table of their activities are shown in Figure 7 and Table 4.

The *in vitro* cell-free assays antioxidant properties of several of these compounds have been tested in different methods including free radical scavenging activity assays (i.e. 2,2-diphenyl-1-picrylhydrazyl (DPPH), 2,2- azinobis (3-ethyl-benzothiazoline-6-sulfonic acid)

(ABTS), oxygen radical absorption capacity (ORAC) assays) as well as ferric reducing antioxidant power.

"Mushroom of immortality" (*Ganoderma lucidum* (Fr.) P. Karst) as panacea to cure cancer, cardiovascular diseases, chronic inflammation and an endless number of diseases in the Chinese traditional medicine, attributed its properties to antioxidant mechanisms related with triterpenes. The total triterpenes fraction isolated from *G. lucidum* possessed good scavenging properties against the radicals DPPH ($IC_{50}$ 41.45 μg/mL), ABTS ($IC_{50}$ 30 μg/mL), superoxide ($IC_{50}$ 25 μg/mL) and ORAC ($IC_{50}$ 3515 μmoles/Trolox per gram) as well as high reducing ferric ion antioxidant potential (FRAP assay) ($IC_{50}$ 6 μg/mL) [63]. Moreover, authors found that the oral treatment of total triterpenes fraction in mice increased the activity of the antioxidant enzymes CAT, SOD and GPx and the level of the non-enzymatic antioxidant GSH in blood [63]. Among the different triterpenes containing in *G. lucidum*, the most abundant ones and mainly responsible for the antioxidant activity of this fungus are the lanostane triterpenes *ganoderic acids A, B, C* and *D*, *lucidenic acid B* and *ganodermanontriol* [64]. However, other types of lanostane triterpene compounds such as *3-oxo-5α-lanosta-8,24-dien-21-oic acid* and *3β-hydroxylanosta-9,24-dien-24-oic acid*, isolated from the stem bark of *Protorhus longifolia* (Benrh.) Engl., have shown weak DPPH and ABTS scavenging properties and low reducing ferric ion antioxidant potential [65].

The measurement of free radical-scavenging properties of the quinonemethide triterpenes *salacin*, *pristimerin*, *maytenin*, *20α-hydroxymaytenin* and *netzahualcoyene*, isolated from the root barks of *Salacia campestris* have been assessed using DPPH method. The most active triterpene was netzahualcoyene at a concentration of 50 μM (ΔA=55%), followed by maytenin (ΔA=39%), 20α-hydroxymaytenin (ΔA=28%), pristimerin (ΔA=20%) and salacin (ΔA=19%). The dienone-phenol system presented in the structure of all the six tested quinonemethide triterpenes seems to be a necessary structural requirement for the antioxidant properties of these compounds. Moreover, as shown for netzahualcoyene, the higher number of conjugated systems, higher antioxidant potency [66].

The antioxidant properties of a series of ursane triterpenes isolated from the Italian varietal "*Annurca*" apple fruits has also been tested by DPPH assay. Among the different pure triterpenes, the higher antioxidant activity based on scavenging properties corresponded to the compounds *ursolic acid* and *uvaol*, both more potent antioxidants than the reference substance α-tocopherol. Moreover, the capacity of inhibiting the autoxidation of methyl linoleate was also evaluated for determining the antioxidant properties of these ursane triterpenes. In this assay, *posmolic acid* was the most active one among all tested compounds (inhibition of autooxidation by 75%) [67].

Ethanolic extracts obtained from the leaves of *Centella asiatica* (L) Urban possessed as good scavenger DPPH free radicals properties than grape seed extract, which is a well-recognized potent antioxidant rich in proanthocyanidins. *Asiaticoside*, one of the major terpene compounds found in *Centella asiatica*, could account for the higher antioxidant activity of this plant species [68].

The antioxidant activity of the oleanene-type triterpenes *salsolin A* and *salsolin B* and the ursene-type triterpene *2α,3β,23,24 tetrahydroxyurs-12-en-28-oic acid*, isolated from the whole plant *Salsola baryosma* (Schult.), has been estimated using DPPH method. The three polyoxygenated triterpenes showed good free radical scavenger properties, being salsolin B the most active one ($IC_{50}$ 4.5 μM), followed by salsolin A ($IC_{50}$ 20.4 μM) and 2α,3β,23,24 tetrahydroxyurs-12-en-28-oic acid ($IC_{50}$ 60 μM) [69].

The DPPH and the ß-carotene autoxidation assays revealed that the triterpenes *3β-E-caffeoyl-oleanolic acid* and *3β-E-feruloyl oleanolic acid* found in the stem of *Dioclea lasiophylla* Marth ex Benth. possessed good antioxidant properties via hydrogen atom donation. The chemical groups responsible for this activity are the hydroxyl groups bound to the benzene ring. The higher number of these reactive groups, the greater antioxidant activity [70].

The metal chelating activity of the lupane triterpenes *monogynol A, 3β-acetylmonogynol A, 3β-acetyl,22β-hydroxymonogynol A* and *3β-acetyl,21β,22β-dihydroxymonogynol A*, isolated from the whole plant *Salvia macrochlamys* Boiss and Kotschy resulted to be very similar than that of the reference compounds butylated hydroxytoluene (BHT), butylated hydroxyanisole (BHA) and α-tocopherol (TOC) [71].

The antioxidant properties of *maslinic acid*, a triterpene presented in the olive pomace of *Olea europaea* L., have been evaluated *ex vitro*, *in vitro* and *in vivo* using $Fe^{2+}/H_2O_2$, $Fe^{3+}$/ascorbate and $CCl_4$ as free radical generating systems, respectively. This study revealed that maslinic acid acts inhibiting free radical production and consequently, preventing lipid peroxidation [72].

The triterpenoid glycoside niga-ichigoside F1 (*23-hydroxytormentic acid 28-O-glucoside*; NIF1) and its aglycone *23-hydroxytormentic acid* (23-HTA) isolated from the unripe fruits of *Rubus coreanus* Miq. have shown a protective role via modulating intracellular redox status in several experimental models with animals. Particularly, both compounds were able to protect gastric mucosa from the damage of ethanol and sodium salycilate caused by inhibiting lipid peroxidation, hydroxyl radical production and increasing the activities of superoxide dismutase (SOD) and glutathione peroxidase (GPx) [73]. Similar protective mechanisms of action were found for these compounds in a model of carrageenan-induced edema [74]. Moreover, these triterpenes prevented from the nephrotoxicity originated by the chemotherapeutic agent cisplatin in renal epithelial LLC-PK1 cells by increasing GSH levels, decreasing malondialdehyde (MDA) and reactive oxygen species (ROS) levels and modulating antioxidant enzymes via Nrf2 [75].

The pentacyclic triterpene *lupeol*, isolated from the stem bark *Crataeva nurvala* Buch Ham., mediates cardioprotection in oxidative stress-induced experimental rat models. Sudharsan et al. showed that this triterpene play a beneficial role against the cardiotoxicity caused by the anticancer drug cyclophosphamide through free radical scavenging and enzymatic enhancement mechanisms [76]. Moreover, *lupeol* inhibited oxidative-stress-associated changes under hypercholesterolemic conditions by decreasing the lipid peroxidation and increasing the enzymatic and nonenzymatic antioxidants [77]

Accumulating evidences have demonstrated the antioxidant properties of the *oleanolic acid*. *In vitro* studies on primary rat vascular smooth muscle cells (VSMCs) revealed that oleanolic acid confers protection against oxidative stress involved in the development and progression of atherosclerosis by inducing the expression of the enzyme heme oxygenase-1 (HO-1) through PI3K/ Akt pathways and the transcription factor NF-E2-related factor 2 (Nrf2) [78]. This indirect antioxidant mechanism of oleanolic acid has been also involved in the protection against tert-butyl hydroperoxide (tBHP) injury on normal hepatocyte cell line

QZG [79]. Besides, in an *ex vivo* rat heart model, *oleanolic acid* acted as cardioprotective antioxidant through increasing GSH and α-to
coferol levels in mitochondria [80].

**Table 4. Summary table of the antioxidant activities of triterpenes**

| **TRITERPENES** | | | |
|---|---|---|---|
| **Triterpene type** | **Compound** | **Type of assay** | **Reference** |
| Lanostane | ganoderic acids A | *In vitro*; cell-free assays<br>*In vivo* | [63,64] |
| | ganoderic acids B | *In vitro;* cell-free assays<br>*In vivo* | [63,64] |
| | ganoderic acids C | *In vitro*; cell-free assays<br>*In vivo* | [63,64] |
| | ganoderic acids D | *In vitro*; cell-free assays<br>*In vivo* | [63,64] |
| | ganodermanontriol | *In vitro*; cell-free assays<br>*In vivo* | [63,64] |
| | lucidenic acid B | *In vitro*; cell-free assays<br>*In vivo* | [63,64] |
| Lupane | 3β-acetyl,21β,22β-dihydroxymonogynol A | *In vitro*; cell-free assays | [71] |
| | 3β-acetyl,22β-hydroxymonogynol A | *In vitro*; cell-free assays | [71] |
| | 3β-acetylmonogynol A | *In vitro*; cell-free assays | [71] |
| | bacosine | *In vivo* | [81] |
| | lupeol | *In vivo* | [76,77] |
| | monogynol A | *In vitro*; cell-free assays | [71] |
| Oleanane | 3ß-E-caffeoyl-oleanolic acid | *In vitro*; cell-free assays | [70] |
| | 3ß-E-feruloyl oleanolic acid | *In vitro*; cell-free assays | [70] |
| | maslinic acid | *In vivo*<br>*In vitro*<br>*Ex vivo* | [72] |
| | oleanolic acid | *In vitro*; culture system<br>*In vivo*<br>*Ex vivo* | [78-80, 82] |
| | platycodin A | *In vitro*; culture system | [83] |
| | salsolin A | *In vitro*; cell-free assays | [69] |
| | salsolin B | *In vitro*; cell-free assays | [69] |
| Quinonemethide | 20α-hydroxymaytenin<br>maytenin<br>netzahualcoyene<br>pristimerin<br>salacin | *In vitro*; cell-free assays<br>*In vitro*; cell-free assays<br>*In vitro*; cell-free assays<br>*In vitro*; cell-free assays<br>In vitro; cell-free assays | [66]<br>[66]<br>[66]<br>[66]<br>[66] |

| TRITERPENES | | | |
|---|---|---|---|
| **Triterpene type** | **Compound** | **Type of assay** | **Reference** |
| Ursane | 23-hydroxytormentic acid | *In vivo* | [73-75] |
| | 23-hydroxytormentic acid 28-O-glucoside | *In vivo* | [73-75] |
| | 2α,3β,23,24 tetrahydroxyurs-12-en-28-oic acid | *In vitro*; cell-free assays | [69] |
| | asiaticoside | *In vitro*; cell-free assays | [68] |
| | posmolic acid | *In vitro*; cell-free assays | [67] |
| | ursolic acid | *In vitro*; cell-free assays | [67] |
| | uvaol | *In vitro*; cell-free assays | [67] |
| | α-amyrin | *In vivo* | [84] |

Oxidative stress is also involved in the pathogenesis of diabetes. The triterpenes *bacosine*, presented in the aerial parts of *Bacopa monnieri* Linn. and *oleanolic acid* from the fruits of *Ligustrum lucidum* Ait exerted antihyperglycaemic effects in rats with alloxan-induced diabetes, pointing out its antioxidant properties, particularly modulation of cellular antioxidant defense system and prevention of ongoing lipid peroxidation, as one of the major mechanisms of action involving in this anti-diabetic effect [81, 82].

The triterpene saponin *platycodin A*, isolated from the methanol extract of the root of *Platycodon grandiflorum* (Jacq.) A. DC., "*Platycodi radix*", provided concentration-dependent protection against glutamate-mediated excitotoxicity in primary cultured rat cortical cells. Comparing the structure of platycodin A to other triterpene saponins (platycodin C, platycodin D and deapioplatycodin D) that did not display activity under identical experimental conditions, it is believed that the the acetoxyl group located on the C-2 of the rhamnosyl moiety of glycoside contributes to the neuroprotective activity of platycodin A against glutamate toxicity [83].

The triterpenes *α-amyrin* and *β-amyrin* found in the trunk wood resin of *Protium heptaphyllum* (Aubl.) March. reduced acetaminophen-induced hepatotoxicity in mice. These compounds reduced the formation of toxic byproduct N-acetyl-p-benzoquinone imine (NAPQI) during acetaminophen metabolism as well as they prevented the hepatocellular oxidative stress through antioxidant mechanisms [84].

## Oleanane-type triterpenes

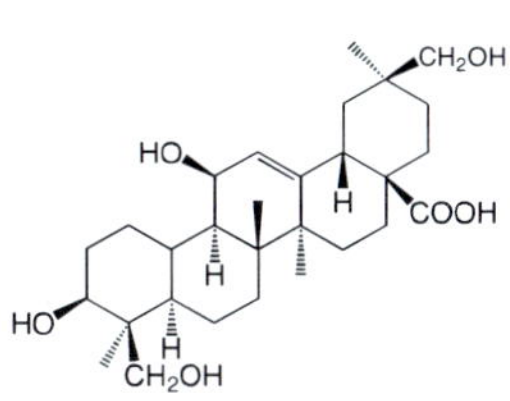


**Salsolin A**

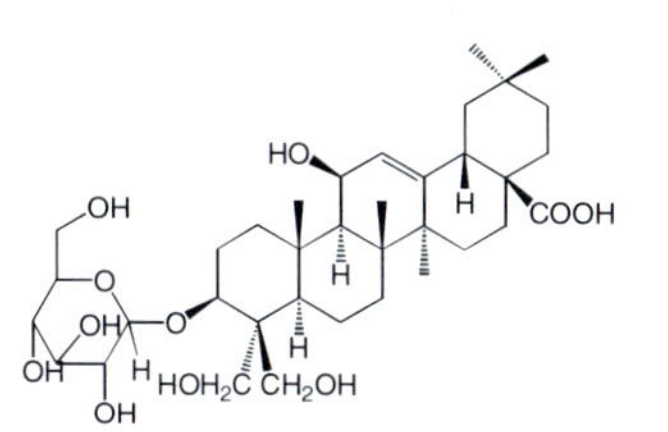


**Salsolin B**

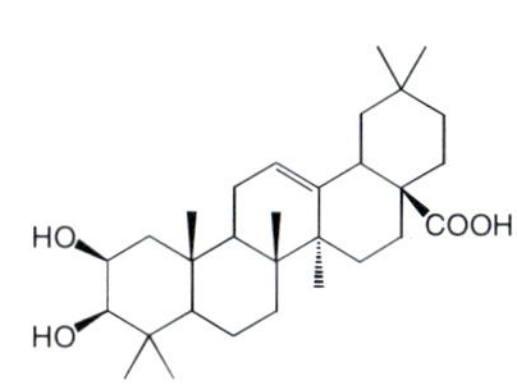


**Maslinic acid**

Figure 7. (Continued)

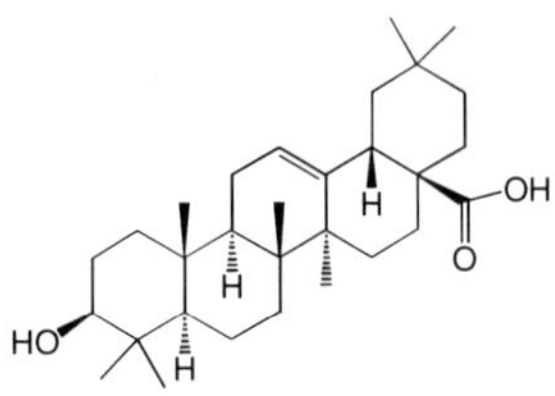

**Oleanolic acid**

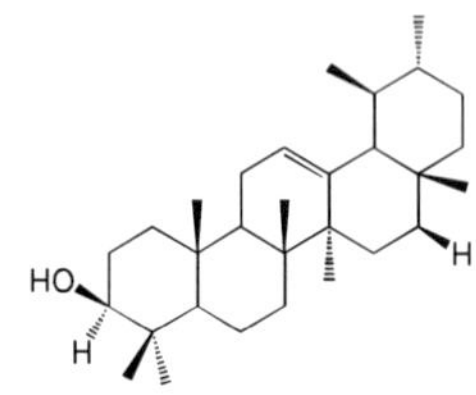

**ß-amyrin**

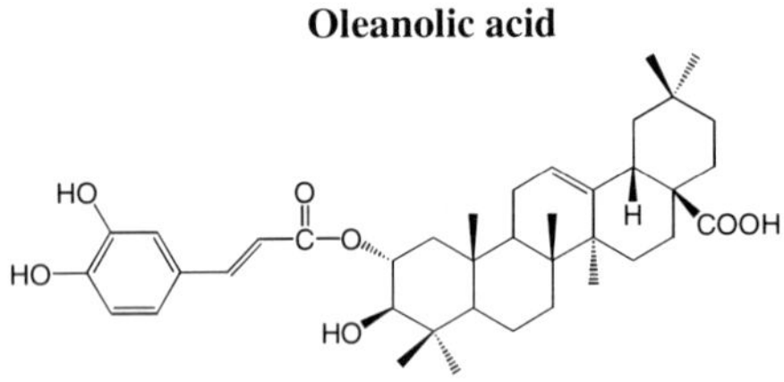

**3ß-E-caffeoyl-oleanolic acid**

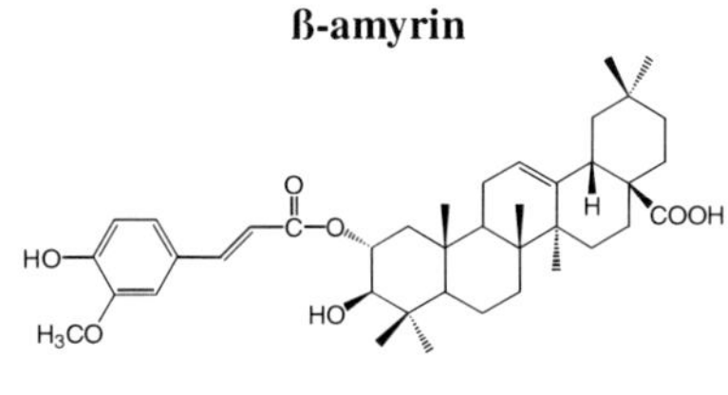

**3ß-E-feruloyl oleanolic acid**

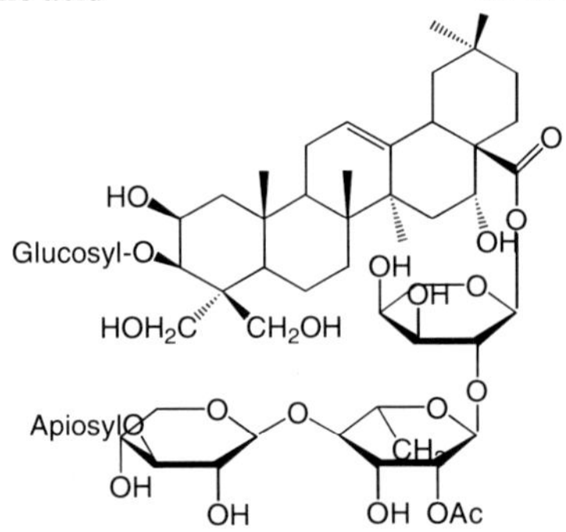

**platycodin A**

## Ursane-type triterpenes

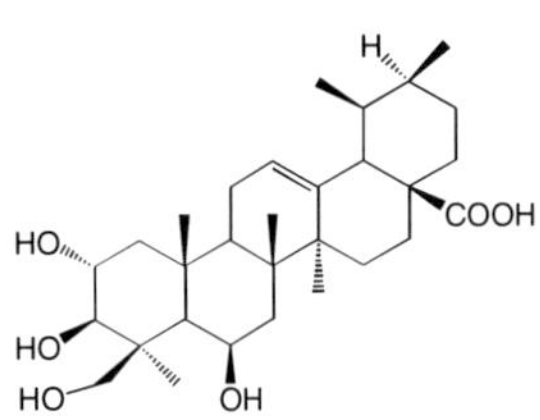

**2α,3β,23,24 tetrahydroxyurs -12-en-28-oic acid**

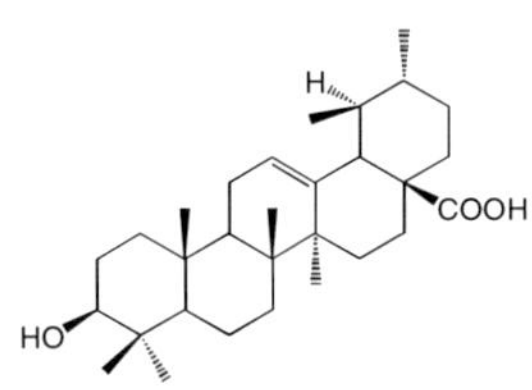

**Ursolic acid**

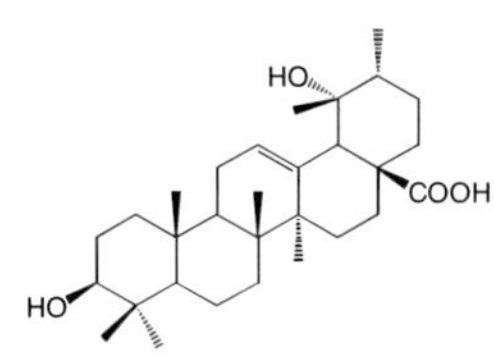

**Posmolic acid**

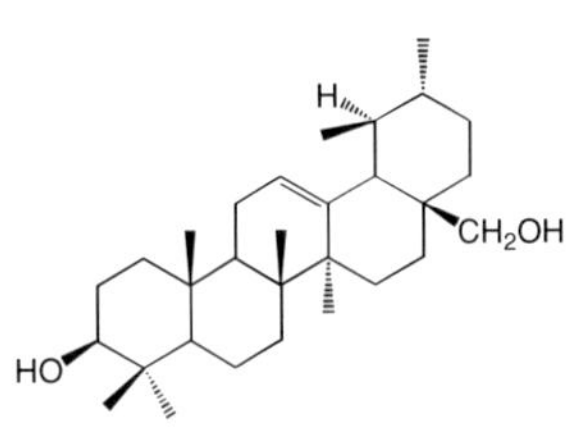

**Uvaol**

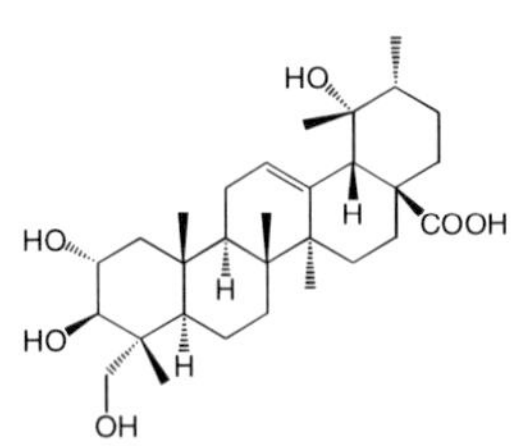

**23-hydroxytormentic acid**

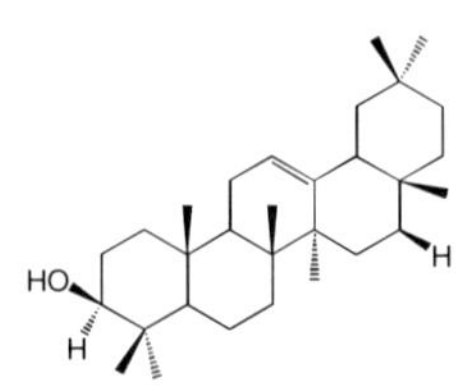

**α-amyrin**

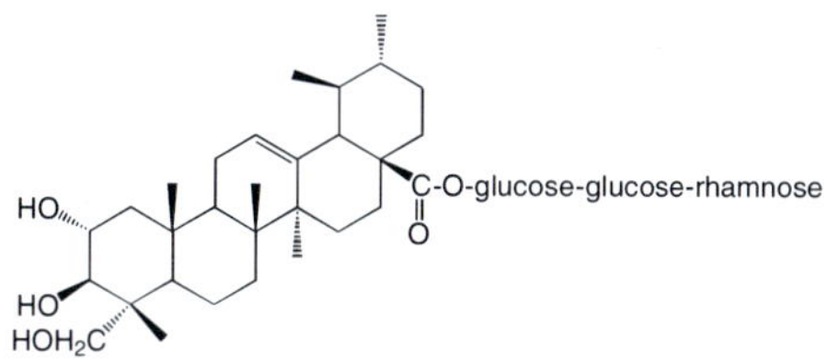

Asiaticoside

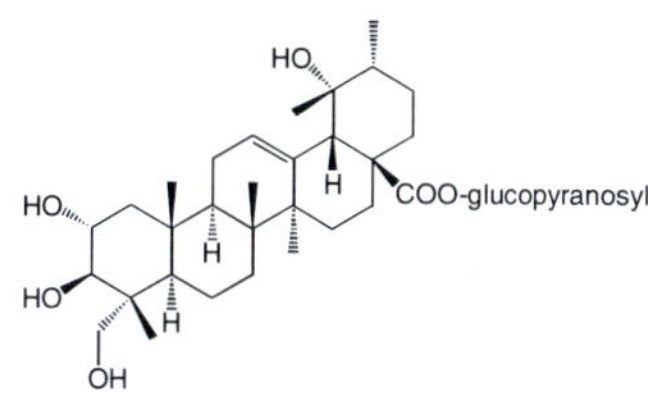

23-hydroxytormentic acid 28-O-glucoside

**Quinonemethide-type triterpenes**

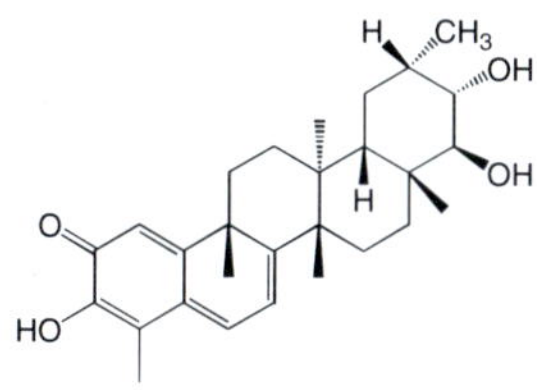

Salacin

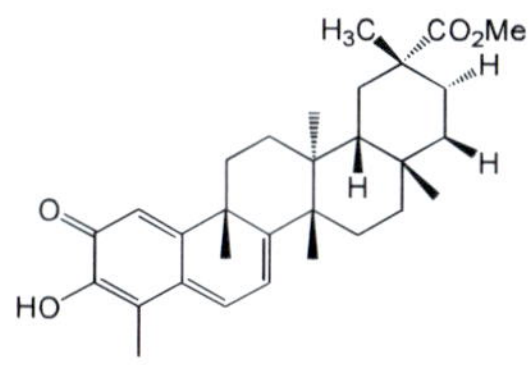

Pristimerin

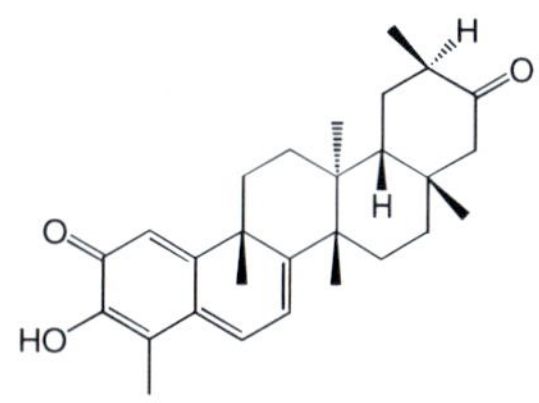

Maytenin

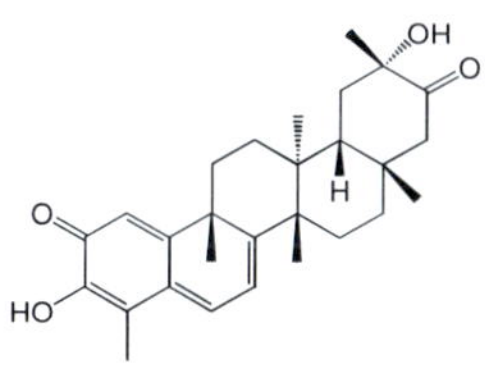

20α-hydroxymaytenin

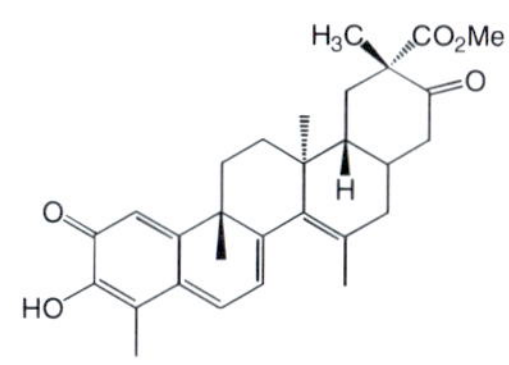

Netzahualcoyene

**Lupane-type triterpenes**

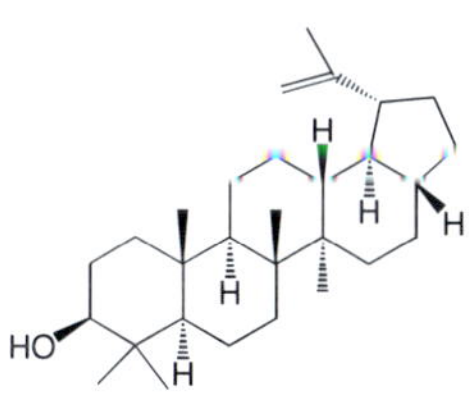

Lupeol

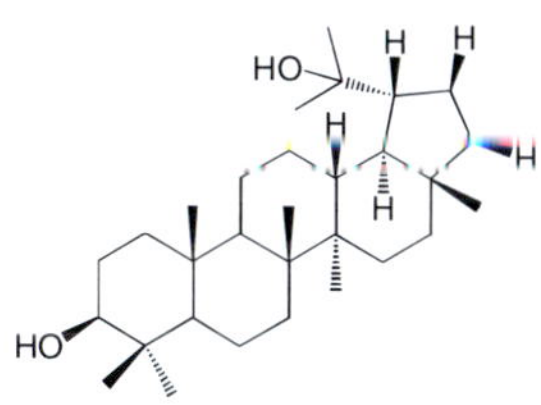

Monogynol A

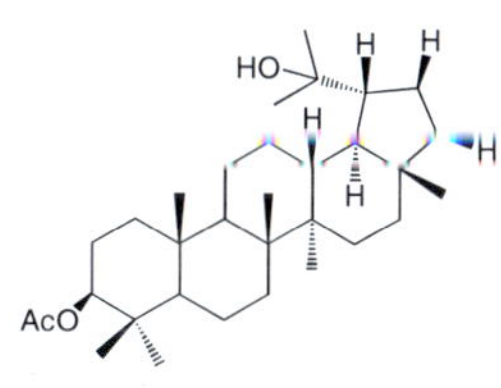

3ß-acetylmonogynol A

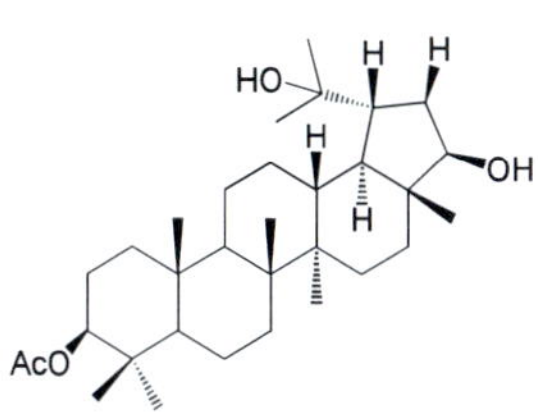

3ß-acetyl-22ß-hydroxymonogynol A

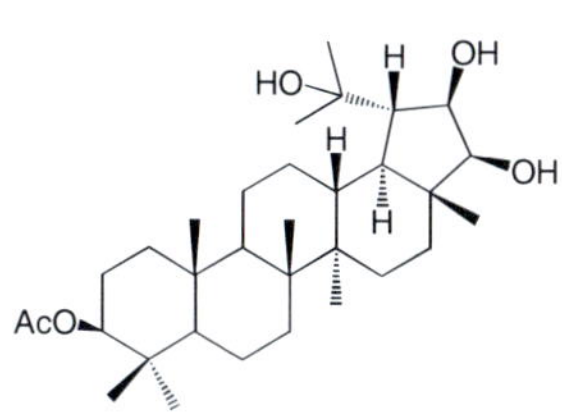

3ß-acetyl-21ß, 22ß-hydroxymonogynol A

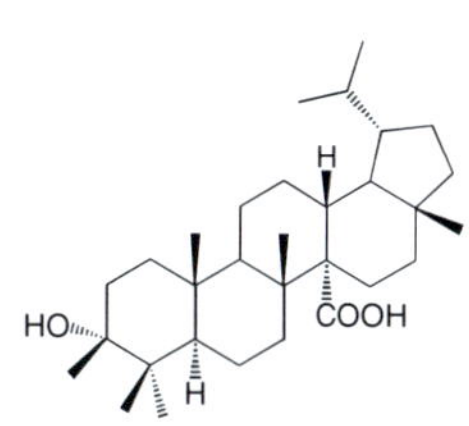

Bacosine

Figure 7. (Continued)

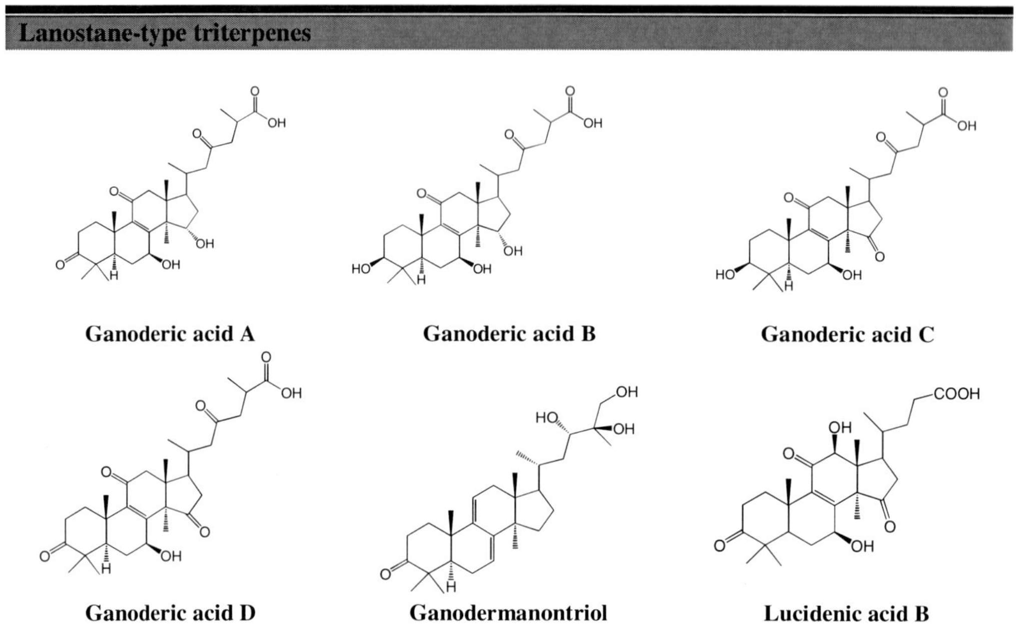


Figure 7. Chemical structure of triterpenes with antioxidant properties.

**Table 5. Summary table of the antioxidant activities of tetraterpenes**

| TETRATERPENES | | | |
|---|---|---|---|
| **Tetraterpene type** | **Compound** | **Type of assay** | **Reference** |
| Carotenoids | α-carotene | *In vitro*; cell-free assays<br>*In vivo* | [63,64] |
| | β-carotene | *In vitro;* cell-free assays<br>*In vitro*; culture system<br>*In vivo* | [87-92] |
| | lycopene | *In vitro*; cell-free assays<br>*In vivo* | [63,64] |
| | lutein | *In vitro*; cell-free assays<br>*In vivo* | [63,64] |
| | zeaxanthin | *In vitro*; cell-free assays<br>*In vivo* | [63,64] |

## V. Tetraterpenes

Carotenoids are the most abundant tetraterpenes ($C_{40}$, 8 isoprene units) found in nature, occurring at least 700 different types. These compounds have an important and significant role as antioxidant attributed to the conjugated double bonds. The higher length of the conjugated systems and the presence of hydroxyl groups on the end rings as in xanthophylls lead to a greater antioxidant potency [85].

Among all these carotenoids, we will focus on β-carotene, α-carotene, lycopene, lutein and zeaxanthin [86]. The chemical structures of carotenoids with antioxidant activity as well as summary table of their activities are shown in Figure 8 and Table 5.

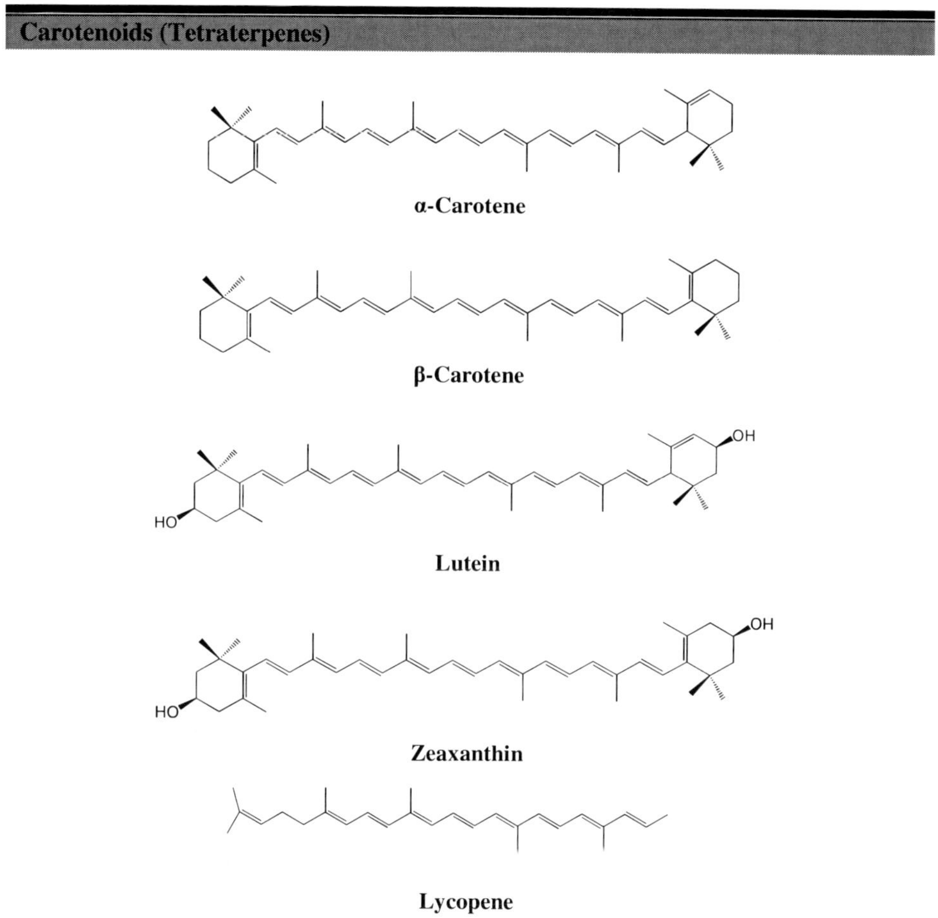


Figure 8. Chemical structure of carotenoids with antioxidant properties.

The behavior of β-carotene in terms of antioxidant activity depends on concentration and oxygen tension ($pO_2$). Oxygen tension above baseline levels of tissue and high concentrations of this natural compound leads to a prooxidant effect [87, 88]. As antioxidant, this carotenoid acts as single oxygen and free radical scavenger. This mechanism of action is related to the formation of β-carotene-radical adducts epoxide type [89, 90]. Based on these scavenger properties, several *in vitro* and *in vivo* studies have demonstrated the protective role of this compound. For instance, in an *in vitro* study, β-carotene protected Chinese hamster ovary cells from oxygen species $H_2O_2$ – induced [91]. Besides, this compound protected non-tumor bone marrow cells from genetic events oxidative stress-related due to the exposition of the anticancer drug doxorubicin in rats [92].

The carotenoid *α-carotene*, has shown to be a better antioxidant than its isomer β-carotene, in multilamellar liposomes exposed to a lipid-soluble radical generator 2,4-dimethyl valeronitrile (AMVN) as deduced from different scavenging assays [93].

*Lycopene*, carotenoid responsible for the red color in tomatoes, watermelons and papayas, is one of the most potent known antioxidants. On one hand, lycopene acts as quencher of singlet oxygen (100 times more efficient than the reference substance α-tocoferol) [94] and other free radicals including nitrogen dioxide ($NO_2^{\cdot}$), thiol ($RS^{\cdot}$) and sulphonyl ($RSO_2^{\cdot}$) [95]. On the other hand, lycopene enhances the activity of the antioxidant enzymes (CAT, SOD, GPx) [96]. Several *in vitro*, *in vivo* and epidemiological studies on the antioxidant properties of lycopene have been reported. As example, lycopene improved those parameters of oxidative stress that are modified by cypermethrin (CYP) in carp (*Cyprinus carpio*) [97], by selenite in cataracts in young rats [98] and by glucose oxidase in pancreatic acinar AR42J cells [99].

Most of the studies focused on the antioxidant *lutein* and *zeaxanthin* are subjected on its protective role in ocular diseases. Both carotenoids are main constituents of the macular region and they are obtained from the diet (fruits and vegetables). *In vitro* studies on rat retinal neurons in culture revealed that the pretreatment with these two xanthophylls protected from the oxidative damage induced by hydrogen peroxide and paraquat [100]. Moreover, lutein prevented from the oxidative stress-related modifications that occurs during acute ischemic in the inner retina and in the retinal ganglion cells [101,102] as well as it prevented the development of cataracts that happens as a complication of the diabetes [103].

## References

[1] J. Buckingham Dictionary of natural products on CD-ROM, version 6.1. London: Chapman & Hall (1998).

[2] R. Paduch, M. Kandefer-Szerszen, M. Trytek, J. Fiedurek. Arch. *Immnol. Ther. Exp. (Warsz).* 55, 315 (2007).

[3] R.J. Fischbach, M. Staudt., I. Zimmer., S. Rambal, J.P. Schnitzler. *Physiol. Plantarum.* 114, 354 (2002)

[4] F. Rapparini, R. Baraldi, O. Facini. *Phytochem.* 57, 681 (2001).

[5] J. Gershenzon, N. Dudareva. *Nat. Chem. Biol.* 3, 408 (2007).

[6] J.C. Sacchettini, C.D. Poulter. *Science.* 277, 1788 (1997).

[7] G. Wang, W. Tang, R.R. Bidigare. *Natural products: Drug discovery and therapeutic medicine.* L. Zhang, A.L. Demain (Eds). pp: 197-277. (2007).

[8] D.J. McGarvey, R. Croteau. *Plant. Cell.* 7, 1015. (1995).

[9] L. Ruzlcka., *Experientia.* 9, 357. (1953).

[10] G. Flesh, M. Rohmer. Eur. J. Biochem. 175, 405 (1988).

[11] M. Rohmer, M. Knani, P. Simonin, B. Sutter, H. Sahm. *Biochem. J.* 295, 517 (1993).

[12] M. Rohmer. *Nat. Prod. Rep.* 16, 565 (1999).

[13] J. Gershenzon, R Croteau, *Lipid Metabolism in Plants.* T.S Moore (Ed.) pp. 339–388 (1993).

[14] J.C. Gray. Advances in Botanical Research. J.A. Callow (Ed.) pp. 26-85 (1987).

[15] W. Eisenreich, A. Bacher, D. Arigoni, F. Rohdich. *Cell. Mol. Life Sci.* 61, 1401 (2004).

[16] P.-H. Liang, T.-P. Ko, A. H.-J. Wang. *Eur. J. Biochem.* 269, 3339 (2002).
[17] M. Valko, C.J. Rhodes, J. Moncol, M. Izakovic, M. Mazur. *Chem. Biol. Interact.* 160, 1 (2006).
[18] S.V. Avery. *Biochem. J.* 434, 201 (2011).
[19] M. Viuda-Martos, Y.R. Navajas, E.S. Zapata, J. Fernandez-Lopez, J.A. Perez-Alvarez. *Flavour Fragance J.* 25, 13 (2010).
[20] G. Ruberto, M.T. Baratta. *Food Chem.* 69, 167 (2000).
[21] J. Grassmann, S. Hippeli, R. Spitzenberger, E.F. Elstner. *Phytomed.* 12, 416 (2005).
[22] M. Foti, M. Piattelli, M.T. Baratta, G.J. Ruberto. *Agric. Food. Chem.* 44, 497 (1996).
[23] G. Li, Z. Liu. *J. Agric. Food Chem.* 57, 3943 (2009).
[24] P.C. Braga, M.D. Sasso, M. Culici, L. Galastri, M.T. Marceca, E.E. Guffanti. *Pharmacol.* 76, 61 (2006).
[25] M.G. Miguel, S.A. Dandlen, A.C. Figueiredo, L.G. Pedro, J.G. Barroso, M.H. Marques. *Acta Hort. (ISHS).* 853, 363 (2010).
[26] J. Mastelić, I. Jerković, I. Blažević, M. Poljak-Blaži, S. Borović, I. Ivančić-Baće, V. Smrečki, N. Zarkovic, K. Brcic-Kostic, D. Vikic-Topic, N. Müller. *J Agric Food Chem.* 11, 3989 (2008)
[27] S. Fujisawa, T. Atsumi, Y. Kadoma, H. Sakagami. *Toxicol.* 177, 39 (2002).
[28] E. Nagababu, J.M. Rifkind, S. Boindala, L. Nakka. *Methods Mol. Biol.* 610, 165 (2010).
[29] R. Aeschbach, J. Loliger, B.C. Scott, A. Murcia, J. Butler, B. Halliwell, O.I. Aruoma. *Food Chem. Toxicol.* 32, 31 (1994).
[30] A.L. Al-Malki. JKAU: *Sci.* 22, 239 (2010).
[31] A.G. Guimarães, G.F. Oliveira, M.S. Melo, S.C.H. Cavalcanti, A.R. Antoniolli, L.R. Bonjardim, F.A. Silva, J.P.A. Santos, R.F. Rocha, J.C.F. Moreira, A.A.S. Araújo, D.P. Gelain, L.J. Quintans-Júnior. *Clin. Pharmacol. Toxicol.*107, 949 (2010).
[32] N. Vidhya, S.N. Devaraj. *Indian J. Exp. Biol.* 37, 1192 (1999).
[33] N. Žarković, K. Brčić-Kostić, D. Vikić-Topić, N. Müller. *J. Agric. Food Chem.* 56, 3989 (2008).
[34] Y. Ding, J.A. Kim, S.Y. Yang, W.-K. Kim, S.H.Lee, H.D. Jang, Y.H. Kim. *Bull. Korean Chem. Soc.* 32, 3493 (2011).
[35] K. Umemura, T. Itoh, N. Hamada, Y. Fujita, Y. Akao, Y. Nozawa, N. Matsuura, M. Iinuma, M. Ito. *Biochem Biophys Res Commun.* 368, 948 (2008).
[36] Y. Nakamura, C. Yoshida, A. Murakami, H. Ohigashi, T. Osawa, K. Uchida. *FEBS Lett.* 572, :245 (2004).
[37] Natural, *Natural products on CD-ROM,* Chapman and Hall, London (1995).
[38] S. Y. Wang, J. H. Wu, L.F. Shyur, Y.H. Kuo, S. T. Chang. *Holzforschung.* 56, 487 (2001).
[39] K.P. Chao, K.F. Hua, H.Y. Hsu, Y.C. Su, S.T. Chang, *Planta Med.* 71, 300 (2005).
[40] J.C. Luis and C.B. Johnson, *Spanish J. Agr. Res.*3, 106 (2005).
[41] A. Kabouche, Z. Kabouche, M. Öztürk, U. Kolak, G. Topçu, *Food Chem.* 102, 1281 (2007).
[42] Y. Narukawa, N. Shimizu, K. Shimotohno, T. Takeda, Chem. Pharm. Bull. 49, 1182 (2001).
[43] P. Rijo, J. Chumbo, A. Oliveira, M.C. Costa, B. Rodríguez, M.F. Simões, *Revista de Fitoterapia.* 9, S1 (2009).

[44] S.R. Ayinampudi, R. Domala, R. Merugu, S. Bathula, M.R. Janaswamy, *Fitoterapia.* 83, 88 (2012).

[45] R.A. Dickson, P.J. Houghton, P.J. Hylands, *Phytochem.* 68, 1436 (2007).

[46] S.E. Ayyad, M.S. Makki, N.S. Al-Kayal, S.A. Basaif, K.O. El-Foty, A.M. Asiri, A.M. Walied, F.A. Badria, *Eur. J. Med. Chem.* 46, 175 (2011).

[47] H. Haraguchi, H. Ishikawa, I. Kubo, *Planta Med.* 63, 213 (1997).

[48] H. Haraguchi, H. Ishikawa, S. Sakai, B-P. Ying, I. Kubo, *Experientia.* 52, 573 (1996).

[49] K. Kunio, K. Chieko, *Frag. J.* 32, 55 (2004).

[50] S.S. Wijeratne, S.L. Cuppett. *J. Agric. Food Chem.* 55, 1193 (2007).

[51] H.H. Zeng, P.F. Tu, K. Zhou, H. Wang, B.H. Wang, J.F. Lu. *Acta Pharmacol. Sin.* 22, 1094 (2001).

[52] J. Xu, P. Guo, C. Liu, Z. Sun, L. Gui, Y. Guo, T. Yamakuni, Y. Ohizumi, *Biosci. Biotechnol. Biochem.*75, 1386 (2011).

[53] J. Xu, Y. Guo, C. Xie, Y. Li, J. Gao, T. Zhang, W. Hou, L. Fang, L. Gui, *J. Nat. Prod.* 74, 2224 (2011).

[54] P. Guo, Y. Li, J. Xu, Y. Guo, D-Q, Jin, J. Gao, W. Hou, T. Zhang, *Fitoterapia.* 82, 1123 (2011).

[55] C.C. Chen, Y.J Shiao, R.D. Lin, Y.Y. Shao, M.N. Lai, C.C. Lin, L.T. Ng, Y.H. Kuo, *J. Nat. Prod.* 69, 689 (2006).

[56] F. Osakada, Y. Kawato, T. Kume, H. Katsuki, H. Sugimoto, A. Akaike, J. Pharmacol. Exp. Ther. 311, 51 (2004).

[57] T. Kume, N. Asai, H. Nishikawa, N. Mano, T. Terauchi, R. Taguchi, H. Shirakawa, F. Osakada, H. Mori, N. Asakawa, M. Yonaga, Y. Nishizawa, H. Sugimoto, S. Shimohama, H. Katsuki, S. Kaneko, A. Akaike, *Proc. Natl. Acad. Sci. U.S.A.* 99, 3288 (2002).

[58] K.J. Lee, J.H. Choi, H.G. Jeong, *Food Chem. Toxicol.* 45, 2118 (2007).

[59] S.Y. Choi, K.J. Lee, H.G. Kim, E.H. Han, Y.C. Chung, N.J. Sung, H.G. Jeong, *Bull. Korean Chem. Soc.* 27, 9 (2006).

[60] I.B. Koul, A. Kapil, *Indian J. Pharmacol.* 26, 296 (1994).

[61] N.P. Trivedi, U.M. Rawal. *Indian J. Exp. Biol.* 39, 41 (2001).

[62] Y.L. Wu, Y.Z. Jiang, X.J. Jin, L.H. Lian, J.Y. Piao, Y. Wan, H.R. Jin, J. Joon Lee, J.X. Nan. *Phytomedicine.* 17, 475 (2010).

[63] T.P. Smina,; J. Mathew,; K.K. Janardhanan,; T.P. Devasagayam. *Environ. Toxicol. Pharmacol.* 32, 438 (2011).

[64] X. Zhu, Z. Lin. *Carbohyd. Polym.* 63, 188 (2006).

[65] R.A. Mosa, A.O. Oyedeji, F. O. Shode, M. Singh, A. R. Opoku. *Afr. J. Pharm. & Pharm.* 5, 2698 (2011).

[66] P.R.F. Carvalho, D.H.S. Silva, V.S. Bolzani, M. Furlan. *Chemistry & Biodiversity.* 2, 367 (2005).

[67] B. D'Abrosca, A. Fiorentino, P. Oriano, P. Monaco, S. Pacifico. *Food Chem.* 98, 285 (2006).

[68] P. Hashim, H. Sidek, M.H.M Helan, A. Sabery, U.D. Palanisamy M. Ilham. *Molecules.* 16, 1310 (2011).

[69] Z. Ahmad, S. Mehmood, I. Fatima, A. Malik, R. Ifzal, N. Afza, L. Iqbal, M. Latif, T.A. Nizami, *Magn. Reson. Chem.* 46, 94 (2008).

[70] J.M. David, A.L.B.S. Barreisors, J.P. David, *Pharm. Biol.* 42, 36 (2004).

[71] G. Topçu, A. Ertas, U. Kolak, M. Ozturk, A. Ulubelen. *ARKIVOC*. 7, 195 (2007).

[72] M.P. Montilla, A. Agil, M.C. Navarro, M.I. Jimennez, A. Garcia-Granados, A. Parra, M.M. Cabo. *Planta Med.* 69, 472 (2003).

[73] J.H. Nam, H.J. Jung, J. Choi, K.T. Lee, H.J. Park. *Biol. Pharm. Bull.* 29, 967 (2006).

[74] J. Choi, K.T. Lee, J. Ha, S.Y. Yun, C.D. Ko, H.J. Jung, H. J. Park. *Biol. Pharm. Bull.* 26, 1436 (2003).

[75] Y.H. Kim, J.H. Choi, H.K. Rim, H.J. Kang, S.G. Chang, J.H. Park, H.J. Park, J.W. Choi, S.D. Kim, K.T. Lee. *Biol. Pharm. Bull.* 34, 906 (2011).

[76] P.T. Sudharsan, Y. Mythili, E. Selvakumar, P. Varalakshmi. *Hum. Exp. Toxicol.* 24, 313. (2005).

[77] V. Sudhahar, S.A. Kumar, P. Varalakshmi. *Life Sci.* 78, 1329 (2006).

[78] J. Feng, P. Zhang, X. Chen, G. He. *J. Cell. Biochem.* 112, 1524 (2011).

[79] X. Wang, X.L. Ye, R. Liu,; H.L. Chen,; H. Bai, X. Liang, X.D. Zhang, Z. Wang, W.L. Li, C.X. Hai. *Chem. Biol. Interact.* 184, 328 (2010).

[80] Y. Du, K.M. Ko. *Planta Med.* 72, 222 (2006).

[81] T. Ghosh, T.K. Maity, J. Singh. *Planta Med.* 77, 804 (2010).

[82] D. Gao, Q. Li, Y. Li, Z. Liu, Y. Fan, H. Zhao, J. Li, Z. Han. *Phytother. Res.* 23, 1257 (2009).

[83] I.H. Son, Y.H. Park, S.I. Lee, H.D. Yang, H. Moon. *Molecules.* 12, 1147 (2007).

[84] F.A. Oliveira, M.H. Chaves, F.R.C. Almeida, R.C.P. Lima, R.M. Silva, J.L. Maia, G.A. Brito, F.A. Santos, V.S. Rao. *J. Ethnopharmacol.* 98, 103 (2005).

[85] A. Jimenez-Escrig, I. Jimenez-Jimenez, C. Sanchez-Moreno, F. Saura-Calixto. *J. Sci. Food Agric.* 80, 1686 (2000).

[86] K.H. Wagner, I. Elmadfa. *Ann. Nutr. Metab.* 47, 95 (2003).

[87] G.W. Burton, K.U. Ingold. *Sci.* 224, 569 (1984).

[88] S.A.R. Paiva, R.M. Russell. *Am. Coll. Nutr.* 18, 426 (1999).

[89] D.C. Liebler, T.D. McClure. *Chem. Res. Toxicol.* 9, 8 (1996).

[90] K. Fukuzawa, Y. Inokami, A. Tokumura, J. Terao, A. Suzuki. *Biofactors.* 7, 31 (1988).

[91] R. Cozzi, R. Ricordy, T. Aglitti, V. Gatta, P. Perticone, R. De Salvia. *Carcinogenesis.* 18, 223 (1997).

[92] A. El-makawy, H.M. EL-Ashmaoui. *Arab. J. Biotech.* 6, 289 (2003).

[93] E.O. Farombi, G. Britton. *Redox Report.* 4, 61 (1999).

[94] P. DiMascio, S. Kaiser, H. Sies. *Arch. Biochem. Biophys.* 274, 532 (1989).

[95] A. Mortensen, L.H. Skibsted. *FEBS Lett.* 417, 261 (1997).

[96] C. Luo, X-G. Wu. *Int. J. Mol. Sci.*, 12, 3340 (2011).

[97] M.E. Yonar. *Environ. Toxicol.* In Press (DOI: 10.1002/tox.20757).

[98] S.K. Gupta, D. Trivedi, S. Srivastava, S. Joshi, N. Holder, S.D. Verma. *Nutrition.* 19, 794 (2003).

[99] Y. Seo, A. Masamune, T. Shimosegawa, H. Kim. *Ann. NY Acad. Sci.* 1171, 570 (2009).

[100] A.J. Chucair, N.P. Rotstein, J.P. Sangiovanni, A. During, E.Y. Chew, L.E. Politi, *Invest. Ophthalmol. Visual Sci.* 48, 5168 (2007).

[101] S.Y. Li, Z.J. Fu, H. Ma, W.C. Jang, K.F. So, D. Wong, A.C. Lo. *Invest. Ophthalmol. Vis. Sci.* 50, 836 (2009).

[102] S.-Y. Li, A.C.Y. Lo. *Int. J. Mol. Sci.* 11, 2109 (2010).

[103] E. Arnal, M. Miranda, I. Almansa, M. Muriach, J.M. Barcia, F.J. Romero, M. Díaz-Llopis, F. Bosch Moreer. *Arch. Clin. Exp. Ophthalmol.* 247, 115 (2009).

In: Phytochemicals
Editor: Hai-Xue Kuang

ISBN: 978-1-62417-354-7


*Chapter VII*

# Supercritical Extraction of Phloroglucinols from St. John's Wort (*Hypericum Perforatum* L.) – Process Optimization, Mathematical Modeling and Antimicrobial Activity

***Sandra B. Glisic* and Aleksandar M. Orlovic***
Faculty of Technology and Metallurgy,
University of Belgrade, Belgrade, Republic Serbia

## Abstract

St. John's Wort (*Hypericum perforatum* L.) has been widely used as an anti-inflammatory and healing agent in traditional medicine. Some para-pharmaceutical products, prepared from the extract of St. John's Wort, are frequently used because of their anti-depressive and antiviral properties. Moreover, the extract prepared from St. John's Wort (*H. perforatum* L.) has been also reported to have an antimicrobial, anti-inflammatory, and healing effect.

Thus, extraction of a specific component from St. John's Wort which gives a high yield of total extract and high content of desired pharmacological active compounds is obviously in focus of many research activities today. Despite the evident clinical effects, there is still a controversy regarding the active principle of the extracts. Some authors attribute the antidepressant activity to hypericin and its derivatives, whereas, other studies are in the favor of hyperforin.

* Phone: +381 11 3303707; Fax: +381 11 3370387; Email: sglisic@tmf.bg.ac.rs.

Hyperforin, as a phloroglucinol derivative is a relatively lipophilic compound which is susceptible to rapid thermal, light-induced and oxidative decomposition. Adhyperforin presents its methyl derivative.

The supercritical fluid extraction (SFE) is an outstanding method for the extraction of natural agents and valuable components from plant materials, because heat-sensitive compounds can be extracted without any degradation, and in addition, it is an environmentally acceptable technology. Supercritical carbon dioxide ($SCCO_2$) is the most useful supercritical fluid in the food industry, because it is non-toxic, non-explosive, and available in high (food grade) purity and can be removed from the extracted products without leaving any residue.

The main goal of this study was oriented to the investigation of process optimization of the best extraction method applicable to intergradient in food and pharmaceutical industry – supercritical fluid extraction as well as the antimicrobial activity of obtained extracts. The first part of study will represent the literature review of St. John's Wort occurrence in nature, their health effects and medical and pharmaceutical use.

# 1. Introduction: Occurrence in Nature, Health Effects and Medical Use

*Hypericum* (*Guttiferae*, syn. *Clusiaceae*) is a large genus of herbs which grows widely in temperate regions. The genus Hypericum contains, by the most recent count, 469 species that have been classified into 36 taxonomic sections [1]. Hypericum perforatum L. (common St. John's wort), the most thoroughly studied taxon of the genus, is well known for its historical and contemporary use as a medicinal plant.

Figure 1. The flower and life of the *Hypericum perforatum* L.

*Hypericum perforatum* is a yellow-flowering, stoloniferous or sarmentose, perennial herb (Figure 1) native to Europe, which has been introduced to many temperate areas of the world and grows wild in many meadows. Short description of St. John's Wort could be: perennial rhizomatous herb with erect stems to 1.2 m high; leaves hairless, paler on the lower surface, ovate to narrow-oblong to about 3 cm long; stems with opposite weakly pronounced longitudinal ridges and with occasional black glands; fruit sticky, containing many pitted seeds about 1 mm long; roots to 1 m deep.

Distinguished features are numerous translucent glands on the leaves and yellow flowers often with black glands on the petal margins (Figure 1) [1,2,3].

The common name comes from its traditional flowering and harvesting on Saint John's day, 24 June. The genus name *Hypericum* is derived from the Greek words *hyper* (above) and *eikon* (picture), in reference to the traditional use of the plant to ward off evil, by hanging plants over a religious icon in the house during Saint John's day. The species name perforatum refers to the presence of small oil glands in the leaves that look like windows, which can be seen when they are held against the light [2]. Secretory canals in the floral and vegetative tissue in H. perforatum are of three types. Type A canals generally have a narrow diameter (although this can be variable in the pistil and root tissue) and are delimited by four or more polygonal cells. Type B and C canals are structurally similar to the translucent glands, but have different patterns of ontogeny. Hyperforin is identified in sepal tissue containing Type B while hypericin and pseudohypericin could be found in translucent and dark glands [4].

Most of these species have been used for a long time for the treatment of external wounds and gastric ulcers, and also as sedative, antiseptic, and antispasmodic herbs in folk medicine [3].

The plant has a wide range of medicinal applications, including skin wounds, eczema, burns, diseases of the alimentary tract and psychological disorders [3,4]. St. John's Wort has been widely used as an anti-inflammatory and healing agent in traditional medicine. Some para-pharmaceutical products, prepared from the extract of St. John's Wort, are frequently used because of their anti-depressive and antiviral properties [3,4].

Moreover, the extract prepared from St. John's Wort has been also reported to have an antimicrobial, anti inflammatory, and a healing agent in traditional medicine [3-7]. Despite the evident clinical effects, there is still a controversy regarding the active principle of the extracts.

Some authors attribute the antidepressant activity to hypericin and its derivatives, whereas, other studies are in the favour of hyperforin [6].

Primary production areas for H. perforatum in Europe include Germany, Italy and Romania, and the majority of the harvest goes toward the production of the crude drug [8]. A small fraction of the plant material is sold for the production of *Oleum hyperici*, which is also used as carrier oil in aromatherapy, and for the distillation of the essential oil [9]. The steam-distilled essential oil, which has only recently become available on a broader scale on the market, is largely produced by small companies in Serbia, Croatia, Poland, Bulgaria, France, Canada and the United States [9].

Thus, extraction of specific component from St. John's Wort which gives a high yield of total extract and high content of desired pharmacological active compounds is obviously in focus of many research activities, today.

## 1.1. Main Components and Composition of Essential Oil and Extract

Species of *Hypericum* contain many bioactive constituents, including proanthocyanins, flavonoids, biflavonoids, xanthones, phenylpropanes and naphthodianthrones that are characterized by their relative hydrophilicity, as well as acylphloroglucinols and essential oil components that are more hydrophobic in nature [3,4].

### *1.1.1. Essential Oil - Oleum Hyperici*

Hypericum species are generally classed as essential oil-poor plants (generally oil yield <1%, w/w). The content of essential oils in healthy H. perforatum plants is highest during the full-bloom stage versus pre-bloom or fruiting stage (0.35% versus 0.12 and 0.18%, respectively) [3,4,9].

**Table 1. Content of essential oil in different part of *Hypericum perforatum L.* [3]**

| Plant part | Content, mass% |
|---|---|
| Whole plant | > 0.07-0.33 |
| Blosom plant | 0.059-0.11 |
| Pre-blosom plant | 0.26 |
| Plant without petals | up to 0.35 |
| Fresh flowers | 0.25 |

**Table 2. Essential oil composition (*Oleum hyperici*) [4]**

| Class of component | Component | mass% |
|---|---|---|
| Monoterpenes | α-Pinene | 4.2-16 |
| | β-Pinene | 2.2 |
| | α-Terpinene | 1.7 |
| | β-Myrcene | 2.6 |
| | p-Cymene | 3.3 |
| Hydrocarbons | 2-Me-octane | 13-15 |
| | n-Nonane | 1.4-5.6 |
| | 3-Me-nonane | 3.1 |
| | 2-Me-decane | 6.3-7.9 |
| | n-Undecane | 2.1-3.9 |
| | n-Tridecane | 1.3 |
| | 2-Me-dodecane | 4.5 |
| | 2-Me-tetradecane | 1.7 |
| | iso-Tridecane | 1.2 |
| Alcohols | Dodecanol | 5.0 |
| | Terpineol-4 | 0.7 |
| | 4-Tetradecanol | 4.1 |
| | n-Tetradecanol | 2.2 |
| Terpenoid oxides | Cariofilene oxide | 2.5-8.69 |
| | 1,8-Cineole | 2.5 |
| Sesquiterpenes | β-Cariofilene<br>β -Farnesene Germacrene-D<br>δ-Cadinene | 3-20<br>3.2<br>18-26<br>1.5 |
| Ketones | 6-Me-5-hepten-2-on | 2.1 |

Research has shown that essential oil extraction efficiency with liquid or supercritical $CO_2$ (yield 1.00%, w/w) is higher than with steam distillation (0.06%, w/w) [4]. Hyperforin and derivatives can be extracted using liquid or supercritical $CO_2$, but not with steam distillation [4,9].

Typical essential oil constituents for this species include the monoterpenes α- and β-pinene, limonene and myrcene; the sesquiterpenes β-caryophyllene and caryophyllene oxide; and hydrocarbons such as n-decane, $C_{16}$- and $C_{29}$ alkanes (nonakosan) and $C_{24}$-, $C_{26}$-, and $C_{28}$-alkanols (1-tetrakosanol, 1-oktakosanol and 1-triakontanol) [3,4,9]. The content of oil in a different part of plant could be significant as it is shown in Table 1 and depending also on the period of harvesting and extraction.

Main components in essential oil as well as their variation, depending on diversity and plant origin, are summarized in Table 2.

### *1.1.2. Main Bioactive Components in Extract*

Main bioactive constituents include proanthocyanins, flavonoids, biflavonoids, xanthones, phenylpropanes, naphthodianthrones and acylphloroglucinols (Table 3) [3,4]. The different content of main bioactive components in the flowers is summarized in Table 4 [3,4].

Naphthodianthrones are typical of the genus *Hypericum* with an intense red color and phototoxic properties. The major components of this group of compounds are the hypericins.

Two so-called protoderivatives, protohypericin and pseudoprotohypericin, were isolated from the plant but due to their unstable nature are efficiently converted to the stable products, i.e., hypericin and pseudohypericin, respectively. The naphthodianthrones show a limited solubility in almost all solvents. The pure compounds, especially hypericin, are almost insoluble in water at ambient temperature [4].

Phloroglucinol derivatives are widely distributed in the genus Hypericum. Two closely related compounds found in *H. perforatum* are hyperforin and adhyperforin which contains an additional methyl group. Hyperforin is quite unstable – especially in aqueous solutions and when exposed to light and heat.

The hyperforins present very interesting compounds from a pharmacological standpoint and it was shown to inhibit or modulate several neurotransmitter systems in vitro. It is a potent uptake inhibitor of serotonin, dopamine and noradrenaline. This supports the possible role of hyperforin in the antidepressant activity of the herb [3,6].

Flavonoids comprise the major group of biologically active compounds in *H. perforatum* (2–4%). The flavonol aglycones identified so far include kaempferol, luteolin, myricetin and quercetin. Hyperoside (hyperin) and rutin usually dominate among the glycosides of H. perforatum followed by quercitrin and isoquercitrin [4].

Biflavones are a rare group of dimeric flavones found in some vegetable sources. Three biflavones detected in H. perforatum are I3,II8-Biapigenin, amentoflavone and I6,II8-diquercetin. The therapeutic significance of these biflavones in *H. perforatum* is still unknown. Amentoflavone, however, was shown to possess anti-inflammatory and analgesic activities [4].

Other typical constituents are xanthones and kielcorin C, acids (isovalerianic, nicotinic, myristic, palmitic, and stearic), carotenoids, choline, nicotinamide, pectin, β-sitosterol, pectin, fatty acids, amino acids, vitamin C, tannins, hydroperoxycadiforin and bisanthraquinone glycosides [3,4].

**Table 3. The chemical structure of main bioactive components in *Hypericum perforatum* L. [3,4]**

| | | |
|---|---|---|
| Naphtodianthrones | 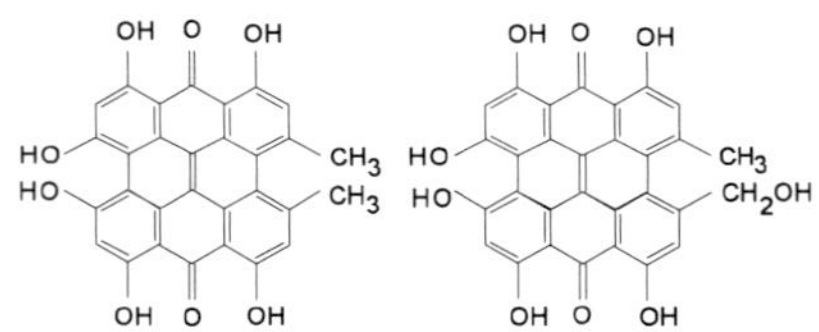 Hipericin Psudohipericin | 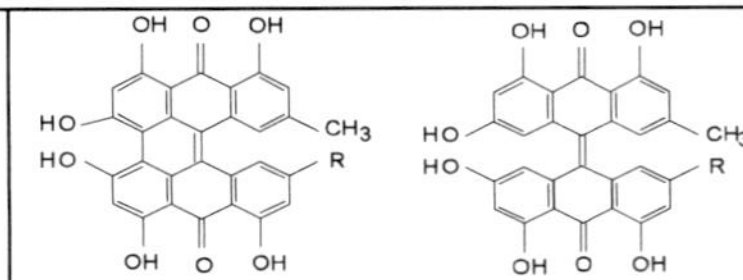 Protopseudohypericin R=$CH_2OH$<br>Pseudohyperikodehidrodiantron R=$CH_2OH$<br>Protohypericin R=$CH_3$<br>Hiperikodehidrodiantron R=$CH_3$ |
| Flavons | 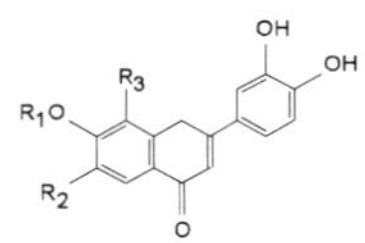 | Homoorjentin $R_2$ = glukozil<br>Leutin $R_1 = R_2 = R_3$ = H |
| Flavonoides | 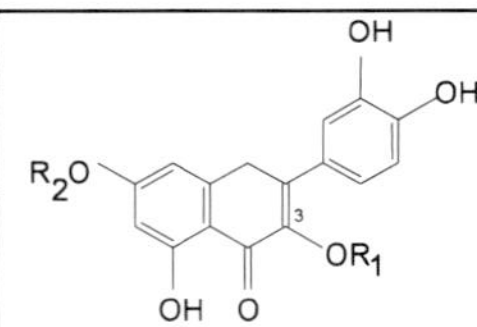 | Kvercetin $R_1$=$R_2$=H<br>Hyperozid $R_1$=galaktozil. R2=H<br>Izokvercetin R1=glukozil. $R_2$=H<br>Kvercitrin $R_1$=ramnozil. $R_2$=H<br>Rutin $R_1$=ramnglukozil. $R_2$=H |
| Biflavonoides | 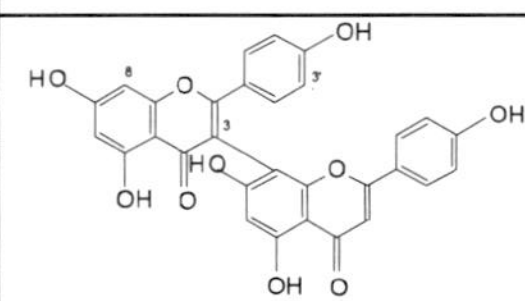 I3.II8-Biapigenin | 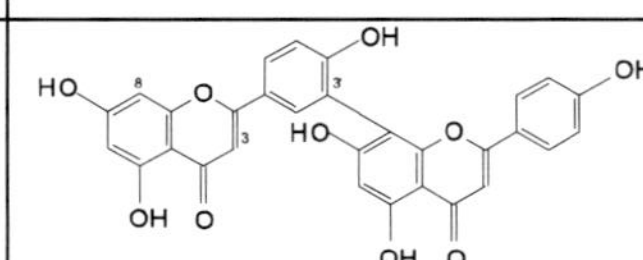 I3`.II8-Biapigenin (amentoflavon) |

| | | |
|---|---|---|
| Flavonols | 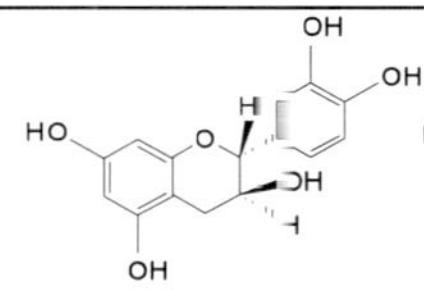 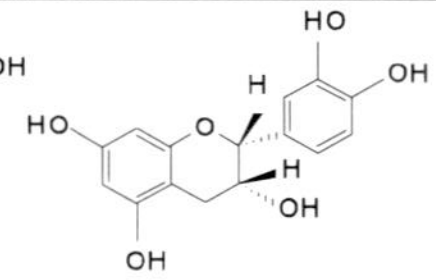 Katehin Epikatehin | 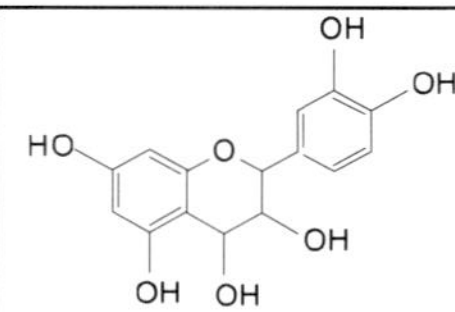 Leukoantocijanidin |
| Ksantons | 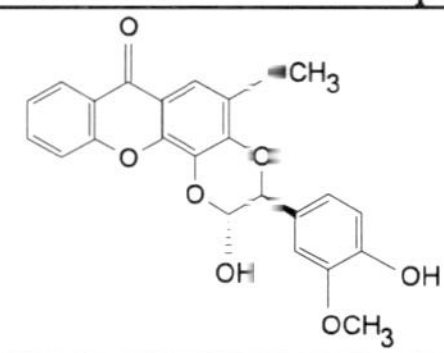 | Klikorin |
| Cumarines | 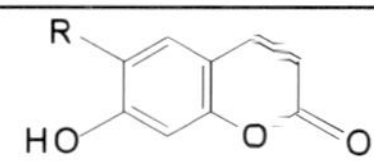 | Skoleptin R=$OCH_3$ Umbeliferon R=H |
| Phloroglucinols | 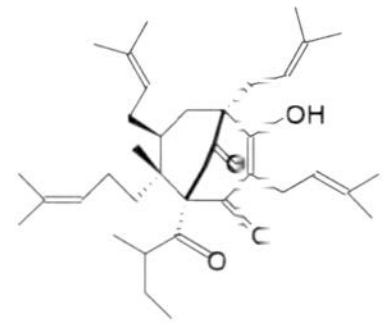 Adhyperforin | 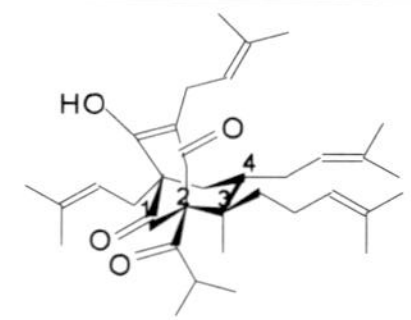 Hyperforin |

**Table 4. The content of main bioactive components [3]**

| Components | Content in flowers, μg | | |
|---|---|---|---|
| | min | max | average |
| Hypericin | 2.9 | 22.8 | 11.2 |
| Pseudohypericin | 11.2 | 63.5 | 29.6 |
| Hyperforin | 206.0 | 606.8 | 363.0 |
| Amentoflavone | 11.2 | 71.4 | 34.1 |
| Rutine | 19.3 | 61.0 | 36.4 |
| Hyperoside | 43.9 | 140.0 | 85.4 |
| Isoquercitrin | 15.1 | 106.7 | 39.9 |
| Quercitrin | 21.0 | 112.2 | 41.9 |

# 2. Supercritical $CO_2$ Extraction – Comparison to Conventional Extraction Techniques

The supercritical fluid extraction (SFE) is an outstanding method for the extraction of natural agents and valuable components from plant materials, because heat-sensitive compounds can be extracted without any degradation, and in addition it is an environmentally acceptable technology. Supercritical carbon dioxide ($SCCO_2$) is the most useful supercritical fluid in the food industry, because it is non-toxic, non-explosive, and available in high (food grade) purity and can be removed from the extracted products without leaving any residue.

Supercritical extraction and fractionation of natural matter is one of the early and most studied applications in the field of supercritical fluids. In the last 10 years, studies on the extraction of classical compounds like essential and seed oils from various sources: seeds, fruits, leaves, flowers, rhizomes, etc., with or without the addition of a co-solvent have been published. Supercritical extraction of antioxidants, pharmaceuticals, colouring matters, and pesticides has also been studied [3,4,10-23]. Indeed, supercritical fluids extraction (SFE) has significant advantages over traditional extraction techniques: it is a flexible process due to the possibility of continuous modulation of the solvent power/selectivity of the supercritical fluids, allows the elimination of polluting organic solvents and of the expensive post-processing of the extracts for solvent elimination. Supercritical fluid extraction of essential oils with carbon dioxide has certain advantages over steam distillation. Steam distillation can lead to thermal degradation and partial hydrolysis of some essential oil compounds. SFE with carbon dioxide can be performed at temperatures around 313 K, thereby preserving original oil composition and properties [23].

The extraction of specific components from St. John's Wort, as well as the high yield of total extract, are therefore of considerable research interest [1,2]. To date, there are a few literature data available, presenting results of supercritical carbon dioxide extraction of St. John's Wort [3,10-18]. A review on the supercritical extraction of St. John's Wort has been published recently [3,10]. According to data recently published in literature, hypericin was not obviously detected as constituent of $SCCO_2$ extract, even when ethanol was used as a co-solvent [11,12]. Extraction performed with liquid $CO_2$ (subcritical condition) gave the highest amount of hyperforin and adhyperforin [10].

Recent investigation reported that increased amount of extract is followed by decrease of hyperforin concentration in extract as a consequence of increased yield of other compounds. Internal mass transfer or diffusion of hyperforin from the plant material controls the extraction process and that effect of supercritical extraction is primarily limited by solubility of hyperforin in the supercritical carbon dioxide [11].

## 2.1. Supercritical $CO_2$ Extraction

Extractions of dry St. John's Wort were carried out using $SCCO_2$ at 313 and 323 K and 10, 15 and 20 MPa (Figure 2 representing the extractions curves at different conditions of temperatures and pressures for the exact $CO_2$ consumption) by the experimental procedure described in previously published paper [10,19-22].

The investigated pressure and temperature conditions as well as obtained yield and hyperforin and adhyperforin content are summarised in Table 5.

LC-MS/MS screening revealed the presence of hyperforin and adhyperforin, while naphthodianthrones (hypericin and pseudohypericin) and flavonoids (rutin, hyperoside, quercetin and quercitrin) were absent in the extracts, which is in accordance with the previously published data [11-14]. Obviously the yield of total extract increases at increasing pressure or decreasing temperature (Table 5) which is in accordance with the change of $SCCO_2$ density.

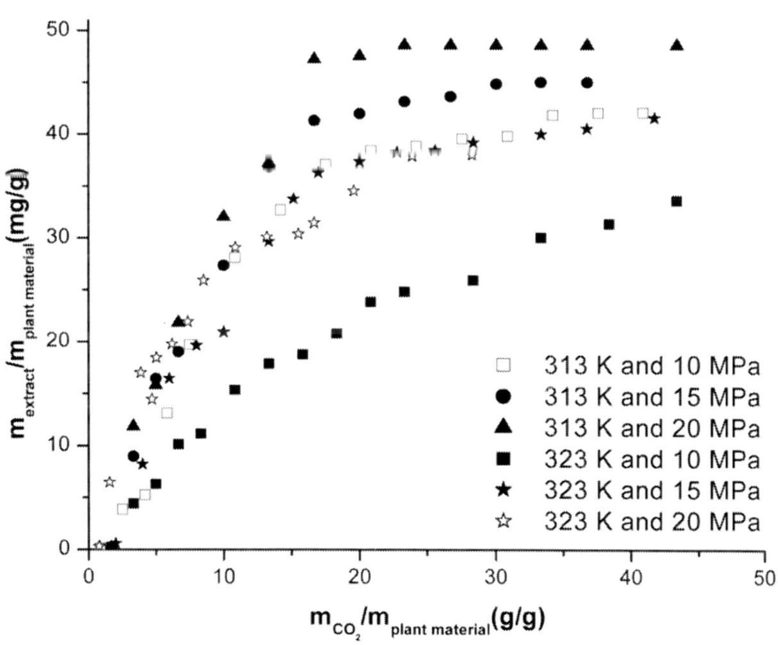


Figure 2. Extractions curves obtained at different conditions of temperatures and pressures [10].

Obtained data, calculated as total extract content (mg/g plant material) versus $SCCO_2$ density is presented on Figure 3 and compared with other data recently published in literature [11,14,16]. The total extract content (mg/g plant material) at the same $SCCO_2$ density depends mainly on the origin of St John Wort used for extraction, but similar trends of increasing yield by increasing $SCCO_2$ density evidently exist (Figure 3).

The lowest content (mg/g plant material) of two phloroglucinol derivatives as well as their concentration (mg/g total extract) was obtained at 10 MPa and 323 K (Table 5) when density of $SCCO_2$ had the lowest value applied in this study (384 kg/m$^3$). Even the highest extraction yield was obtained at 20 MPa and 323 K the qualitative content of extract is not obtained (because of higher extraction of some havy components as tannins and waxes.

**Table 5. Content of hyperforin and adhyperforin in $SCCO_2$ extract [10]**

| Extraction conditions | | | Total extract[a] | Hyperforin | | Adhyperforin[b] | |
|---|---|---|---|---|---|---|---|
| P (MPa) | T (K) | Density (kg/m$^3$) | (mg/g plant material) [a] | (mg/g plant material) | (mg/g total extract) | (mg/g plant material) | (mg/g total extract) |
| 10 | 313 | 629 | 41.8 ± 0.5 | 19.0 | 450 | 2.9 | 68 |
| 15 | 313 | 780 | 45.4 ± 0.6 | 18.9 | 420 | 2.8 | 63 |
| 20 | 313 | 839 | 49.2 ± 0.6 | 19.4 | 399 | 2.8 | 58 |
| 10 | 323 | 384 | 35.3 ± 0.4 | 14.7 | 427 | 2.1 | 61 |
| 15 | 323 | 670 | 40.9 ± 0.5 | 19.3 | 459 | 2.8 | 67 |
| 20 | 323 | 784 | 32.6 ± 0.4 | 18.0 | 438 | 2.7 | 65 |

[a] Mean ± standard deviation (n=3).
[b] Results estimated according to response of hyperforin.

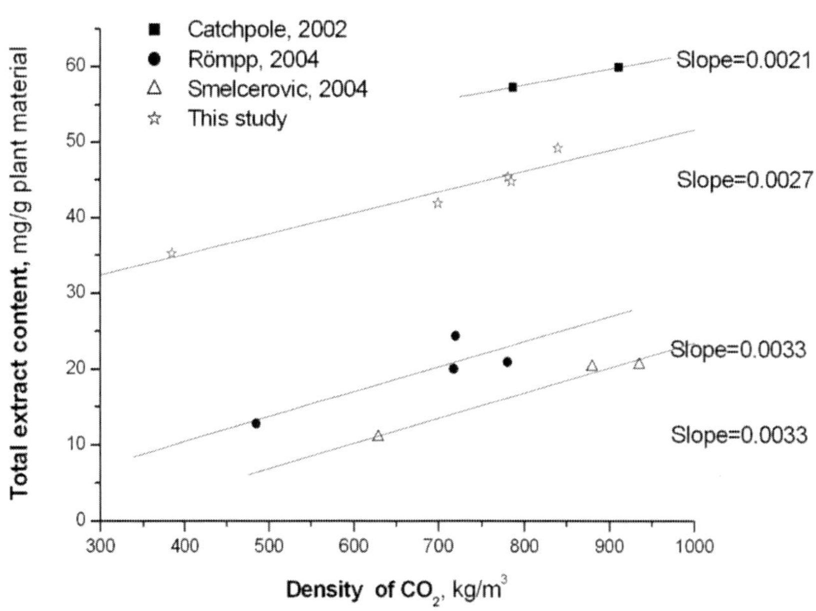


Figure 3. Yield of extract versus density of $SCCO_2$ [10].

Such small density poses the smallest extraction power of $SCCO_2$ indicating that extraction power mainly depends on supercritical fluid density. Obtained data in this study show that high extraction efficiency of hyperforin could be achieved when the fluid density is above 620 kg/m$^3$, what was also concluded by Manilla and co-workers [13].

### 2.1.1. The Rate of Extraction of Hyperforin and Adhyperforin

$SCCO_2$ extraction at 313 K and 10 MPa was performed by collecting the extract for some period from the beginning to the end of the experiment (0-3 hours). Content of hyperforin and adhyperforin in fraction collected during extraction/fractionation as well as the amount of total extract is shown in Table 6.

Determined amount and content of hyperforin and adhyperforin in collected fractions (Fraction 1-4) are used for the calculation of the extraction rate of total extract (i.e. amount of extract per consumed mass of $SCCO_2$ per time, average value of each period) versus time as well as the extraction rates of hyperforin and adhyperforin (Figure 4).

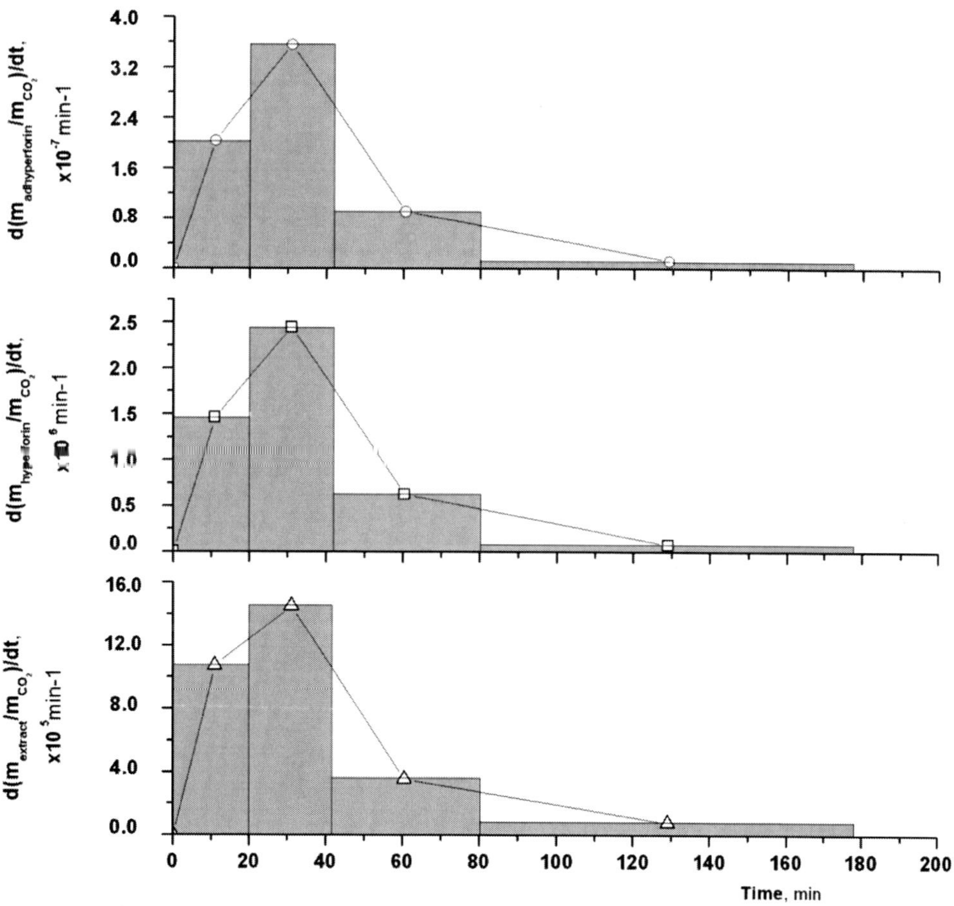


Figure 4. The rate of extraction for: adhyperforin (a), hyperforin (b) and total extract (c) as a function of time [10].

**Table 6. Content of hyperforin and adhyperforin in fraction collected during SC $CO_2$ extraction at 10MPa and 313K**

| Fraction | | Total extract | Hyperforin | | Adhyperforin[b] | |
|---|---|---|---|---|---|---|
| No | Time interval (min) | (mg/g plant material)[a] | (mg/g plant material) | (mg/g total extract) | (mg/g plant material) | (mg/g total extract) |
| 1 | 0–22 | 11.9 ± 0.2 | 4.8 | 408 | 0.7 | 57 |
| 2 | 22–40 | 9.3 ± 0.1 | 4.4 | 506 | 0.6 | 73 |
| 3 | 40–81 | 9.0 ± 0.1 | 5.1 | 534 | 0.7 | 77 |
| 4 | 81–177 | 11.0 ± 0.2 | 3.6 | 330 | 0.5 | 47 |

Performed analysis of extraction rates indicated that consumption of about 7.0 - 8.5 g of $SCCO_2$ per 1 g of dry St John Wort gave the highest extraction rate of total extract, hyperforin and adhyperforin at 313 K and 10 MPa [10].

Extraction of hyperforin and adhyperforin from St. John's Wort by $SCCO_2$ depends strongly on the $CO_2$ density and this study indicated that favourable $CO_2$ density for the best extraction results (desirable amount of hyperforin and adhyperforin in extract which is around 50%) is between 600 kg/m$^3$ and 700 kg/m$^3$. Such conclusions indicate that liquid $CO_2$ at 8 MPa and 303 K (having density of 700 kg/m$^3$) might give the same extraction yield of desired components (hyperforin and adhyperforin) as $SCCO_2$ extraction [10]. The mild working condition (a lower temperature) used for sub-critical extraction have many advantages: technical convenience, reduced extraction of unwanted products, and decreased risk of hyperforin and adhyperforin decomposition during extraction.

## 2.2. Extraction of Hyperforin and Adhyperforin Performed by Ultrasound Extraction with Methanol

The generally accepted explanation for ultrasound enhancement is related to the effect of ultrasonic waves on the plant material and breaks of cells and releases of cells' contents into the extraction medium. It is well known that an ultrasonic device is an adequate unit for the disruption of plant cells, but again no detailed information exists for the effect of ultrasound on plant tissues. Furthermore, the extraction from dried material could be described as a two-step process involving: (i) swelling and solvation of plant materials in corresponding solvent; (ii) the mass transfer of soluble constituents from the plant material to solvent by diffusion and osmosis [24]. The possible benefits of ultrasound application in extraction are most probably an intensification of mass transfer, improved solvent penetration into the plant tissue and capillary effects. All of these effects make an easier access of the solvent to the plant cell. The collapse of cavitation bubbles near the cell walls is expected to produce the cell disruption together with a good penetration of the solvent into the cells, through the ultrasonic jet [24].

From a botanical standpoint, the plant tissues are composed of cells surrounded by middle lamella which consists of pectinous compounds, thus the plant cells are surrounded by membranes and the cell walls.

**Table 7. Content of total extract, hyperforin, adhyperforin and hypericin (mg/g plant material) in samples obtained by ultrasound extraction with (sample 1) and without (sample 2) pre-treatment (313 K and 10 MPa for 2 hours)**

| Sample no. | Total extract [a] | Hyperforin | Adhyperforin[b] | Hypericin |
|---|---|---|---|---|
| 1 | 158 ± 24 | 18.4 | 2.3 | 1.5 |
| 2 | 143 ± 20 | 13.3 | 1.6 | 0.8 |

[a] Mean ± standard deviation (n=3).
[b] Results estimated according to response of hyperforin.

In a live cell, the wall is a network of crystalline cellulose microfibrils glued together by various cross-linked polysaccharides, phenolic compounds and proteins into a coherent ‘‘matrix” that allows diffusion to take place. During the drying process of the plant tissues, a series of physical and chemical modifications take place, producing major differences between the fresh and the dried tissue. Thus, it is known that the dried cells lose their capacity for diffusion and osmosis as a result of cell wall and middle lamella desiccation. The rehydration process which leads to swelling during the steeping stage of extraction, is the result of the contact between the large number of hydroxyl groups, the cell wall and middle lamella. The general effect of ultrasonic treatment is to enhance the process of softening via the hydration of pectinous material from the middle lamella which then becomes more malleable. This leads to the break-up of the plant tissues with ultrasonic vibration in which the primary benefits of sonication during solvent extraction are an enhanced hydration process, which takes place simultaneously with plant (vegetal) material fragmentation [24].

The extract's yield depends on used solvent and increases with increasing its polarity. It is well known that the plant material contains hydrophilic and hydrophobic compounds but the content of the water soluble compounds is much higher. The comparative analysis of combined extraction techniques (the ultrasound-assisted solvent extraction followed by, e.g. supercritical extraction) and supercritical extraction, which might fulfil the main task to obtain the extract with desired concentration of naphtodiantrones is the main task. The influence of the pre-treatment (323 K and 10 MPa for 2 hours; plant material as a charge; without flow of $SCCO_2$) on the yield of total extract, hyperforin and adhyperforin obtained by ultrasonic extraction with methanol was investigated. The main idea was to give enough time for penetration of $SCCO_2$ into the tissue structure of plant material and than destruct it [19]. Thus bioactive components of St. John's Wort are better available for contact with methanol used in ultrasonic extraction process. The plant material was then treated by methanol in ultrasound bath and the content of the hyperforin and hypericin was compared with those obtained by sole ultrasonic extraction.

Obtained results (Table 7) indicated that a higher content (mg/g dry plant material) of total extract and three pharmacological compounds under study could be achieved if pre-treatment is applied.

Approximately 3-4 time higher content of total extract might be achieved using ultrasound extraction with pre-treatment than those obtained by $SCCO_2$ extraction. The content of hyperforin in the extract obtained by ultrasonic extraction followed after pre-treatment is almost the same as those obtained by $SCCO_2$, with the exception of the $SCCO_2$ extract reached at 313 K and 10 MPa which had a lower content of hyperforin [4,10,22].

The content of adhyperforin obtained by ultrasonic extraction followed after pre-treatment is slightly lower than those obtained by $SCCO_2$. Combined method of extraction based on pre-treatment (pressurized the plant materials for 2 hours) which is followed by ultrasonic extraction of St John's Wort with methanol as solvent indicated that extracts so obtained contain higher content of hyperforin, adhyperforin and hypericin than those obtained by ultrasonic extraction without pre-treatment.

# 3. Mathematical Modelling

Mathematical modelling can provide important process data for optimization of supercritical fluid extraction process. The model with high accuracy and good description of underlying physical phenomena can be used for determination of the process design parameters: optimum operating conditions like temperature and pressure, equipment design and sizing, optimum solvent flow rate and optimum size of solid particles. Also, mathematical modeling makes possible estimation of the viability of SFE processes on industrial scale, through the simulation of overall extraction curves (OECs). The optimization of operating conditions used in supercritical fluid extraction can make this technique economically viable in spite of high operating pressures. This can be achieved by minimization of the total cost through the application of optimum operational conditions for each type of extraction method performed in a single unit. Genetic algorithm (GA) is the optimization technique developed by Holland [25] and applied in the optimization problems for both its accuracy and global searching ability.

Over the last 15 years several mathematical models of supercritical extraction using carbon dioxide were developed and published [26-32]. Meireles and co-workers [26] applied a mathematical modelling on the supercritical extraction from the vetiver root. The influences on the extraction yield were investigated for the following process parameters: the supercritical fluid flow rate, the ratio of extractor diameter to extractor length, the vetiver root particle diameter and the bed void fraction. Hatami and co-authors [27] applied a mathematical modelling study for the extraction of oil from clove buds using supercritical $CO_2$. They identified the optimum temperature and pressure that maximized the extraction yield and reported that the mass of extract increases with an increase in the fluid flow rate and decreases with an increase in particle diameter. Deterministic models based on type and structure of plant material was developed by Zizovic et al. [29,30]. Zahedi et al. [31] studied modelling and optimization of the extraction of nimbin from neem seeds using supercritical carbon dioxide with methanol as co-solvent. A novel correlation for Sherwood number as a function of Reynolds number and Schmidt number was developed using a Genetic Algorithm (GA) technique. Hatami et al. [28] extracted the volatile oil from khoa using pressurized $CO_2$ with and without ethanol as co-solvent. Kinetic experiments were performed at pressures of 6.5 and 7 MPa and temperature of 294.15 K. Based on the literature survey, no work has been reported on the optimization of St. John's Wort extraction with supercritical carbon dioxide.

Two partial differential equations were obtained for solute concentration: one for fluid and one for solid phase. The lumped-system assumption has been applied for the particles. The only adjustable parameter in the model is the equilibrium constant of the extract between the fluid and solid phases, which was obtained by genetic algorithm. The complete model and solving procedure are described in detail in previously published paper [33].

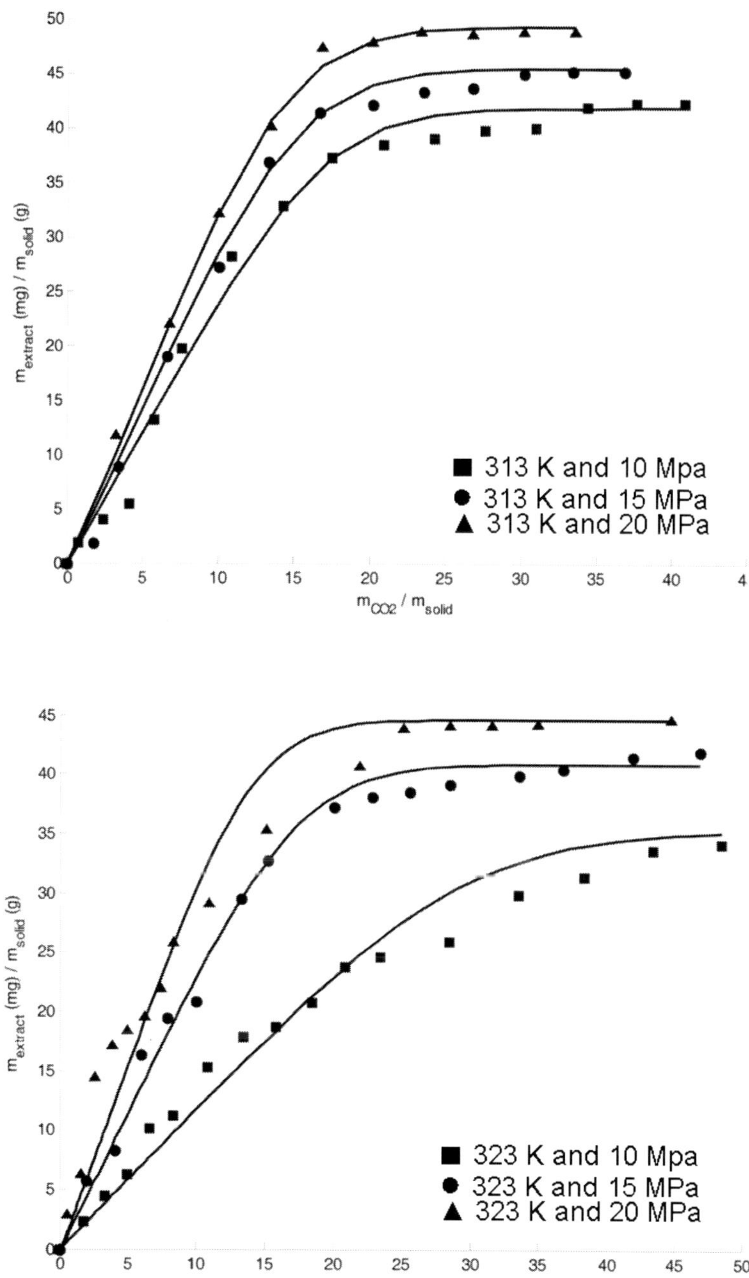


Figure 5. Comparison between model results (solid line) and experimental data of SFE from St. John's Wort [33].

The solving of proposed model and optimisation is based on Genetic Algorithm optimisation technique by using solubility information and the simulation results showed excellent agreement between model and experimental data [33] (Figure 5). The results of model prediction were compared with the experimental data for each of the six performed experiments. The results are presented in Figure 5.

These results show that good agreement was obtained between the modelling and experimental data at the initial time of extraction, but the model prediction deviated from the experimental data at higher moderate or higher extraction time. The reason for this discrepancy is that the concentration profile of the extract in the St. John's Wort seeds is not flat at higher extraction time, as it is at the lower extraction time. As a result, the assumption of the lumped system, which is employed for the particles here, is more reliable at the lower extraction time than at the higher extraction time [33].

# 4. Process Optimization

A mathematical model based on mass conservation has been employed [33]. The main process conditions that have to be determined in order to maximize the amount of extract are: particle diameter, solvent flow rate, temperature and pressure. The optimal particle size should be as small as possible, but not to the extent that it leads to sticking of small particles with consequent solvent channelling in the extractor and excessive pressure drop across the bed. The optimal solvent flow rate should be as large as possible, but keeping a minimum necessary solvent residence time so that the solvent can interact with the raw material; the channelling problem also has to be considered. Regarding the optimum temperature and pressure, the problem is more complex because both the equilibrium constant and the amount of extractable material change with the system operating conditions [33].

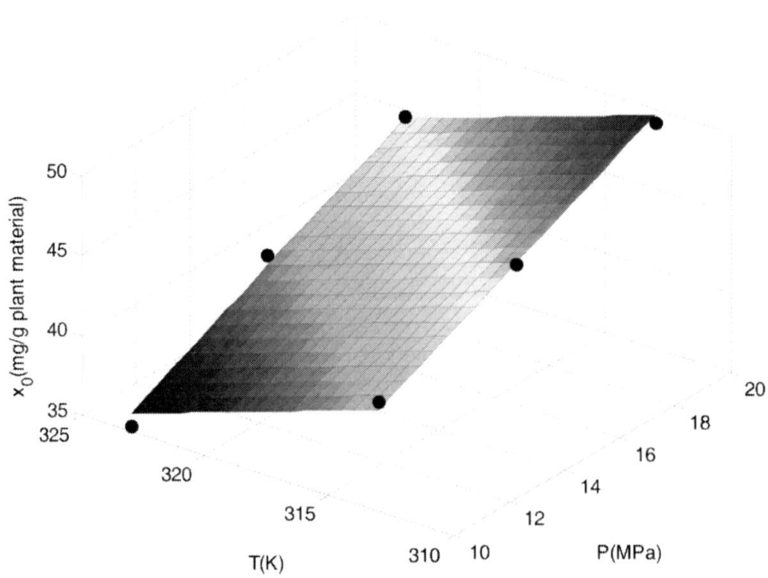


Figure 6. Influence of temperature and pressure on the extract yield [33].

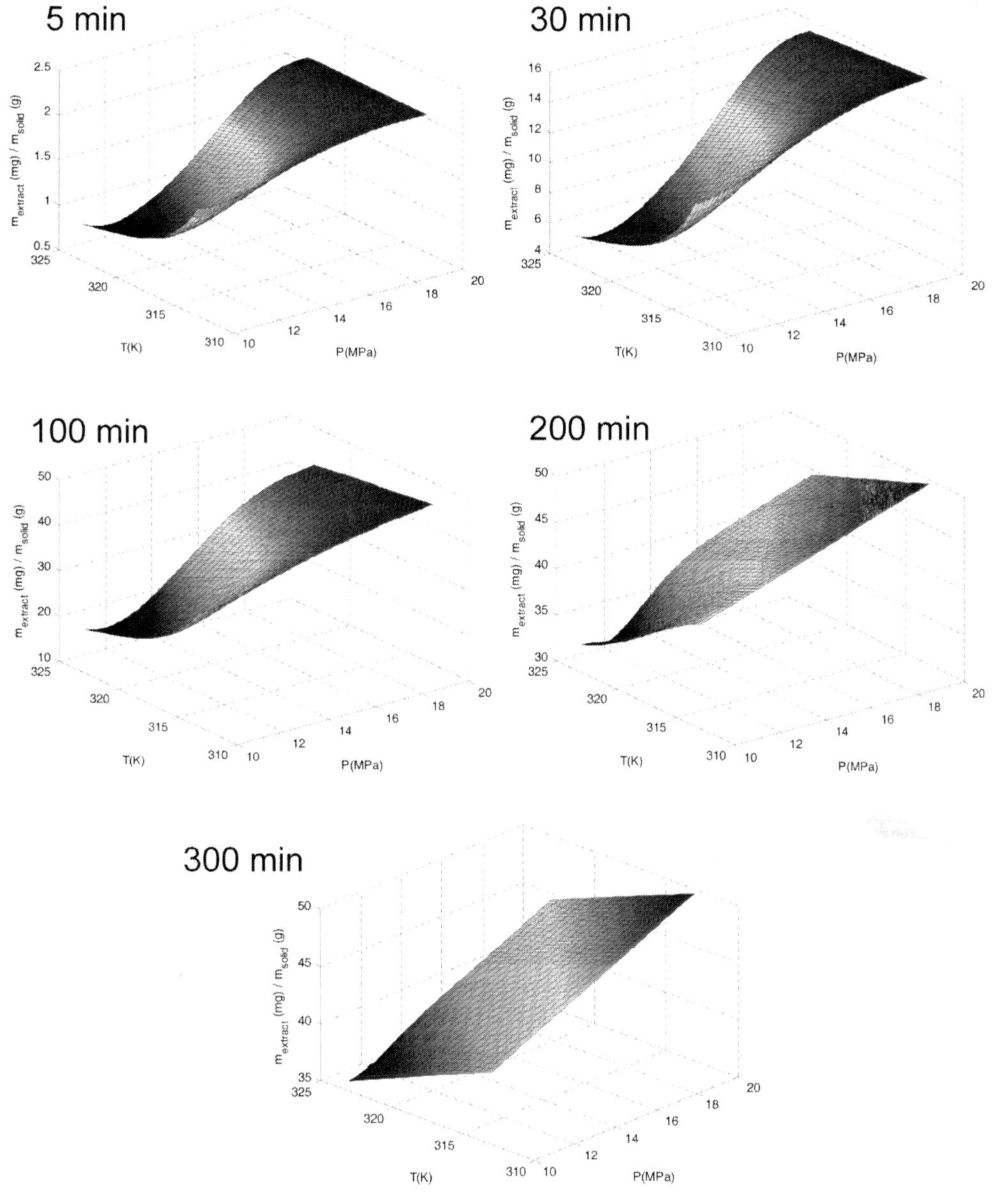


Figure 7. Influence of temperature and pressure on the extract yield (mg of extract per g of plant material) during the extraction time of 5. 30. 100. 200 and 300 min (Q=5 g/min) [33].

The influence of temperature and pressure on the amount of obtained extract is shown in Figure 6. The amount of extract is strongly correlated to equilibrium constant ($K$) and initial mass fraction of extractable oil in solid phase ($x_0$).

Actually, $K$ is the dynamic parameter and indicates the kinetics of extraction, and $x_0$ is the static parameter and indicates the mass of extract at final extraction time. At initial time of extraction $K$ is the key parameter and higher values of $K$ lead to lower extraction yield. However, $x_0$ is the key parameter at final extraction time and higher value of $x_0$ leads to higher extraction yield. From the Figure 5 it is clear visible that the amount of extract increases with pressure and decreases with temperature.

In particular, the amount of extract is a strong function of temperature in the low pressure region, and it is a weak function of temperature at high pressure. The amount of extract is strongly correlated to equilibrium constant ($K$) and initial mass fraction of extractable oil in solid phase ($x_0$). Actually, $K$ is the dynamic parameter and indicates the kinetics of extraction, and $x_0$ is the static parameter and indicates the mass of extract at final extraction time. At initial time of extraction $K$ is the key parameter and higher values of $K$ lead to lower extraction yield. However, $x_0$ is the key parameter at final extraction time and higher value of $x_0$ leads to higher extraction yield. Furtermore, the influence of temperature and pressure on the amount of obtained extract is shown in Figure 7 for different values of extraction time. At the beginning of the extraction the same amount of extract could be obtained at both analyzed temperatures (313 and 323 K) but very high extraction pressure is required. As the extraction process is approaching its final stage the influence of density is becoming more important (capability of the dense $CO_2$ to dissolve less soluble, "heavier", components) and the high amount of final extract is obtained at lower temperature. Thus, the optimal conditions of St. John's Wort supercritical extraction, as determined by the genetic algorithm within investigated range of operating conditions, are temperature of 313 K and pressure of 20 MPa.

Moreover, the procedure for SFE could be probably optimized regarding $CO_2$ consumption taking into account recently published data related to the $SCCO_2$ extraction efficiency and its dependence on the plant micro-structure [29,30]. Namely, SFE can be performed following proposed procedure of operation: the first 1 hour of operation the plant material could be kept under the pressurized conditions and without flowing of $SCCO_2$ through an extractor; after 1 hour from the beginning of extraction the continuous flow of $SCCO_2$ through the extractor can be established. The proposed time of 1 hour is necessary for achieving complete penetration of $CO_2$ inside the plant tissue structure when plant material is only under the pressure of $CO_2$ at certain temperature. That will result in faster extraction rate in the next period of continuous mode of extraction. Such procedure of plant material pre-treatment and its advantage is proven in the case of ultrasonic extraction to.

# 5. Antimicrobial Activity

## 5.1. Short Review

Several antifungal, antibiotic, antiviral and anticancer components are isolated from the *Hypericum* species [3,4]. Most of these bioactive components have phenolic compaunds with the base fluoroglycinol structure. The most important fluoroglycinol compaunds isolated from the *Hypericum perforatum* are hyperforin and adhyperforin [3,4]. Fluoroglycinol derivates in the lipophilic fraction poses antifungal and antimicrobial activity against the *Staphylococcus aureus*, *Bacillus cereus*, *Bacillus subtilis* and *Nocardia gardenen*. Moreover the other components, benzopiranes, ksantones, flavonoides and tanines, founded in the *Hypericum perforatum* extract also show the antimicrobial activity against strains of bacteria and fungus [3,4].

Hyperforin demonstrate an excellent effect towards strains of methicillin-restistant and penicillin-restistant *Staphylococcus aureus* (MRSA and PRSA) [3,4] with a MIC value of 0.1 to 1.0 µg/ml. A butanol fraction of St. John's Wort revealed anti-Helicobacter pylori activity with MIC values ranging between 15.6 and 31.2 µg/ml. The hypericins are the most interesting compounds of *Hypericum perforatum* and many pharmacological effects have been described for this class of compounds. Hypericin and pseudohypericin specifically inhibit protein kinase C with IC50 values of 1.7mgL−1 and 15mgL−1, respectively, and show antiproliferative activity against mammalian cells. This explains the antiretroviral activity of Hypericin and pseudohypericin, which can be attributed to the inhibition of phosphorylation by protein kinase C during the viral infection of cells. Hypericin was also found to have potential for the treatment of T cell-mediated diseases due to its inhibitory effect on the CD8+ T cell-mediated cytotoxicity reaction [3,4]. Hypericin is the major photosensitizing constituent of *Hypericum perforatum* (causing hypericism) and has been proposed as a photosensitizer for photodynamic cancer therapy. Pseudohypericin has no phototoxicity. Several studies have revealed antidepressant, antiviral, wound healing, antioxidant and antimicrobial activity of various extracts of *Hypericum perforatum* and all data are summarised in our previously published papers [3,4]. Antiviral and anticancer activity of this extract could be prescribed to hypericin which also poses strong anksiolitic and antidepresiv activity. Hypericin shows toxic activity to human carcinomic cells [3]. The mechanism of antimicrobial action of terpeness and other phytochemicals is not fully understood but it is speculated to involve membrane disruption by the lipophilic compounds. All the bacterial strains demonstrated some degree of sensitivity of the plant volatile oils, although the growth of a number of bacteria was uninhibited by specific volatile oils. There are many views which state that every plant volatile oil shows antimicrobial activity on some type of microorganisms, Gram-positive bacteria, Gram-negative bacteria, fungi or yeast. The activity of the oils would be expected to relate to the respective composition of the plant volatile oils, the structural configuration of the constituent components of the volatile oils and their functional groups and possible synergistic interactions between the components [35]. Investigations of the effects of terpenoids upon isolated bacterial membranes suggest that their activity is a function of the lipophilic properties of the constituent terpenes, the potency of their functional groups and their aqueous solubility. Their site of action appeared to be at the phospholipid bilayer, caused by a biochemical mechanism catalyzed by the phospholipid bilayers of the cell. These processes include the inhibition of electron transport, protein translocation, phosphorylation steps and other enzyme-dependent reactions. Although a similar tendency of water solubility was observed, specific statements on the action of single terpenoids in vivo have to be assessed individually, taking into account not only the structure of the terpenoid, but also the chemical composition of the cell wall. The plant extracts clearly demonstrate antibacterial properties, although the mechanistic processes are poorly understood.

## 5.2. Antimicrobial Activity of Obtained $SCCO_2$ Extract and Fractions

The in vitro antimicrobial activities of the different extracts were tested against a panel of laboratory control strains as: gram-positiv bacteria (*Staphylococcus aureus* ATCC 6538 P,

*Enterococcus faecalis* ATCC 29212, *Bacillus subtilis* ATCC 6633 BB, *Bacillus cereus* ATCC 11778, *Listeria monocytogenes* ATCC 19115, *Rhodococcus equi* ATCC 6939), gram-negativ bacteria (*Escherichia coli* ATCC 25922, *Salmonella enteritidis* ATCC 13076, *Pseudomonas aeruginosa* ATCC 27853) and yeast (*Candida albicans* ATCC 10231).

Antimicrobial effects were investigated by the agar dilution method and the broth dilution method described in previously published paper [4,35]. The preparation of the investigated bacterial suspension was performed according to prescribed references by NCCLS for bacterial sensitivity to antibiotics investigation [36].

Results of the antimicrobial activity obtained by the agar dilution method and broth dilution method were identical and they are presented in Tables 8, 9, 10 and 11 along with Table 12 the MIC values for referent antibiotics and medicaments.

The all tested samples obtained by supercritical extraction with $CO_2$ (Table 5) show very strong antimicrobial activity against Gram-positive microorganism with the MIC value around 0.04 mg/ml to 1.28 mg/ml.

**Table 8. MIC values and zone of inhibition for the samples obtained bu $SCCO_2$ extraction (samples described in Table 5 along with content of hyperforin and adhyperforin)**

| Microorganism | 1 | 2 | 3 | 4 | 5 | 6 |
|---|---|---|---|---|---|---|
| | MIC μg/ml | MIC μg/ml | MIC μg/ml | MIC μg/ml | MIC μg/ml | MIC μg/ml |
| Gram-positive bacteria | | | | | | |
| *Staphylococcus aureus* ATCC 6538 P | 320 | 40 | 640 | 40 | 160 | 40 |
| *Enterococcus faecalis* ATCC 29212 | 320 | 320 | 1280 | 1280 | 640 | 640 |
| *Bacillus subtilis* ATCC 6633 BB | 320 | 1280 | 1280 | >1280 | >1280 | 1280 |
| *Bacillus cereus* ATCC 11778 | 320 | 1280 | 640 | 640 | 1280 | 640 |
| *Listeria monocytogenes* ATCC 19115 | 320 | 160 | 640 | 40 | 40 | 40 |
| *Rhodococcus equi* ATCC 6939 | 320 | 320 | 1280 | 40 | 40 | 40 |
| Gram-negative bacteria | | | | | | |
| *Escherichia coli* ATCC 25922 | >1280 | >1280 | >1280 | >1280 | >1280 | >1280 |
| *Salmonella Enteritidis* ATCC 13076 | >1280 | >1280 | >1280 | >1280 | >1280 | >1280 |
| *Pseudomonas aeruginosa* ATCC 27853 | >1280 | >1280 | >1280 | >1280 | >1280 | >1280 |
| Yest | | | | | | |
| *Candida albicans* ATCC 10231 | >1280 | >1280 | 1280 | >1280 | 1280 | 1280 |

**Table 9. MIC values and zone of inhibition for the samples obtained by fractionation with $SCCO_2$ (samples described in Table 6 along with content of hyperforin and adhyperforin)**

| Microorganism | VF1 | | VF2 | | VF3 | | VF4 | |
|---|---|---|---|---|---|---|---|---|
| | MIC µg/ml | Zone 100 µl. mm | MIC µg/ml | Zone 100 µl. mm | MIC µg/ml | Zone 100 µl. mm | MIC µg/ml | Zone 100 µl. mm |
| Gram-positiv bacteria | | | | | | | | |
| *Staphylococcus aureus* ATCC 6538 P | 640 | 32 | 320 | 40 | 320 | 36 | 640 | 30 |
| *Enterococcus faecalis* ATCC 29212 | 1280 | 30 | 1280 | 30 | 1280 | 30 | 1280 | 30 |
| *Bacillus subtilis* ATCC 6633 BB | 640 | 40 | 640 | 42 | 640 | 42 | 640 | 40 |
| *Bacillus cereus* ATCC 11778 | 320 | 24 | 320 | 26 | 320 | 24 | 320 | 24 |
| *Listeria monocytogenes* ATCC 19115 | 640 | 40 | 160 | >60 | 640 | 40 | 640 | 40 |
| *Rhodococcus equi* ATCC 6939 | 640 | 36 | 320 | 36 | 640 | 36 | 640 | 38 |
| Gram-negativ bacteria | | | | | | | | |
| *Escherichia coli* ATCC 25922 | >1280 | 0 | >1280 | 0 | >1280 | 0 | >1280 | 0 |
| *Salmonella enteritidis* ATCC 13076 | >1280 | 0 | >1280 | 0 | >1280 | 0 | >1280 | 0 |
| *Pseudomonas aeruginosa* ATCC 27853 | >1280 | 0 | >1280 | 0 | >1280 | 0 | >1280 | 0 |
| Yeast | | | | | | | | |
| *Candida albicans* ATCC 10231 | 1280 | 0 | 640 | 0 | 640 | 0 | 1280 | 0 |

The strongest inhibition is visible at *Staphylococcus aureus* ATCC 6538 P, *Listeria monocytogenes* ATCC 19115 and *Rhodococcus equi* ATCC 6939.

From the other side, no growth-inhibitory effect was seen in gram-negative bacteria or in *Candida albicans* what was also concluded in the study of the Schempp and co-authors [38] who are determined the hyperforin activity to several bacterial strains.

The content of hyperforin in tested samples is in accordanc with there antimicrobial activity (higher content shows stronger inhibition with smaller MIC value) what is also concluded in previously published paper [37,38].

The sample 6 has a 19.418 mg/g of hyperforin in extract and shows the best results agains Gram-positive organisms. The interesting observation is that the samples with the smallest content of hyperforin and adhyperforin shows stronger inhibition of *Bacillus cereus* ATCC 11778 and *Bacillus subtilis* ATCC 6633 BB. More over the antimicrobial activity of the different fractions obtained by $SCCO_2$ (Table 6) are summarised in Table 9. The same antimicrobial activity for the specific microorganisms (especially gram-positive bacteria) shows these extracts but with the lower MIC values.

That trend is in accordance with the lower content of hyperforin and adhyperforin in obtained fractions then in the $SCCO_2$ extracts. Interesting observations is that these fractions shows the inhibition of the *Candida albicans* ATCC 10231 strains. The some significant conclusion could not be derived for this observation so the further investigation of fraction composition should be performed.

**Table 10. MIC values and inhibitons zones for the ultrasound obtained extracts (Table 7)**

| Microorganism | UM1 | | UM2 | | UM3 | | Methanol |
|---|---|---|---|---|---|---|---|
| | MIC µg/ml | Zone 100 µl. mm | MIC µg/ml | Zone 100 µl. mm | MIC µg/ml | Zone 100 µl. mm | Zone 100 µl. mm |
| Gram-positive bacteria | | | | | | | |
| *Staphylococcus aureus* ATCC 6538 P | >1280 | 30 | >1280 | 30 | >128 0 | 30 | 44 |
| *Enterococcus faecalis* ATCC 29212 | >1280 | 40 | >1280 | 40 | >128 0 | 40 | >50 |
| *Bacillus subtilis* ATCC 6633 BB | 40 | 40 | 40 | 40 | 40 | 40 | 40 |
| *Bacillus cereus* ATCC 11778 | 40 | 34 | 40 | 34 | 40 | 34 | 40 |
| *Listeria monocytogenes* ATCC 19115 | 40 | 50 | 40 | 50 | 40 | 50 | >50 |
| *Rhodococcus equi* ATCC 6939 | 160 | >50 | 160 | >50 | 80 | >50 | >50 |
| Gram-negative bacteria | | | | | | | |
| *Escherichia coli* ATCC 25922 | >1280 | 0 | >1280 | 0 | >128 0 | 0 | 36 |
| *Salmonella enteritidis* ATCC 13076 | >1280 | 0 | >1280 | 0 | >128 0 | 0 | 30 |
| *Pseudomonas aeruginosa* ATCC 27853 | >1280 | 0 | >1280 | 0 | >128 0 | 0 | 20 |
| Yeast | | | | | | | |
| *Candida albicans* ATCC 10231 | 40 | 40 | 40 | 40 | 40 | 40 | 40 |

**Table 11. Inhibition zone (mm) of control samples and some referent antibiotics and medicaments**

| Microorganism | Floram fenicol 100 μl c=40 μg/ml | Nistatin 100 μl c=30 μg/ml | Gentamycin 100 μl c=40 μg/ml | Ethanol 100 μl 96 % | Methanol 100 μl 98 % | Distiled water |
|---|---|---|---|---|---|---|
| Gram-positive bacteria | | | | | | |
| *Staphylococcus aureus* ATCC 6538 P | 44 | - | - | 0 | 44 | 0 |
| *Enterococcus faecalis* ATCC 29212 | 40 | - | - | 0 | >50 | 0 |
| *Bacillus subtilis* ATCC 6633 BB | 48 | - | - | 0 | 40 | 0 |
| *Bacillus cereus* ATCC 11778 | 48 | - | - | 0 | 40 | 0 |
| *Listeria monocytogenes* ATCC 19115 | 46 | - | - | 0 | >50 | 0 |
| *Rhodococcus equi* ATCC 6939 | 42 | - | - | 0 | >50 | 0 |
| Gram-negative bacteria | | | | | | |
| *Escherichia coli* ATCC 25922 | 40 | - | - | 0 | 36 | 0 |
| *Salmonella Enteritidis* ATCC 13076 | 40 | - | - | 0 | 30 | 0 |
| *Pseudomonas aeruginosa* ATCC 27853 | 0 | - | 40 | 0 | 20 | 0 |
| Yeast | | | | | | |
| *Candida albicans* ATCC 10231 | - | 26 | - | 0 | 40 | 0 |

Regarding the antimicrobial activity of the ultrasound obtained extract of St. John's Wort using methanol as a solvent (Table 7) the zone of inhibitions and MIC values are presented in Table 10. All tested samples were used together with methanol so some inhibitory effect should be prescribed to methanol to. The antimicrobial activity of the pure solvents as well as the some referent antibiotics is shown in Table 11 and 12. The inhibition of *Candida albicans* ATCC 10231 should be full prescribed to methanol which is already known as a very good antifungicid. The same imact of methanol is also visible at *Bacillus subtilis* and *Bacillus cereus* inhibition. Gram-negative bacteria are totally tolerant to this extracts obtained by ultrasound extraction with methanol.

Fractions obtained by $SCCO_2$ (samples 2, 4, 5 and 6, Table 5 and Table8) as well as the fractions obtained by fractionation with $SCCO_2$ (Table 6 and Table 9) shows the same inhibitory effect as the referent (commercial) mediacmants, genthamycin, floramfenicol and nistatin. The most siginificant conclusion is that fractions obtained by fractionation with $SCCO_2$ (Table 6 and Table 9) are very effective against gram-negative *Pseudomonas aeruginosa* ATCC 27853 which is totally resistant to the floramfenikol as referent antibiotic.

# References


[1] Robson NKB, Hypericum botany, In: Ernst E. editor. Hypericum: The genus Hypericum, Taylor and Francis; New York. USA: 2003. pp. 1–22.

[2] St John's wort. http://en.wikipedia.org/wiki/St_John's_wort#Botanical_ description.

[3] S. Glisic, S. Popadic, D. Skala, St. John's Wort (*Hypericum perforatum* L.)-Supercritical extraction, antimicrobial and antidepressant activity of extract and their components. *Chemical Industry* 60 (2006) 61-72.

[4] S. Glisic, MSc Thesis: "Isolation and fractionation of St. John's (*Hypericum perforatum* L.) and Juniper's (*Juniper communis* L.) essential oils with supercritical $CO_2$ – Antimicrobial activity of the essential oils and their fractions", Faculty of Technology and Metallurgy, University of Belgrade, University of Belgrade, Serbia, June 2006.

[5] J. Reichling. A. Weseler. R. Saller. A current review of the antimicrobial activity of *Hypericum perforatum* L., *Pharmacopsychiatry* 34 (2001) 116-118.

[6] S.S. Chatterjee. M. Noldner. E. Koch. C. Erdelmeier. Antidepressant activity of *Hypericum perforatum* and hyperforin: the neglected possibility, *Pharmacopsychiatry* 31 (1998) 7–15.

[7] Z. Saddiqe, I. Naeem, A. Maimoona, A review of the antibacterial activity of *Hypericum perforatum* L., *Journal of Ethnopharmacology*, 131(3) (2010) 511-521.

[8] European Herb Growers Association (EUROPAM) *Production of medicinal and aromatic plants in Europe.* 2010. http://www.europam.net/index.php.

[9] S.L. Crockett. Essential Oil and Volatile Components of the Genus Hypericum (Hypericaceae), *Natural Product Communications* 5 (9)(2010) 1493–1506.

[10] S. Glisic. A. Smelcerovic. S. Zuehlke. M. Spiteller. D. Skala. Extraction of hyperforin and adhyperforin from St. John's Wort (*Hypericum perforatum* L.) by supercritical carbon dioxide. *The Journal of Supercritical Fluids* 45 (2008) 332-337.

[11] H. Römpp. C. Seger. C.S. Kaiser. E. Haslinger. P.C. Schmidt. Enrichment of hyperforin from St. John's Wort (*Hypericum perforatum* L.). by pilot scale supercritical carbon dioxide extraction. *Eur. J. Pharm. Sci.* 21 (2004) 443–451.

[12] Y. Cui, C.Y.W. Ang, Supercritical fluid extraction and high performance liquid chromatographic determination of phloroglucinols from St. John's Wort (*Hypericum perforatum* L.), *Journal of Agricultural and Food Chemistry* 50 (2002) 2755–2757.

[13] M. Mannila, K. Hakwon. C. Isaacson, C.M. Wai, Optimization of supercritical fluid extraction for the separation of hyperforin and adhyperforin in St. John's wort (*Hypericum perforatum* L.). *Green Chemistry* 4 (2002) 331-336.

[14] O.J. Catchpole, N.B. Perry, B.M.T. da Silva, J.B. Grey, B.M. Smallfield, Supercritical extraction of herbs I: Saw Palmetto, St John's Wort, Kava Root and Echinacea, *Journal of Supercritical Fluids* 22 (2002) 129-138.

[15] C. Seger. H. Römpp. S. Sturm. E. Haslinger. P. C. Schmidt. F. Hadacek. Characterization of supercritical fluid extracts of St. John's Wort (*Hypericum perforatum* L.) by HPLC–MS and GC–MS, *European Journal of Pharmaceutical Sciences*, 21 (2004) 453.

[16] Smelcerovic, Z. Lepojevic, S. Djordjevic, Sub-and supercritical $CO_2$-extraction of *Hypericum perforatum* L. *Chemical Engineering and Technology*, 27 (2004) 1327-1329.

[17] E. Reverchon, I. De Marco, Review: Supercritical fluid extraction and fractionation of natural matter, *Journal of Supercritical Fluids* 38 (2006) 146–166.

[18] E. Dimitrieska-Stojkovic, Z. Zdravkovski, Supercritical fluid extraction of quercetin and rutin from *Hyperici* herba. *Journal of Liquid Chromatography and Related Technologies* 26 (2003) 2517-2533.

[19] S. Glisic, D. Mišić, M. Stamenić, I. Zizovic, R. Ašanin, D. Skala, Supercritical Carbon Dioxide Extraction of Carrot Fruit Essential Oil – Chemical Composition and Antimicrobial Activity, *Food Chemistry*, 105 (1) 2007, 346-352.

[20] S. Glisic, J. Ivanovic, M. Ristic, D. Skala, Extraction of Sage *(Salvia officinalis L.)* by Supercritical $CO_2$: Kinetic Data, Chemical Composition and Selectivity of Diterpenes, *The Journal of Supercritical Fluids*, 52 (2010) 62–70.

[21] S. Glisic, M. Ristic, D. Skala, The combined extraction of sage (*Salvia officinalis* L.): ultrasound followed by supercritical $CO_2$ extraction, *Ultrasonics Sonochemistry* 18 (2011) 318-326.

[22] S. Glisic, A. Šmelcerović, S. Zuehlke, M. Spiteller, D. Skala, Supercritical carbon dioxide extraction and fractionation of hyperforin and adhyperforin from St. John's Wort *(Hypericum perforatum L.)*, 17th International Congress of Chemical and Process Engineering - CHISA 2006, Praha, Czech Republic, Summaries 2, P3.053, 488. Full Text CD Rom (6 pages).

[23] Žižović, M. Stamenić, A. Orlović, D. Skala, Supercritical carbon dioxide extraction of essential oils from plants with secretory ducts: Mathematical modelling on the micro-scale. *Journal of Supercritical Fluids* 39 (2007) 338-346.

[24] S. Glisic, M. Ristic, D. Skala, The combined extraction of sage (*Salvia officinalis* L.): ultrasound followed by supercritical $CO_2$ extraction, *Ultrasonics Sonochemistry* 18 (1) (2011) 318-326.

[25] J.H. Holland. Adaptation in Natural and Artifical Systems. University of Michigan press. *Ann Arbor* (1975).

[26] A.A. Meireles. G. Zahedi. T. Hatami. Mathematical modeling of supercritical fluid extraction for obtaining extracts from vetiver root. *The Journal of Supercritical Fluids* 49 (2009) 23-31.

[27] T. Hatami. M.A.A. Meireles. G. Zahedi. Mathematical modeling and genetic algorithm optimization of clove oil extraction with supercritical carbon dioxide. *The Journal of Supercritical Fluids* 51 (2010) 331-338.

[28] T. Hatami. M. Rahimi. P.C. Veggi. R. Portillo-Prieto. M.A.A. Meireles. Near-critical carbon dioxide extraction of khoa (Satureja boliviana Benth Briq) using ethanol as a co-solvent: Experiment and modelling. *The Journal of Supercritical Fluids* 55 (2011) 929-936.

[29] Zizovic. M. Stamenić. A. Orlović. D. Skala. Supercritical carbon dioxide essential oil extraction of Lamiaceae family species: Mathematical modelling on the micro-scale and process optimization. *Chemical Engineering Science* 60 (2005) 6747-6756.

[30] Zizovic. M. Stamenić. A. Orlović. D. Skala. Mathematical modelling of essential oil SFE on the micro-scale-Classification of plant material. *The Journal of Supercritical Fluids* 46 (2008) 285-292.

[31] G. Zahedi. A. Elkamel. A. Lohi. T. Hatami. Optimization of supercritical extraction of nimbin from neem seeds in presence of methanol as co-solvent. *The Journal of Supercritical Fluids* 55 (2010) 142-148.

[32] G. Brunner. Gas Extraction: An Introduction to Fundamentals of Supercritical Fluid and Applications to Separation Processes. Steinkopff Darmstadt Springer. New York. 1994.

[33] T. Hatami, S.B. Glisic, A.M. Orlovic, Modelling and optimization of supercritical $CO_2$ extraction of St. John's Wort (*Hypericum perforatum* L.) using genetic algorithm, *Journal of Supercritical Fluids*, 62 (2012) 102-108.

[34] S.B. Glišić. D.U. Skala. Phase transition at subcritical and supercritical conditions of triglycerides methanolysis. *The Journal of Supercritical Fluids* 54 (2010) 71-80.

[35] The National Committee for Clinical Laboratory Standards (NCLLS). (2003). Methods for dilution antimicrobial susceptibility test for bacteria that grow aerobically (6th ed.) Wayne, Pennsylvania, USA, vol. 23, pp. 9–18.

[36] S. Glisic, S. Milojević, S. Dimitrijević, A. Orlović, D. Skala, Antimicrobial activity of the essential oil and different fractions of *Juniperus communis* L. and a comparison with some commercial antibiotics, *Journal of the Serbian Chemical Society*, 72 (4) 2007, 311-320.

[37] Reichling, A. Weseler, R. Saller, A current review of the antimicrobial activity of. Hypericum perforatum L. *Pharmacopsychiatry* 34 (2001) S116-118.

[38] C. M. Schempp, K. Pelz, A. Wittmer, E. Schopf, J. C. Simon, Antibacterial activity of hyperforin from St. John_s wort, against multiresistant *Staphylococcus aureus* and Gram-positive bacteria, *Lancet*. 353 (1999) 2129-2135.

In: Phytochemicals
Editor: Hai-Xue Kuang

ISBN: 978-1-62417-354-7


Chapter VIII

# Natural Antioxidants in Food

***Gui-Fang Deng*[1], *Xiang-Rong Xu*[2], *Sha Li*[1], *Fang Li*[1], *Shan Wu*[1] *and Hua-Bin Li*[1,*]**

[1]Guangdong Provincial Key Laboratory of Food, Nutrition and Health, School of Public Health, Sun Yat-Sen University, Guangzhou, China

[2]Key Laboratory of Marine Bio-resources Sustainable Utilization, South China Sea. Institute of Oceanology, Chinese Academy of Sciences, Guangzhou, China

## Abstract

The oxidative stress imposed by reactive oxygen species plays an important role in many chronic and degenerative diseases, such as cardiovascular diseases, cancer, diabetes mellitus, ageing and neurodegenerative diseases. The scavenging of reactive oxygen species is thought to be an effective measure to depress the level of oxidative stress of organism for prevention and treatment of some chronic and degenerative diseases. It has been reported that intake of vegetables and fruits was inversely associated with the risk of many chronic diseases, such as cardiovascular diseases and cancer.

Natural antioxidants in vegetables and fruits are considered to be responsible for these health benefits. In recent years, there is considerable interest in studying antioxidant activities of food. In this chapter, natural antioxidants in food are reviewed, and special attention is paid to antioxidant activities of vegetable, fruit and spice, including extraction methods of antioxidants, evaluation methods of antioxidant activity, antioxidant capacities and total phenolic contents, separation and identification of antioxidants as well as structure-activity relationship of antioxidant. The review provides an account on our research work combined with a reference of the literature published in the past ten years.

* Corresponding author. Tel.: +86-20-87332391; fax: +86-20-87330446. Email address: lihuabin@mail. sysu.edu.cn (H.-B. Li).

# Introduction

Reactive oxygen species (ROS) are constantly generated in the body as a result of oxidative metabolism. Certain disease, alcohol, smoking, environmental poisons and ionizing radiation promote the generation of free radicals. When increased ROS are produced and go beyond the scavenging capacity of endogenous antioxidant defence system, oxidative stress occurs [1]. This imbalance leads to oxidative damage to various biological macromolecules including proteins, lipids, lipoproteins and DNA. Compelling evidence has emerged in the last two decades demonstrating that oxidative damage is closely involved in the pathophysiology of many chronic and degenerative diseases, such as cardiovascular disease, cancer, diabetes, neurodegenerative disease and ageing [2-6]. In addition to endogenous antioxidant defence, dietary components may also help counteract oxidation stress either by providing redox active compounds that can directly scavenge or neutralize ROS, or by providing compounds that can induce the gene expression of the endogenous enzymatic antioxidants (e.g., superoxide dismutase, glutathione peroxidase, and catalase) [7,8]. Epidemiological studies have established a positive correlation between consumption of fruits and vegetables and low risk of chronic diseases like cancer, cardiovascular disease, and diabetes. Natural antioxidants in fruits and vegetables, such as vitamins and polyphenols, are considered to be responsible for these health benefits [9-13]. The increasing attention has been paid to identify safe and inexpensive antioxidants of natural origin due to potential health risks, toxicity and carcinogenicity of some synthetic antioxidants, such as butylated hydroxytoluene and butylated hydroxyanisole [14, 15].

In recent years, evaluation on antioxidant activities of food have increased remarkably due to increased interest in their potential of being used as a rich and natural source of antioxidant compounds [16-21]. On the basis of the high oxidative radical absorbance capacity (ORAC) values and antioxidant benefits, various products are marketed, including whole foods and beverages as well as isolated substances primarily as dietary supplements or added to foods [22]. In this chapter, natural antioxidants in food are reviewed, and special attention is paid to antioxidant activities of vegetable, fruit and spice, including extraction methods of antioxidants, evaluation methods of antioxidant activity, antioxidant capacities and total phenolic contents, separation and identification of antioxidants as well as structure-activity relationship of antioxidant.

# Extraction Methods of Antioxidants

The extraction of antioxidants from food is the first step in the evaluation of antioxidant activities. Antioxidants can be extracted from fresh, frozen or dried food samples. Before extraction, food samples are usually treated by milling and homogenization, and may be preceded by air-drying or freeze-drying. So far as we know, several extraction methods, including solvent extraction, supercritical fluid extraction, subcritical water extraction, enzyme extraction, microwave-assisted extraction and ultrasonic-assisted extraction have been developed for the extraction of antioxidants from food.

Solvent extraction is a conventional technique and has been used to extract antioxidants from different materials such as fruits, vegetables, legumes, spices, and other foodstuffs. It is applied in different solvent systems with respect to organic solvents (such as acetone, methanol and ethanol) and water in certain proportion as extraction solvents [23-25]. The yield of solvent extraction depends on the type of solvents with varying polarities, extraction time and temperature, sample-to-solvent ratio as well as on the chemical composition and physical characteristics of the samples [25]. Among extraction solvents, methanol is usually the most efficient, but also toxic [26, 27]. Additionally, the nature of both food material and antioxidant should be considered in order to obtain high extraction efficiency. Lipophilicity or hydrophilicity of antioxidant affects its solubility in the extraction solvent [28].

Supercritical fluid extraction has been extensively applied in the extraction of natural antioxidants. Usually, carbon dioxide ($CO_2$) is used as an extraction solvent of this extraction technique. Generally, organic solvents are added as modifiers during the extraction of ionic or polar compounds. Alternatively, the compounds are first derivatized to decrease their polarities [29-31]. Extraction with supercritical $CO_2$ could be used as an alternative method to replace the non-polar organic solvents and results in solvent-free products and reduces degradation processes, due to low process temperatures [32].

In enzyme extraction, enzymatic treatment was used to degrade cell walls components of plant food material and promote phenolic compounds release [33-35]. The main limitations for the application of enzyme extraction in industrial application are the high cost of the enzymes, low conversion rate and time-consuming. Additionally, microwave-assisted and ultrasound-assisted extractions have been alternative methods for the extraction of bioactive components from natural sources. The main advantages of both techniques are better extraction efficiency and the reduction in extraction time, solvent volume and energy [36,37].

## Evaluation Methods of Antioxidant Activities

Antioxidants in food include several classes: (1) simple phenols and phenolic acids, (2) flavonoids, and (3) others such as ascorbic acid, α-tocopherol, β-carotene, and lutein. Each class of antioxidants has its own unique chemical properties [38]. Because natural antioxidants are often multifunctional, a reliable antioxidant protocol requires the measurement of more than one property relevant to foods. Inconsistent results have been obtained because non-specific one-dimensional methods were used to evaluate natural antioxidant activities [39]. Therefore, it is necessary to perform more than one type of antioxidant capacity measurement to take into account the various mechanisms of antioxidant action.

Many methods have been developed to evaluate antioxidant activity. Depending on the chemical reaction involved, antioxidant capacity assays can be mainly divided in two categories [40]. The first category is hydrogen atom transfer (HAT) based methods which measure the ability of an antioxidant to quench free radicals by hydrogen donation, including oxygen radical absorbance capacity (ORAC), total radical-trapping antioxidant parameter (TRAP), and inhibition of auto-oxidation of induced low-density lipoprotein (LDL). The second category is electron transfer (ET) methods to measure the ability of a potential antioxidant to transfer electrons to reduce radicals, metals or carbonyls, including 2,2'-azinobis-(3-ethylbenzothiazoline-6-sulfonic acid) (ABTS), ferric reducing antioxidant power

(FRAP), 2,2'-diphenyl-l-picrylhydrazyl (DPPH), and cupric reducing antioxidant capacity (CUPRAC) methods.

In spite of the diversity of methods, there is no universal method presently able to measure the total antioxidant capacity of a complex food system accurately and quantitatively and a comparative evaluation of antioxidant efficacy is also difficult to perform because the antioxidant activity are influenced by notable factors including the substrate, the reaction medium, the oxidation conditions, interfacial phenomena and the antioxidant partitioning properties between phases [39, 41]. However, there is a great need to standardize antioxidant testing protocol to minimize the present chaos in the methodologies used to evaluate antioxidant activities [42]. Three methods including the Folin-Ciocalteu assay, the TEAC assay, and the ORAC assay, could be considered for standardization. The Folin-Ciocalteu assay is based on electron transfer and gives reducing ability, which has usually been expressed as phenolic contents [28]. The TEAC assay and the ORAC assay take into consideration the major redox reactions that commonly happen in human body [43]. Additionally, there are some assays developed to evaluate the scavenging capacity to other ROS causing oxidative damage in the human body such as $O_2$ •– scavenging capacity assay, $H_2O_2$ scavenging capacity assay, hydroxyl radical scavenging assay, singlet oxygen scavenging capacity assay, and peroxynitrite scavenging capacity assay, etc [40].

# Antioxidant Capacities and Total Phenolic Contents of Food

A literature search revealed that the number of publications on antioxidants and oxidative stress has nearly quadrupled in the past decades [40]. It is of considerable interest to the general public as well as food science researchers to know the antioxidant activities and constituents in the foods we consume. Mass of researches were carried out to provide data regarding the antioxidant activities of plant food materials such as fruits, vegetables, spices, grains, and nuts, etc [1, 16, 44-47], and their positive function to human health and important role as primary dietary sources of natural antioxidants were confirmed.

## Antioxidant Capacities and Total Phenolic Contents of Fruits

Antioxidant capacities and total phenolic contents of edible portions of 62 fruits marketed in China were systematically evaluated by ferric reducing antioxidant power (FRAP) and Trolox equivalent antioxidant capacity (TEAC) as well as the Folin-Ciocalteu assays, respectively [47]. Generally, high antioxidant capacities were detected in those fruits, the values of which ranged from 0.11 to 72.11 μmol Fe (II)/g fresh weight (FW) and 0.84 to 80.68 μmol Trolox/g FW. Additionally, high total phenolic contents were also found in those fruits analyzed, the values of which ranged from 11.88 to 585.52 mg gallic acid equivalent (GAE)/100g FW. Seven fruits, Chinese date, pomegranate, guava, sweetsop, persimmon, Chinese wampee and plum, possessed the highest antioxidant capacities and total phenolic contents and could be potential dietary sources of natural antioxidants for prevention of oxidative stress-related diseases. A very weak correlation ($R^2$= 0.0337) between the FRAP value and TEAC value suggested that the compounds capable of reducing oxidants could be different from those scavenging free radicals in these fruits. Generally, a high correlation ($R^2$ = 0.8416 to 0.9305) between antioxidant capacities and their total phenolic contents was found, which indicated phenolic compounds were a major contributor of antioxidant capacities of these fruits. Twenty-four exotic Colombian fruits were measured for antioxidant capacities and total soluble phenolic compounds (edible part, peel and seed) and ascorbic acid content (edible part) [48]. The antioxidant capacities were evaluated by ABTS and FRAP assays, respectively. The edible part of banana passion fruits (*P. mollisima and P. tarminiana*) showed the highest levels of antioxidant capacity as well as total phenolic compounds and the edible parts of guava, apple and cashew were found to possess the highest levels of ascorbic acid. Further more, the highest values of antioxidant capacity and total phenolic compounds were detected in the seeds of cashew, algarrobo, araza and coastal sapote as well as the peel of coastal sapote and algarrobo, suggesting that fruit residues principal peels and seeds could be fully utilized as source of natural antioxidants for food additives and ingredients use. This will not only help bringing about commercial benefits, but also help alleviating environmental pollution problems caused by the poor disposal of such residues.

To the Aboriginals of Australia, edible native Australian fruits have been very important sources of food and medicine for centuries [49], as well as novel rich source of antioxidant compounds. In a investigation to twelve native Australian fruits, brush cherry, Burdekin plum, Davidson's plum, Kakadu plum, Illawarra plum, Cedar Bay cherry, finger lime (red and yellow), Molucca raspberry, riberry, muntries and Tasmanian Pepper, five fruits possess stronger radical scavenging activities (3.1 to 5.2-fold in the TEAC assay and 1.2 to 4.2-fold in the PCL assay, respectively) than blueberry. Additionally, six fruits showed higher levels of phenolic compounds (2.5 to 3.9-fold in the Folin–Ciocalteu assay) than blueberry. A high correlation between total phenolics and antioxidant capacity was observed, indicating that phenolic compounds were a major contributor as a source of antioxidants in the analyzed native Australian fruits. Various phenolic compounds such as flavonoids (anthocyanins, flavan-3-ols, etc.) and phenolic (benzoic and cinnamic) acids were considered to contribute to this activity [50]. In addition, a survey to fourteen fresh fruits commonly consumed in India reported a range of 0.879-19.817 μmol Trolox/g of TEAC values as well as 26-374 mg/100 g of total phenolic contents [51].

Similar to other tropical countries, Ecuador is rich in a wide range of delicious fruits that are reaching international markets. Seventeen fruits from Ecuador were evaluated. Total soluble phenolic compounds content were measured by the Folin-Ciocalteu method and the antioxidant capacities were analyzed using DPPH, FRAP and ABTS assays, respectively. Several fruits andean blackberry, capuli cherry peel and banana passion fruit appeared the highest total phenolics with contents of 2167, 1494 and 1010 mg of GAE/100 g FW, respectively [52].

## Antioxidant Capacities and Total Phenolic Contents of Vegetables

As an important source of plant food, vegetables are commonly consumed in their fresh state or after processing and cooking with the advantage of low calories. Recently, they were hypothesized to help counteracting oxidation stress and reducing the risk of oxidative stress-related diseases, synergizing with food and other nutrient-rich plant foods [53, 54]. Worldwide interest increased in investigating the antioxidant activities and phenolic compounds of vegetables [20, 55-61].

A total of 38 commonly consumed vegetable samples grown in Colorado, including 3 broccoli, 5 carrot, 2 green bean, 3 kale, 11 potato, 1 rhubarb, 3 spinach, and 10 tomato samples, were investigated for their total phenolic contents (Folin–Ciocalteu assay) and antioxidant properties (ABTS, DPPH, ORAC assays, etc.). The total phenolic contents and the measured antioxidant properties were positively correlated to each other. Among the 38 tested vegetable samples, kale, spinach, broccoli, and rhubarb were the better dietary sources of natural antioxidants and phenolic compounds [60].

The contents of bioactive compounds and related antioxidant activity in fruits and vegetables are influenced by geographical region, climatic and storing conditions, and degree of ripeness.

A study evaluated the antioxidant and antiproliferative activities of collected vegetables including raw white cabbage (*Brassica oleracea*), white and yellow garlic (*Allium sativum*), red and green peppers (*Capsicum annuum*), and red onions (*Allium cepa*), all of which were harvested in the same year and same geographical and climatic conditions [61]. The results showed that red onion possessed the highest phenolic content (2519 ± 220 mg GAE/100g) and antioxidant activity (41.32 ± 3.9, 31.05 ± 2.7, 59.17 ± 5.2 and 58.94 ± 5.1 μmol Trolox/g for DPPH, FRAP, CUPRAC and ABTS, respectively). In another study, ten Mauritian vegetables, broccoli, cauliflower, white cabbage, Chinese cabbage, carrot, chilli pepper, lettuce, mugwort, onion, and tomato were analyzed for their total phenolic compounds, flavonoids, and proanthocyanidins and antioxidant capacity. The TEAC and FRAP values were detected at ranges of 0.43-3.68 μmol Trolox/g FW and 0.60-8.47 μmol Fe (II)/ g FW, respectively.

Levels of total phenolic compounds in the vegetables varied from 13.2 to 118.9 mg GAE/100g FW and those of total flavonoids were between 4.5 and 94.4 mg Quercetin/100g FW, while proanthocyanidins were detected at very low levels in only a few vegetables. Several vegetables possessed the highest antioxidant capacities, including Chinese cabbage, onion, mugwort and broccoli, thus represented an important source of phenolic antioxidants, with quercetin derivatives being most abundant [55].

Additionally, in a comprehensive antioxidant study on the common vegetables, a total of 927 freeze-dried vegetable samples, including 25 beets, 130 broccoli, 59 carrots, 57 cauliflower, 111 white cabbages, 50 green peppers, 33 white onions, 88 peas, 48 purple onions, 18 red peppers, 51 snap beans, 88 spinach, and 169 tomatoes, were analyzed. According to the results of the ORAC assay, green pepper, spinach, purple onion, broccoli, beet, and cauliflower were found to be the leading sources of antioxidant activities against the peroxyl radicals [62].

## Antioxidant Capacities and Total Phenolic Contents of Spices

Spices are common ingredients in cuisine and have been used as flavoring, seasoning, and coloring agents and sometimes as preservatives worldwide for thousands of years, especially in India, China, and many other southeastern Asian countries [63]. Owing to health benefits for people, spices have become an important source of traditional medicines in the eastern countries. In recent years, like fruits and vegetables, spices cultivated in the different regions of the world, have been extensively studied on their antioxidant properties and total phenolic contents, and massive reports have been published [64-71].

Antioxidant capacity and total phenolic content of 26 common spice extracts from 12 botanical families were investigated by TEAC assay and Folin-Ciocalteu method, respectively. The results showed that the ranges of TEAC and TPC values of spice extracts varied from 5.5 to 1687 μmol Trolox/g dry weight (DW) and 40 to 14380 mg of GAE/100 g DW, respectively. A highly positive linear relationship ($R^2 = 0.95$) between TEAC values and total phenolic content was calculated, indicating that phenolic compounds were the dominant antioxidant components in the tested spices. Major types of phenolic constituents identified in the spice extracts were phenolic acids, phenolic diterpenes, flavonoids, and volatile oils (e.g., aromatic compounds). Among the 26 spices analyzed, several spices, including clove, cinnamon, oregano, sage, thyme, and rosemary, exhibited stronger antioxidant capacities and higher levels of phenolics and could be used as potential sources of natural antioxidants [65]. Six spices consumed in Central Italy (the Marches region), including cumin, cardamom, coriander, fresh ginger, as well as two spice products seasoned salt 1 and seasoned salt 2, were investigated for their phenolic, flavonoid and antioxidant capacity (ORAC assay). The results showed that cumin exhibited the most significant antioxidant capacity, with values of 750 mg caffeic acid equivalents (CAE)/100 g, 740.1 mg CAE /100 g, 768 μmol Trolox/g in total phenols, flavonoids, and ORAC assays, respectively [58]. In another evaluation to five Portuguese spices, fennel, mint, pennyroyal, rosemary and wild thyme, the water extracts of mint and pennyroyal showed the highest radical-scavenging activity (DPPH test) values ($IC_{50}$ = 5.7 and 8.9 μg/ml DW respectively). The ethanol extracts of mint, wild thyme and fennel showed the highest antioxidant activities measured by the β-carotene/linoleic acid assay, $IC_{50}$ = 36.9, 41.2 and 68.7 μg/ml DW, respectively. The ethanol extracts of wild thyme and mint exhibited the highest total phenolic compounds by Folin-Ciocalteu method, with TPC values of 113 and 81.2 mg pyrogallol equivalents/g, respectively [67]. Additionally, an antioxidant measurement (FRAP assay) of 425 spices and herbs collected worldwide indicated that clove had the highest mean antioxidant value, followed by peppermint, allspice, cinnamon, oregano, thyme, sage, rosemary, saffron and estragon, with mean values ranging from 440 to 2770 μmol Fe (II)/ g DW [20].

Because plant based food such as fruits, vegetables and spices are wide accepted to be main sources of natural antioxidants, as confirmed in above researches, and common consumed in the diet worldwide, thus in this chapter antioxidant activities and total phenolic contents of plant food including vegetable, fruit and spice were detailedly introduced. In addition to the food introduced above, there are some other food materials got their antioxidant capacities and phenolic contents analyzed, such as cereals, legumes, mushrooms, teas and nuts, animal based food as well as mixed food [16, 20, 72-75].

It is difficult to directly compare the results from different investigations because the wide varieties of extraction and evaluation methods have been used. In this chapter, the antioxidant capacities and total phenolic contents of some fruits, vegetables as well as spices are summarized in Table 1, which were from the studies that applied the similar extraction condition and measurement methods [47, 48, 55, 65, 76, 77].

**Table 1. Antioxidant capacities and total phenolic contents of fruits, vegetables and spices**

| Plant food | Antioxidant capacities | Total phenolic contents |
|---|---|---|
| *Fruits* | *(μmol Trolox/g FW)* | *(mg GAE/ 100g FW)* |
| Abiu | 21.0 ± 2.52 | 83.0 ± 0.45 |
| Aguaje | 70.2 ± 3.62 | 281 ± 2.25 |
| Algarrobo | 26.7 ± 1.89 | 97.2 ± 2.69 |
| American oil palm | 16.4 ± 1.31 | 80.5 ± 3.44 |
| Apple (green, delicious) | 4.98 ± 0.07 | 68.29 ± 2.12 |
| Apple (organic) | 4.25 ± 0.27 | 58.12 ± 3.98 |
| Apple (red, delicious) | 4.62 ± 0.03 | 73.96 ± 3.52 |
| Apple (rose red) | 5.47 ± 0.04 | 70.57 ± 1.85 |
| Arrayana | 48.8 ± 1.40 | 187 ± 3.87 |
| Avocado | 1.16 ± 0.05 | 21.86 ± 1.25 |
| Banana (cooking) | 1.74 ± 0.13 | 29.07 ± 2.06 |
| Banana (general) | 3.44 ± 0.29 | 57.13 ± 3.64 |
| Banana (royal) | 1.95 ± 0.05 | 25.55 ± 0.60 |
| Banana passion fruit (*P. mollissima*) | 131 ± 0.64 | 635 ± 2.71 |
| Banana passion fruit (*P. tarminiana*) | 144 ± 1.88 | 1018 ± 14.5 |
| Blueberry | 1.26 ± 0.02 | 46.24 ± 2.03 |
| Borojo | 6.29 ± 0.86 | 41.8 ± 1.54 |
| Brazilian guava | 44.8 ± 1.98 | 192 ± 11.5 |
| Cantaloup | 2.56 ± 0.03 | 31.50 ± 1.29 |
| Carambola | 2.80 ± 0.04 | 36.04 ± 1.63 |
| Cashew | 115 ± 15.2 | 445 ± 15.2 |
| Cassabanana | 6.49 ± 0.47 | 15.7 ± 1.13 |
| Cherries | 5.48 ± 0.10 | 114.56 ± 4.72 |
| Cherry tomato | 3.94 ± 0.11 | 73.51 ± 2.03 |
| Chinese date | 8.33 ± 0.12 | 585.52 ± 18.59 |
| Chinese wampee | 8.19 ± 0.09 | 116.10 ± 7.48 |

| Plant food | Antioxidant capacities | Total phenolic contents |
|---|---|---|
| Citrus (green) | 3.65 ± 0.05 | 53.73 ± 0.20 |
| Citrus (South Africa) | 4.03 ± 0.06 | 63.14 ± 2.06 |
| Citrus (Thailand) | 3.71 ± 0.16 | 69.91 ± 3.21 |
| Coastal sapote | 8.56 ± 0.07 | 23.9 ± 0.09 |
| Cupuacu | 9.59 ± 0.25 | 40.3 ± 0.57 |
| Durian | 4.98 ± 0.17 | 79.15 ± 3.05 |
| Flat peach | 6.02 ± 0.06 | 74.85 ± 3.13 |
| Giant granadilla | 16.3 ± 1.39 | 70.7 ± 2.27 |
| Grape (green) | 1.27 ± 0.03 | 23.20 ± 1.58 |
| Grape (jufeng) | 3.34 ± 0.06 | 49.43 ± 0.17 |
| Grape (red) | 3.95 ± 0.11 | 80.28 ± 4.32 |
| Grape (USA) | 1.23 ± 0.11 | 19.32 ± 0.67 |
| Grapefruit | 3.04 ± 0.23 | 60.35 ± 3.68 |
| Guava | 15.18 ± 0.81 | 194.11 ± 7.01 |
| Guava apple | 50.2 ± 0.42 | 309 ± 6.81 |
| Jackfruit | 2.37 ± 0.08 | 60.96 ± 3.69 |
| Kiwifruit | 4.60 ± 0.07 | 87.54 ± 1.22 |
| Lemon | 2.54 ± 0.17 | 61.47 ± 0.57 |
| Litchi | 4.63 ± 0.01 | 59.77 ± 1.22 |
| longan | 5.88 ± 0.34 | 80.32 ± 5.38 |
| Macadamia nut | 18.6 ± 0.08 | 38.7 ± 1.24 |
| Mango (Hainan) | 3.87 ± 0.01 | 35.02 ± 1.43 |
| Mango (shuixian) | 4.01 ± 0.09 | 37.03 ± 2.06 |
| Mangosteen | 3.34 ± 0.01 | 43.68 ± 1.34 |
| Mountain papaya | 7.63 ± 1.36 | 36.8 ± 0.76 |
| Muskmelon (white pulp) | 1.32 ± 0.01 | 20.36 ± 1.09 |
| Muskmelon (yellow pulp) | 1.62 ± 0.10 | 25.06 ± 2.01 |
| Naranjilla | 12.2 ± 0.85 | 58.3 ± 2.39 |
| Navel orange (China) | 2.78 ± 0.12 | 51.09 ± 1.70 |
| Navel orange (USA) | 3.72 ± 0.01 | 61.08 ± 3.66 |

**Table 1. (Continued)**

| Plant food | Antioxidant capacities | Total phenolic contents |
|---|---|---|
| Netted melon | 2.29 ± 0.03 | 28.72 ± 1.88 |
| Olive | 80.68 ± 2.11 | 21.68 ± 1.12 |
| Orange (South Africa) | 4.90 ± 0.29 | 77.23 ± 0.08 |
| Papaya | 2.92 ± 0.06 | 47.13 ± 2.09 |
| Passionfruit | 1.83 ± 0.11 | 37.91 ± 0.14 |
| Peach | 2.38 ± 0.05 | 27.58 ± 1.57 |
| Peach tomato | 15.2 ± 0.49 | 70.9 ± 2.57 |
| Pear (Australian) | 4.30 ± 0.06 | 68.06 ± 3.45 |
| Pear (fragrant) | 1.51 ± 0.02 | 18.65 ± 0.96 |
| Pear (honey) | 0.84 ± 0.03 | 11.88 ± 0.11 |
| Pear (royal) | 3.58 ± 0.02 | 34.84 ± 1.46 |
| Pear apple | 8.92 ± 0.82 | 40.8 ± 1.74 |
| Pejibaye | 14.1 ± 0.16 | 65.7 ± 2.07 |
| Persimmon | 9.38 ± 0.25 | 112.09 ± 4.60 |
| Pineapple | 5.93 ± 0.05 | 94.04 ± 1.47 |
| Pitaya | 2.18 ± 0.13 | 27.52 ± 2.48 |
| Plum (black) | 6.48 ± 0.57 | 88.28 ± 1.88 |
| Plum (red) | 5.22 ± 0.12 | 73.04 ± 5.63 |
| Plum (sanhua) | 7.23 ± 0.02 | 102.43 ± 2.83 |
| Pomegranate | 40.61 ± 0.11 | 146.94 ± 0.04 |
| Pomelo | 3.07 ± 0.14 | 57.43 ± 2.25 |
| Prune | 3.61 ± 0.04 | 65.06 ± 0.14 |
| Sweetsop | 23.60 ± 0.06 | 405.41 ± 16.70 |
| Watermelon (jintong) | 1.69 ± 0.08 | 23.15 ± 1.47 |
| Watermelon (red pulp) | 2.64 ± 0.06 | 24.66 ± 1.04 |
| Watermelon (yellow pulp) | 0.91 ± 0.03 | 18.62 ± 1.32 |
| Wax apple | 3.52 ± 0.14 | 40.69 ± 2.65 |
| Yellow mombin | 8.60 ± 0.57 | 52.6 ±1.55 |
| *Vegetables* | *(μmol Trolox/g FW)* | *(mg GAE/ 100g FW)* |

| Plant food | Antioxidant capacities | Total phenolic contents |
|---|---|---|
| Broccoli | 2.85 ± 0.56 | 82.2 ± 8.9 |
| Capsicum | 19.9 ± 1.4 | 286.7 ± 5.1 |
| Carrot | 4.8 ± 1.1 | 35.2 ± 2.0 |
| Cauliflower | 2.09 ± 0.95 | 27.8 ± 1.5 |
| Celery leaves | 113.5 ± 6.1 | 605.6 ± 9.4 |
| Celery root | 15.3 ± 1.2 | 43.8 ± 2.3 |
| Chilli pepper | 36.1 ± 6.5 | 298.6 ± 1.8 |
| Chinese cabbage | 3.68 ± 1.10 | 118.9 ± 12.5 |
| Cucumber | 1.2 ± 0.2 | 24.2 ± 2.4 |
| Eggplant | 16.2 ± 2.0 | 102.9 ± 4.7 |
| Goathorn pepper | 30.6 ± 1.4 | 260.5 ± 5.0 |
| Green bean | 14.5 ± 1.2 | 101.5 ± 3.3 |
| Green onion | 14.7 ± 1.5 | 92.7 ± 0.9 |
| Green pepper | 5.6 ± 0.3 | 80.7 ± 3.6 |
| Gumbo | 14.6 ± 0.8 | 89.8 ± 4.0 |
| Lettuce | 1.14 ± 0.21 | 13.4 ± 2.8 |
| Lovage | 57.3 ± 5.0 | 267.0 ± 2.1 |
| Mugwort | 2.86 ± 0.86 | 95.6 ± 7.1 |
| Onion | 3.57 ± 0.30 | 101.0 ± 7.5 |
| Parsley leaves | 108.6 ± 13.1 | 599.7 ± 0.4 |
| Potato | 10.3 ± 1.3 | 65.5 ± 1.1 |
| Radish | 23.6 ± 1.7 | 89.9 ± 1.7 |
| Red beet | 12.6 ± 1.6 | 81.5 ± 1.6 |
| Red pepper | 9.3 ± 0.9 | 115.7 ± 1.5 |
| Tomato | 5.4 ± 0.3 | 41.3 ± 2.2 |
| Tomato | 1.00 ± 0.06 | 35.0 ± 1.1 |
| Vegetable marrow | 2.9 ± 0.3 | 20.0 ± 1.7 |
| White cabbage | 1.73 ± 0.06 | 15.3 ± 2.1 |
| *Spices* | *(μmol Trolox/g DW)* | *(mg GAE/ 100g DW)* |

**Table 1. (Continued)**

| Plant food | Antioxidant capacities | Total phenolic contents |
|---|---|---|
| Angelica dahurica root | 41.44 ± 2.73 | 786 ± 21 |
| Bay | 342.9 ± 0.01 | 41700 ± 5 |
| Black pepper | 45.6 ± 0.13 | 300 ± 2 |
| Caraway | 55.0 ± 0.08 | 610 ± 17 |
| Chilli | 60.5 ± 0.03 | 860 ± 4 |
| Chinese prickly ash | 369.2 ± 0.05 | 3130 ± 4 |
| Cinnamon | 617.5 ± 0.14 | 6340 ± 21 |
| Cinnamon stick | 1076.9 ± 0.10 | 11900 ± 4 |
| Clove | 1686.6 ± 0.24 | 14380 ± 6 |
| Coriander | 70.2 ± 0.04 | 880 ± 7 |
| Cumin | 66.1 ± 0.02 | 230 ± 5 |
| Dill | 63.6 ± 0.06 | 980 ± 9 |
| Dried tangerine peel | 43.00 ± 1.98 | 1004 ± 55 |
| Fennel | 55.77 ± 1.61 | 676 ± 11 |
| Galangal | 593.90 ± 15.51 | 5825 ± 189 |
| Ginger | 78.9 ± 0.09 | 630 ± 9 |
| Green cardamom | 75.3 ± 0.04 | 460 ± 9 |
| Green peppercorn | 111.5 ± 0.07 | 380 ± 3 |
| Green prickleyash | 405.01 ± 12.32 | 4637 ± 108 |
| Lemon grass | 44.1 ± 0.03 | 250 ± 9 |
| Mint | 338.3 ± 0.16 | 5150 ± 25 |
| Mustard | 39.21 ± 1.28 | 762 ± 35 |
| Nutmeg | 200.1 ± 0.17 | 1610 ± 1 |
| Oregano | 1006.7 ± 0.09 | 10170 ± 10 |
| Parsley | 63.1 ± 0.05 | 970 ± 2 |
| Poppy | 5.5 ± 0.02 | 40 ± 4 |
| Rosemary | 378.0 ± 0.21 | 5070 ± 36 |
| Sage | 518.9 ± 0.06 | 5320 ± 6 |
| Sichuan pepper | 368.47 ± 8.92 | 3886 ± 234 |

| Plant food | Antioxidant capacities | Total phenolic contents |
|---|---|---|
| Star anise | 203.0 ± 0.08 | 2020 ± 14 |
| Sweet basil | 295.9 ± 0.04 | 3640 ± 14 |
| Thyme | 380.7 ± 0.03 | 4520 ± 6 |
| Tsaoko amomum fruit | 187.83 ± 5.16 | 2209 ± 26 |
| Villous amomum fruit | 83.47 ± 2.08 | 899 ± 9 |
| White pepper | 89.7 ± 0.07 | 780 ± 4 |

# Separation and Identification of Antioxidants from Food

Various different techniques could be employed for separation and identification of antioxidants from different foods [78-83]. Considering there are various kinds of plant food and related studies, in this chapter, only fruits, vegetables and spices with strong antioxidant capacities and high levels of phenolic compounds were introduced about the isolation and identification of their predominant phenolic compounds.

## Separation and Identification of Antioxidants from Fruits

Jujube fruits are commonly consumed fresh, dried and processed (jams, loaf, cakes, jelly, etc.) throughout the world [84]. They have been used in Chinese and Korean traditional medicine for antifungal, antibacterial, antiulcer, anti-inflammatory, sedative, antispastic, cardiotonic, immunity regulating, and wound healing properties [85-87]. According to Fu et al [47], jujube possessed strong antioxidant capacities and rich total phenols. In the edible part of jujube, a level of 585.5 mg GAE/100 g total phenolic compounds was detected. Recently, the phenolic composition of four promising jujubes (*Ziziphus jujuba* Miller) were analyzed with an HPLC device equipped with a diode array detector ($\lambda_{max}$ = 278) and on an Agilent Eclipse XDB-C18 column. The phenolic compounds were eluted using 3% acetic acid (A) and 100% methanol (B) at a gradient mode with a flow rate of 0.8 mL/min. Seven phenolic compounds, catechin, caffeic acid, epicatechin, ferulic acid, rutin, *p*-hydroxybenzoic acid and chlorogenic acid, were isolated from the jujubes. Predominant phenolic compounds were catechin and rutin for the jujube fruits, the levels of which were ranged from 2.46 to 3.74 mg/100 g and 0.88 to 3.60 mg/100 g, respectively [88].

The phenolics in the peels and pulps of 6 Tunisian pomegranate ecotypes were analyzed by an HPLC-UV system on a C18 column with methanol/acetonitrile (50/50, v/v) (A) and acetic acid in ultrapure water, pH 3.2 (B) as mobile phase at a flow rate of 1.0 mL/min. Four phenolic compounds, 2 hydroxybenzoic acids (gallic and ellagic acids) and 2 hydroxycinnamic acids (caffeic and p-coumaric acids) were identified and quantified. Antioxidants capacities of pomegranate fractions (pulp, peel, and seed oil) were confirmed by FRAP and ORAC assays. The highest values were detected in peels with 256.3 μmol

Trolox/g and 220.8 μmol Trolox/g for FRAP and ORAC assays, respectively. The antioxidant capacity of pomegranate extracts was positively correlated with their phenolic compound content. In particular, the highest correlation was found in peels. High correlations were also found between peel hydroxybenzoic acids and antioxidant capacities. The results implied that the pomegranate peel might be used as potential resources of antioxidant dietary food [89].

Persimmon fruit (*Diospyros kaki L.*) contains abundant polyphenols, including condensed tannin and other phenolic compounds. Proanthocyanidin from persimmon pulp comprised a heteropolyflavanol series with flavan-3-O-galloylated extenders, flavan-3-ol and flavonol terminal units, and A-type interflavan linkages [90]. HPLC–MS analysis showed that the thiolysis degradation products of high molecular weight tannin (HMWT, >10,000 Da) consisted of (epi) gallocatechin, epigallocatechin-3-O-gallate, and epicatechin-3-O-gallate. The antioxidant capacities of persimmon tannins were evaluated with regard to their ability in scavenging hydroxyl radical (2-deoxyribose oxidation system and salicylic acid system) and superoxide anion, and inhibiting linoleic acid lipid peroxidation, respectively. HMWT exhibited excellent antioxidant activities in all tested systems in a dose-dependent manner. The antioxidant activity of HMWT was more potent than that of low molecular weight tannin (<10,000 Da) and grape seeds proanthocyanidins, suggesting that high molecular weight condensed tannins were the major antioxidant composition in persimmon pulp [91].

At present, although there are some studies on isolation and identification of bioactive compounds like phenolic compounds from sweetsop seeds, very few references involved in that of sweetsop edible part. The isolation and characterization of eleven flavan-3-ols from sweetsop pulp were reported [92]. Among them, four major flavan-3-ols were epicatechin, epicatechin-(4,8)-epicatechin-(4,8)-epicatechin, catechin-(4,8)-epicatechin, epicatechin-(4,6)-catechin.

The hydroxycinnamoylquinic acid esters were isolated from European plums (*Prunus domestica*). In particular, three main isomers of caffeoylquinic acid were neochlorogenic acid, cryptochlorogenic acid, and chlorogenic acid [93-95]. Additionally, compounds from four classes of flavonoids, anthocyanins, flavonols, flavanols, and proanthocyanidins have been identified from plums (*Prunus domestica*) cultivars. Cyanidin 3-rutinoside was the major anthocyanin, followed by peonidin 3-rutinoside, cyaniding 3-glucoside and cyanidin 3-xyloside [95, 96]. Minor amounts of flavonols (quercetin 3-rutinoside and quercetin 3-glucoside) have been characterized from several plum cultivars by HPLC-MS [93, 95]. Three dimeric flavanols, procyanidins B3, B4 and B7, have recently been isolated in some plum cultivars [97-99].

## Separation and Identification of Antioxidants from Vegetables

The polyphenols of a common red variety of curly kale (*Brassica oleracea*) were analyzed by HPLC-DAD-ESI-MS and a total of 47 different acylated and nonacylated flavonoid glycosides and complex hydroxycinnamic acids were characterized and quantified. Those compounds included mono- to tetraglycosides of quercetin, kaempferol, and cyanidin and derivatives of *p*-coumaric, ferulic, sinapic, and caffeic acid. Among the compounds characterized, several novel flavonols (quercetin-3-sinapoyl-triglucoside-7-diglucoside, kaempferol-3-feruloyl-triglucoside-7-diglucoside, kaempferol-3-feruloyl-diglucoside-7-gluco side, and kaempferol-3-sinapoyl-diglucoside-7-glucoside), anthocyanins (cyanidin-3-

triglucoside-5-glucoside, cyanidin-3-hydroxy-feruloyl-feruloyl-digluoside-5-glucoside, and cyanidin-3-diferuloyl-diglucoside-5-glucoside), and phenolic acids (feruloyl-triglucoside and diprotocatechuic acid-diglucoside) were identified in the Brassica family. The major flavonol, anthocyanin, and phenolic acid were kaempferol-3-sinapoyl-diglucoside-7-diglucoside, cyanidin-3-sinapoyl-feruloyl-diglucoside-5-glucoside, and disinapoyl-diglucoside, respecttively [100, 101]. In addition, three novel acylated isorhamnetin glycosides (isorhamnetin-3-O-sinapoyl-sophoroside-7-O-D-glucoside, isorhamnetin-3-O-feruloyl-sophoroside-7-O-diglucoside and isorhamnetin-3-O-disinapoyl-triglucoside-7-O- diglucoside) and seven novel non-acylated isorhamnetin glycosides were identified from kale (*Brassica oleracea*) by high-performance liquid chromatography diode-array detection/electrospray ionization multi-stage mass spectrometry [102].

A study analyzed the phenolic components of collard greens, kale, and Chinese broccoli. The results showed that similar phenolic component profiles were detected in the three vegetables, in which a total of 45 flavonoids and 13 hydroxycinnamic acid derivatives were identified. For each, kaempferol glycosides and acylgentiobiosides were the major phenolic compounds and most of the flavononol glycosides existed in their acylated forms. In addition, the structures caffeoyl-, p-coumaroyl-, and feruloylquinic acid monomers with a 3-position derivative as the dominant isomer were characterized in each of the vegetable [103].

Phenolic compounds of fourteen broccoli cultivars (*Brassica oleracea*) grown under uniform cultural conditions were identified and quantified by an HPLC-UV system on a LiChroCART column using water/formic acid (95:5 v/v) (A) and methanol (B) as mobile phase at a gradient mode with a flow rate of 1.0 mL/min. The predominant glucosinolates in all broccoli cultivars were 4-methylsulphinylbutyl glucosinolate and 3-indolylmethyl glucosinolate. Other glucosinolates such as 4-hydroxyl-glucobrassicin and N-methoxy-glucobrassicin were detected in lower amounts [104].

Using a high-performance liquid chromatographic method with photodiode array and mass spectrometric detection, eighteen flavonoids representing glucuronides and acylated di- and triglycosides of methylated and methylene dioxide derivatives of 6-oxygenated flavonols were isolated, identified and quantified from spinach extracts (methanol-water-formic acid)of five genotypes [105]. Previously, natural water soluble antioxidant (NAO) isolated from spinach leaves extracts (water-acetone, 1:9, v/v) were reported to be capable of preventing lipid peroxidation both in plant and animal systems [106]. Using HPLC, NMR, and CMR spectroscopy, the main components were identified as glucuronic acid derivatives of flavonoids, *trans*- and *cis*- isomers of *p*-coumaric acid and *meso*-tartarate derivatives of *p*-coumaric acid [107]. In addition, three phenolic acids were identified and quantified from the spinach (*Spinacia oleracea*) using mass spectroscopy and UV–vis spectrophotometry, including *ortho*-coumaric acid, ferulic acid and *para*-coumaric acid [108]. Compared the antioxidant capacities of glucurinated flavonoid (GF) and NAO to well-known antioxidants, the results showed that GF and NAO were more effective in inhibiting ROS formation in the auto-oxidation system of linoleic acid and in fibroblasts exposed to metal oxidation than EGCG, NAC and Trolox [109].

High-performance liquid chromatography-diode array detector coupled with electron spray mass spectrometry was used to analyze the flavonol profile in southern Italian red onions (*Allium cepa*). Seven flavonols were recognized, including two major flavonols quercetin 4-glucoside and quercetin 3, 4'-diglucoside, and several minor flavonols, quercetin 3-glucoside, quercetin 7,4'-diglucoside, quercetin 3,7,4'-triglucoside, and isorhamnetin 4'-

glucoside [110]. In another study, five quercetin derivatives and kaempferol were identified and quantified from 5 traditional onion cultivars (*Allium cepa*) from Tenerife by HPLC coupled with a diode array detector [111]. Compared with other plant food, the onion is a very important source of quercetin and its derivatives. In a survey to 28 vegetables and 9 fruits, the onion was found to rank the highest in quercetin content [112]. Amounts of this flavonoid in onions vary with bulb color and type, mostly existing in the outer skins and rings [110, 113].

## Separation and Identification of Antioxidants from Spices

Cloves have been traditionally used as a culinary spice and a flavoring agent in the manufacturing of cigarettes, as well as in traditional medicine as a carminative and antiemetic, or as a topical toothache remedy and counter-irritant [114]. Chemical composition and antioxidant properties of a commercial rectified clove leaf essential oil (*Eugenia caryophyllus*) were analyzed using GC-MS method. A total of 23 constituents were identified in this essential oil, among which the main components were found to be eugenol (76.8%), followed by α-caryophyllene (17.4%), humulene (2.1%), and eugenylacetate (1.2%). The essential oil from clove exhibited higher scavenging activity against DPPH radical than eugenol, butylated hydroxytoluene (BHT), and butylated hydroxyanisole (BHA) [115].

Acetone extract of aerial parts of Greek oregano (*Origanum vulgare*) were analyzed using an on-line LC-UV-SPE-NMR-MS system. Combining the data provided by the UV, MS, and NMR spectra, five flavonoids (taxifolin, aromadendrin, eriodictyol, naringenin, and apigenin), one phenolic acid (rosmarinic acid), and one monoterpene (carvacrol) were identified [116]. In another study, polyphenols in aqueous methanol extracts from Greek oregano (*Origanum vulgare*) were quantified by RP-HPLC-PDA and identified by LC-PDA-MS and NMR spectroscopy. Six major polyphenols were eriodictyol 6,8-di-C-glucoside, apigenin 6,8-di-C-glucoside, luteolin 7-O-glucuronide, epi-lithospermic acid B, lithospermic acid B and rosmarinic acid [117]. Flavonoids and phenolic acids of sage samples in the North of Portugal (both cultivated and commercially available) were determined using reversed-phase-HPLC-DAD system. Six phenolic compounds were identified and quantified, including caffeic acid, luteolin 7-O-glucoside, rosmarinic acid, apigenin, hispidulin, and cirsimaritin. Among those compounds, rosmarinic acid ranked the highest in the cultivated samples, while caffeic acid and luteolin 7-O-glucoside were found to be the minor compounds in all the samples, commercially available or cultivated [118].

Rosemary has been traditionally used as a culinary spice to modify or to improve food flavors, as well as in folk medicine. In deed, rosemary and sage both belong to the Labiatae family. Their chromatographic profiles appear to be close [119]. Twenty-two compounds were identified in the Labiatae family extracts (rosemary and sage) by HPLC-MS-APCI and HPLC-PDA, including phenolic acids, carnosol derivatives, and flavonoids. According to the previous studies, the presence of three phenolic compounds carnosic acid, carnosol, and rosmarinic acid contributed the most for the antioxidant capacities of rosemary and sage [120-123] and this was further confirmed by Cuvelier and his colleagues [119].

The antioxidants isolated and identified from some fruits, vegetables and spices are summarized in Table 2.

**Table 2. Antioxidants isolated and identified from fruits, vegetables and spices**

| Plant food | Antioxidant components | References |
|---|---|---|
| Blueberry | chlorogenic acid | [124] |
| | delphinidin 3-*O*-arabinoside | |
| | delphinidin 3-*O*-galactoside | |
| | delphinidin 3-*O*-glucoside | |
| | petunidin 3-*O*-galactoside | |
| | petunidin 3-*O*-glucoside | |
| Blackberry | cyanidin-3-glucoside | [125] |
| | cyanidin-3-rutinoside | |
| | cyanidin-3-malonyl glucoside | |
| | lambertianin C | |
| | sanguiin H-6, quercetin | |
| | ellagic acid, kaempferol | |
| | lambertianin A | [126] |
| | lambertianin C isomer | |
| Chinese hawthorn | chlorogenic acid | [127] |
| | cyanidin-3-*O*-galactoside | |
| | epicatechin, procyanidin B2 | |
| | quercetin-3-*O*-galactoside | |
| | quercetin-3-*O*-glucoside | |
| Cranberry | benzoic acid, myricetin | [128] |
| | quercetin | |
| Grapefruit | ferulic acid, naringin | [129] |
| | vanillic acid | |
| Mulberry | caffeic acid | [130] |
| | cyanidin 3-rutinoside | |
| | hydroxycinnamic acids | |
| | cyanidin 3-glucoside | |
| | quercetin 3-glucoside | |
| | quercetin 3-rutinoside | |
| | ellagic acid derivative | |
| Papaya | caffeic acid, rutin, ferulic acid | [131] |
| Persimmon | catechin | [90] |
| | (*epi*)gallocatechin-3-*O*- | |

**Table 2. (Continued)**

| Plant food | Antioxidant components | References |
|---|---|---|
| | gallate benzylthioether, | |
| | epicatechin benzylthioether | |
| | (*epi*)catechin-3-*O*- | |
| | gallate benzylthioether, | |
| | benzyl mercaptan | |
| | epigallocatechin benzylthioether | |
| | epigallocatechin-3-*O*-gallate | |
| | myricetin, ursolic acid | |
| | oleanolic acid | |
| Plum | cyanidin 3-rutinoside | [96] |
| | peonidin 3-rutinoside | |
| | cyanidin 3-glucoside | |
| | cyanidin 3-xyloside | |
| | peonidin 3-glucoside | |
| | catechin, procyanidin B1 | [99] |
| | 3-caffeoylquinic acid | |
| | 3-O-feruloylquinic acid | |
| | quercetin-3-*O*-rutinoside | |
| Pomegranate | delphinidin 3,5-diglucoside | [132] |
| | cyanidin 3,5-diglucoside | |
| | delphinidin 3-glucoside | |
| | pelargonidin 3,5-diglucoside | |
| | ellagic acid | |
| Quince | 3-*O*-caffeoylquinic acid | [133] |
| | 4-*O*-caffeoylquinic acid | |
| | 3, 5-*O*-caffeoylquinic acid | |
| | 3,5-dicaffeoylquinic acids | |
| | rutin | |
| Strawberry | ellagic acid, ellagitannin | [134] |
| | malonic acid, *p*-coumaric acid | |
| | pelargonidin-3-glucoside | |
| | proanthocyanidin B3 | |
| | quercetin-3-glucuronide | |
| Sweetsop | quercetin-3-*O*-rutinoside | [135] |

| Plant food | Antioxidant components | Refcrences |
|---|---|---|
| | myricetin-3-*O*-galactoside | |
| | quercetin-3-*O*-glycoside | |
| | quercetin-3,7-*O*-dimethyl ether | |
| Ziziphus | alphitolic acid, betulinic acid | [136] |
| | betulonic acid, ceanothic acid | |
| | epiceanothic acid | |
| | maslinic acid, oleanolic acid | |
| | pomonic acid, ursolic acid | |
| Broccoli | sulforaphane | [137] |
| | progoitrin | [138] |
| | 4-hydroxyglucobrassicin | |
| | neoglucobrassicin | |
| Curly kale | caffeic acid, ferulic acid | [100] |
| | sinapic acid | |
| | kaempferol, quercetin | |
| Onion | quercetin, tryptophan | [139] |
| | isorhamnetin 4'-glucoside | |
| | isorhamnetin 3,4'-glucoside | |
| Red onion | quercetin 4'-glucoside | [110] |
| | quercetin 3,4'-diglucoside | |
| | isorhamnetin 3-glucoside | |
| Spinach | jaceidin, patuletin | [105] |
| | spinatoside, spinacetin | |
| Wampee | 8-hydroxypsoralen | [140] |
| Clove | caffeic acid, gentisic acid | [141] |
| | *p*-hydroxybenzoic acid | |
| | vanillic acid, quercetin | |
| | ferulic acid, *p*-coumaric acid | |
| | protocatechuic acid | |
| | synapic acid, syringic acid | |
| | eugenol, kaempferol | |
| Mint | apigenin, caffeic acid | [142] |
| | eriocitrin, eriodictyol | |
| | isorhoifolin, luteolin | |
| | luteolin-7-*O*-glucoside | |
| | rosmarinic acid | |

**Table 2. (Continued)**

| Plant food | Antioxidant components | References |
|---|---|---|
| Oregano | eriodictyol 6,8-di-*C*-glucoside | |
| | apigenin 6,8-di-*C*-glucoside | [117] |
| | luteolin 7-*O*-glucuronide | |
| | *epi*-lithospermic acid B | |
| | lithospermic acid B | |
| | rosmarinic acid | |
| Rosemary | carnosic acid, rosmarinic acid | [143] |
| | carnosol, cirsimaritin | [144] |
| | genkwanin, gallocatechin | |
| | homoplantaginin, scutellarein | |
| | rosmadial, rosmanol | |
| Sage | caffeic acid | [118] |
| | apigenin, rosmarinic acid | |
| | luteolin 7-*O*-glucoside | |
| | cirsimaritin, hispidulin | |
| Thyme | arbutin, caffeic acid | [145] |
| | eriocitrin, eriodictyol | |
| | eriodictyol-7-*O*-b-glucoside | |
| | hesperidin, luteolin | |
| | luteolin-7-*O*-b-glucuronide | |
| | luteolin-7-*O*-rutinoside | |
| | methyl rosmarinate | |
| | naringenin, narirutin | |
| | naringenin-7-*O*-b-glucoside | |
| | rosmarinic acid | |

## Structure-Activity Relationship of Antioxidant

In recent years, the structure-activity relationships of some phenolic compounds such as flavonoids, anthocyanins, flavonols, flavanols, and proanthocyanidins were also studied. The data obtained could be used to modify structure of antioxidant as well as to design and synthesize novel antioxidant with special function.

The DPPH radical-scavenging activities and structure-activity relationships of thirty-eight phenolic compounds (including 31 flavonoids) were studied *in vitro*. Obviously, the presence of *ortho*-dihydroxyl substituent in phenolics might be an important determinant in antioxidant effects of the compounds. 3-Hydroxyl was also essential for their radical-scavenging activities. Other structural features including an increasing number of hydroxyls on flavones with a 3',4'-dihydroxyl basic structure, the presence of a third hydroxyl group at C-5', glycosylation and methylation of the hydroxyls, and some other hydroxyls such as 5-, and 7-hydroxyl in ring A, played a modified role in the radical-scavenging activities of flavonoids and other phenolics [146].

The total peroxyl radical-scavenging capacity (TOSC) of a series of flavonoids was determined. The flavonol quercetin exhibited the highest TOSC, followed by its 3-glycoside derivative rutin, which were 7 and 5 times higher than Trolox respectively, thus found to be better scavengers of peroxyl radicals. Among the flavones with both hydroxy and methoxy substitution, thymonin (5,6,4'-trihydroxy-7,8,3'-trimethoxyflavone) was the most effective, with a TOSC value 1.5 times greater than that of Trolox. The TOSC of the remaining flavones decreased in the following order: 5,4'-dihydroxy-6,7,8,3-tetramethoxyflavone>5-hydroxy-3,6,7,3',4'-pentamethoxyflavone>5,4'-dihydroxy-3,6,7-trimethoxyflavone> 5,6,7,8,2',3',4',5'-octamethoxyflavone [147].

The antioxidant activities of phenolics anthocyanidins, anthocyanins, and catechins were investigated in a liposomal model system. Hydroxyl group at the B-ring and position 3 played a significant part in the potent antioxidant activity of anthocyanidins. Whereas an increasing number of hydroxyl substituents on the B-ring and glycosyl units at position 3 decreased the antioxidant activity of cyanidins. Likewise, the substitution of a galloyl group at position 3 of the flavonoid moiety resulted in significantly decreased antioxidant activity for the catechins. To catechins, cis-trans isomerism, epimerization, and racemization did not influence their overall antioxidant activity [148]. The ranking of superoxide radical scavenging activity of anthocyanins decreased in the following order: delphinidin > petunidin > malvidin approximate to cyanidin > (+)-catechin > peonidin > pelargonidin. The order of scavenging activity toward ONOO- decreased as follows: delphinidin > cyanidin approximate to petunidin > malvidin approximate to (+)-catechin > peonidin > pelargonidin. Methylation of 4'-OH had a negative effect on the antioxidant activity of anthocyanin. In addition, synergistic action was observed in both O-2(.-)- and ONOO--scavenging activities among the mixed anthocyanins [149].

Generally, among the phenolic acid analogues, the antioxidant capacity was dependent on the number of hydroxy groups and methoxy substitutions on the aromatic ring. Cinnamic acid derivatives and their efficacy of hydrogen peroxide scavenging activity ($SA_{HP}$) were tested. The orders of $SA_{HP}$ among benzoic acid derivatives and cinnamate derivatives were hydrocaffeic acid > caffeic acid > *L*-phenylalanine > ferulic acid > *p*-coumaric acid > *m*-coumaric acid, and vanillic acid > gallic acid > *m*-hydroxybenzoic acid > *p*-hydroxybenzoic acid > protocatechuic acid, respectively. Benzoic acid derivatives were shown to be far more effective than cinnamate derivatives. The scavenging potency of cinnamate derivatives seemed to be negatively affected by the side chain unsaturation structure [150,151]. Free radical scavenging activities of hydroxycinnamic acid derivatives were tested as reflected in the ability to inhibit *in vitro* LDL peroxidation. Caffeic acid, chlorogenic acid, sinapic acid, ferulic acid and *p*-coumaric acid exhibited inhibition from both AAPH- and $Cu^{2+}$-induced LDL peroxidation. Among the tested compounds, caffeic acid and chlorogenic acid showed the strongest activities in both measurements. It was worthy noting that *ortho*-dihydroxyl and

4-hydroxy-3-methoxyl play a remarkable role in the protection effects of hydroxycinnamic acid derivatives against LDL peroxidation [152].

The flavanones are mainly found in citrus. The antioxidant or prooxidant behaviors of some common flavanones and their activity–structure relationships as antioxidant were studied using the crocin bleaching inhibition assay. With comparison of heridictyol and naringenin, the substitution of 3', 4'-di-hydroxy in the aglycone form did not obviously influence the antioxidant activity. On the other hand, comparison of neoeriocitrin and naringenin showed that the 3',4'-catechol structure noticeably increased the antioxidant power in the flavanones glycosylated with a neohesperidose of the 7th hydroxy group. However, replaced with a neohesperidoside molecule in the 7th position, a methoxylation in the 4th position markedly decreased the antioxidant potential, suggesting that the sugar molecule in the 7th position interacted with the methoxyl group in the 4th position and decreased the antioxidant potential (comparison between Neohesperidin and Neoeriocitrin) [153]. Flavanones and flavones were reported to be broadly effective [154]. Though having the same arrangement of OH groups substituted into the flavonoid skeleton, flavanone naringenin having 4-oxo group but no 2, 3-double bond had higher antioxidant capacity than flavone apigenin with its 4-oxo group and 2, 3-double bond in the C ring. The flavone rhoifolin (apigenin 7-*O*-neohesperidoside) also had weaker antioxidant capacity than the flavanone naringin (naringenin 7-*O*-neohesperidoside) [150].

Curcumin and its analogues were effective antioxidants against AAPH- and $Cu^{2+}$-induced LDL peroxidation, and the H-atom abstraction from the phenolic group were found to be responsible for the activity. The *ortho*-diphenoxyl functionality significantly increased the antioxidant activity of curcumin analogues. Similarly, 4-hydroxy-3-methoxyphenyl group also played an important role in the antioxidant activity. On the other hand, in the *in vitro* LDL peroxidation model, synergistic antioxidant interaction between curcumin and its analogues and a-tocopherol was not observed [155].

The level of free radical scavenging activity for tannins was in decreasing order as follows: EGCG > ECG > EGC > EC, and digallic acid > ellagic acid. The calculation results from electron transfer mechanism and H-abstraction mechanism indicated that galloyl groups played a pivotal role on the free radical scavenging activity of condensed tannins [156].

# Conclusion

Epidemiological studies have established a positive correlation between consumption of fruits and vegetables and low risk of chronic diseases like cancer, cardiovascular disease, and diabetes. Natural antioxidants in fruits and vegetables, such as vitamins and polyphenols, are considered to be responsible for these health benefits. Recently, studies on antioxidant activities of food have increased remarkably due to increased interest in their potential of being used as a rich and natural source of antioxidant compounds. The extraction of antioxidants from food is the first step in the evaluation of antioxidant activities. At present, several extraction methods, including solvent extraction, supercritical fluid extraction, subcritical water extraction, enzyme extraction, microwave-assisted extraction and ultrasonic -assisted extraction have been applied as alternative methods for antioxidants extraction from food. Due to the multifunctional property of natural antioxidants, it is necessary to perform more than one type of antioxidant capacity measurement to take into account the various mechanisms of antioxidant action. In spite of the diversity of methods, there is a great need to standardize antioxidant testing protocol to minimize the present chaos in the methodologies used to evaluate antioxidant activities. Three assays including the oxygen radical absorbance capacity assay, the Folin-Ciocalteu method, and the Trolox equivalent antioxidant capacity assay, could be considered for the standardization. Mass of researches were carried out to provide data regarding the antioxidant activities of plant food materials such as fruits, vegetables, spices, grains, and nuts, etc. and their positive function to human health and important role as primary dietary sources of natural antioxidants were confirmed. The isolation, identification and quantification of natural antioxidants from different plant food have been carried out using various different techniques. The structure-activity relationships of natural antioxidant especially the phenolic compounds such as flavonoids, anthocyanins, flavonols, flavanols as well as proanthocyanidins were also studied. The data obtained could be used to modify structure of antioxidant as well as to design and synthesize novel antioxidant with special function.

## References

[1] R. Blomhoff, M. H. Carlsen, L. F. Andersen, and D. R. Jr. Jacobs, *Brit. J. Nutr.* 96 (Suppl. 2), S52 (2006).

[2] A.P. Wickens, *Resp. Physiol.* 128, 379 (2001).

[3] O.A. Azizova, *Biol. Membrany.* 19, 451 (2002).

[4] K. J. Barnham, C. L. Masters, and A. I. Bush, Nat. *Rev. Drug Discov.* 3, 205 (2004).

[5] M. Quintero, P. A. Brennan, G. J. Thomas, and S. Moncada, *Cancer Res.* 66, 770 (2006).

[6] R. Nagler, A. Reznick, Y. Shafir, and N. Shehadeh, *Free Radic. Res.* 40, S156 (2006).

[7] R. Blomhoff, *Curr. Opin. Lipidol.* 16, 47 (2005).

[8] J. O. Moskaug, H. Carlsen, M. C. Myhrstad, and R. Blomhoff, *Am. J. Clin. Nutr.* 81, 277 (2005).

[9] M.V. Eberhardt, C.Y. Lee, and R. H. Liu, *Nature.* 405, 903 (2000).

[10] P. Van't Veer, M. C. Jansen, M. Klerk, and F. J. Kok, *Public Health Nutr.* 3, 103 (2000).

[11] C. La Vecchia, A. Altieri, and A. Tavani, *Europ. J. Nutr.* 40, 261 (2001).

[12] WHO. Technical Report Series 916. Geneva (2003).

[13] C. Ruxton, E. Gardner, and D. Walker, *Int. J. Food Sci. Nutr.* 57, 249 (2006).

[14] N. Ito, M. Hirose, S. Fukushima, H. Tsuda, T. Shirai, and M. Tatematsu, *Food Chem Toxicol*, 24, 1071 (1986).

[15] A. M. Safer, and A. J. Al-Nughamish, *Histol Histopathol*, 14, 391 (1999).

[16] K. K. Adom, R. H. Liu, *J. Agric. Food Chem.* 50, 6182 (2002).

[17] B. L. Halvorsen, K. Holte, M. C. Myhrstad, I. Barikmo, E. Hvattum, S. F. Remberg, A. B. Wold, K. Haffner, H. Baugerød, L. F. Andersen, Ø. Moskaug, D. R J. Jacobs, and R. Blomhoff, *J. Nutr.* 132, 461 (2002).

[18] B.L. Halvorsen, M. H. Carlsen, K. M. Phillips, S. K. Bøhn, K. Holte, D. R. R. Jacobs, and R. Blomhoff, *Am. J. Clin. Nutr.* 84, 95 (2006).

[19] R. Tzulker, I. Glazer, I. Bar-Ilan, D. Holland, M. Aviram, and R. Amir, *J. Agric. Food Chem.* 55, 9559 (2007).
[20] M. H. Carlsen, B. L. Halvorsen, K. Holte, S. K. Bøhn, S. Dragland, L. Sampson, C. Willey, H. Senoo, Y. Umezono, C. Sanada, I. Barikmo, N. Berhe, W. C. Willett, K. M. Phillips, D. R. J. Jacobs, and R. Blomhoff, *Nutr. J.* 9, 3 (2010).
[21] L. Fu, B. T. Xu, X. R. Xu, X. S. Qin, R. Y. Gan, and H. B. Li, *Molecules.* 15, 8602 (2010).
[22] J. W. Finley, A. N. Kong, K. J. Hintze, E. H. Jeffery, L. L. Ji, and X. G. Lei, *J. Agric. Food Chem.* 59, 6837 (2011).
[23] J. Perez-Jimenez, S. Arranz, M. Tabernero, M, Elena Diaz-Rubio, J. Serrano, I. Goni, and F. Saura-Calixto, *Food Res. Internat.* 41, 274 (2008).
[24] D. R. Pompeu, E. M. Silva, and H. Rogez, *Bioresource Technol.* 100, 6076 (2009).
[25] J. Dai, and R. J. Mumper, *Molecules.* 15, 7313 (2010).
[26] P. G. Kapasakalidis, R. A. Rastall, and M. H. Gordon, *J. Agric. Food Chem.* 54, 4016 (2006).
[27] A. Marston, and K. Hostettmann, *Separation and quantification of flavonoids.* Andersen and Markham, Boca Raton: CRC Press, New York, pp. 1–36 (2006).
[28] H. B. Li, D. Li, Y. Zhang, R. Y. Gan, F. L. Song, and F. Chen, *Reactive Oxygen Species and Antioxidants in Higher Plants.* Dutta Gupta, S. pp. 331–362. Science Publishers (2010).
[29] M. C. Lin, M. J. Tsai, and K. C. and Wen, *J. Chromatogr. A.* 830, 387 (1999).
[30] M. C. Pascual-Martí, A. Salvador, A. Chafer, and A. Berna, *Talanta.* 54, 735 (2001).
[31] M. A. Rostagno, J. M. A. Araújo and D. Sandi, *Food Chem.*78, 111 (2002).
[32] V. Tunde, S. Mojca, and K. Zeljko, *J. Food Eng.* 90, 246 (2009).
[33] R. T. Threlfall, J. R. Morris, and A. Mauromoustakos, *Am. J. Enol. Viticul.* 50, 57 (1999).
[34] T. Sun, J. Tang, and J. R. Powers, *J. Agric. and Food Chem.* 53, 42 (2005).
[35] J. M. Koponen, A. M. Happonen, S. Auriola, H. Kontkanen, J. Buchert, K. S. Poutanen, and A. R. Törrönen, *J. Agric. Food Chem.* 56, 3136 (2008).
[36] X. J. Pan, H. Z. Liu, G. H. Jia, and Y. Y. Shu, *Biochem. Eng. J.* 5, 173 (2000).
[37] J. Wang, B. G. Sun, Y. P. Cao, Y. A. Tian, and X. H. Li, *Food Chem.* 106, 804 (2008).
[38] T. Sun, and S. A. Tanumihardjo, *J. Food Sci.* 72, R159 (2007).
[39] E. N. Frankel, and A. S. Meyer, *J. Sci. Food Agric.* 80, 1925 (2000).
[40] D. J. Huang, B. X. Ou, and R. L. Prior, *J. Agric. Food Chem.* 53, 1841 (2005).
[41] S. C. Litescu, S. Eremia, and G. L. Radu, *Adv. Exp. Med. Biol.* 698, 241 (2010).
[42] E. N. Frankel, and N. S. Meyer, *J. Sci. Food Agric.* 80, 1925 (2010).
[43] N. Pellegrini, Salvatore. S, S. Valtueña, G. Bedogni, M. Porrini, V. Pala, D. Del Rio, S. Sieri, C. Miglio, V. Krogh, I. Zavaroni, and F. Brighenti, *J. Nutr.* 137, 93 (2007).
[44] H. Zielinski, and H. Kozlowska, *J. Agric. Food Chem.* 48, 2008 (2000).
[45] T. Stangeland, S.F. Remberg, and K. A. Lye, *Food Chem.* 113, 85 (2009).
[46] A. Ghasemzadeh, H. Z. E. Jaafar, and A. Rahmat, *Molecules.* 15, 4324 (2010).
[47] L. Fu, B. T. Xu, X. R. Xu, R. Y. Gan, Y. Zhang, E. Q. Xia, and H. B. Li, *Food Chem.* 129, 345 (2011).
[48] J. Contreras-Calderón, L. Calderón-Jaimes, E. Guerra-Hernández, and B. García-Villanova, *Food Res. Int.* 44, 2047 (2011).
[49] R. G. Roberts, R. Jone, and M. A. Smith, *Nature.* 345:153 (1990).

[50] M. Netzel, G. Netzel, Q. Tian, S. Schwartz, and I. Konczak, *Innov Food Sci. Emerg. Technol.* 8, 339 (2007).
[51] C. Vijaya Kumar Reddy, D. Sreeramulu, and M. Raghunath, *Food Res. Int.* 43, 285 (2010).
[52] C. Vasco, J. Ruales, and A. Kamal-Eldin, *Food Chem.* 111, 816 (2008).
[53] E. Riboli, and T. Norat, *Am. J. Clin. Nutr.* 78, 559S (2003).
[54] S. A. Stanner, J. Hughes, C. N. Kelly, and J. Buttriss, *Public Health Nutr.* 7, 407 (2004).
[55] T. Bahorun, A. Luximon-Ramma, A. Crozier, and O. I. Aruoma, *J. Sci. Food Agric.* 84, 1553 (2004).
[56] A. Ismail, Z. M, Marjan, and C. W. Foong, *Food Chem.* 87, 581 (2004).
[57] J. Nilsson, D. Pillai, G. Onning, C. Persson, A. Nilsson, and B. Akesson, *Mol. Nutr. Food Res.* 49, 239 (2005).
[58] P. Ninfali, G. Mea, S. Giorgini, M. Rocchi, and M. Bacchiocca, *Brit. J. Nutr.* 93, 257 (2005).
[59] P. Stratil, B. Klejdus, and V. Kubán, *J. Agric. Food Chem.* 54, 607 (2006).
[60] K. Q. Zhou, and L. L. Yu, *LWT.* 39, 1155 (2006).
[61] S. Gorinstein, ·Y. S. Park, B. G. Heo, J. Namiesnik, H. Leontowicz, M. Leontowicz, K. S. Ham, J. Y. Cho, and S. G. Kang, *Eur. Food Res. Technol.* 228, 903 (2009).
[62] B. Ou, D. Huang, M. Hampsch-Woodill, J. A. Flanagan, and E. K. Deemer, *J. Agric. Food Chem.* 50, 3122 (2002).
[63] K. Srinivasan, *Food Res. Int.* 21, 167 (2005).
[64] T. H. Tsai, P. J. Tsai, and S. C. Ho, *J. Food Sci.* 70, C93 (2005).
[65] B. Shan,, Y. Z. Cai,, M. Sun, and H. Corke, *J. Agric. Food Chem.* 53, 7749 (2005).
[66] I. Hinneburg, H. J. D. Dorman, and R. Hiltunen, *Food Chem.* 97, 122 (2006).
[67] A. T. Mata, C. Proenca, A. R. Ferreira, M. L. M. Serralheiro, J. M. F. Nogueira, and M. E. M. Araújo, *Food Chem.* 103, 778 (2007).
[68] L. Su, J. J. Yin, D. Charles, K. Zhou, J. Moore, and L. L. Yu, *Food Chem.* 100, 990 (2007).
[69] A. Kırca,and E. Arslan, *Int. J. Food Sci. Technol.* 43, 2038 (2008).
[70] L. Galvez-Ranilla, Y. I. Kwon, E. Apostolidis, and K. Shetty, *Bioresource Technol.* 101, 4676 (2010).
[71] M. Viuda-Martos, Y. Ruiz-Navajas, E. Sanchez-Zapata, J. Fernández-López, and J. A. Pérez-Álvarez, *Flavour Fragr. J.* 25, 13 (2010).
[72] C. M. Liyana-Pathirana, and F. Shahidi, *J. Sci. Food Agric.* 86, 477 (2006).
[73] R. Y. Yang, S. C. S. Tsou, T. C. Lee, W. J. Wu, P. M. Hanson, G. Kuo, L. M. Engle, and P. Y. Lai, *J. Sci. Food Agric.* 86, 2395 (2006).
[74] J. Bertoncelj, U. Doberšek, M. Jamnik, and T. Golob, *Food Chem.* 105, 822 (2007).
[75] I. Orhan, and O. Üstün, *J. Food Compost. Anal.* 24, 386 (2011).
[76] M. Číž, H. Čížová, P. Denev, M. Kratchanova, A. Slavov, and A. Lojek. *Food Control.* 21, 518 (2010).
[77] M. Lu, B. Yuan, M. Zeng, and J. Chen. *Food Res. Int.* 44, 530 (2011).
[78] N. Thitilertdecha, A. Teerawutgulrag, J. D. Kilburn, and N. Rakariyatham, *Molecules.* 15, 1453 (2010).
[79] U.Vrhovsek, D. Masuero, L. Palmieri, and F. Mattivi, *J. Food Compos Anal.* 25, 9 (2012).

[80] L. C. Kerio, F. N. Wachira, J. K.Wanyoko, and M. K. Rotich, *Food Chem.* 131, 31 (2012).
[81] A. Busing, and W. Ternes, *Anal Bioanal Chem.* 401, 2843 (2011).
[82] E. L. Regalado, S. Tolle, J. A. Pino, P. Winterhalter, R. Menendez, A. R. Morales, and J. L. Rodríguez, *J. Chromatogr A.* 1218, 7358 (2011).
[83] V. Verardo, A. M. Gomez-Caravaca, A. Segura-Carretero, M. F. Caboni, and A. Fernandez-Gutierrez, *Electrophoresis.* 32, 669 (2011).
[84] O.P. Pareek, Ber-Ziziphus mauritania. 2004 At: http://www.civil.soton.ac.uk/icuc/ cd_icuc_ber_tamarind/content/ber/ ber_book_html.
[85] J. G., Jiang, X. J. Huang, J. Chen, and Q. S. Lin, *Nat. Prod. Res.* 21, 310 (2007).
[86] R.T. Mahajan, and M. Z. Chopda, *Pharmacogn Rev.* 3, 320 (2009).
[87] G. P. Mill, *Phytother.* 7, 31 (2009).
[88] B. San, and A. N. Yildirim, *J. Food Compos. Anal.* 23, 706 (2010).
[89] W. Elfalleh, N. Tlili, N. Nasri, Y. Yahia, H. Hannachi, N. Chaira, M. Ying, and A. Ferchichi, *J. Food Sci.* 76, C707 (2011).
[90] C.M. Li, R. Leverence, J. D. Trombley, S. F Xu, J. Yang, Y. Tian, J. D. Reed, and A. E. Hagerman, *J. Agric. Food Chem.* 58, 9033 (2010).
[91] H. F. Gu, C. M. Li, Y. J. Xu, W. F. Hu, M. H. Chen, and Q. H. Wang, *Food Res. Int.* 41, 208 (2008).
[92] S. de Pascual-Teresa, C. Santos-Buelga, and J. C. Rivas-Gonzalo, *J. Agric. Food Chem.* 48, 5331 (2000).
[93] J. L. Donovan, A. S. Meyer, and A. L. Waterhouse, *J. Agric. Food Chem.* 46, 1247 (1998).
[94] O. K. Chun, D. O. Kim, H.Y. Moon, H. G. Kang, and C. Y. Lee, *J. Agric. Food Chem.* 51, 7240 (2003).
[95] R. Slimestad, E. Vangdal, and C. Bredo, *J. Agric. Food Chem.* 57, 11370 (2009).
[96] V. Usenik, F. Štampar, and R.Veberic, Food Chem. 114, 529 (2009).
[97] J. Los, J. Wilska-Jeszka, and M. Pawlak, *Pol. J. Food Nutr. Sci.* 9, 35 (2000).
[98] Y. Kimura, H. Ito, M. Kawaji, T. Ikami, and T. Hatano, *Biosci., Biotechnol., Biochem.* 72, 1615 (2008).
[99] C. Nunes, S. Guyot, N. Marnet, A. S. Barros, J. A. Saraiva, C. M. Renard, and M. A. Coimbra, *J. Agric. Food Chem.* 56, 5188-5196 (2008).
[100] H. Olsen, K. Aaby, and G. I. Borge, *J. Agric. Food Chem.* 57, 2816 (2009).
[101] H. Olsen, K. Aaby, and G. I. Borge, *J. Agric. Food Chem.* 58, 11346 (2010).
[102] S. Schmidt, M. Zietz, M. Schreiner, S. Rohn, L. W. Kroh, and A. Krumbein, *Rapid. Commun. Mass Spectrom.* 24, 2009 (2010).
[103] L. Z. Lin, and J. M. Harnly, *J. Agric. Food Chem.* 57, 7401 (2009).
[104] F. Vallejo, F. A. Tomás-Barberán, and C. García-Viguera, *J. Sci. Food Agric.* 82, 1293 (2002).
[105] M. J. Cho, L.R. Howard, R.L. Prior, and T. Morelock. *J. Sci. Food Agric.* 88, 1099 (2008).
[106] S. Grossman, R. Reznik, T. Tamari, and M. Albeck, *New plant water soluble antioxidant (NAO) from spinach.* Asada Toshikawa (Eds.), Elsevier Science, Amsterdam, The Netherlands, pp. 343–344 (1994).
[107] M. Bergman, L. Varshavsky, H.E. Gottlieb, and S. Grossman, *Phytochemistry.* 58, 143 (2001).

[108] A. Bunea, M. Andjelkovic, C. Socaciu, O. Bobis, M. Neacsu, M. Verhe, and J. V. Camp, *Food Chem.* 108, 649 (2008).

[109] M. Bergman, A. Perelman, Z. Dubinsky, and S. Grossman, *Phytochemistry.* 62, 753 (2003).

[110] P. Bonaccorsi, C. Caristi, C. Gargiulli, and U. Leuzzi, *J. Agric. Food Chem.* 53, 2733 (2005).

[111] G. B. Rodríguez, R. E. M. Rodríguez, and R. C. Díaz, *J. Food Sci.* 73, C599 (2008).

[112] M. G. L. Hertog, P. C. H. Hollman, and M. B. Katan, *J. Agric. Food Chem.* 40, 2379 (1992).

[113] A. Crozier, M. E. J. Lean, M. S. McDonald, and C. Black, *J. Agric. Food Chem.* 45, 590 (1997).

[114] J. Barnes, L. A. Anderson, and J. D. Phillipson, *Herbal Medicines,* 2nd Edn. Pharmaceutical Press, A Guide for Healthcare Professionals. London. pp.139–140 (2002).

[115] L. Jirovetz, G. Buchbauer, I. Stoilova, A. Stoyanova, A. Krastanov, and E. Schmidt, *J. Agric. Food Chem.* 54, 6303 (2006).

[116] V, Exarchou, M, Godejohann, T. A. van Beek, I. P. Gerothanassis, and J. Vervoort, *Anal. Chem.* 75, 6288 (2003).

[117] K. Grevsen, X. C. Fretté, and L. P. Christensen. *Europ. J. Hort. Sci,* 74, S193–203 (2009).

[118] F. Areias, P. Valentão, P. B. Andrade, F. Ferreres, and R. M. Seabra. *J. Agric. Food Chem.* 48, 6081 (2000).

[119] M. E. Cuvelier, H. Richard, and C. Berset, *J. Am. Oil Chem. Soc.* 73, 645 (1996).

[120] C. Brieskorn, and H. J. Demling, *Z. Lebensm. Unters. Fo rsch.* 41, 10 (1969).

[121] J. Lö1iger, *Natural Antioxidants, in Rancidity in Food*, Allen and Hamilton (edis), New York, pp. 105-124 (1989).

[122] P. Schuler, *Natural Antioxidants Exploited Commercially, in Food Antioxidants*, B. Hudson (edit), New York, pp. 99-170 (1990).

[123] J. L. Lamaison, C. Petitjean-Freytet, and A. Carnat, *Pharm. Acta Helv.* 66, 185 (1991).

[124] S. M. Kim, and B. H. Um, *J. Med. Plants Res.* 5, 5008 (2011).

[125] C. Mertz, V. Cheynier, Z. Günata, and P. Brat, *J. Agric. Food Chem*, 55, 8616 (2007).

[126] T. J. Hager, L. R. Howard, R. Liyanage, J. O. Lay, R. L. Prior. *J. Agric. Food Chem.* 56, 661 (2008).

[127] P. Liu, B. Yang, and H. Kallio, *Food Chem.* 121, 1188 (2010).

[128] C. J. Wang, and Y. G. Zuo. *Food Chem.* 128, 562 (2011).

[129] H. Kelebek. *Ind. Crop. Prod.* 32, 269 (2010).

[130] M. R. Pérez-Gregorio, J. Regueiro, E. Alonso-González, L. M. Pastrana-Castro, and J. Simal-Gándara, *LWT - Food Sci. Technol.* 44, 1793 (2011).

[131] D. M. Rivera-Pastrana, E. M. Yahia, and G. A. González-Aguilar. *J. Sci. Food Agric.* 90, 2358 (2010).

[132] G. Mousavinejad, Z. Emam-Djomeh, K. Rezaei, and M. H. H. Khodaparast, *Food Chem.* 115, 1274 (2009).

[133] B. M. Silva, P. B. Andrade, P. Valentão, F. Ferreres, R. M. Seabra, and M. A. Ferreira, *J. Agric. Food Chem.* 52, 4705 (2004).

[134] K. Aaby, D. Ekeberg, and G. Skrede. *J. Agric. Food Chem.* 55, 4395–4406 (2007).

[135] H. M. Kotkar, P. S. Mendki, S. V. Sadan, S. R. Jha, S. M. Upasani, and V. L. Maheshwari, *Pest Manag. Sci.* 58, 33 (2002).

[136] S. Guo, J. A. Duan, Y. P. Tang, N. Y. Yang, D. W. Qian, S. L. Su, and E. X. Shang. *J. Agric. Food Chem.* 58, 6285 (2010).

[137] D. D. Han, and K. H. Row. *Int. J. Mol. Sci.* 12, 1854 (2011).

[138] J. Borowski, A. Szajdek, E. J. Borowska, E. Ciska, and H. Zieliński. *Eur Food Res Technol.* 226, 459 (2008).

[139] S. U. Lee, J. H. Lee, S. H. Choi, J. S. Lee, M. Ohnisi-Kameyama, N. Kozukue, C. E. Levin, and M. Friedman, *J. Agric. Food Chem.* 56, 8541 (2008).

[140] K. N. Prasad, H. H. Xie, J. Hao, B.Yang, S. X. Qiu, X. Y.Wei, F. Chen, and Y. M. Jiang. *Food Chem.* 118, 62 (2010).

[141] P. S.Variyar, C. Bandyopadhyay, and P. Thomas. *J. Food Sci. Technol*, 33,533 (1998).

[142] H. J. Dorman, M. Koşar, K. Kahlos, Y. Holm, and R. Hiltunen, *J. Agric. Food Chem.* 51, 4563 (2003).

[143] N. Troncoso, H. Sierra, L. Carvajal, P. Delpiano, and G. Günther, *J. Chromatogr A.* 1100, 20 (2005).

[144] I. Borrás Linares, D. Arráez-Román, M. Herrero, E. Ibáñez, A. Segura-Carretero, and A. Fernández-Gutiérrez, *J. Chromatogr. A.* 1218, 7682 (2011).

[145] I. Fecka, and S. Turek. *Food Chem.* 108, 1039 (2008).

[146] C. D. Zheng, G. Li, H. Q. Li, X. J. Xu, J. M. Gao, and A. L. Zhang. *Nat. Prod. Commun.* 5, 1759 (2010).

[147] A. J. J. Dugas, J. Castañeda-Acosta, G. C. Bonin, K. L. Price, N. H. Fischer, and G. W. Winston. *J. Nat. Prod.* 63, 327 (2000).

[148] N. P. Seeram, and M. G. Nair, *J. Agric. Food Chem.* 50, 5308 (2002).

[149] M. M. Rahman, T. Ichiyanagi, T. Komiyama, Y. Hatano, and T. Konishi, *Free Radical Res.* 40, 993 (2006).

[150] D. O. Kim, and C. Y. Lee, *Crit. Rev. Food Sci. Nutr.* 44, 253 (2004).

[151] A. Mansouri, D. P. Makris, and P. Kefalas, *J. Pharmaceut Biomed. Anal.* 39, 22 (2005).

[152] J. C. Cheng, F. Dai, B. Zhou, L.Yang, and Z. L. Liu. *Food Chem.*104, 132 (2007).

[153] D. D. Majo, M. Giammanco, M. L. Guardia, E.Tripoli, S.Giammanco, and E. Finotti. *Food Res. Int.* 38, 1161 (2005).

[154] B. J. F. Hudson, and J. I. Lewis, *Food Chem.* 10, 47 (1983).

[155] W. F. Chen, S. L. Deng, B. Zhou, L. Yang, and Z. L. Liu. *Free Radical Bio. Med.* 40, 526 (2006).

[156] W. H. Zhang, and B. Shi. *J. Soc. Leath Tech. Ch.* 89,137 (2005).

*Reviewed by* Professor Feng Chen in The University of Hong Kong.

In: Phytochemicals
Editor: Hai-Xue Kuang

ISBN: 978-1-62417-354-7


*Chapter IX*

# Regulation of Inflammation by Dietary Phytochemicals: Current Perspectives

***Marta González Castejón***
***and Arantxa Rodriguez Casado****
IMDEA Food Institute, Madrid, Spain

## Abstract

Acute respiratory distress syndrome, sarcoidosis, psoriasis, inflammatory bowel diseases, rheumatoid arthritis, osteoarthritis, and atherosclerosis are examples of inflammatory diseases. Also, chronic inflammation is associated to other serious chronic illnesses - including heart disease, obesity, diabetes, Alzheimer's disease, allergy, asthma, and many forms of cancer. Inflammation is a key feature of autoimmune diseases and other conditions that affect the immune system. Whether inflammation is the root cause of chronic illness, merely accompanies them, or occurs as a result of the disease, different dietary choices play significant roles in inflammation development. Intake of specific foods that influence the inflammatory process is an easy daily way to prevent and reduce long-term risks. In recent years, increasing attention has been paid to medicinal plants, due to many epidemiological and clinical studies that have demonstrated how a proper diet rich in micronutrients reduces chronic inflammation and, simultaneously, helps controlling and managing chronic health conditions. For example, high levels of cytokines cause many of the detrimental consequences of autoimmune disease. Many plants have effects on cytokines, e.g., interleukins, interferons, and chemokines, naturally-produced proteins that modulate the immune response and trigger inflammatory responses to toxins, injury, viruses, and bacteria. Some medicinal plants known to reduce inflammation by modulation of cytokines secretion include *Zingiber officinale* (ginger), *Allium sativum* (garlic), *Taraxacum officinale* (dandelion), *Echinacea purpurea* (purple coneflower), and *Silybum marianum* (milk thistle). Plants exert their beneficial effects

* E-mail address: arantxa.rodriguez@imdea.org.

through the additive or synergistic action of an immense variety of biologically-active, non-nutritive compounds known as phytochemicals acting at single or multiple target sites associated with physiological processes. Moreover, dietary spice phytochemicals, such as curcurmin from turmeric (*Curcuma longa*) and capsaicin from red pepper (*Capsicum baccatum*) are now emerging as agents that can not only prevent but may even treat inflammation.

Many edible plants with anti-inflammatory properties can be used as the sole therapy in inflammatory disease or as complementary of conventional treatments including corticosteroids, allowing patients to take smaller doses or shorter courses of drugs. In addition to influencing inflammation, phytochemicals might provide steady protective effects against different diseases accompanying the inflammatory response. This chapter assesses the anti-inflammatory effects of certain plants, as well as the functional benefits of notable groups of phytochemicals. In it we discuss how spice-derived phytochemicals suppress inflammatory pathways and what their molecular targets are.

# Introduction

Inflammation is the first response of body tissues to injury, infection, toxic chemicals, allergens, or other harmful stimuli. Clinical features of acute inflammation include pain, heat, swelling, and redness at the site of injury. This type of inflammation is normally a localized, defensive, and protective response that follows trauma or infection. Conversely, chronic inflammation results from the persistent activation of inflammatory molecules, viral or microbial infection, environmental antigen, or autoimmune reaction. Although many inflammatory diseases seem to have a genetic component, some of them are triggered by other processes. Inflammation involves a complex biological cascade of molecular and cellular signals that alter physiological responses at the site of the injury: dilation of blood vessels, increased blood flow, increased vascular permeability, exudation of fluids containing proteins like immunoglobulins (antibodies), and invasion by leukocytes (white blood cells). Moreover, an uncontrolled and persistent inflammation may act as an etiologic factor for many of the above-mentioned chronic illnesses [1] because the mediators involved in the inflammatory reaction can easily be induced [2]. These mediators are commonly used as biomarkers of inflammation (Figure 1). As an example, C-reactive protein (CRP), an acute-phase reactant protein synthesized in the liver, is used as general marker of inflammation and also as prototypic biomarker of cardiac related diseases. Clinically, the most common tests to diagnose inflammation include measuring erythrocyte sedimentation rate (ESR), white blood cell count, and albumin levels. High ESR, high white cell counts, and low albumin indicate, generally, states of inflammation. Also, various cytokines and adhesion molecules are frequently used in basic scientific studies to investigate the cellular and molecular processes involved in the pathogenesis of inflammation-related diseases.

There are over a hundred human inflammatory diseases. The most common ones include asthma, Crohn's disease, rheumatoid arthritis, polymyalgia rheumatic, tendonitis, bursitis, laryngitis, gingivitis, gastritis, otitis, celiac disease, dermatitis, diverticulitis, hepatitis, lupus erythematous, nephritis, ulcerative colitis, and inflammatory bowel disease. Additionally, a number of chronic diseases, such as atherosclerosis, obesity, diabetes, cancer, and Parkinson's and Alzheimer's diseases share inflammatory components.

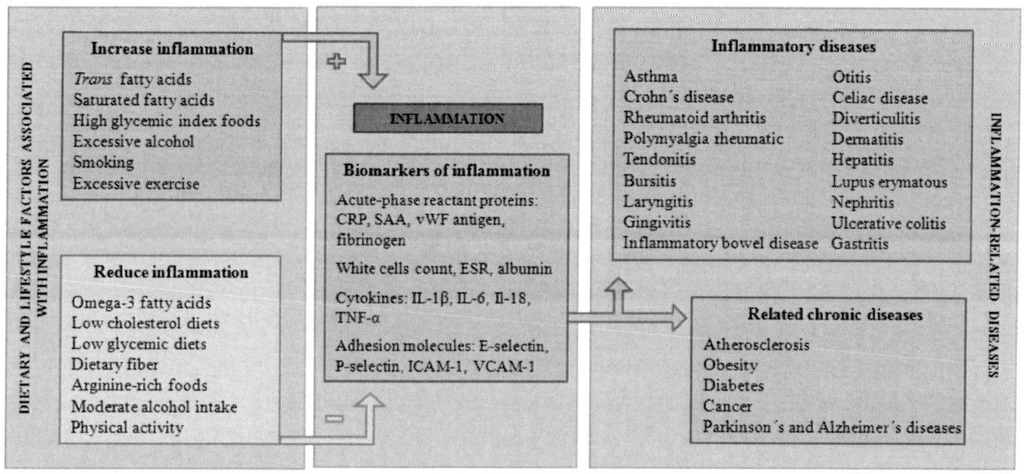


Figure 1. Selected dietary components with important impact on inflammatory processes. Highlighted pro-inflammation mediators commonly used as biomarkers of inflammation.

For instance, it is known that the adipose tissue secretes several inflammatory factors and that obesity is associated with macrophage infiltration in the adipose tissue; however, the exact role of inflammation in the pathogenesis of obesity is currently unknown. Elevated CRP levels have been linked to obesity and animal and human studies have found that physical activity decreases both acute and chronic inflammation, as measured by reductions in CRP and particular cytokines.

Non-steroidal anti-inflammatory drugs (NSAIDs) such as aspirin and ibuprofen are drugs commonly used to control inflammation and relieve pain. Most NSAIDs inhibit one or more isoforms of cyclooxygenase (COX), an enzyme that catalyzes the formation of pro-inflammatory prostaglandins and thromboxanes. Studies indicate increased risk of cardiovascular events associated with the use of these drugs, because the nonselective nature of NSAIDs causes adverse upper gastrointestinal effects. Accordingly, NSAID users experience decreased blood clotting ability, which could increase risk of hemorrhagic stroke. These side effects could be reduced by slightly modifying dietary habits, in order to modulate inflammatory responses. Certain dietary components have an important impact on inflammatory processes (Figure 1).

Overall, studies suggest that diets rich in saturated fats, trans-fats (hydrogenated or partially hydrogenated oils), and high glycemic index foods stimulate inflammation. In contrast, adherence to a Mediterranean-style diet has been shown to reduce inflammation. A Mediterranean-style diet is rich in monounsaturated fatty acids found in olive oil, fruits and vegetables, nuts, beans, and whole grains. Essential fatty acids play a role in the body's inflammatory. In addition, anti-inflammatory protein sources include lean poultry, fish, and seafood. Despite the lack of concluding research findings, some patients experience relief of pain and inflammation when some Solanaceae plants such as potatoes, tomatoes, and eggplants are removed from their diet. These vegetables contain a chemical alkaloid called solanine, which can increase pain sensations in some people. Some vitamins, minerals, and phytochemicals may also have therapeutic benefits in inflammatory diseases. Likewise, nutraceuticals derived from medicinal herbs have been long used in traditional medicine and many of them have wide-ranging biologic effects that suggest novel mechanisms of action.

Some of these have been proven effective in controlled clinical trials of inflammation. Whether these approaches offer safe symptom relief and possibly mediate beneficial joint remodeling in early disease are possibilities that merit further investigation [3].

Recent research has rejuvenated centuries-old traditional herbs to cure various ailments, by using modern tools like diet-based therapy and other regimens. In recent years, we witnessed an increasing preference for the use of natural products - especially those derived from plants - in treating and preventing medical complications. Many currently available drugs such as taxol, vinblastine, vincristine, topotecan, and irinotecan are derived from plants [4, 5]. The search for novel cost-effective anti-inflammatory agents with minimal side effects from natural products continues.

This chapter briefly presents some medicinal plants with remarkable anti-inflammatory effects. The focus is limited to dietary spice curcurmin from turmeric (*Curcuma longa*), capsaicin from red pepper (*Capsicum baccatum*) and ginger, as well as the herb *Allium sativum* (garlic), *Taraxacum officinale* (dandelion), *Echinacea purpurea* (purple coneflower), and *Silybum marianum* (milk thistle).

## Examples of Medicinal Plants as a Source of Anti-inflammatory Agents

(1) *Zingiber officinale*, commonly known as ginger, has been extensively used worldwide for medicinal purposes in numerous human ailments, i.e. common colds, fever, gastrointestinal complications, indigestion, arthritis, rheumatism, muscular aches, pains, hypertension, and infectious diseases. Consumed worldwide as dietary spice and beverage component, ginger possesses antioxidant, anti-inflammatory, and antitumor potential. Ginger is an excellent source of several bioactive phenols, volatile constituents like sesquiterpenes and monoterpenes, and nonvolatile pungent compounds such as gingerols, paradols, shogaols, zerumbone, and gingerones, all attributed with health-enhancing properties.

*In vitro* studies have been published on the anti-inflammatory response to ginger components, including inhibition of prostaglandin, thromboxane and leukotriene biosynthesis and suppression of 5-lipoxygenase activity [6, 7]. Additionally, evidence exists of the strong effects of ginger constituents on proinflammatory signaling and cytokine expression, in validated assays in human whole blood. Ginger constituents inhibit IL-1β expression in monocytes/macrophages [8]. Furthermore, inhibition of phosphorylation of three mitogen-activated protein kinases (MAPKs), extracellular signal-regulated kinases 1 and 2 (ERK1/2), and c-Jun N-terminal kinase (JNK) have been also demonstrated. Similarly, molecular targets involved in inflammatory prevention have been observed via inhibition of NF-κB, suppression of cIAP1 expression, activation of caspase-3/7, up-regulation Cytochrome-C, Apaf-1, and activation PI3K/Akt/I kB kinases IKK [9]. Moreover, a study with rats fed with high-fat diets reported the suppression of hepatic inflammation markers by decreased hepatic cytokine gene expression and NF-κB activation, following treatment for six weeks with an ethanolic ginger extract (400 mg/kg). Moreover, ginger supplementation (100 μg/mL for 24 h) was beneficial against induced inflammation in human hepatocytes, because of decreased cytokine genes expression. Similarly, ginger supplementation decreased NF-κB-target inflammatory gene expression of IL-6, IL-8 and serum amyloid A1 (SAA1) via suppression of NF-κB and IκB kinase (IKK) activities, and IκBα degradation [10, 11].

Anti-inflammatory effects of ginger have been proven in animals and humans suffering of chronic severe inflammatory. Arthritic rats treated with ginger oil exhibited less swelling and inflammation symptoms [12]. Adult volunteers who ate 5 gr of raw ginger daily witnessed a 25% reduction in thromboxane concentrations. A case series of seven patients with rheumatoid arthritis reported improved symptoms following supplemental ginger [13]. Another study on 56 patients with rheumatoid arthritis, osteoarthritis, and muscular discomfort who were given powdered ginger reported that more than three-quarters of the arthritis patients experienced relief in pain and swelling; all the patients with muscular discomfort reported symptoms relief. None of the patients reported adverse effects during the period of ginger consumption, i.e. 2.5 years [14].

Shogaol is one of the most bioactive components of ginger rhizomes. Studies assessed the [6]-shogaol's ability to protect primary rat astrocytes against lipopolysaccharide (LPS)-induced inflammation. This ginger component suppresses the release of pro-inflammatory cytokines and decreases the level of inducible nitric oxide syntheses (iNOS), cyclooxygenase-2 (COX-2), and phospho-NF-kB in LPS-treated astrocytes. Furthermore, [6]-shogaol treatment up-regulates histone acetylation and suppresses histone deacetylase expression, which prevents inflammatory processes. Additionally, [6]-shogaol treatment also increases the expression of heat-shock proteins that are key players in the modulation of inflammation [15, 16]. Another component with anti-inflammatory effects isolated from ginger is [4]-shogaol. A recent report determined that [4]-shogaol suppresses the activation of NF-κB and cell migration and invasion induced by TNF-α. Moreover, [4]-shogaol inhibits the phosphorylation of IκB and the translocation of NF-κB/Snail *in vitro*. Evidence shows that RKIP, an inhibitory molecule of IKK, is up-regulated after 4-shogaol treatment and prolongs the inhibitory effects of [4]-shogaol [17].

These examples of protection and efficacy toward degenerative diseases make ginger a new promising and natural anti-inflammatory agent that may be exploited therapeutically and needs to be paid close attention to by future studies.

(2) *Allium sativum* (garlic) has been traditionally used as a medicinal plant to treat heart disease, headaches, and certain tumors [18]. Moreover, many of garlic‘s health-promoting properties i.e. anti-thrombotic, lipid-lowering, antioxidant, and hypocholesterolemic actions [19] have been demonstrated in scientific studies. Besides these biological activities, garlic has been demonstrated in an *in vitro* model to possess anti-inflammatory properties with modulator effect on multiple cytokines levels. It lowers the production of pro-inflammatory cytokine IL-1, TNF-α, and IL-8, and stimulates IL-10 secretion (an antagonist of pro-inflammatory cytokines) [18]. IL-10 is able to modulate the immunopathology of brain inflammatory diseases such as Alzheimer's disease, providing another potential use for garlic as a cytokine modulator [20]. Garlic suppresses leukocyte inflammatory cytokine production *in vitro*, which abilities with potential therapeutic use in treatment of inflammatory bowel disease [21]. Additionally, garlic acts as a hypoglycemic agent and ameliorates alloxan-induced diabetes in murine models [18, 21, 22]. Garlic powder extracts from plants fertilized with sulphate during cultivation had stronger anti-inflammatory effects than unfertilized garlic, which suggests that organosulfur compounds, rather than other phytochemicals such as flavonoids, are responsible for the anti-inflammatory effects of garlic [19].

As an example, the sulfur compound thiacremonone isolated from garlic has anti-inflammatory effects in models of neuro-inflammation and amyloidogenesis. Namely, thiacremonone decreases LPS-induced memory impairment, glial activation, pro-inflammatory mediators' expression, and amyloidogenesis. In *in vitro* and *in vivo* studies, similar results were obtained with different doses of thiacremonone (1, 2, 5 μg/ml) and (1, 3, 10 mg/kg) in drinking water, respectively. It has been demonstrated that thiacremonone inhibits neuro-inflammation and amyloidogenesis through inhibition of NF-κB activity - a critical transcriptional factor regulating inflammation and amyloid-β generation - via blocking phosphorylation of IκBα in mice brain as well as cultured astrocytes and microglial BV-2 cells. This suggests that thiacremonone could be applied for intervention of inflammation-related neurodegenerative disease including Alzheimer's disease [23, 24].

(3) *Taraxacum officinale* commonly known as dandelion has many culinary and medicinal uses, despite being generally regarded as a weed. Most of the potentially bioactive components of dandelion have been isolated and identified as bitter substances and phenolic compounds with antioxidant and anti-inflammatory activities.

Dandelion influences the expression of certain pro-inflammatory mediators. The acute anti-inflammatory activity of dandelion was assessed by investigating the protective effects of a root methanolic extract in terms of inhibitory activity toward leukotriene B4 produced by leukocytes in response to inflammatory mediators. The observed 86% inhibition was attributed to two sesquiterpene glucosides [25]. Other authors have shown an anti-inflammatory activity of leaf extracts (0.1 and 1.0 μg/mL) in a central nervous system model, as indicated by a significant suppression of both TNF-α and IL-1 production by primary cultures of rat astrocytes stimulated with substance P and LPS [26]. Dandelion polyphenols also mediate the down-regulation of NO, prostaglandin E2 (PGE2), and pro-inflammatory cytokines (TNF-α, IL-1) production in LPS-stimulated macrophages (RAW264.7 cells). Some researchers observed that luteolin and luteolin-7-O-glucoside (< 20 μM) from dandelion flower extracts down-regulate both iNOs and COX-2 expression, without a concomitant reduction of enzymatic activities [27]. Furthermore, recent studies report that the effects of dandelion leaves on iNOs and COX-2 expressions and on inflammatory mediators are produced via inactivation of mitogen-activated protein kinases (MAPK) signal, in a dose-dependent manner [28]. Recently, it has been reported that luteolin and chicoric acid, two foremost constituents of dandelion, synergistically inhibit inflammatory responses via inactivation of PI3K-Akt pathways and impairment of NF-κB translocation in LPS-stimulated RAW264.7 cells [29].

The first animal study on the anti-inflammatory activity of dandelion reported a positive effect of an ethanolic extract of roots, administered per os (100 mg/kg body weight) to carrageenan-induced paw oedema rats [30, 31]. Notably, the observed inhibition was significant higher with methanolic dandelion extracts: flowers (95%), leaf (69%), and roots (51%) [32, 33]. *In vivo*, dandelion ameliorates the severity of induced pancreatitis and reduces IL-6 and TNF-α production during the course of the disease. The anti-inflammatory effects of aqueous extracts of dandelion leaves toward acute pancreatitis (cholecystokinin-octapeptide-induced) have also been evaluated in rats. Protective effects were shown in terms of significant decreases of pancreatic wet weight, cytokines (IL-6, TNF-α) secretion, and increased expression of heat shock proteins in the pancreas [34]. Recent animal studies also reported protective anti-inflammatory effects of dandelion on acute lung injury (ALI) induced by LPS in mice. Dandelion inhibits the production of inflammatory cytokines (IL-6, TNF-α) in the bronchoalveolar lavage fluid, six hours after damage induction, and decreases the number of neutrophils and the wet weight of lungs after 24 hours [35]. In addition, dandelion decreases LPS-induced myeloperoxidase (MPO) activity and increased SOD activity in the lungs [35]. This effect has been attributed to luteolin, which blocks mitogen-activated protein kinase (MEK), extracellular signal-regulated kinases (ERK), and Akt-related signaling cascades, and attenuates neutrophil chemotaxis and respiratory burst, in turn exerting beneficial effects in ALI [36].

In humans, a study of a seven-herb combination that included dandelion found significantly less rectal bleeding and lower disease activity in inflammatory bowel disease (IBD) patients [37]. The etiology of this disease is still unclear: it damages the mucosal lining of the gastrointestinal tract and current treatment is mostly aimed at systemic immunosuppression. Of note, several combined mechanisms of action of herbs might lead to decreased leukocyte infiltration and mucosal ulceration, in turn ameliorating the development of acute colonic inflammation. The tested herbal remedy - whose various constituents work in synergistic fashion - inhibited activity of the pro-inflammatory transcription factor nuclear factor kappa B (NF-kB), down-regulated COX-2, TNF-α, IL-1, IL-6, and iNOs, and also decreased NO, ROS, leukotriene and prostaglandin production, suggesting its potential use ad adjunct therapy for IBD [37].

(4) *Echinacea purpurea* also known as the purple coneflower - is an herbal medicine that has been used for centuries as a treatment for common colds, respiratory infections, and inflammatory conditions [38]. Studies suggest that *Echinacea purpurea* also may help preventing certain types of cancer [39, 40]. Endowed with potent antiviral and anti-microbial activities, *Echinacea purpurea* is best known for its ability to reduce inflammation caused by respiratory problems, bronchitis and strep throat, and also to kick-start the immune system, stimulating the production of white blood cells [38, 41].

Research also suggests that *Echinacea purpurea* can help reduce inflammation associated with arthritis. Steroids are commonly prescribed for inflammatory diseases such as arthritis. It is noteworthy that *Echinacea purpurea* is twice as effective as steroids in treating the symptoms of arthritis, yet no side effects have been reported to date. An added benefit is that, in addition to reducing inflammation, *Echinacea purpurea* enhances immunity, whereas steroids actually suppress the immune system [42, 43]. Because steroid drugs have so many negative side-effects, *Echinacea purpurea* may prove a valuable treatment in immune system reactions that cause conditions such as arthritis.

Studies have determined the inhibitory effects of alcohol extracts and individual fractions of such alcohol extracts of *Echinacea purpurea* on nitric oxide (NO) production in lipopolysaccharide (LPS)-activated the RAW 264.7 macrophage cell line, TNF-α and IL-1β, and explored the mechanism underlying the pharmacological anti-inflammatory activity [44]. Arginase activity of RAW 264.7 cells stimulated with 8-bromo-cAMP is significantly increased by *Echinacea purpurea* alcohol extracts. The polar fraction containing caffeic acid derivatives enhances arginase activity, while the lipophilic fraction containing alkamides inhibits NO production and iNOS expression [44]. This suggests that the anti-inflammatory activity of *Echinacea purpurea* might be due to multiple active metabolites, which work together to switch macrophage activation from their "classical" activation state towards alternative activations.

The most common constituents found in *Echinacea purpurea* include alkylamides, caffeic acid derivatives, polysaccharides, and lipoproteins [45]. Anti-inflammatory principles of *Echinacea purpurea* have been identified in high-molecular weight polysaccharides (molecular weight between 30,000 and 100,000), echinacoside (caffeoyl derivative), alkylamides (dodeca-2Z,4E-diene-8,10-diynoic acid isobutylamide, 2Z,4E-undeca-2,4-dien-8,10-diynoic acid isobutyl amide, 2Z,4E-undeca-2,4-dien-8,10-diynoic acid isobutyl amide) [45].

(5) *Silybum marianum* (milk thistle) is a well-known medicinal plant, of the Mediterranean region and found in many other countries. Its medicinal effects help liver diseases (cirrhosis, jaundice and hepatitis), gallbladder disease, and bile-related disorders [46, 47]. Milk thistle has hepatoprotective and antioxidant properties, which stabilize and protect the membrane lipids of hepatocytes (liver cells) [48]. This medicinal plant inhibits the enzymes peroxidase and lipoxygenase. A recent *in vitro* study also showed activity against liver cancer cells [49].

The most powerful *Silybum marianum* extract is used in herb medicine under the name silymarin (a flavonolignane complex consisting of silibinin A and B/silybin/silymarin I, isosilibinin A and B, silicristin/silymarin II, silidianin), which contains high proportions (approximately 20 to 35%) of oil [50, 51]. Hence, *Silybum marianum* oil has been suggested as being suitable edible and vitamin E-rich oil [52]. Silibinin is used in cerebral edema and acute hepatitis therapy [53].

(6) *Curcuma longa* ((turmeric) is a plant of Indian origin widely used as dietary spice. In folk medicine, turmeric is often used to prevent or treat many diseases [54]. Tumeric possesses antibacterial, antifungal and anti-parasitic properties. Beneficial effects have also been shown in skin, liver, and the gastrointestinal and respiratory tracts. Numerous studies have shown the capacity of turmeric to stabilize membranes and to prevent lipid peroxidation, which is a key process in the onset and progression of many health disorders such as hepatic, renal, cardiovascular and degenerative neurological diseases, diabetes, and cancer [55]. While Curcuma longa has been attributed with numerous pharmacological activities, we focus on one of its best-explored actions, i.e. its anti-inflammatory effects. Turmeric has anti-inflammatory effects mediated through the modulation of the metabolism of ecosanoids and by its immunomodulating capacity, mainly altering the profile of Thl cytokines of T helper cells [54, 56].

Turmeric contains three active major curcuminoids, namely curcumin (diferuloylmethane), demethoxycurcumin, and bisdemethoxycurcumin, as well as volatile oils (tumerone, atlantone, and zingiberone), sugars, proteins, and resins.

Curcumin interacts with numerous molecular targets involved in inflammation. It modulates the inflammatory response by down-regulating the activity of COX-2, lipoxygenase, and iNOs enzymes, by inhibiting the production of the inflammatory cytokines TNF-α, IL-1, -2, -6, -8, and -12, monocyte chemoattractant protein (MCP) and migration inhibitory protein, and by down-regulating mitogen-activated and Janus kinases [56, 57].

It has been shown that turmeric works in acute, subacute, and chronic inflammation in mice and rats. Even tough it has strong analgesic and antipyretics properties; it does not produce significant gastric irritation and has no effect on the CNS [58]. Many studies have demonstrated the anti-inflammatory ability of curcuminoids, proven by modulation of the metabolism of arachidonic acid via inhibition of the activity and synthesis of cyclooxygenase, lipoxygenase, prostaglandins and thromboxanes [59, 60], in turn limiting the development of inflammatory processes and aggregation of platelets [61]. Furthermore, it was observed that the induction of the enzymes responsible for the detoxification of electrophilic products of lipid peroxidation (glutathione transferase and glutathione peroxidase) may contribute to the anti-inflammatory action of curcumin. Another anti-inflammatory mechanism that curcumin can exert is through its antiproliferative and apoptotic effects in arterial smooth muscle cells, both mediated by inhibition of protein tirosin kinase and protein kinase C and other mechanisms molecular [62]. Curcumin inhibits the incorporation of arachidonic acid in lipid membranes, thus avoiding the release of the eicosanoid mediators of inflammation prostaglandin E2, leukotriene B4 and C4, and hydrolytic enzymes: collagenase, elastase and hyaluronidase secreted by macrophages [60]. Curcumin inhibits the production of TNF-α produced by monocytic cell lines and human macrophages [63]. Inflammatory responses play critical roles in development of pancreatitis. In animal models of pancreatitis, curcumin alleviates inflammation by decreasing the activation of NF-κB and AP-1, as well as by inhibiting IL-6, TNF-α, and iNOS expression in the pancreas. Curcumin's inhibitory effect on inflammatory mediators results in improvement of diseases, as measured by histology, serum amylase, pancreatic trypsin, and neutrophil infiltration [64, 65]. *In vitro* studies indicate that curcumin's inhibition of inflammatory cytokines is achieved through regulation of activation of certain transcription factors such as activating protein-1 (AP-1) and NF-κB in stimulated monocytes and alveolar macrophages, thereby blocking cytokine gene expression [66]. In an animal model of inflammation, curcumin inhibits arachidonic acid metabolism and inflammation in skin epidermis via down-regulation of COX-2 and lipoxygenase activities [66].

Turmeric´s anti-inflammatory properties and therapeutic potential have been demonstrated in a variety of pathological conditions, including irritable bowel syndrome (IBS). The most common symptoms in IBS patients are abdominal pain, bloating, altered bowel habits, and increased stool frequency accompanied by low-grade inflammation of the intestinal mucosa (the latter being responsible for some of the symptomology) [67]. In a pilot study of IBS patients, either 72 mg or 144 mg of a standardized turmeric extract was administered to a group of 102 or 105 subjects, respectively. After four weeks, those in the 72-mg group experienced a 53% reduction in IBS prevalence, while the 144-mg group experienced a 60% decrease. Additionally, abdominal pain and discomfort scores were reduced by 22% in the 72-mg group and 25% in the 144-mg group [68]. Crohn's disease (CD) and ulcerative colitis (UC) are the two primary forms of inflammatory bowel disease (IBD). Both diseases bring about abdominal pain, vomiting, diarrhea, bloody stools, weight loss, and clinical manifestations such as arthritis, pyoderma gangrenosum, and primary sclerosing cholangitis. In an ad-hoc study, the effect of curcumin (360 mg three times daily) was examined in five patients with CD and in five patients with UC (550 mg curcumin twice daily). After two months, all UC patients improved and exhibited normal ESR, CRP, and inflammation indices. In the CD group, CDAI scores decreased by an average of 55 points, whereas CRP and ESR decreased in four of five patients [69, 70]. Another clinical trial was conducted to assess the efficacy of curcumin (1 g twice daily) as a maintenance therapy in 82 patients with UC. The authors conclude that curcumin plus standard therapy is more effective in maintaining remission than placebo plus standard UC treatment [70, 71]. A turmeric extract devoid of essential oils was given to Wistar female rats suffering rheumatoid arthritis. Intraperitoneal injection of the extract containing 4 mg total curcuminoids/kg/day for four days prior to arthritis induction inhibited joint inflammation in both the acute (75%) and chronic (68%) phases. Oral administration of a 30-fold higher dose (to compensate for low gastrointestinal absorption) of the curcuminoid preparation, given to rats four days prior to arthritis induction, reduced joint inflammation by 48% on the third day of administration [72]. Moreover, in a preliminary double-blind, randomized, controlled trial, curcumin was compared to phenylbutazone in human patients with rheumatoid arthritis. Curcumin given at 1200 mg daily was effective in improving joint swelling, morning stiffness, and walking time [73]. One pilot study examined the effect of curcumin for tropical pancreatitis in 15 patients. The effects on pain patterns as well as on erythrocyte malonylaldehyde (MDA; an indicator of lipid peroxidation) and glutathione (GSH) were assessed at baseline and after six weeks. A significant reduction in MDA levels was reported in the curcumin group [74, 75].

Results from ongoing clinical trials of curcumin in inflammatory conditions should clarify its role as therapeutic agent and elucidate some of the mechanisms responsible for its efficacy.

(7) *Capsicum baccatum* (red pepper) is the major source of capsaicinoids, which include capsaicin, dihydrocapsaicin, nordihydrocapsaicin, homodihydrocapsaicin, and homocapsaicin, among others [76]. Numerous pharmacological properties have been attributed to capsaicinoids, including induction of analgesia, chemoprevention, anti-inflammation, as well as antioxidant and anti-obesity actions [76-79]. Moreover, capsaicinoids might be beneficial in cardiovascular and gastrointestinal systems [76].

Capsaicin is the main capsaicinoid of chili peppers and is a proven anti-inflammatory phytochemical used by arthritis sufferers to relieve pain and inflammation [80-82]. Capsaicin is believed to be synthesized in the interlocular septum of chili peppers by the addition of a branched-chain fatty acid to vanillylamine [83, 84]. Biosynthesis depends on the gene AT3, which resides at the pun1 locus, and which encodes a putative acyltransferase [85]. Sweet peppers contain an analog non-pungent form of capsaicin called capsiate, which is endowed with potent anti-inflammatory properties and diverse healthful potential [86, 87]. Moreover, sweet peppers are rich in capsinoids, which share similar structure with capsaicinoids and consist of capsiate, dihydrocapsiate, nordihydrocapsiate, and other related molecules. Compared with capsaicinoids, capsinoids are less pungent and easily break down in aqueous solutions [76]. Capsinoids are being attributed with antitumor, antioxidant, and anti-obesity actions. Notably, because capsinoids are less toxic than capsaicinoids, they may have advantages over the latter in chemoprevention and weight loss [76]. Animal studies with *Capsicum baccatum* juice (0.25-2.0 g/kg) have been carried out. The main constituent of this juice (analyzed following chloroform extraction) is capsaicin, which is able to inhibit neutrophil migration towards inflammatory foci. This juice significantly decreased leucocyte and LDH concentration in exudates of a pleurisy model and reduced vascular permeability in a murine model of carrageenan-induced peritonitis. Additionally, the juice reduced pro-inflammatory cytokines (TNF-α, IL-1β) production during mouse inflammatory immune peritonitis [88]. Capsaicin reduces pain associated with rheumatoid arthritis as well as joint or muscle pain deriving from fibromyalgia or other disorders [89]. Capsaicin is being tested as component of novel analgesic drugs, to treat post-surgical and osteoarthritis pain [90].

# Conclusion

Traditional herbal medicine is shifting toward the field of nutraceuticals or functional foods. Numerous phytochemicals found in plants and herbs have interesting pharmacological properties that, through different pathways, modulate inflammation. This chapter reviews few notable examples of the several natural products that have been shown beneficial effects on inflammatory processes. Mechanistically, spices such as *Zingiber officinale* (ginger), *Allium sativum* (garlic), *Taraxacum officinale* (dandelion), *Echinacea purpurea* (purple coneflower), and *Silybum marianum* (milk thistle) reduce systemic inflammation mostly by modulation of cytokines secretion. Selected spice components such as curcurmin from turmeric (*Curcuma longa*) and capsaicin from red pepper (*Capsicum baccatum*) also reduce inflammation by influencing arachidonic acid metabolism, the secretion of lysosomal enzymes by macrophages, and by controlling the release of pro-inflammatory mediators.

In summary, the use of well-characterized spices or of the bioactive principles isolated from them might prove useful in preventing or treated inflammation and associated disorders.

## Acknowledgments

This work has been supported by the Spanish Ministry of Science and Innovation (R&C2007-01920) and the Regional Government of Madrid (CPI/0631/2008). The authors thank Prof. Visioli for his very helpful and constructive comments during the preparation of this review.

## References

[1] V. Kumar, A. Abbas and N. Fausto, *Robbins and Cotran pathologic basis of disease*, Philadelphia (Pennsylvania), p. 47 (2004).

[2] S. Sosa, M.J. Balick, R. Arvigo, R.G. Esposito, C. Pizza, G. Altinier and A. Tubaro, *J. Ethnopharmacol.*, 81, 211 (2002).

[3] N.J. Olsen, *Minerva. Med.*, 102, 33 (2011).

[4] D. Leveque and F. Jehl, *J. Clin. Pharmacol.*, 47, 579 (2007).

[5] H.J. Choi, B.C. Cho, S.J. Shin, S.H. Cheon, J.Y. Jung, J. Chang, S.K. Kim, J.H. Sohn and J.H. Kim, *Cancer. Chemother. Pharmacol.*, 61, 309 (2008).

[6] F. Kiuchi, M. Shibuya and U. Sankawa, *Chem. Pharm. Bull. (Tokyo)*, 30, 754 (1982).

[7] D.L. Flynn, M.F. Rafferty and A.M. Boctor, *Prostaglandins. Leukot. Med.*, 24, 195 (1986).

[8] A. Nievergelt, J. Marazzi, R. Schoop, K.H. Altmann and J. Gertsch, *J. Immunol.*, 187, 4140 (2011).

[9] M.S. Butt and M.T. Sultan, *Crit. Rev. Food. Sci. Nutr.*, 51, 383 (2011).

[10] X.H. Li, K.C. McGrath, S. Nammi, A.K. Heather and B.D. Roufogalis, *Basic. Clin. Pharmacol. Toxicol.*, 110, 238 (2012).

[11] E.A. Al-Suhaimi, N.A. Al-Riziza and R.A. Al-Essa, *Am. J. Chin. Med.*, 39, 215 (2011).

[12] J.N. Sharma, K.C. Srivastava and E.K. Gan, *Pharmacology.*, 49, 314 (1994).

[13] K.C. Srivastava and T. Mustafa, *Med. Hypotheses.*, 29, 25 (1989).

[14] K.C. Srivastava and T. Mustafa, *Med. Hypotheses.*, 39, 342 (1992).

[15] S. Shim, S. Kim, D.S. Choi, Y.B. Kwon and J. Kwon, *Food. Chem. Toxicol.*, 49, 2734 (2011).

[16] H. Chen, L. Lv, D. Soroka, R.F. Warin, T.A. Parks, Y. Hu, Y. Zhu, X. Chen and S. Sang, *Drug. Metab. Dispos.*, (2012).
[17] Y.L. Hsu, C.Y. Chen, I.P. Lin, E.M. Tsai, P.L. Kuo and M.F. Hou, *J. Agric. Food. Chem.*, 60, 852 (2012).
[18] M. Ebadi, *Pharmacodynamic Basis of Herbal Medicine*, FL: CRC Press, Boca Raton (Florida) (2002).
[19] H.P. Keiss, V.M. Dirsch, T. Hartung, T. Haffner, L. Trueman, J. Auger, R. Kahane and A.M. Vollmar, *J. Nutr.*, 133, 2171 (2003).
[20] A.M. Szczepanik and G.E. Ringheim, *Journal of Alzheimer's disease : JAD*, 5, 105 (2003).
[21] G. Hodge, S. Hodge and P. Han, *Cytometry.*, 48, 209 (2002).
[22] K. Huang and W. Williams, *The Pharmacology of Chinese Herbs*, CRC Press, Boca Rato (FLorida) (1999).
[23] G.H. Lin, Y.J. Lee, D.Y. Choi, S.B. Han, J.K. Jung, B.Y. Hwang, D.C. Moon, Y. Kim, K.W. Oh and H.S. Jung, *Journal of Alzheimer's disease : JAD*, (2012).
[24] K. Spelman, J. Burns, D. Nichols, N. Winters, S. Ottersberg and M. Tenborg, *Altern. Med. Rev.*, 11, 128 (2006).
[25] Y. Kashiwada, K. Takanaka, H. Tsukada, Y. Miwa, T. Taga, S. Tanaka and Y. Ikeshiro, *J. Asian. Nat. Prod. Res.*, 3, 191 (2001).
[26] H.M. Kim, H.Y. Shin, K.H. Lim, S.T. Ryu, T.Y. Shin, H.J. Chae, H.R. Kim, Y.S. Lyu, N.H. An and K.S. Lim, *Immunopharmacol. Immunotoxicol.*, 22, 519 (2000).
[27] C. Hu and D.D. Kitts, *Mol. Cell. Biochem.*, 265, 107 (2004).
[28] Y.J. Koh, D.S. Cha, J.S. Ko, H.J. Park and H.D. Choi, *J. Med. Food.*, 13, 870 (2010).
[29] C.M. Park, K.S. Jin, Y.W. Lee and Y.S. Song, *Eur. J. Pharmacol.*, 660, 454 (2011).
[30] N. Mascolo, F. Capasso, A. Menghini and M.P. Fasulo, *Phytother. Res.*, 1, 28 (1987).
[31] B. Tita, U. Bello, P. Faccendini, R. Bartolini and P. Bolle, *Pharmacol. Res.*, 27, 23 (1993).
[32] K. Yasukawa, T. Akihisa, H. Oinuma, Y. Kasahara, Y. Kimura, S. Yamanouchi, K. Kumaki, T. Tamura and M. Takido, *Biol. Pharm. Bull.*, 19, 1329 (1996).
[33] K. Yasukawa, T. Akihisa, Y. Inoue, T. Tamura, S. Yamanouchi and M. Takido, *Phytother. Res.*, 12, 484 (1998).
[34] S.W. Seo, H.N. Koo, H.J. An, K.B. Kwon, B.C. Lim, E.A. Seo, D.G. Ryu, G. Moon, H.Y. Kim, H.M. Kim and S.H. Hong, *World journal of gastroenterology : WJG*, 11, 597 (2005).
[35] L. Liu, H. Xiong, J. Ping, Y. Ju and X. Zhang, *J. Ethnopharmacol.*, 130, 392 (2010).
[36] J.P. Lee, Y.C. Li, H.Y. Chen, R.H. Lin, S.S. Huang, H.L. Chen, P.C. Kuan, M.F. Liao, C.J. Chen and Y.H. Kuan, *Acta. Pharmacol. Sin.*, 31, 831 (2010).
[37] L.N. Jackson, Y. Zhou, S. Qiu, Q. Wang and B.M. Evers, *Am. J. Chin. Med.*, 36, 953 (2008).
[38] J.B. Hudson, *J. Biomed. Biotechnol.*, 2012, 769896 (2012).
[39] L. Tenney, *The Encyclopedia of Natural Remedies* Woodland Publishing, Utah (1995).
[40] D. Mowrey, *The Scientific Validation of Herbs* Keats Publishing, New Canaan (Connecticut) (1986).
[41] P. Ody, *The Complete Medicinal Herbal* Dorling-Kindersley, New York (1993).
[42] *Echinacea*, Woodland Publishing, Pleasant Grove (UT) (1995).

[43] C. Steinmuller, J. Roesler, E. Grottrup, G. Franke, H. Wagner and M.L. Lohmann-Matthes, *Int. J. Immunopharmacol.*, 15, 605 (1993).
[44] Z. Zhai, A. Solco, L. Wu, E.S. Wurtele, M.L. Kohut, P.A. Murphy and J.E. Cunnick, *J. Ethnopharmacol.*, 122, 76 (2009).
[45] R. Bauer and L. Lawson, *Phytomedicines of Europe*, American Chemical Society, Washington, DC (1998).
[46] F. Fraschini, G. Demartini and D. Esposti, *Clin. Drug. Invest.*, 22, 51 (2002).
[47] V. Kurkin, *Pharmaceut. Chem. J.*, 37, 189 (2003).
[48] C. Mulrow, V. Lawrence, B. Jacobs, C. Dennehy, J. Sapp, G. Ramirez, C. Aguilar, K. Montgomery, L. Morbidoni, J.M. Arterburn, E. Chiquette, M. Harris, D. Mullins, A. Vickers and K. Flora, *Evid. Rep. Technol. Assess. (Summ)*, 1 (2000).
[49] J.J. Lah, W. Cui and K.Q. Hu, *World journal of gastroenterology : WJG*, 13, 5299 (2007).
[50] K. Ramasamy and R. Agarwal, *Cancer. Lett.*, 269, 352 (2008).
[51] M. Malekzadeh, S. Mirmazloum, S. Mortazavi, M. Panahi and H. Angorani, *J. Med. Plants Res.*, 5, 1485 (2011).
[52] M.H. El-Mallah, S. El-Shami and M.M. Hassanein, *Grasas. Aceites.*, 54, 397 (2003).
[53] R. Saller, R. Brignoli, J. Melzer and R. Meier, *Forsch. Komplementmed.*, 15, 9 (2008).
[54] G. Ramadan, M.A. Al-Kahtani and W.M. El-Sayed, *Inflammation.*, 34, 291 (2011).
[55] A. Shehzad, T. Ha, F. Subhan and Y.S. Lee, *Eur. J. Nutr.*, 50, 151 (2011).
[56] J.S. Jurenka, *Altern. Med. Rev.*, 14, 141 (2009).
[57] A. Goel, A.B. Kunnumakkara and B.B. Aggarwal, *Biochem. Pharmacol.*, 75, 787 (2008).
[58] R. Srimal and D. BN., *Development of unani drugs from herbal sources and the role of elements in their mechanism of action*, Hamdard National Foundation New Delhi, p. 131 (1985).
[59] G. Began, E. Sudharshan and A.G. Appu Rao, *Lipids.*, 33, 1223 (1998).
[60] B. Joe and B.R. Lokesh, *Lipids.*, 32, 1173 (1997).
[61] K.C. Srivastava, A. Bordia and S.K. Verma, *Prostaglandins. Leukot. Essent. Fatty. Acids.*, 52, 223 (1995).
[62] M.M. Chan, H.I. Huang, M.R. Fenton and D. Fong, *Biochem. Pharmacol.*, 55, 1955 (1998).
[63] M.M. Chan, *Biochem. Pharmacol.*, 49, 1551 (1995).
[64] I. Gukovsky, C.N. Reyes, E.C. Vaquero, A.S. Gukovskaya and S.J. Pandol, *Am. J. Physiol. Gastrointest. Liver. Physiol.*, 284, G85 (2003).
[65] E. Vaquero, I. Gukovsky, V. Zaninovic, A.S. Gukovskaya and S.J. Pandol, *Am. J. Physiol. Gastrointest. Liver. Physiol.*, 280, G1197 (2001).
[66] M.T. Huang, T. Lysz, T. Ferraro, T.F. Abidi, J.D. Laskin and A.H. Conney, *Cancer. Res.*, 51, 813 (1991).
[67] G. Barbara, R. De Giorgio, V. Stanghellini, C. Cremon and R. Corinaldesi, *Gut.*, 51, i41 (2002).
[68] R. Bundy, A.F. Walker, R.W. Middleton and J. Booth, *J. Altern. Complement. Med.*, 10, 1015 (2004).
[69] P.R. Holt, S. Katz and R. Kirshoff, *Dig. Dis. Sci.*, 50, 2191 (2005).
[70] R.A. Taylor and M.C. Leonard, *Altern. Med. Rev.*, 16, 152 (2011).

[71] H. Hanai, T. Iida, K. Takeuchi, F. Watanabe, Y. Maruyama, A. Andoh, T. Tsujikawa, Y. Fujiyama, K. Mitsuyama, M. Sata, M. Yamada, Y. Iwaoka, K. Kanke, H. Hiraishi, K. Hirayama, H. Arai, S. Yoshii, M. Uchijima, T. Nagata and Y. Koide, *Clin. Gastroenterol. Hepatol.*, 4, 1502 (2006).

[72] J.L. Funk, J.N. Oyarzo, J.B. Frye, G. Chen, R.C. Lantz, S.D. Jolad, A.M. Solyom and B.N. Timmermann, *J. Nat. Prod.*, 69, 351 (2006).

[73] S.D. Deodhar, R. Sethi and R.C. Srimal, *Indian. J. Med. Res.*, 71, 632 (1980).

[74] S. Durgaprasad, C.G. Pai, Vasanthkumar, J.F. Alvres and S. Namitha, *Indian. J. Med. Res.*, 122, 315 (2005).

[75] M. Mesa, M. Ramírez-Tortosa, C. Aguilera, A. Ramírez-Boscá and A. Gil, *Ars. Pharmaceutica.*, 41, 307 (2000).

[76] X.J. Luo, J. Peng and Y.J. Li, *Eur. J. Pharmacol.*, 650, 1 (2011).

[77] G.J. He, X.L. Ye, X. Mou, Z. Chen and X.G. Li, *Eur. J. Med. Chem.*, 44, 3345 (2009).

[78] S. Handler and D. Rorvik, *PDR for Nutritional Supplements*, Thomas Reuters (2008).

[79] A.G. Dulloo, *Obes. Rev.*, 12, 866 (2011).

[80] A.R. Zimmer, B. Leonardi, D. Miron, E. Schapoval, J.R. Oliveira and G. Gosmann, *J. Ethnopharmacol.*, 139, 228 (2012).

[81] A.A. Oyagbemi, A.B. Saba and O.I. Azeez, *Indian. J. Cancer.*, 47, 53 (2010).

[82] L. Reyes-Escogido Mde, E.G. Gonzalez-Mondragon and E. Vazquez-Tzompantzi, *Molecules.*, 16, 1253 (2011).

[83] H. Fujiwake, T. Suzuki, S. Oka and K. Iwai, *gricultural and Biological Chemistry*, 44, 2907 (1980).

[84] I. Guzman, P. Bosland and M. O'Connell, *Recent Advances in Phytochemistry* Springer, p. 117 (2011).

[85] C. Stewart, Jr., B.C. Kang, K. Liu, M. Mazourek, S.L. Moore, E.Y. Yoo, B.D. Kim, I. Paran and M.M. Jahn, *Plant. J.*, 42, 675 (2005).

[86] E.J. Lee, M.S. Jeon, B.D. Kim, J.H. Kim, Y.G. Kwon, H. Lee, Y.S. Lee, J.H. Yang and T.Y. Kim, *Free. Radic. Biol. Med.*, 48, 1133 (2010).

[87] R. Sancho, C. Lucena, A. Macho, M.A. Calzado, M. Blanco-Molina, A. Minassi, G. Appendino and E. Munoz, *Eur. J. Immunol.*, 32, 1753 (2002).

[88] F. Spiller, M.K. Alves, S.M. Vieira, T.A. Carvalho, C.E. Leite, A. Lunardelli, J.A. Poloni, F.Q. Cunha and J.R. de Oliveira, *J. Pharm. Pharmacol.*, 60, 473 (2008).

[89] L. Fraenkel, S.T. Bogardus, Jr., J. Concato and D.R. Wittink, *Arch. Intern. Med.*, 164, 1299 (2004).

[90] *Chronic Pain (Pain Management Series)* Informa Healthcare, New York (2007).

In: Phytochemicals
Editor: Hai-Xue Kuang

ISBN: 978-1-62417-354-7


*Chapter X*

# Antioxidative Phytochemicals and Skin Care

***Frédéric Bonté*** [*] ***and Gallic Beauchef***
LVMH Recherche, Saint Jean de Braye, France

## Abstract

Skin is one of the most important parts of the body and represents approximately two square meters in the adult. Day by day, skin undergoes internal and external aggressions. Daily emotional stress, ultraviolet radiations and abrupt temperature variations make it a target of choice for the oxidative stress. The biological targets are numerous: DNA, proteins, lipids mitochondria... During ageing or in case of strong aggressions, the cutaneous defense systems are overflowed, and the skin will develop wrinkles, spots or pathologies. How natural substances can increase the systems of natural detoxification and defence is one of the current scientific questions. This article proposes to focus on these targets and phytochemicals, especially the main antioxydant natural substances used in the dermatological care of the skin. The article also relates to the recent discoveries in this field. The main biological targets and the expected efficiency of antioxidants phytochemicals used in cosmetics are detailed.

## Introduction

With a surface area of 2 $m^2$ and making up nearly 5-8 % of the body's weight, the skin is one of our most important organs. It is not only important for social interaction but also essential for the protection of human life itself. It ensures multiple functions: protection against external aggressions (mechanical, chemical, thermal and infectious types of stress); conservation of fluids; thermoregulation; regulation of exchanges; metabolic functions (vitamin D synthesis, lipid synthesis) and sensory functions (touch, pain) [1]. Aging and

[*] E-mail address: fredericbonte@research.lvmh-pc.com.

stress modify the skin's appearance, and skin care products have been developed to fight the appearance of aging skin and to help skin preserve its youthful qualities. These products very often contain antioxidant phytochemicals. This paper, after presenting the causes and the biological targets of aging and stress in the skin, will focus on the main antioxidative phytochemicals used. It will show that thanks to the complexity of the mechanism involved, only combinations of molecules are able to reverse or delay the signs of aging skin and are able to maintain its functional longevity.

# I. Skin Aging and Oxidative Stress

All of the skin's functions are ensured by a complex structure, associating various tissue elements (epithelial, conjunctive, vascular, muscular and nervous), organized in four distinct compartments, which are, from surface to depth (Figure 1):

- the horny layer also called stratum corneum
- the epidermis;
- the dermis, separated from the epidermis by the epidermal dermal junction (EDJ);
- the hypodermis.

The epidermis consists of four cellular types: mostly keratinocytes (90%) but also melanocytes, which play a role in pigmentation and the defense of the skin against ultraviolet rays (UV), Merkel cells, which act as mechanoreceptors, and Langerhans cells, which collect environmental antigens in order to present them to local T lymphocytes and to stimulate a natural immune response [1].

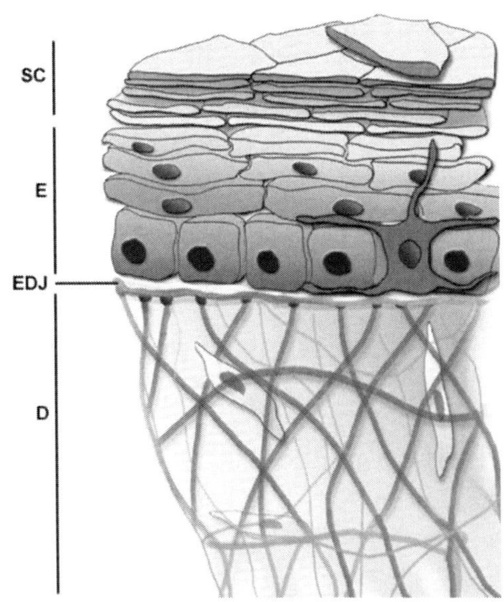


Figure 1. Human skin architecture: SC stratum corneum, E epidermis, EJD epidermal dermal junction, D dermis.

The skin thus makes it possible to implement this double role: to protect the skin from external threats and to maintain internal hydration of the tissue [2], in particular via its uppermost surface layer, the horny layer, also called the stratum corneum.

The dermis is a fibrous, elastic component, much thicker than the epidermis, and is in charge of biomechanical skin properties. The dermis and epidermis are connected by an adhesion zone centered around the EJD, a very efficient, complex interface, which transfers signals and nutrients [3]. The dermis contains the cutaneous vascular system (which the epidermis does not). The dermis is made up of fibroblasts, dendrocytes and immune system cells, which can be recruited if necessary following stress or UV damage. The various components of the extracellular matrix (ECM), synthesized by the fibroblasts, are primarily collagen (mainly types I and III), proteoglycans (PG), glycosaminoglycans (GAG), hyaluronic acid (HA) and elastin. The dermis, thanks to this complex cell-interactive matrix, is firm, extensible and elastic. The hypodermis, or fat tissue, composed of mesenchyme cells and adipocytes in a fatty and fibrous environment and acts primarily as a reservoir of hormones, growth factors, heat and mechanical insulator.

Everybody knows that the skin's appearance changes with aging. The skin becomes thinner and loses its biomechanical properties. During the aging process, the number of epidermal cells, as well as their renewal rates, fall [4-6]. The cells of the basal membrane increase in size [7], the keratinocytes become shorter [6] and the corneocytes become larger, owing to the reduction of their renewal rate [8]. The Langerhans cells decrease in number, which leads to a weakening of cutaneous immunity [9]. Old skin is thinner than young skin and looks drier due to a lower sebum level and sometimes altered water content. In this regard, it was recently shown that UV exposure provoked downregulation of epidermal water transport protein named aquaporin 3, associated with skin dehydration [10]. With stress and aging, skin pigment disorders, brown spots, a blotchy complexion, lentigines and liver spots appear, especially on exposed skin. The EJD is flattened, leading to a reduced exchange surface and less nutrition of the epidermis. The renewal rate of various matrix proteins decreases: collagen (approximately 2% *per annum* [4]), GAGs and HA [5, 11], due to the reduction in the number, strength and metabolic activity of the fibroblasts. This loss of the structural integrity of the dermis leads to a significant decrease in its biomechanical properties, especially extensibility [5] and elasticity, facilitating the appearance of wrinkles [12]. Furthermore, some studies show a relationship between hormone balance disruption and cutaneous aging, the erosion of the properties mentioned above being faster in women than in men [2]. The global cutaneous metabolism is also impacted during aging. Protective and defense capacities decrease, oxidized proteins or lipids accumulate, glycation increases, free radicals or reactive oxygen species (ROS) and ROS-induced damage increase [13]. Oxidative stress also impacts the functioning of the mitochondria, the cell's bioenergy organelle, forcing the cells to compensate [14]. Moreover, often visible at the surface of old skin, the cutaneous blood network decreases because of modifications in angiogenesis.

## The Main Theories Related to Skin Aging

Three major theories of aging are often cited in regard to skin aging: replicative senescence, or telomeric theory; mitochondrial dysfunction; and the third theory, which connects the previous two, free radical theory.

Telomeres are non-coding regions of chromatin located at each 3' extremity of the chromosomes, allowing for replication without the loss of nucleotide coding information. Formed by the repetition of the sequence TTAGGG extending over several hundred base-pairs (bp), they ensure the stability of the chromosome ends [15, 16]. According to telomeric theory, somatic cells have too weak a telomerase activity to maintain these telomeric repetitions. The shortening of the telomeres is partly attributed to the accumulation of DNA breaks induced by oxidative stress or UV rays. Furthermore, telomerase was also shown to be a target of UV rays in skin keratinocytes and fibroblasts [17].

The mitochondria are complex intracellular organelles, fuel factories that convert the food we eat into biological cellular energy. They generate biological energy in the form of adenosine triphosphate (ATP) during an oxidative phosphorylation process, which occurs in the internal mitochondrial membrane. Five multiprotein complexes are responsible for the electron transport chain (ETC) necessary for ATP synthesis. Most electrons cross the protein complexes successfully, but some electrons can "escape" from the ETC and react with available oxygen to form ROS, in particular, the radical superoxide. It is supposed that approximately 2% of mitochondrial oxygen ($O_2$) is used to form ROS. Mitochondria, thus damaged by ROS, see their function rapidly decline and in return generate more ROS, thus creating a vicious circle where the ROS are at the same time cause and consequence of the protein, lipid and mitochondrial DNA (mtDNA) damage [18-20]. The outcome is lower bioenergy production as confirmed in human fibroblasts of 51 old donors from 1 to 103 years of age in a recent study [21]. It is important to note that mtDNA synthesis takes place close to the mitochondrial membrane, site of the formation of ROS, and is therefore weakened compared to the nuclear one. As a matter of fact, unlike nuclear DNA, it neither profits from mechanisms of repair by excision or recombination [22] nor has protective histones. Thus, it has been highlighted that the cumulated mtDNA damage could play an important role in cutaneous aging via ROS and that messages between mt and nuclear DNA exist [23, 24].

Free radicals and oxidative stress have been recognized as one of the most important factors in the process of aging [25]. Cutaneous cells have natural repair mechanisms, including many detoxifying enzymes, like superoxide dismutase (SOD), catalase (CAT), glucose-6-phosphate dehydrogenase (G6PD), heme oxygenase-1 (HO-1) and glutathione peroxidase (GPx). They also are rich in natural non-enzymatic antioxidants, like tocopherols, ascorbic acid, pyruvate, nicotinamide adenine dinucleotide (NADH), glutathione, etc., molecules, which can give an electron by reduction and thus scavenge ROS. Many studies describe a diminution in antioxidant potential in old skin cells, suggesting redox imbalance during aging and an excess of production of ROS leading to an accumulation of cellular damage [26-28]. Oxidation of the lipid cellular membranes thus leads, for instance, to cellular fluidity and significant modifications in phospholipid architecture. The consequences are that all these attacks modify signaling transduction pathways and internal or external molecular traffic. Polyunsaturated fatty acids present in the cellular membrane are extremely sensitive to oxidation and form very reactive peroxyl radicals (ROO), which will ultimately from very reactive and oxidative products, i.e., malondialdehyde (MDA) and hydroxynonenal (HNE). The complex lipids of the horny layer, responsible for skin barrier properties, are also the targets of ROS [29-31]. Proteins present in skin and its appendages, such as hair, are also major oxidative stress targets. Oxidation of lysine, arginine, proline and threonine leads to carbonyl residue. Hydroxyl radicals, in particular, can react with the protein residues cysteine and methionine present in skin keratins to form disulfide mixtures. Some parts of the skin, face, hair and hands are more exposed, more visible in our social contacts and sensitive to stresses. As a consequence, all of these oxidation processes bring about visible modifications of the skin and its appendages, which is why scientists have developed active ingredients to fight them.

## Cutaneous Aging and Oxidative Stress

A distinction was made in the past between an *intrinsic* aging, depending on time, internal or emotional stress and genetic background, and an *extrinsic* aging, resulting from the cumulative exposure of the skin to external sources of aging like UV and environmental stress [32]. Clinically, these two types of aging were presented differently, with intrinsic aging responsible for dry, pale, finely wrinkled skin with the appearance of benign epithelial neoplasms (seborrheic keratosis) after the age of 40, independent of solar exposure [33, 34]. By contrast, extrinsic aging is related to deep wrinkles, redness, dryness, laxity and atrophy, with the skin having a leathery appearance and pigment disorders like the actinic lentigines (solar lentigos). From a cellular and biochemical point of view, during intrinsic aging, on the one hand, all the cellular layers of the skin are finer, the corneocytes are less adherent between them and the EJD is flattened. The epidermis is atrophied with a cellular loss in melanocytes and Langerhans cells [4, 35]. The dermis present a loss of ECM and a decline of the vascular network of the papillary dermis [36]. Dermal collagen becomes more scattered and elastin is slowly degraded [37]. Aging is also characterized in particular by a marked modification of the hormone balance (estrogen, testosterone, dehydroepiandrosterone [DHEA], etc.) [38] and cellular receptors (those for vitamin D, neurotransmitters, interleukin-1 [IL-1], platelet-derived growth factor [PDGF] and epidermal growth factor [EGF]) [39-44] and accumulation of an undegradable intralysosomal material [45-47].

Aging due to extrinsic factors, on the other hand, is characterized by an abundant inflammatory infiltration made up of mononuclear mastocytes, histiocytes and other mononuclear cells [48] and an increase in the proliferation and dysplasia of the keratinocytes associated with a loss of Langerhans cells and sebaceous glands. Also, the rates of some GAG and dermal PG are increased, as well as a dysfunctional and insoluble form of elastin (a phenomenon called elastosis), while the collagen fibers are over-degraded by matrix metalloproteinases (MMPs) [48, 49]. Moreover, partially degraded collagen can amplify collagen degradation; indeed, the half-life of collagens is estimated at 17 years, and consequently fragments of natural collagen accumulate with the years [50]. Indeed, the mechanisms behind the two types of aging are due to mechanisms implying ROS [51-53].

### *Intrinsic Aging and ROS Generated by Endogenous Stress*

Apart from mitochondrion, the sources of "endogenous" ROS are numerous: Nicotinamide adenine dinucleotide phosphate (NADPH) oxidases of the plasmic membrane during the inflammation, hydrogen peroxide ($H_2O_2$) resulting from the degradation of the fatty acids in the peroxisomes, lipoxygenase, COX, NADH dehydrogenase, cytochrome P450 during the detoxification of xenobiotics, cytochrome b5, etc. [54]. However, the escapes of electrons during the mitochondrial metabolism constitute approximately 90% of the "endogenous" ROS. They then damage the DNA (mitochondrial and nuclear), most frequently by oxidation of a guanine base to form the 8-oxo-guanine (8-oxoG), which often disturbs the machinery of replication of the DNA. It is worthwhile to note that free 8-oxoG is more susceptible to oxidation than in its nucleoside form and in the presence of molecular oxygen can lead to the transformation of the highly reactive hydroperoxide free base 8-oxoG*. These changes can irreversibly accumulate if repair mechanisms are not undertaken. Because oxidative stress prevents the repair of DNA [55], ROS, moreover, can also affect directly and indirectly other cellular components, like lipids and proteins, as mentioned above. The pigment lipofuscin can also be a direct target of ROS: indeed, lipofuscin is a highly oxidized aggregate that consists of covalently cross-linked proteins or lipids and can inhibit the proteasome by binding surface motifs [56]. Moreover, the accumulation of lipofuscin during aging process was also recently identified as a robust marker of human aging in brain and muscle and should be, in the future, the focus of more research on skin [45,47].

### *Extrinsic Aging and ROS Generated by Exogenous Stress*

Environmental aging implies chronic exposure to various factors, primarily: UV, pollution and infra-red A (IR-A) [57]. Only UVA (315-400 nm) and -B (280-315 nm) reach the Earth's surface; UVC is absorbed by atmospheric ozone. In particular, enhanced ultraviolet radiation exposure, inflammation and oxidative stress and telomere attrition during chronological aging may induce severe DNA damage and genomic instability in the skin-resident stem/progenitor cells and their progeny [58]. Epidermal progenitor cells, which include stem cells and transitory amplifying cells, due to their fundamental role in epidermal construction and regeneration, require the greatest attention. Although UVA strength is 1,000 times less than that of UVB, because of its wavelength, it penetrates the skin more deeply than UVB to reach the dermis layers [59]. UVB penetrates the epidermis and is strong enough to cause direct damage to the DNA without the intermediate formation of ROS (even if UVB, like UVA, can also produce ROS) [60]. Thus, UVB makes it possible to bind pyrimidine bases together (thymine and cytosine), cyclobutane pyrimidine dimers (CPD) being the most common form of DNA damage; CPD, moreover, being formed ten times more frequently than 8-oxo-G (generated on the DNA via ROS, also generators of single- and double-strand breaks of DNA) [61]. To a lesser extent, another type of bases fusion, the 6-4 photoproduct (<6-4>PP), can also be formed. CPD is formed by all of the ultraviolet wavelengths, even if UVA, which is less energetic, is more deleterious due to the production of ROS both in the epidermis and dermis [62-65]. In this respect, the UVA is absorbed by natural cellular chromophores, like urocanic acid [66], riboflavin [67] and precursors of melanin [68], which also are able to act as photosensitizers [64, 69], leading to the formation of ROS through reaction with cellular oxygen. Glycation is a non enzymatic reaction between reducing sugars (generally glucose) and the lysine residue of the proteins. The formation of cross-links between matrix molecules, especially dermal collagen, leads to the stiffening of cutaneous tissue. Advanced glycation end-products (AGE), which accumulate with age, also act as chromophores with UVA and become photosensitizers, thus affecting primarily the dermis via ROS [70-72]. Note, finally, that intense exposure to visible light can also produce some DNA damage via the formation of ROS [73].

Mitochondrial DNA is damaged by UV radiation: greater accumulations of damaged mtDNA were indeed found on photo-exposed skins in comparison with photo-protected skins [74-77]. The UV-induced mtDNA breaks are mediated by ROS, mainly superoxide radicals $O_2^{\bullet -}$ [78]. The deletions of mtDNA 4,977-bp and 3,895-bp are among those most linked to skin photo ageing [75, 76, 79]. The common deletion 4,977-bp constitutes the most widespread marker of damaged mtDNA in humans. The deletion 3,895-bp, which is less known, can also play a part in skin photo ageing [76]. UVs induce DNA damage in fibroblasts and keratinocytes. Moreover, the exposure of keratinocytes to UVB induces two mtDNA deletions: 4,977-bp and a new deletion, 5,128-bp [80]. The induction of the common deletion in human fibroblasts is concurrent with a reduction in oxygen uptake or in ATP content and to an increase in MMP-1, while the naturals inhibitors of these MMPs (called TIMPs) are not upregulated; this imbalance is known to play a major part in cutaneous photo ageing.

It should be noted that UVB affects the immune response, leading to a reduction in antigen presentation and the formation of T cells, favoring the propagation of cancer cells [81, 82]. In this respect, it has been suggested that UVA, by inducing lipid peroxidation, stimulates the exit of immunizing cells from the epidermis, contributing to immunosuppression [81]. Moreover, UVB induces the secretion of epidermal cytokines, such as TNF-α and IL-10, which play a major role in immunosuppression [81]. UVs also activate many cellular receptors for cytokines and growth factors, such as the EGF receptor, the TNF-α receptor, the platelet-activated factor (PAF) receptor, the insulin receptor, the IL-1 receptor and the PDGF receptor [83]. This simultaneous increase in cytokines and their receptors induces signaling pathway cascades, which imply an increase in the expression of p21Ras, extracellular signal-regulated kinase (ERK), c-jun aminoterminal kinase (JNK) and p38. Increased C-jun, with c-fos, leads to an augmentation in the activation of activator protein-1 (AP-1). Irradiation with UV thus makes it possible to activate transcription factors like AP-1 but also nuclear factor-kappa B (NF-κB) [84]. AP-1 inhibits the expression of collagen [85] and increases that of MMPs, degrading matrix proteins, thus contributing to the formation of cutaneous wrinkles. Its action is joined with that of NF-κB, which then controls positively its target genes, such as IL-1beta, IL-6, IL-8 and TNF-alpha. These cytokines then bind to their membrane receptors, both in the dermis and epidermis, activating AP-1 and NF-κB and amplifying the UVs' effects and the skin's inflammatory response. NF-κB has also been previously shown to be an inhibitor of collagen synthesis [86, 87]. In addition, the accumulation of nonfunctional elastin and the degradation of collagen by ROS characterize photo ageing [53, 88]. The exacerbated catabolism of aging is also at the origin of the degradation of glycosaminoglycans, such as hyaluronic acid, a target of hyaluronidases activated by the UVB [89].

There are many other environmental types of stress involved in the process of extrinsic aging (or premature cutaneous aging), like the tobacco smoke, IR rays or ozone and other xenobiotics. Almost all of them imply the same molecular mechanisms, with the activation of MMP-1 and non regulation of TIMP-1, activation of p38 MAP kinase (MAPK), ERK (extracellular signal-regulated kinase) signaling pathway and activation of transcription factor AP-1, leading to the diminution of collagen synthesis and to its simultaneous degradation. This is the case in particular with people who smoke [90, 91]. The increased formation of wrinkles in smokers seems to be simultaneously of molecular origin, through the activation of MMPs and increase in elastosis [90, 92], and of behavioral origin, through the repeated pinching of the lips and the contraction of the facial muscles, explaining, thus, partly the formation of the wrinkles around the mouth [93]. Studies also show that young children exposed to cigarette smoke are likelier to develop atopic eczema than non-exposed children

[94]. The activation of MMP-1 is also true for IR-A [95-98]. Infrared radiation includes IR-A (760-1,440 nm), IR-B (1,440-3,000 nm) and IRC (3,000 nm-1 mm). They are not as strong as UVs but constitute more than half of the solar energy that reaches the skin (compared to 7% for UVs [99]). IR-A (near IR) is most penetrating type of infrared radiation—compared with IR-B and -C, which affect only the superficial part of the skin—and can penetrate the dermis more deeply than UVA. Like UVs, IR-A initially leads to an increase in ROS rates, by disturbing the ETC [57]. The skin reacts by increasing the secretion of ferritin, a cellular antioxidant; but in the long term, IR-A is linked to a reduction in epidermal proliferation and density of Langerhans cells [100, 101].

Atmospheric ozone constitutes one of the principal components of "smog" and one of the sources of air quality problems in urban and rural areas. Ozone is a very powerful oxidizing agent, initially only affecting the superficial layers of the skin and taking part in the reduction of the rate of natural antioxidants of the epidermis, like vitamin E, tocopherol and vitamin C [102]. Lipid ozonation products (LOP), which are small-size molecules, and very diffusible ones, crossing the surface layers of the skin to reach the dermis, act as second messengers of lipid peroxidation by ROS and generate an inflammatory response [102-104]. In particular, the heat shock proteins (HSP) 27 and 70 and HO-1, stress proteins activated by many endogenous and exogenous stimuli, are induced. Moreover, a potential pro-inflammatory role of ozone and its derivatives is linked to an increase in the rates of cyclooxygenase-2 (COX-2) and inducible nitric oxide synthase (iNOS) and to activation of NF-κB. In addition, studies showed that a thermal shock, generating HSP, reduced the damage caused by UVs, which can, moreover, prove direct inductors of HSP [105]. In this regard, Dams et al. recently showed that a thermal shock in human dermal fibroblasts led to an increase in the rates of HSP and type I and III collagens [106]. In addition, various types of other stress, like heat, cold and osmotic stress, lead to a common implication of the AP-1 transcription factor, activating MMPs at the origin of the formation of the wrinkles [107]. Recently, it was even shown that "airborne particulate matter" (PM), a concentrate of many radicals, causes wrinkles, pigment disorders and spots at the cutaneous level [108].

# II. Oxygen and ROS, a Subtile Equilibrium

In contrast to the previous theory of ROS described as solely toxic and responsible of damages, ROS are necessary for the normal physiology and are now recognized having a regulatory role in a wide range of metabolic functions. Oxygen is necessary for many biological reactions, and the use of oxygen by mammal cells is an essential point of basic redox biology and, in this regard, hypoxia seems likely to induce MMP secretion via the protein kinase C (PKC) pathway [109]. Indeed, aerobic breathing, oxidases and some extracellular stimuli lead to the natural production of ROS. The oxidation of various cysteine residues, like those of redox-sensitive transcription factors, genes, NF-κB, hypoxia inducible factors (HIF), the vascular endothelial growth factor (VEGF), MMP, COX-2 and inducible NO synthase, will lead to the specific activation of signaling pathways. The local rate of free radicals is fundamental, beneficial within physiological amounts and likely to induce a tolerance to stress, while overproduced ROS will be deleterious for skin cells or will even generate pathological phenomena.

Oxidative stress is defined as a substantial modification of the balance redox of a cell resulting from an imbalance between the results of its natural defense mechanisms and the rate of ROS and reactive nitrogen species present inside the cells or in their very close environment. The ROS act at a short distance and have a very short lifespan but induce chain reactions, which propagate and amplify their action. The main free radicals encountered in skin are the reactive derivatives of oxygen, i.e., mainly singlet oxygen $^{1}0_2$, superoxide anion radical $O_2^{\bullet -}$, hydroxyl radical $OH^{\bullet}$, alkoxyle radical ($RO^{\bullet}$, where R is an aliphatic chain), radicals derived from an unsaturated fatty acid (lipid peroxidation), such as peroxyl radical ($ROO^{\bullet}$) or peroxinitrite $ONOO^{\bullet}$ formed from a reaction between nitric oxide NO and $O_2^{\bullet -}$. Schrek et al. were among the first to regard hydrogen peroxide as a second messenger involved in various signaling pathways of NF-κB [110-114]. ROS play a functional role in gene regulation, cell signaling, senescence, turnover, cell differenciation and apoptosis. More precisely, ROS impact other *trans* factors, like p53, phosphoinositide 3 kinase (PI3K), Akt, Erk, JNK and p38, thus targeting key genes for apoptosis, proliferation, differentiation, growth arrest and senescence [115-119]. Moreover, the addition or withdrawal of a methyl group (hyper- or hypo-methylation) of cytosines from DNA cytosines can impact gene expression, reinforce genotoxicity of free radicals and increase the deterioratation of cell functioning [120- 122]. This confirms the hypothesis that aging results primarily from epigenetic changes leading to distortions of gene expression. Many other age-related metabolic alterations, such as accumulation of damaged proteins, are "secondary" and may be reversed by the restoration of the physiologically optimal (young) state of "top" epigenetic regulators.

Let us shortly focus on proteins and lipids the consequences of redox imbalance. Proteins, owing to their abundance in all the skin compartments, are preferential ROS targets, in particular certain amino-acids like cysteine, methionine are very present in keratin skin appendages. A "frank" oxidation of proteins lead to their carbonylation, cross-linking and denaturation with many impacts on the visible skin appearance [123]. Recent results on the interaction of hydroxyl radicals on the repeat motif Val-Gly-Val-Ala-Pro-Gly of the skin protein elastin show nonrandom hydroperoxide formation and confirm the occurrence of chain reactions [124]. Oxidative stress, by modifying certain proteins, can also activate, via oxidized danger signals, the innate immune system, thereby triggering inflammatory responses. ROS also react with the unsaturated fatty-acids, leading to the hydroperoxide formation. Furthermore, AGE resulting from glycation reactions are oxidative stress catalysts. Addition of oxidative damage of different origins at different levels and the imbalance of free radical scavenger enzymes are factors causing premature aging, primarily at the dermal level [125, 126].

On the cellular surface (hydrolipidic film, sebum) or at the more external horny layer of skin, lipids responsible for the skin barrier are also targets of ROS, the lipid peroxidation leading to the formation of reactive derivatives such as aldehydes or indirectly pro-inflammatory isoprostane-like compounds [127, 128]. The lipid rafts, cholesterol and sphingolipid-rich microdomains of the cellular lipid plasmatic mosaic, seem increasingly involved in the local production of ROS. These domains are also low in docosahexaenoic acid (DHA), essential for the biological activity of close membrane proteins and thus modulating events such as cell death, cell proliferation and skin sensitivity [129-131].

### Inflammation and Physiological Consequences

The inflammation resulting from these various modified targets, as many infra-clinical inflammations sources, will take a major part in aging via the activation of NF-κB and AP-1 leading to the implication of pro-inflammatory COX-2, MMPs, cytokines such as the IL-6, TNF-α or IL1-β and direct and indirect modifications of the immune response [132-134]. All these elements contribute to what many authors today call "inflamm-aging" [135-139]. Inflamm-aging becomes a target for scientists who developed active ingredients for skincare.

### Redox Equilibrium and Longevity

Another topical question is the degree of implication of ROS and antioxidants in longevity. The idea of the existence of natural extremely powerful systemic mechanisms of epigenetic rejuvenation explains vast differences of lifespans between different species. Augmentation of antioxidative defense fail to extend longevity, and it seems today difficult to explain the limited lifespan by only the accumulation of structural damages. Very recently, Cabreiro and collaborators showed on the model of the nematode *C. elegans* that SOD acts rather by a promoter effect of transcription factors related to longevity than by its role in the detoxification of free radicals [140]. A few years ago, sirtuins were recognized as pivotal mediators in quality programs that control cellular homeostasis and possible targets to increase functional longevity, even for skin [141, 142]. Stem cell differentiation is under nitric oxide-cyclic GMP signaling, and the maintenance of the stem cell pool requires regulation of redox homeostasis and Daf-16 from the transcriptional factor FoxO family (FoxO3), which was also shown to regulate *C. elegans*' lifespan [143]. It is now clear that sirtuin affects lifespan, especially the mitochondrial SIRT3, are complex but also involves their ability to regulate the systems that control the cellular environment [144]. Nevertheless, up to now, it has not been possible to directly reverse aged tissue through antioxidative properties only, to protect and restore the structure of a tissue/organ before it suffered heavy damage and acquired aged features.

## III. Antioxidant Phytochemicals for Skin Care

The antioxidants are compounds that protect from damage due to free radicals and electrophilic compounds attacks. Free radicals are very unstable molecules from which an electron is missing and which destabilize surrounding molecules (damage) in taking the electron they lack. The cellular damaging effects of free radicals depend on their concentration and act either directly through molecular attacks or as a second messenger in various redox pathways. They can also have some positive effects when they intervene in adaptive phenomena with stress. In general, antioxidants collect free radicals before they attack other molecules. Reaction times are determined by their redox properties, which take into account their dissociation energy and their ionization potential. The effectiveness of antioxidants depends on their chemical reactivity, the number and the kinetics of the collected free radicals, what becomes of the free radical derivatives, their interactions with other antioxidants, their concentration, their mobility in the close environment, their absorption, distribution and metabolism. To stop oxidative stress, it seems necessary to help the cells to increase their pool of natural intracellular antioxidant defenses. The most effective way is to use cocktails that act at various levels of the oxidative phenomena.

How can we judge the relevance of a presumed antioxidative molecule ? It is important to have broad, reliable and recognized evaluation methods to determine the real antioxidant capacity and target domains of the studied phytochemicals. In general, *in vitro* methods—testing a phytochemical's reaction with stable radicals or methods in which cell cultures are subjected to various stresses—as well as the *in vivo* use of biomarkers are employed [145-148]. In *in vitro* models, cell types (lines or strains), origin (human or not), culture condition (additives), pre- or post-treatment (preventive, curative), the concentration used and the nature of stress are of prime importance to correctly evaluate the real antioxidative capacity of a phytochemical. The physicochemical characteristics of the molecule, its bioavailability and its chemical structure can influence the results. Only a large combination of methods allows a confident evaluation of the real antioxidative power of a phytochemical. Genomic influence of natural antioxidants on different skin cell types shows that the receptivity of keratinocytes to antioxidants seems broader than fibroblasts and that gene sets responding to a treatment are often the same [149]. As we will see, in describing the different antioxidative phytochemicals used, the most important element in improving skin appearance is the nature of the biological target(s) and the direct and indirect power level of the biologically active ingredients rather than its chemical structure. Within the framework of cutaneous photoprotection in anti-aging or whitening skin care products, they can also complement chemical and mineral UV filters [150].

## Powerful Plant Antioxidants and Cosmetics

The Federal Food, Drug, and Cosmetic Act defines cosmetics by their intended use, as "articles ...applied to the human body...for cleansing, beautifying, promoting attractiveness, or altering the appearance." Among the products included in this definition are skin moisturizers, perfumes, lipsticks, fingernail polishes, eye and facial makeup ... In fact, cosmetics address healthy skin only. Antioxidants neutralize free radicals and also are able to calm minor inflammation by sacrificing one of their electrons without adverse effect. They are widely used in skin care to prevent, protect and cure skin aging and related disorders. Usually plants having vitamins C, E, carotenoids, flavonoids (flavones, flavanones, anthocyanins and catechins), polyphenols (ellagic acid, gallic acid, chlorogenic acid, tannins, etc.) show some antioxidant and/or free-radical-scavenging properties. Very often, their specificity of action remains to be elucidated [151, 152]. A wide variety of phytochemicals are able, by various mechanisms, to exert a chemo-preventive effect in inducting antioxidant and detoxifying enzymes. The benefits of topical phytochemicals cannot always be explained only by their ability to prevent free radical stress, to counteract cell oxidative stress or to help natural repair systems. Some of them exhibit hormetic properties. The best-known hormetic plant antioxidants are sulforaphane, resveratrol, curcumin, flavonoids and green tea catechins, which are some phytochemicals classically used in skin care products. Hormetic molecules act as mild or low-dose stressors in order to prepare cells to adapt to and gradually resist more severe stress. This innovative view has been recently adapted and opens a very interesting approach to cosmetic skin care [153, 154]. If the majority of antioxidant phytochemicals is used topically, a few are also used as nutritional supplements in order to improve skin aspect.

## A Short Alphabetical Listing of Interesting Topical Antioxidants

*Anthocyanidins and pro-anthocyanidins:* Anthocyanins are water-soluble pigments derived from glycosylated molecules called anthocyanidins and are naturally present in many vegetals (Figure 2).

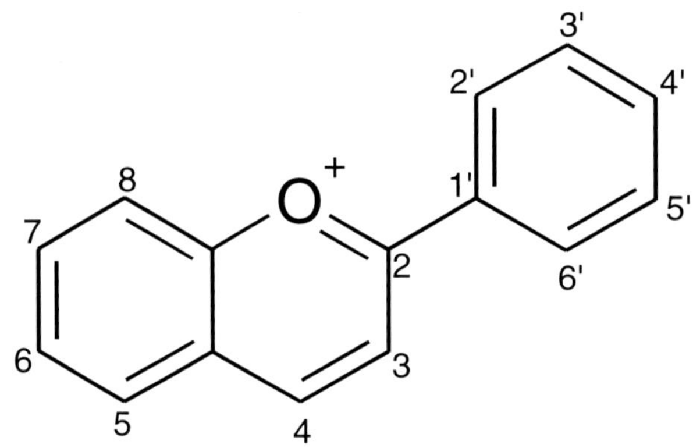


Figure 2. Anthocyanins.

They belong to the flavonoid family and are synthesized in plants via the phenylpropanoid pathway. These molecules, present in many vegetal materials (tropical fruits and different foods), are described as very effective carbonyl scavengers, sometimes enhancer of estrogen receptor-α expression and prevent the formation of AGE.

Possible interesting sources of natural anthocyanins can be purple sweet potato or even the purple black rice cultivar, which contain cyaniding 3-O-beta-D-glucoside [155, 156].

Proanthocyanidin oligomers contained in French maritime pine (flavangenol) could be also an interesting way to prevent chronic UVB radiation-induced skin damage. Proanthocyanidins include oligomeric proanthocyanidins (OPC) and polymeric proanthocyanidins (PPC); some have been used in sun care products but due to their instability have to be carefully formulated or even used after interfacial polymerisation.

*Baicalein and Sanguisorba officinalis extracts:* The roots of *Scutellaria baicalensis* are a traditional remedy for inflammatory diseases, used for centuries in China and Japan. *Sanguisorba officinalis* extracts and Baicalein (5,6,7-trihydroxyflavone), a known nitric oxide synthase (NOS) inhibitor already described for its interest in protective skin care, have reignited interest in light of new pharmacological data [157]. Baicalein offers high protection against lipid peroxidation damage, and the phenolic moity of the molecule has been shown to be responsible for its scavenging property. Baicalein also acts as a chain-breaking antioxidant and protects mitochondrial membranes through a double action, i.e., in scavenging primary and secondary generated radicals [158].

It also protects mitochondrial cellular function, reduces antiapoptotic Bcl-2 family proteins, protein carbonyl formation, and induces manganese and superoxide dismutase [159, 160].

*Berries:* Berries contain a wide range of polyphenols and have a dietary role as natural antioxidants. Glycosylated, polyhydroxylated and methylated anthocyanins are often found in berries. Flavonols, lignan, stilbenes, phenolic acid, condensed or hydrolysable tannins, carotenoids and sterols are, with vitamins, the antioxidants most present in berries. Wolfberry or goji (*Lycium barbarum*), cultivated in China, acerola (*Malpighia punicifolia*) from Central America, acai (*Euterpe oleracea*) from Brazil and blackberry (*Rubus fruticosus*) leaf extracts are the new ingredients used in some cosmetics and are promoted for marketing reasons, but more *in vitro* and *in vivo* studies to assess their real benefits to the skin need to be performed. Oily seed extracts from cranberry, cloudberry, strawberry and Finnish oats are rich in functional lipids, especially phystosterols, tocotrienols and tocopherols, and have been used in surface relipidation and protection of aging skin. Goldenberry (*Physalis peruviana*), a fruit known to be rich in vitamins and carotenoids, has been recently offered in cosmetics for its ability to increase microcirculation and collagen synthesis. In order to improve their antioxidant properties and to respect a sustainable development approach, supercritical carbon dioxide extracts were developed [161].

*Carotenoids:* Even though they are powerful antioxidants and a precursor of vitamin A, due to their strong yellow-orange color, power and instability, natural carotenoid mixtures are now rarely used in skin care (Figure 3). Astaxanthin is normally used in food coloring, is an antioxidant with a slightly lower antioxidant activity in some model systems than other carotenoids and is coming back in cosmetics.

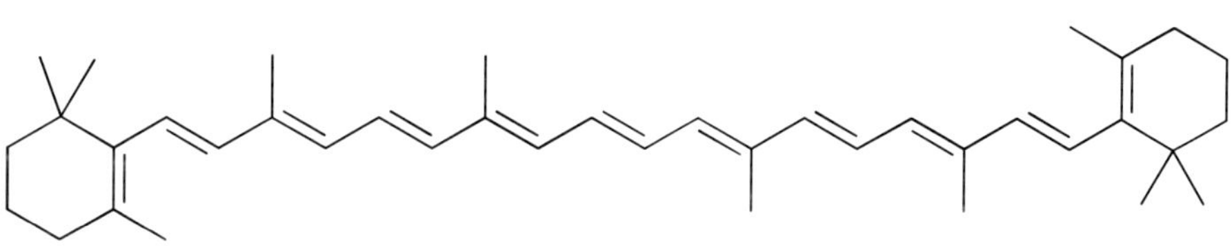

Figure 3. β- carotene.

*Curcumin:* Curcumin, or diferuloylmethane, a small-molecular-weight yellow compound isolated from the rhizome of turmeric (*Curcuma longa*), zingiberaceae, has been used traditionally for centuries in Asia, particularly India, for medicinal, culinary and other purposes (Figure 4).

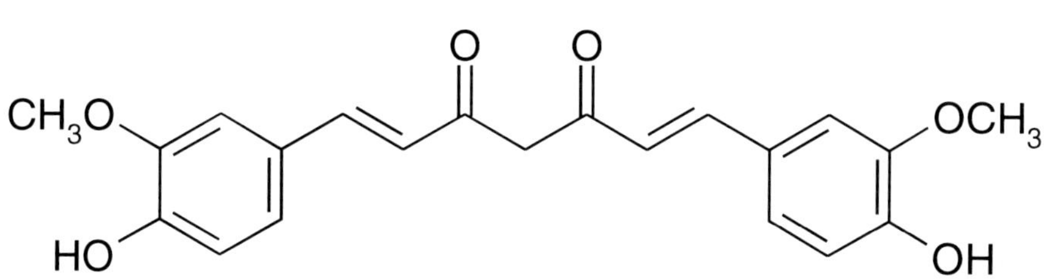


Figure 4. Curcumin.

First, curcuminoids have been shown to protect normal human keratinocytes from hypoxanthine/xanthine oxidase injury. Curcumin, presented to be able to chelate iron, is also able to induce cellular stress responses in normal human skin fibroblasts through phosphatidylinositol 3-kinase/Akt pathways and redox signaling, supporting the view that curcumin induces also hormetic stimulation of cellular antioxidant defenses. Curcumin induces *in vitro* and *in vivo* Nrf2/ARE pathways, Nrf2 (associated with detoxification) being highly expressed in skin exposed to a stressful external environment. As with other Chinese medicinal plants, tumeric extracts, curcumin and some derivatives such as bisdemethoxy-curcumin, inhibit glyoxalase-1, an enzyme linked to the glycation process. Other studies show that curcumin has a beneficial effect as a proangiogenic agent in wound-healing by inducing transforming growth factor-beta, angiogenesis, accumulation and the tension of the extracellular matrix architecture. A curcumin gel has been shown to clinically accelerate perfect skin regeneration following surgical procedures. More stable tetrahydrocurcuminoids used to treat vitiligo via phototherapy seem disappointing in repigmentation, perhaps due to their metabolism, original curcumin being known to decrease melanogenesis in human melanocytes [162-167]. Loaded in different vesicular vehicles for topical use, *Curcuma longa* extract showed more photoprotective properties in transferosomes than in liposomes or ethosomes [168].

*Epigallocatechin gallates and green tea polyphenols: Green* tea is manufactured from the fresh leaves of the *Camellia sinensis* plant, originally from China, and used in skin care products [169-172].

The leaves of this plant are fermented through a process that, for the most part, prevents oxidation and polymerization of the plant's polyphenols. Green tea contains alkaloids and four major polyphenols: (-epicatechin (EC), -epicatechin-3-gallate (ECG), -epigallocatechin (EGC), and -epigallocatechin-3-gallate (EGCG) (Figure 5)). It also contains other agents that have chemo-preventive activities.

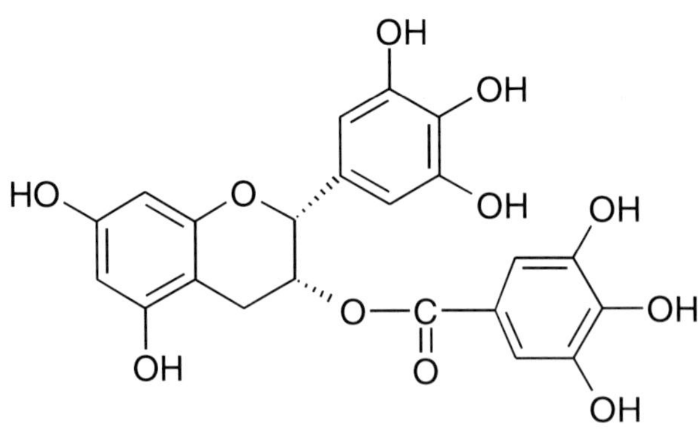


Figure 5. EGCG: epigallocatechin-3-gallate.

Green tea polyphenols have also been shown to inhibit the formation of AGE [173]. Catechins have recently garnered a lot of attention thanks to their strong antioxidative activity due to their ability to chelate metals ions such as pro-oxidative $Fe^{++}$ and $Cu^{++}$ and to protect from DNA modifications, suggesting a real role in protecting the population from the risk of skin cancer. [174].

Recent studies also reveal that EGCG can also be pro-oxidant and can induce proapoptic properties for cancer cells via the NADPH-dependent reduction of the thiorredoxin/ thioredoxin reductase pathway [175]. Interestingly, through the stimulation of interleukine 12 and a nucleotide excision repair mechanism, green tea polyphenols or EGCG rapidly remove or repair UVB-induced DNA damage such as cyclobutane pyrimidine dimers, known to play a role in photocarcinogenesis and immunosuppression in skin [176]. Epigallocatechin gallates and green tea polyphenols are widely used, and their efficiency is recognized.

*Flavonoids:* Flavonoids (a class of more than 4,000 molecules) are polyphenolic compounds that are ubiquitous in nature and are categorized according to chemical structure. The antioxidant capacity of flavonoids depends upon their molecular structure. The position of hydroxyl groups in the chemical structure of flavonoids is important for their antioxidant and free radical scavenging activities (see polyphenols).

*Lycopene:* Lycopene is a natural pigment synthesized by plants and microorganisms. It belongs to the carotenoid family and, thanks to its high number of conjugated dienes, is probably one of their most effective free radical scavengers, especially of single oxygen. Lycopene is found predominantly in tomatoes and tomato products. In contrast to beta-carotene, it is not a precursor of vitamin A in humans. Lycopene has been recognized for its dietary photoprotective properties. Recently, supplementation in type I and II phototype women over 12 weeks with a tomato paste rich in lycopene demonstrated skin protection against UV-induced acute photodamage, erythema, matrix changes and mtDNA damage [177]. Lycopene has been proposed but little used in skin care due to its tendency to tint skin and its controversial pro-oxidative properties under ultraviolet radiation [178, 179]. Nevertheless, recent *in vivo* data reveal its potential to protect skin against photodamage by acting on erythema and matrix changes (MMP1, fibrillin, mtDNA deletion following ultraviolet radiation) [177].

*Orchid extracts:* Orchidaceae, with more than 30,000 species, is one of the most important plant families. Some orchids have traditional European and Chinese medicinal uses, and a few have been subjected to phytochemical analysis. Recent studies on orchids have revealed stilbenoids, triterpenoids, alkaloids and anthocyanins with pharmacological properties. The stilbenoid skeleton comprises two aromatic rings joined by an ethylene bridge (C6–C2–C6). A large array of compounds has been synthesized from plants, and some found in Chinese epiphyte orchids, such imbricatin, gigantol, coelonine have generated interest due to their pharmacological activity (Figure 6).

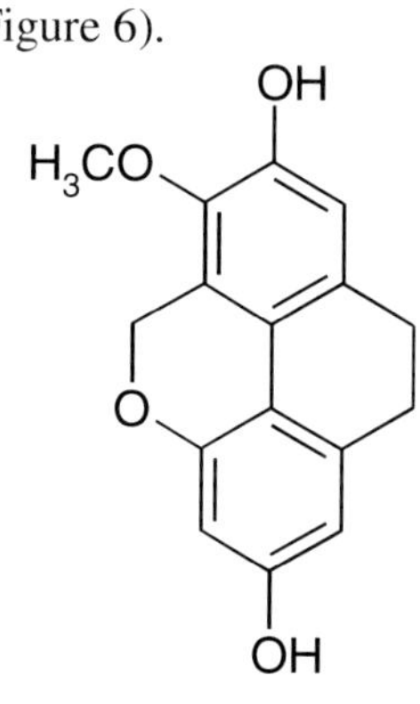


Figure 6. Imbricatin.

Recent studies on Vanda sp. stem or Brassocattleya extracts have shown anti-inflammatory, anti-senescence, whitening or mitochondrial respiratory properties appropriate for cosmetic uses [180-183]. Phenolic secondary metabolite compounds present in different parts of terrestrial and freshwater orchids seem responsible for their antioxidant properties and exhibit a possible classical "anti-aging" cell maintenance interest [184-187].

*Polyphenols:* They are a large and diverse family of secondary metabolites synthesized by plants. A benzoic ring linked with a hydroxyl group characterizes them (Figure 7). The three double bonds decrease the bond strength between hydrogen and oxygen of the linked hydroxyl group, revealing antioxidant properties due to the freedom of the hydrogen atom.

Flavonoids, natural benzopyrone derivatives, are responsible for the colorful appearance of fruits and vegetables. With more than 4,000 flavonoids having been identified, they are the most common antioxidants synthesized by photosynthezing cells.

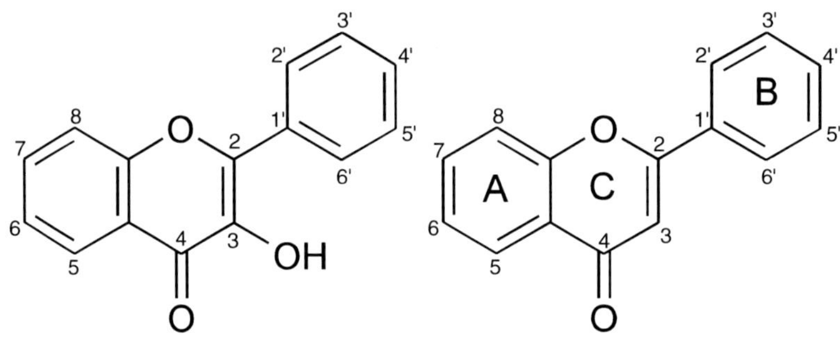


Figure 7. Flavones Flavonols.

P. Mladenka et al. remind us in a recent review article that the catecholic B ring is necessary for scavenging activity; hydroxyl groups in an ortho position, 3-hydroxy-4-keto group or 5-hydroxy-4-keto group, enable iron chelation and other properties required for their principal biological activities [188]. Iron is known to exacerbate the decomposition of lipid hydroperoxides and to catalyze oxidative reactions. The ability of some flavonoids to strongly bind intracellular iron ions is responsible for their potential to protect cells against $H_2O_2$-induced DNA damage [189]. The reaction rate constants of plant antioxidants are essential for their scavenging abilities. Compounds with pyrogallol or catechol structures in ring B are the most rapid superoxide scavenger against superoxide, but *in vivo*, biologically high concentrations are necessary to compete with superoxide anion radical [190]. Due to their structure and ability to absorb UV, they can also have a sunscreen-like and photoprotective or even anti-photocarcinogenic effect when repeatedly topically applied on skin. Some Thai plants, rich in polyphénols, such as *Phyllantus emblica*, *Terminalia chebula* and *Houttunya cordata*, have strong antioxidant properties, potential antiglycation properties, and they have been proposed for cosmetic uses [191].

*Pomegranate:* Recently, pomegranate (*Punica granatum*) seed oil, fruit and peel extracts, often rich in polyphenols such as ellagic acid, have been used in cosmetics as anti-inflammatories or as wound-healing enhancers. They are said to have more antioxidant properties than green tea. The juice and peel contain anthocyanins and other antioxidants molecules such as tannins. Punicalagins, a hydrolizable bioavailable ellagitannin composed of or 2,3-(S)-hexahydroxydiphenoyl-4,6-(S,S)-gallagyl-D-glucose, also present in numerous species of the genus Terminalia (combretaceae), are of great interest for skin care, having demonstrated an ability to protect human skin fibroblasts from UV (especially UVB) damage, reducing the pro-inflammatory factor NF-K B, proapoptotic caspase3, MMPs and cyclobutane pyrimidine thymers and increasing G0-G1 phase, associated with DNA repair. [192,193]. Pomegranate flower extract has recently been demonstrated to reduce MMP-1mRNA fibroblast gene expression following UVA irradiation, a result that must be confirmed with more data and analysis of its phytochemical composition [194].

*Pycnogenol:* Pycnogenol is the registered trade name of standardized extract from the bark of the French maritime pine (*Pinus pinaster*). It is widely used as a nutritional supplement, a phytochemical remedy or applied topically. It has proved antioxidative properties and as an nutritional supplement, has recently been shown to repair damaged skin in increasing its hydration, elasticity and in exhibiting anti-inflammatory properties [195].

*Resveratrol:* Resveratrol (3,5,4'-trihydroxystilbene) is a polyphenol produced in plants by the enzyme stilbene synthase and recognized as a phytoalexin (Figure 8). It is found naturally in two geometric isomers; cis-(Z) and trans-(E). (See below, Vitis vinifera).

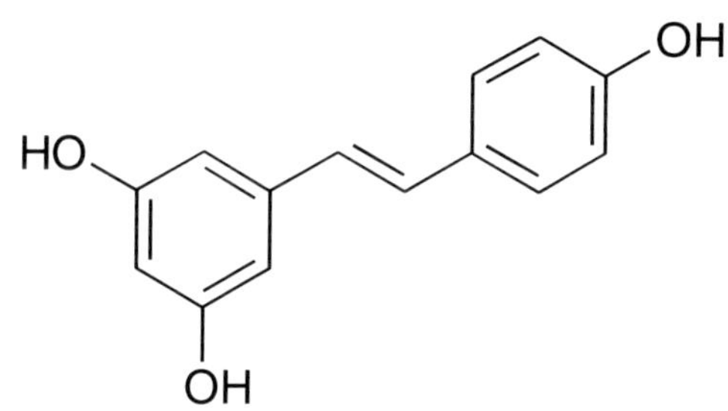


Figure 8. Resveratrol.

*Rosmarinic acid:* A phenolic acid derived from cinnamic acid, from the phenylpropanoid family, especially present in the lamiaceae and boraginaceae plant families, in *Rosmarinus officinalis, Prunella vulgaris, Perilla frutescens, Melissa officinalis* and Java tea (orthosiphon); rosmarinic acid is a natural hydroxylated ester of caffeic acid (3,4-hydroxycinnamic acid) used in cosmetics. Psotova et al. proved that rosmarinic acid and *Prunella vulgaris* extract exhibit a concentration-dependant UVA photoprotection potential. Later, rosmarinic acid was shown to exhibit its photoprotective effect via induction of melanin synthesis and in an emulsion at 0.3% to improve xerosis and pruritis in atopic dermatitis. Further work is needed to link its antioxidant role with its pharmacological properties [196, 197].

*Ruscogenins:* Ruscogenin, a major steroidal sapogenin from *Ophiopogon japonicus*, has no direct antioxidative properties, yet it is known to act as an anti-elastase, anti-hyaluronidase and anti-inflammatory due to its inhibition of intercellular adhesion molecules, of ROS induced NF-KB on endothelial cells, and it helps to restore microcirculation in photo-exposed areas of aged skin [198, 199]. This product is often used in eye contour cosmetics with antioxidants to fight aging signs of this thin and very fragile zone.

Salicin: The willow family of plants, Salix, contains notable amounts of endogenous salicylate compounds. Willow bark has been used in Europe and China since ancient times due to the anti-inflammatory beta-glycoside salicin. Very recent data on 3D in-vitro skin culture reveal its wide interesting skin architecture, firmness and elasticity genes regulation and its possible rejuvenation potential [200]. Thanks to the antioxidant content in willow bark, new crude extracts will likely be proposed soon on the market for topical anti-aging applications.

*Silymarin:* Silymarin, a standardized flavonoid extract of flavonolignans from the seeds of *Silybum marianum* (asteraceae), has been demonstrated to protect normal human fibroblasts [201]. Difficult to introduce in a stable skin care formula, this compound has regained interest due to new formulation approaches. Combined with methylsulfonilmethane (MSM) over a one-month period, the silymarin/MSM treatment has reduced redness and itching and has improved symptoms related to rosacea. Silymarin is mainly a mixture of silybin, silidianin, silychristin and isosilybin. Silybin is considered to be the principal biologically active constituent, with antioxidant and anti-inflammatory properties. Silymarin also repairs UVB-induced DNA fragmentation, as demonstrated by comet assay, but needs to be stabilized in formulation to retain its effectiveness [202].

*Sulforaphane:* Contained in the cruciferae family, isolated from broccoli (*Brassica oleacera*) or cress (*Lepidium sativum*) sprouts, this natural isothiocyanate molecule is a potent inducer of antioxidant response enzymes, which mobilize cellular defenses, activate Nrf2 in keratinocytes, inhibit mitochondrial permeability transition pore, induce phase -2 antioxidant enzymes, increase resistance and protect skin against UV damage and inflammation. Acting as a plant hormetic antioxidant, it contributes to skin adaptation to environmental stress [203-205]. Due to its poor stability, it has to be formulated carefully in oxygen-deprived formula.

*Tocopherols and tocotrienols:* Tocopherols are used in skin care products because of their amphiphilic/lipophilic nature and their antioxidant protective skin properties. Tocopherol is the most important natural chain-breaking oxidative compound due to its presence in the natural cellular membrane.

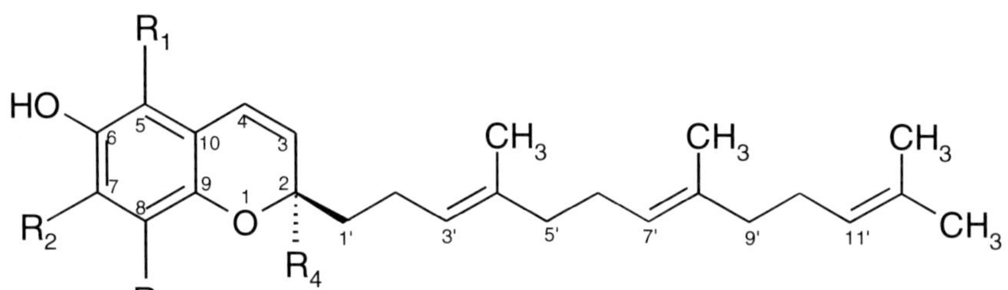

Tocotrienols

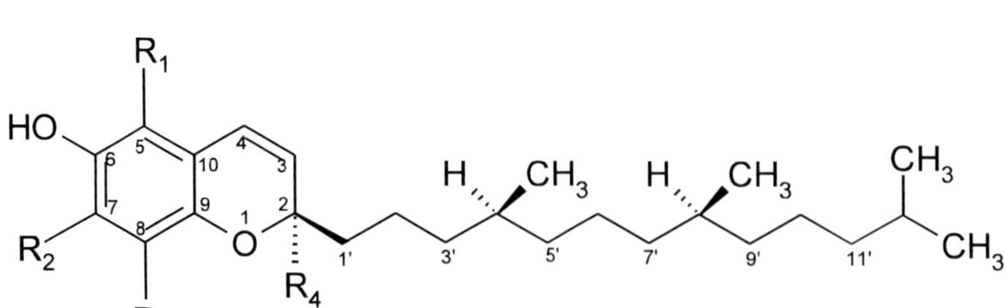

Tocopherols

Figure 9. Tocopherols and Tocotrienols.

Since the 90s, topical DL-alpha-tocopherol (vitamin E) has been shown to reduce ultraviolet-induced damage to the epidermis [206]. Tocopherol acetate or the water-dispersible vitamin E derivative, sodium dl-α-tocopheryl-6-O-phosphate, are also used. Sodium dl-α-tocopheryl-6-O-phosphate possesses anti-allergic properties and has recently been shown to inhibit PGE2 production in keratinocytes induced by UVB, IL-1β and peroxidants [207, 208]. Tocotrienols, like tocopherols, are members of the vitamin E family. Tocotrienols can be found in monocotyledons, especially palm, rice bran, barley and annatto beans. Tocotrienols and tocopherols differ in the unsaturated side chain, having three double bonds in the farnesyl isoprenoid side chain. Various isometric forms of tocotrienol are linked to the number and location of the methyl groups and the chromanol rings. The tocotrienols (T3s), especially delta tocotrienol, have been shown to have the most neuroprotective, anti-oxidant activity that are often not exhibited by tocopherols.

Tocotrienols are often claimed more valuable than tocopherols in protecting the interior cell membranes, such as those that surround the cell nucleus and mitochondria, because of their greater fluidity and ease in being incorporated into membranes.

*Vitis vinifera flavonoids:* Flavonoids present in grape leaves and seed have been widely studied and recognized as being rich in anthocyanins and proanthocyanins. They display a large spectrum of activity. Resveratrol (3,5,4-trihydroxystilbene) is a unique, multifaceted biologically active stilbenoid molecule present in more than 72 plant species, with antioxidant, chemoprevention and anti-inflammatory properties, upregulating stress resistance via mitochondrial superoxide dismutase. The involvement of resveratrol in sirtuin activation and sirtuin's role in longevity and life extension are controversial and still under debate [209, 210]. Nevertheless, these molecules, or extracts from plants containing it (*Polygonum sp*), glycosylated or oligomer derivatives (piceid, viniferin), are used extensively in anti-aging cosmetic products [211-214]. Another derivative used in anti-aging cosmetic skincare, claimed to be a stabilized form of resveratrol named Resveratrate™, exerts protective effects against repetitive solar simulator UVR-induced sunburn and suntan on human volunteers [215]. Miyabenol A, a stilbenoid isolated from Vitis sp., has also been shown to decrease inflammation and NO production by interfering in the MAP kinase pathway [216]). *Vitis vinifera* shoot extracts containing Miyabenol C, in combination with ectoine (an amphoteric, organic molecule discovered in extreme halophilic bacteria) and hydroxyectoine, have been shown to have some antioxidant potential and clinically to be very useful in improving skin appearance [217].

## Other Natural Antioxidants of Potential Interest

In cosmetics, flavonoid rutin-rich extracts or its glycoside are used for their action on the dermal microvascular system. Rutin found in many plants (especially the buckwheat plant *Fagopyrum esculentum*) metabolites are novel AGE inhibitors and associated with other molecules can be of prime interest for skincare. Edelweiss, *Leontopodium alpinum*, is an asteraceae originally from Siberia and has been proposed as an additive to whitening skin care products. Thanks to new data, leontopodic acid, a novel glucaric acid glycoside, has been linked to some antioxidative and DNA protection properties (Fig 10) [218].

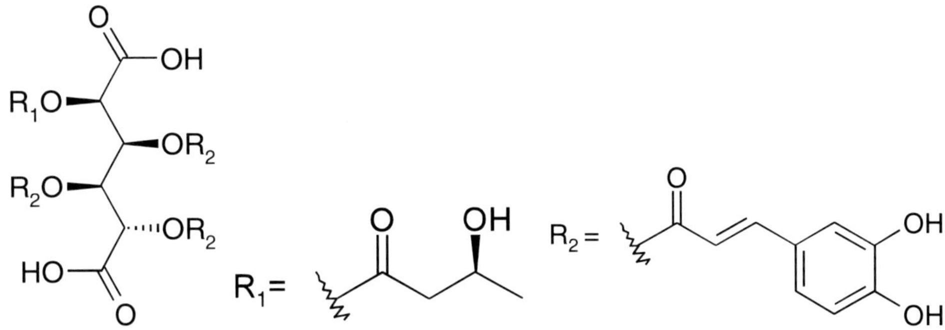


Figure 10. Leontopodic acid.

Another study emphasizes the possible antioxidative and skin-detoxifying role of glucaric acid and its derivatives, compounds found in many healthy foods such as pumpkins and Brussels sprouts. [219]. Sunflower (*Helianthus annuus*) seed extract, *Jasminum officinale* flower extract, *Centaurea cyanus*, *Epilobium angustifolium*, *Terminalia chebula* fruit extract and *Hippophae* extract have also been studied and have been shown to combat photo-induced free radicals, protect natural enzymes and help in the recovery of intercellular glutathione levels. Interestingly, *Rhodiola rosea*, a crassulaceae used as an adaptogen in Eastern Europe and Asia, known to increase the nematode *C. elegans'* longevity, induces translocation of the DAF-16 transcription factor from the cytoplasm into the nucleus, suggesting a reprogramming of transcriptional activities favoring the synthesis of proteins involved in longevity and stress resistance. [220]. Other studies demonstrated that *Rhodiola* increases cell resistance against oxidative stress, probably acting through an antioxidant (due to its polyphenolic content) or hormetic mechanism, but this remains to be documented [221]. Much research on antioxidants reveals that antioxidant plant extracts are able to inhibit tyrosinase and may be useful as a depigmenting agent. Nevertheless, the specific inhibiting activity on melanin-synthesis enzymes has rarely been demonstrated, and the observed *in vitro* action is often nonspecific. Despite the large number of *in vitro* tyrosinase "inhibitors" described in the literature, only a very few are able to produce real visible results in clinical trials [222, 223].

## Conclusion

Independent of its complex architecture, human skin is an important and biologically active fragile envelope exposed to a multitude of internal and external stressors. If reactive oxygen species are vital for normal cell metabolism, the disruption of defense and redox-regulated signaling mechanisms is linked to aging. In fact, recent data strongly suggest that only epigenetic modifications can be reversed, and compounds with some antioxidant properties are good candidates. Natural phytochemical antioxidants are now used to protect skin in topical applications. Many extracted molecules or standardized complex mixtures are used in cosmetics and dermatological products. More and more research is being performed to understand the way in which these phytochemicals act on skin cells *in vitro* and/or *in vivo* after formulation. Thanks to laboratory chemical modifications, some natural compounds undergo structural modifications to improve their hydrogen or electron-giving ability and to reinforce their antioxidant potential. In addition, in the future, pre- and probiotics, natural products of biofermentation, new molecules derived from tropical plants, archeobacteria or even fungi will probably also be interesting sources of new and powerful antioxidative chemicals.

Today, only careful combinations of different antioxidants at specific doses and ratios can have a synergistic antioxidative effect and be effective for skin protection and repair. Much recent data confirm that the link between antioxidative properties and longevity is not a simple one. To increase the longevity of living organisms, a combination of modes of action—hormetic, antioxidant, anti-inflammatory, detoxifying, energetic and maintenance—is necessary. Therefore, anti-inflammatory and antioxidant treatments could accompany pharmaceutical therapies and thus offer great promise to treat diverse pathological skin disorders with fewer side effects. If UV protection and a healthy lifestyle are necessary to delay skin aging, some antioxidative phytochemicals present in dermatological skin care and cosmetics are very important. They will fight direct or indirect damage of skin proteins, lipids, more or less severe metabolic perturbations and the degradation of the skin's natural barrier, which make the skin more sensitive and susceptible to new stress, accelerating the signs of aging. In protecting, repairing, neutralizing free radicals, regenerating and keeping the skin maintenance level efficient, it is possible to rejuvenate skin and look younger. Crow's feet, dullness, fine lines, wrinkles, pimples and sunspots will fade away within a few weeks; tone, complexion, texture will improve following skin care treatment. As we have seen, skin is a fantastic communicator, often revealing our emotions to others. Skin reactions are also very often linked to internal stress, anxiety or anger, and an everyday cosmetic treatment will help to moderate superficial reactions, improving our social image and well-being.

## References


[1] Z. D. Draelos and P. T. Pugliese, *Physiology of the Skin,* Allured Pub. Corp., Carol Stream (2011).

[2] R. McCallion and A. Li Wan Po, *J. Clin. Pharm. Ther.* 18, 15 (1993).

[3] R. A. Briggaman and C. E. Wheeler, *J. Invest. Dermatol.* 65, 71 (1975).

[4] N. A. Fenske and C. W. Lober, *J. Am. Acad. Dermatol.* 15, 571 (1986).

[5] F. H. Martini, *Fundamentals of Anatomy and Physiology*, Benjamin-Cummings Pub. Company, San Francisco (2004).

[6] J. Suter-Widmer and P. Elsner, *Age and Irritation*, The irritant contact dermatitis syndrome (CRC Press, Boca Raton, 257, 1994).

[7] F. Brégégère, Y. Soroka, J. Bismuth, B. Friguet, and Y. Milner, *Exp. Gerontol.* 38, 619 (2003).

[8] K. Sauermann, S. Jaspers, U. Koop, and H. Wenck, *BMC Dermatol.* 4, 13 (2004).

[9] T. Phillips and L. Kanj, *Clinical Manifestations of Skin Aging*, The Effect of aging in oral mucosa and skin (CRC Press, Boca Raton, 25, 1994).

[10] F. Bonté, *Ann. Pharm. Fr.* 69, 135 (2011).

[11] K. M. Südel, K. Venzke, H. Mielke, U. Breitenbach, C. Mundt, S. Jaspers, U. Koop, K. Sauermann, E. Knussman-Hartig, I. Moll, G. Gercken, A. R. Young, F. Stäb, H. Wenck, and S. Gallinat, *Photochem. Photobiol.* 81, 581 (2005).

[12] S. Verdier-Sévrain, F. Bonté, and B. Gilchrest, *Exp. Dermatol.* 15, 83 (2006).

[13] R. Widmer, I. Ziaja, and T. Grune, *Free Radic. Res.* 40, 1259 (2006).

[14] S. Prahl, T. Kueper, T. Biernoth, Y. Wöhrmann, A. Münster, M. Fürstenau, M. Schmidt, C. Schulze, K.-P. Wittern, H. Wenck, G.-M. Muhr, and T. Blatt, *Biofactors* 32, 245 (2008).

[15] E. H. Blackburn, *Cell* 106, 661 (2001).

[16] T. de Lange, *Oncogene* 21, 532 (2002).

[17] I. Botchkina, R. Kurfurst, F. Bonté, and E. Blackburn, *Mol. Biol. Cell* 13, 434 (2002).

[18] G. Barja, *Ann N Y Acad Sci* 854, 224 (1998).

[19] J. J. Chen and B. P. Yu, *Free Radic. Biol. Med.* 17, 411 (1994).

[20] S. Agarwal and R. S. Sohal, *Proc. Natl. Acad. Sci. USA* 91, 12332 (1994).

[21] M. Greco, G. Villani, F. Mazzucchelli, N. Bresolin, S. Papa, and G. Attardi, *Faseb J.* 17, 1706 (2003).

[22] C. Richter, *Int. J. Biochem. Cell Biol.* 27, 647 (1995).

[23] Y. Michikawa, F. Mazzucchelli, N. Bresolin, G. Scarlato, and G. Attardi, *Science* 286, 774 (1999).

[24] J. Miquel, *Exp. Gerontol.* 33, 113 (1998).

[25] T. M. Callaghan and K.-P. Wilhelm, *Int. J. Cosmet Sci* 30, 313 (2008).

[26] R. Kohen, *Biomed. Pharmacother* 53, 181 (1999).

[27] R. S. Sohal and R. Weindruch, *Science* 273, 59 (1996).

[28] K. Hensley and R. A. Floyd, *Arch. Biochem. Biophys.* 397, 377 (2002).

[29] A. Nicolaou, S. M. Pilkington, and L. E. Rhodes, *Chem. Phys. Lipids* 164, 535 (2011).

[30] F. Bonté, A. Saunois, P. Pinguet, and A. Meybeck, *Arch. Dermatol. Res.* 289, 78 (1997).

[31] F. Bonté, M. Dumas, G. Deby-Dupont, C. Deby, and A. Meybeck, *CR Soc. Biol.* 190, 667 (1996).

[32] J. Uitto, *Dermatol Clin* 4, 433 (1986).

[33] H. Nakamura, S. Hirota, S. Adachi, K. Ozaki, H. Asada, and Y. Kitamura, *J. Invest. Dermatol.* 116, 506 (2001).

[34] S. G. Silver and V. C. Y. Ho, *Benign Epithelial Tumors*, Fitzpatrick's Dermatology in General Medicine (McGraw-Hill, New York, p. 774, 2003).

[35] N. A. Fenske and C. B. Conard, *Am. Fam. Physician* 37, 219 (1988).

[36] M. Yaar, M. S. Eller, and B. A. Gilchrest, *J. Investig. Dermatol. Symp. Proc.* 7, 51 (2002).

[37] J. M. Waller and H. I. Maibach, *Skin Res Technol* 11, 221–235 (2005).

[38] M. A. Farage, K. W. Miller, P. Elsner, and H. I. Maibach, *Int. J. Cosmet Sci.* 30, 87 (2008).

[39] T. J. Phillips, Z. Demircay, and M. Sahu, *Clin. Geriatr. Med.* 17, 661 (2001).

[40] N. H. Bell and G. J. Jackson, *Endocr. Pract.* 1, 44 (1995).

[41] A. Peters, *Neurosci. Biobehav. Rev.* 26, 733 (2002).

[42] C. O. Barland, E. Zettersten, B. S. Brown, J. Ye, P. M. Elias, and R. Ghadially, *J. Invest. Dermatol.* 122, 330 (2004).

[43] E. J. Yeo, I.-S. Jang, H.-K. Lim, K.-S. Ha, and S. C. Park, *Exp. Gerontol.* 37, 871 (2002).

[44] W. R. Reenstra, M. Yaar, and B. A. Gilchrest, *Exp. Cell Res.* 209, 118 (1993).

[45] E. Hütter, M. Skovbro, B. Lener, C. Prats, R. Rabøl, F. Dela, and P. Jansen-Dürr, *Aging Cell* 6, 245 (2007).

[46] T. von Zglinicki, E. Nilsson, W. D. Döcke, and U. T. Brunk, *Gerontology* 41 Suppl 2, 95 (1995).

[47] K. L. Double, V. N. Dedov, H. Fedorow, E. Kettle, G. M. Halliday, B. Garner, and U. T. Brunk, *Cell Mol. Life Sci.* 65, 1669 (2008).

[48] M. Yaar and B. A. Gilchrest, *Aging of Skin*, Fitzpatrick's Dermatology in General Medicine (McGraw-Hill, New York, pp. 1386 2003).

[49] S. E. G. Fligiel, J. Varani, S. C. Datta, S. Kang, G. J. Fisher, and J. J. Voorhees, *J. Invest. Dermatol.* 120, 842 (2003).

[50] G. J. Fisher, T. Quan, T. Purohit, Y. Shao, M. K. Cho, T. He, J. Varani, S. Kang, and J. J. Voorhees, *Am. J. Pathol.* 174, 101 (2009).

[51] D. Harman, *J. Gerontol.* 11, 298 (1956).

[52] J. H. Chung, J. Y. Seo, H. R. Choi, M. K. Lee, C. S. Youn, G. Rhie, K. H. Cho, K. H. Kim, K. C. Park, and H. C. Eun, *J. Invest. Dermatol.* 117, 1218 (2001).

[53] E. C. Naylor, R. E. B. Watson, and M. J. Sherratt, *Maturitas* 69, 249 (2011).

[54] B. N. Ames, M. K. Shigenaga, and T. M. Hagen, *Proc. Natl. Acad. Sci. USA* 90, 7915 (1993).

[55] W. Eiberger, B. Volkmer, R. Amouroux, C. Dhérin, J. P. Radicella, and B. Epe, *DNA Repair (Amst.)* 7, 912 (2008).

[56] A. Höhn, T. Jung, S. Grimm, B. Catalgol, D. Weber, and T. Grune, *Free Radic. Biol. Med.* 50, 585 (2011).

[57] P. Schroeder, J. Haendeler, and J. Krutmann, *Exp. Gerontol.* 43, 629 (2008).

[58] M. Mimeault and S. K. Batra, *J Cell Mol Med* 14, 116 (2010).

[59] M. S. Kaminer, *Photodamage: Magnitude of the Problem*, Photodamage (Blackwell, Cambridge, 295, 1995).

[60] M. Yaar and B. A. Gilchrest, *Br. J. Dermatol.* 157, 874 (2007).

[61] W. J. Schreier, T. E. Schrader, F. O. Koller, P. Gilch, C. E. Crespo-Hernández, V. N. Swaminathan, T. Carell, W. Zinth, and B. Kohler, *Science* 315, 625 (2007).

[62] J. H. Yoon, C. S. Lee, T. R. O'Connor, A. Yasui, and G. P. Pfeifer, *J. Mol. Biol.* 299, 681 (2000).

[63] R. M. Tyrrell, *Biochem. Soc. Symp.* 61, 47 (1995).

[64] E. Kvam and R. M. Tyrrell, *Carcinogenesis* 18, 2379 (1997).

[65] N. S. Agar, G. M. Halliday, R. S. Barnetson, H. N. Ananthaswamy, M. Wheeler, and A. M. Jones, *Proc. Natl. Acad. Sci. USA* 101, 4954 (2004).

[66] E. L. Menon and H. Morrison, *Photochem. Photobiol.* 75, 565 (2002).

[67] K. Sato, H. Taguchi, T. Maeda, H. Minami, Y. Asada, Y. Watanabe, and K. Yoshikawa, *J. Invest. Dermatol.* 105, 608 (1995).

[68] C. Kipp and A. R. Young, *Photochem. Photobiol.* 70, 191 (1999).

[69] K. Scharffetter-Kochanek, M. Wlaschek, P. Brenneisen, M. Schauen, R. Blaudschun, and J. Wenk, *Biol. Chem.* 378, 1247 (1997).

[70] G. T. Wondrak, M. J. Roberts, M. K. Jacobson, and E. L. Jacobson, *J. Invest. Dermatol.* 119, 489 (2002).

[71] G. T. Wondrak, M. J. Roberts, D. Cervantes-Laurean, M. K. Jacobson, and E. L. Jacobson, *J. Invest. Dermatol.* 121, 578 (2003).

[72] G. T. Wondrak, M. J. Roberts, M. K. Jacobson, and E. L. Jacobson, *J. Biol. Chem.* 279, 30009 (2004).

[73] B. H. Mahmoud, C. L. Hexsel, I. H. Hamzavi, and H. W. Lim, *Photochem. Photobiol.* 84, 450 (2008).

[74] J. H. Yang, H. C. Lee, K. J. Lin, and Y. H. Wei, *Arch. Dermatol. Res.* 286, 386 (1994).

[75] M. Berneburg, H. Plettenberg, K. Medve-König, A. Pfahlberg, H. Gers-Barlag, O. Gefeller, and J. Krutmann, *J. Invest. Dermatol.* 122, 1277 (2004).

[76] K. J. Krishnan, A. Harbottle, and M. A. Birch-Machin, *J. Invest. Dermatol.* 123, 1020 (2004).

[77] A. Eshaghian, R. A. Vleugels, J. A. Canter, M. A. McDonald, T. Stasko, and J. E. Sligh, *J. Invest. Dermatol.* 126, 336 (2006).

[78] M. Berneburg, S. Grether-Beck, V. Kürten, T. Ruzicka, K. Briviba, H. Sies, and J. Krutmann, *J. Biol. Chem.* 274, 15345 (1999).

[79] H. Koch, K.-P. Wittern, and J. Bergemann, *J. Invest. Dermatol.* 117, 892 (2001).

[80] F. Ji, Z. Pierre, S. Liu, B. Hwuang, H. Hill, and K. Hubbard, *J. Photochem. and Photobiol.* 184, 340 (2006).

[81] R. D. Granstein, *Photoimmunology*, Fitzpatrick's Dermatology in General Medicine (McGraw-Hill, New York, 378, 2003).

[82] P. R. Bergstresser, *Immediate and Delayed Effects of UVR on Immune Responses in Skin*, Photodamage (Blackwell, Cambridge, 81, 1995).

[83] Y. Matsumura and H. N. Ananthaswamy, *Expert Rev. Mol. Med.* 4, 1 (2002).

[84] G. J. Fisher, S. Kang, J. Varani, Z. Bata-Csorgo, Y. Wan, S. Datta, and J. J. Voorhees, *Arch. Dermatol.* 138, 1462 (2002).

[85] K. Y. Chung, A. Agarwal, J. Uitto, and A. Mauviel, *J. Biol. Chem.* 271, 3272 (1996).

[86] R. A. Rippe, L. W. Schrum, B. Stefanovic, J. A. Solís-Herruzo, and D. A. Brenner, *DNA Cell Biol.* 18, 751 (1999).

[87] G. Beauchef, N. Bigot, M. Kypriotou, E. Renard, B. Porée, R. Widom, A. Dompmartin-Blanchere, T. Oddos, F. Maquart, M. Demoor, K. Boumediene, and P. Galéra, *J. Biol. Chem.* 287, 3462 (2012).

[88] Y. Kawaguchi, H. Tanaka, T. Okada, H. Konishi, M. Takahashi, M. Ito, and J. Asai, *Free Radic. Biol. Med.* 23, 162 (1997).

[89] S. Kurdykowski, S. Mine, V. Bardey, L. Danoux, C. Jeanmaire, G. Pauly, E. Brabencova, Y. Wegrowski, and F.-X. Maquart, *Photochem. Photobiol.* 87, 1105 (2011).

[90] L. Yin, A. Morita, and T. Tsuji, *Arch. Dermatol. Res.* 292, 188 (2000).

[91] C. Lahmann, J. Bergemann, G. Harrison, and A. R. Young, *Lancet* 357, 935 (2001).

[92] A. S. Boyd, T. Stasko, L. E. King, G. S. Cameron, A. D. Pearse, and S. A. Gaskell, *J. Am. Acad. Dermatol.* 41, 23 (1999).

[93] J. B. Smith and N. A. Fenske, *J. Am. Acad. Dermatol.* 34, 717 (1996).

[94] U. Krämer, C. H. Lemmen, H. Behrendt, E. Link, T. Schäfer, J. Gostomzyk, G. Scherer, and J. Ring, *Br. J. Dermatol.* 150, 111 (2004).

[95] M.-S. Kim, Y. K. Kim, K. H. Cho, and J.-H. Chung, *Mech. Ageing Dev.* 127, 875 (2006).

[96] P. Schroeder, J. Lademann, M. E. Darvin, H. Stege, C. Marks, S. Bruhnke, and J. Krutmann, *J. Invest. Dermatol.* 128, 2491 (2008).

[97] S. M. Schieke, P. Schroeder, and J. Krutmann, *Photodermatol. Photoimmunol. Photomed.* 19, 228 (2003).

[98] P. Schroeder, C. Pohl, C. Calles, C. Marks, S. Wild, and J. Krutmann, *Free Radic. Biol. Med.* 43, 128 (2007).

[99] I. E. Kochevar, C. R. Taylor, and J. Krutmann, *Fundamentals of Cutaneous Photobiology and Photoimmunology*, Fitzpatrick's Dermatology in General Medicine (McGraw-Hill, New York, 797 2007).

[100] K. Danno and N. Sugie, *Photodermatol. Photoimmunol. Photomed.* 12, 233 (1996).

[101] L. A. Applegate, C. Scaletta, R. Panizzon, E. Frenk, P. Hohlfeld, and S. Schwarzkopf, *Int. J. Mol. Med.* 5, 247 (2000).

[102] J. J. Thiele, M. G. Traber, K. Tsang, C. E. Cross, and L. Packer, *Free Radic. Biol. Med.* 23, 385 (1997).

[103] R. M. Kafoury, W. A. Pryor, G. L. Squadrito, M. G. Salgo, X. Zou, and M. Friedman, *Am. J. Respir. Crit. Care Med.* 160, 1934 (1999).

[104] S. U. Weber, J. J. Thiele, C. E. Cross, and L. Packer, *J. Invest. Dermatol.* 113, 1128 (1999).

[105] C. Jonak, G. Klosner, and F. Trautinger, *Int. J. Cosmet. Sci.* 28, 233 (2006).

[106] S. D. Dams, M. de Liefde-van Beest, A. M. Nuijs, C. W. J. Oomens, and F. P. T. Baaijens, *Skin Res. Technol.* 17, 167 (2011).

[107] P. Chauhan and M. Shakya, *Indian J. Dermatol. Venereol. Leprol.* 75, 463 (2009).

[108] A. Vierkötter, T. Schikowski, U. Ranft, D. Sugiri, M. Matsui, U. Krämer, and J. Krutmann, *J. Invest. Dermatol.* 130, 2719 (2010).

[109] B. A. Wagner, S. Venkataraman, and G. R. Buettner, *Free Radic. Biol. Med.* 51, 700 (2011).

[110] R. Schreck, P. Rieber, and P. A. Baeuerle, *Embo J.* 10, 2247 (1991).

[111] M. Meyer, R. Schreck, and P. A. Baeuerle, *Embo J.* 12, 2005 (1993).

[112] M. J. Morgan and Z.-G. Liu, *Cell Res* 21, 103 (2011).

[113] G. Gloire, S. Legrand-Poels, and J. Piette, *Biochem Pharmacol* 72, 1493 (2006).

[114] Z.-H. Wu and S. Miyamoto, *J. Mol. Med.* 85, 1187 (2007).

[115] R. van den Berg, G. R. Haenen, H. van den Berg, and A. Bast, *Br. J. Nutr.* 86 Suppl 1, S121 (2001).

[116] A. Vigneron and K. H. Vousden, *Aging* 2, 471 (2010).

[117] A. Barthel and L.-O. Klotz, *Biol. Chem.* 386, 207 (2005).

[118] X. Wang, J. Z. Liu, J. X. Hu, H. Wu, Y. L. Li, H. L. Chen, H. Bai, and C. X. Hai, *Free Radic. Biol. Med.* 51, 539 (2011).

[119] Y. Morel and R. Barouki, *Biochem. J.* 342 Pt 3, 481 (1999).

[120] Y. Morel and R. Barouki, *Med. Sci. (Paris)* 14, 713 (1998).

[121] N. H. Zawia, D. K. Lahiri, and F. Cardozo-Pelaez, *Free Radic. Biol. Med.* 46, 1241 (2009).

[122] S. E. Johnstone and S. B. Baylin, *Nat. Rev. Genet.* 11, 806 (2010).

[123] Z. Draelos and P. Pugliese, *Cosmetics and Toiletries* 126, 438 (2011).

[124] P. E. Morgan, D. I. Pattison, and M. J. Davies, *Free Radic. Biol. Med.* 52, 328 (2012).

[125] H. P. Indo, M. Davidson, H.-C. Yen, S. Suenaga, K. Tomita, T. Nishii, M. Higuchi, Y. Koga, T. Ozawa, and H. J. Majima, *Mitochondrion* 7, 106 (2007).

[126] C. Y. Lu, H. C. Lee, H. J. Fahn, and Y. H. Wei, *Mutat. Res.* 423, 11 (1999).
[127] A. Catalá, Chem. Phys. *Lipids* 157, 1 (2009).
[128] F. H. Greig, S. Kennedy, and C. M. Spickett, *Free Radic. Biol. Med.* 52, 266 (2012).
[129] K. Simons and D. Toomre, *Nat. Rev. Mol. Cell Biol.* 1, 31 (2000).
[130] K. M. Boini, C. Zhang, M. Xia, W.-Q. Han, C. Brimson, J. L. Poklis, and P.-L. Li, *Biochim. Biophys. Acta* 1801, 1294 (2010).
[131] C. Mathay, M. Pierre, M. R. Pittelkow, E. Depiereux, A. F. Nikkels, A. Colige, and Y. Poumay, *J. Invest. Dermatol.* 131, 46 (2011).
[132] H. Y. Chung, E. K. Lee, Y. J. Choi, J. M. Kim, D. H. Kim, Y. Zou, C. H. Kim, J. Lee, H. S. Kim, N. D. Kim, J. H. Jung, and B. P. Yu, *J. Dent Res.* 90, 830 (2011).
[133] J. Caamaño and C. A. Hunter, *Clin. Microbiol. Rev.* 15, 414 (2002).
[134] S. Ghosh, M. J. May, and E. B. Kopp, *Annu. Rev. Immunol.* 16, 225 (1998).
[135] C. Franceschi, M. Bonafè, and S. Valensin, *Vaccine* 18, 1717 (2000).
[136] C. Franceschi, S. Valensin, M. Bonafè, G. Paolisso, A. I. Yashin, D. Monti, and G. De Benedictis, *Exp. Gerontol.* 35, 879 (2000).
[137] A. Salminen, J. Huuskonen, J. Ojala, A. Kauppinen, K. Kaarniranta, and T. Suuronen, *Ageing Res. Rev.* 7, 83 (2008).
[138] E. S. Cannizzo, C. C. Clement, R. Sahu, C. Follo, and L. Santambrogio, *J. Proteomics* 74, 2313 (2011).
[139] S. Giunta and G. Sergio, *Inflamm. Res.* 57, 558 (2008).
[140] F. Cabreiro, D. Ackerman, R. Doonan, C. Araiz, P. Back, D. Papp, B. P. Braeckman, and D. Gems, *Free Radic. Biol. Med.* 51, 1575 (2011).
[141] D. A. Sinclair and L. Guarente, *Sci. Am.* 294, 48–51, 54 (2006).
[142] M. Moreau, M. Neveu, S. Stéphan, E. Noblesse, C. Nizard, N. S. Sadick, S. Schnebert, F. Bonté, M. Dumas, P. Andre, and E. Perrier, *J. Drugs Dermatol.* 6, s14 (2007).
[143] J. Liu, L. Cao, and T. Finkel, *Free Radic. Biol. Med.* 51, 2158 (2011).
[144] B. R. Webster, Z. Lu, M. N. Sack, and I. Scott, *Free Radic. Biol. Med.* 52, 281 (2012).
[145] E. Niki, *Free Radic. Biol. Med.* 49, 503 (2010).
[146] J. Tabart, C. Kevers, J. Pincemail, J. Defraigne, and J. Dommes, *Food Chemistry* 113, 1226 (2009).
[147] R. Gebhardt, *Planta Med.* 66, 99 (2000).
[148] A. Ratz-Lyko, J. Arct, and K. Pytkowska, *Skin Res. Technol.* Epub Nov 14 (2011).
[149] J. V. Gruber and R. Holtz, *Mediators Inflamm.* 2010, 1 (2010).
[150] S. Q. Wang, Y. Balagula, and U. Osterwalder, *Dermatol. Ther.* 23, 31 (2010).
[151] V. Gupta and S. Sharma, *Natural Product Radiance* 5, 326 (2006).
[152] M. E. Obrenovich, N. G. Nair, A. Beyaz, G. Aliev, and V. P. Reddy, *Rejuvenation Res.* 13, 631 (2010).
[153] S. I. S. Rattan, *Ageing Res Rev* 7, 63 (2008).
[154] A. Speciale, J. Chirafisi, A. Saija, and F. Cimino, *Curr. Mol. Med.* 11, 770 (2011).
[155] H. Ichikawa, T. Ichiyanagi, B. Xu, Y. Yoshii, M. Nakajima, and T. Konishi, *J. Med. Food* 4, 211 (2001).
[156] L. Einbond, K. Reynertson, X. Luo, M. Basile, and E. Kennelly, *Food Chemistry* 84, 23 (2004).
[157] Y. Kimura and M. Sumiyoshi, *Eur. J. Pharmacol.* 661, 124 (2011).
[158] S. Adhikari, J. C. Tilak, and T. P. A. Devasagayam, *Indian J. Biochem. Biophys.* 48, 275 (2011).

[159] K. A. Kang, R. Zhang, M. J. Piao, S. Chae, H. S. Kim, J. H. Park, K. S. Jung, and J.-W. Hyun, *Toxicol Ind. Health* Epub Sep 28 (2011).
[160] I. K. Lee, K. A. Kang, R. Zhang, B. J. Kim, S. S. Kang, and J.-W. Hyun, *Environ. Toxicol. Pharmacol.* 31, 233 (2011).
[161] S. J. Wu, J. Y. Tsai, S. P. Chang, D. L. Lin, S. S. Wang, S. N. Huang, and L. T. Ng, *J. Ethnopharmacol.* 108, 407 (2006).
[162] I. Chattopadhyay, K. Biswas, U. Bandyopadhyay, and R. Banerjee, *Current Science* 87, 44 (2004).
[163] F. Bonté, M. S. Noel-Hudson, J. Wepierre, and A. Meybeck, *Planta Med.* 63, 265 (1997).
[164] C. F. Lima, C. Pereira-Wilson, and S. I. S. Rattan, *Mol. Nutr. Food Res.* 55, 430 (2011).
[165] R. L. Thangapazham, A. Sharma, and R. K. Maheshwari, *Adv. Exp. Med. Biol.* 595, 343 (2007).
[166] M. C. Y. Heng, *Int. J. Dermatol.* 50, 1058 (2011).
[167] P. Asawanonda and S.-O. Klahan, *Photomed. Laser Surg* 28, 679 (2010).
[168] C. D. Kaur and S. Saraf, *J. Cosmet Dermatol.* 10, 260 (2011).
[169] S. Hsu, *J. Am. Acad. Dermatol.* 52, 1049 (2005).
[170] N. Ahmad and H. Mukhtar, *Skin Pharmacol Appl. Skin. Physiol.* 14, 69 (2001).
[171] C. Fourneau, A. Laurens, and R. Hocquemiller, *Phytotherapy Research* 10, 529 (1996).
[172] A. E. Chiu, J. L. Chan, D. G. Kern, S. Kohler, W. E. Rehmus, and A. B. Kimball, *Dermatol. Surg.* 31, 855 (2005).
[173] P. Ouyang, W.-L. Peng, D.-L. Xu, W.-Y. Lai, and A.-L. Xu, *Di Yi Jun Yi Da Xue Xue Bao* 24, 247 (2004).
[174] N. Yusuf, C. Irby, S. K. Katiyar, and C. A. Elmets, *Photodermatol. Photoimmunol. Photomed.* 23, 48 (2007).
[175] H. Zhang, D. Cao, W. Cui, M. Ji, X. Qian, and L. Zhong, *Free Radic. Biol. Med.* 49, 2010 (2010).
[176] J. A. Nichols and S. K. Katiyar, *Arch. Dermatol. Res.* 302, 71 (2010).
[177] M. Rizwan, I. Rodriguez-Blanco, A. Harbottle, M. A. Birch-Machin, R. E. B. Watson, and L. E. Rhodes, *Br. J. Dermatol.* 164, 154 (2011).
[178] U. Heinrich, H. Tronnier, W. Stahl, M. Béjot, and J. M. Maurette, *Skin Pharmacol. Physiol.* 19, 224 (2006).
[179] M. Darvin, A. Patzelt, S. Gehse, S. Schanzer, C. Benderoth, W. Sterry, and J. Lademann, *Eur. J. Pharm. Biopharm.* 69, 943 (2008).
[180] F. Bonté, C. Simmler, A. Lobstein, F. Pellicier, and J.-H. Cauchard, *Ann. Pharm. Fr* 69, 177 (2011).
[181] C. Simmler, C. Antheaume, and A. Lobstein, *PLoS ONE* 5, e13713 (2010).
[182] C. Simmler, C. Antheaume, P. Andre, F. Bonté, and A. Lobstein, *J. Nat. Prod.* 74, 949 (2011).
[183] T. Tadokoro, F. Bonté, J. C. Archambault, J. H. Cauchard, M. Neveu, K. Ozawa, F. Noguchi, A. Ikeda, M. Nagamatsu, and S. Shinn, *J. Dermatol.* 37, 522 (2010).
[184] D. Stajner, B. M. Popović, A. Kapor, P. Boza, and M. Stajner, *Phytother Res.* 24, 759 (2010).
[185] R. Perez Gutierrez, A. Neira Gonzalez, E. Baez, and S. Diaz, *Chemistry of Natural Compounds* 46, 554 (2010).
[186] T. Shafazila and P. Lee, *IPCBEE* 7, 19 (2011).

[187] M. K. Johnson, K. E. Alexander, N. Lindquist, and G. Loo, Comp. Biochem. Physiol. C *Pharmacol Toxicol Endocrinol* 122, 211 (1999).
[188] P. Mladenka, L. Zatloukalová, T. Filipský, and R. Hrdina, *Free Radic. Biol. Med.* 49, 963 (2010).
[189] M. Melidou, K. Riganakos, and D. Galaris, *Free Radic. Biol. Med.* 39, 1591 (2005).
[190] D. Taubert, T. Breitenbach, A. Lazar, P. Censarek, S. Harlfinger, R. Berkels, W. Klaus, and R. Roesen, *Free Radic. Biol. Med.* 35, 1599 (2003).
[191] W. Kusirisin, S. Srichairatanakool, P. Lerttrakarnnon, N. Lailerd, M. Suttajit, C. Jaikang, and C. Chaiyasut, *Medicinal Chemistry* 5, 139 (2009).
[192] L. A. Pacheco-Palencia, G. Noratto, L. Hingorani, S. T. Talcott, and S. U. Mertens-Talcott, *J. Agric. Food Chem.* 56, 8434 (2008).
[193] F. Afaq, M. A. Zaid, N. Khan, M. Dreher, and H. Mukhtar, *Exp. Dermatol.* 18, 553 (2009).
[194] J. Smits and N. Herbst, *Personal Care* 2, 25 (2012).
[195] A. Marini, S. Grether-Beck, T. Jaenicke, M. Weber, C. Burki, P. Formann, H. Brenden, F. Schönlau, and J. Krutmann, *Skin Pharmacol. Physiol.* 25, 86 (2012).
[196] J. Psotova, A. Svobodova, H. Kolarova, and D. Walterova, *J. Photochem. Photobiol. B* 84, 167 (2006).
[197] J. Lee, E. Jung, J. Koh, Y. S. Kim, and D. Park, *J. Dermatol.* 35, 768 (2008).
[198] R. M. Facino, M. Carini, R. Stefani, G. Aldini, and L. Saibene, *Arch. Pharm. (Weinheim)* 328, 720 (1995).
[199] J. Song, J. Kou, Y. Huang, and B. Yu, *J. Pharmacol. Sci.* 113, 409 (2010).
[200] R. Gopaul, H. E. Knaggs, and J. Lephart, *Int. J. Cosmet. Sci.* 33, 416 (2011).
[201] M. S. Noel-Hudson, C. de Belilovsky, N. Petit, A. Lindenbaum, and J. Wepierre, *Toxicol. in Vitro* 3, 103 (1989).
[202] S. K. Katiyar, S. K. Mantena, and S. M. Meeran, *PLoS ONE* 6, e21410 (2011).
[203] A. E. Wagner, I. Ernst, R. Iori, C. Desel, and G. Rimbach, *Exp. Dermatol.* 19, 137 (2010).
[204] T. Greco, J. Shafer, and G. Fiskum, *Free Radic. Biol. Med.* 51, 2164 (2011).
[205] P. Talalay, J. W. Fahey, Z. R. Healy, S. L. Wehage, A. L. Benedict, C. Min, and A. T. Dinkova-Kostova, *Proc. Natl. Acad. Sci. USA* 104, 17500 (2007).
[206] E. F. Ritter, M. Axelrod, K. W. Minn, E. Eades, A. M. Rudner, D. Serafin, and B. Klitzman, *Plast Reconstr. Surg.* 100, 973 (1997).
[207] A. Meybeck, M. Dumas, F. Bonté, and C. Maréchal, FR Patent No. 2,657,526 (1997).
[208] E. Kato, Y. Sasaki, and N. Takahashi, *Bioorg Med. Chem.* 19, 6348 (2011).
[209] M. Pacholec, J. E. Bleasdale, B. Chrunyk, D. Cunningham, D. Flynn, R. S. Garofalo, D. Griffith, M. Griffor, P. Loulakis, B. Pabst, X. Qiu, B. Stockman, V. Thanabal, A. Varghese, J. Ward, J. Withka, and K. Ahn, *J. Biol. Chem.* 285, 8340 (2010).
[210] H. Ledford, *Nature* 464, 480 (2010).
[211] F. Orallo, *Curr. Med. Chem.* 15, 1887 (2008).
[212] P. Kovacic and R. Somanathan, *Oxid. Med. Cell Longev* 3, 86 (2010).
[213] E. L. Robb and J. A. Stuart, *Free Radic. Biol. Med.* 50, 821 (2011).
[214] R. A. Baxter, *J. Cosmet. Dermatol.* 7, 2 (2008).
[215] Y. Wu, L.-L. Jia, Y.-N. Zheng, X.-G. Xu, Y.-J. Luo, B. Wang, J. Z. S. Chen, X.-H. Gao, H.-D. Chen, M. Matsui, and Y.-H. Li, *J. Eur. Acad. Dermatol. Venereol.* Epub jan 5 (2012).

[216] K.-T. Ku, Y.-L. Huang, Y.-J. Huang, and W.-F. Chiou, *J. Agric. Food Chem.* 56, 8911 (2008).

[217] S. Cornacchione, N. S. Sadick, M. Neveu, S. Talbourdet, K. Lazou, C. Viron, I. Renimel, D. de Quéral, R. Kurfurst, S. Schnebert, C. Heusèle, P. André, and E. Perrier, *J. Drugs Dermatol.* 6, s8 (2007).

[218] S. Schwaiger, R. Cervellati, C. Seger, E. Ellmerer, N. About, I. Renimel, C. Godenir, P. André, F. Gafner, and H. Stuppner, *Tetrahedron* 61, 4621 (2005).

[219] J. Kolodziejczyk, J. Saluk-Juszczak, and B. Wachowicz, *J. Physiol. Biochem.* 67, 175 (2011).

[220] F. A. C. Wiegant, S. Surinova, E. Ytsma, M. Langelaar-Makkinje, G. Wikman, and J. A. Post, *Biogerontology* 10, 27 (2009).

[221] S. E. Schriner, A. Avanesian, Y. Liu, H. Luesch, and M. Jafari, *Free Radic. Biol. Med.* 47, 577 (2009).

[222] F. Bonté, *Phytothérapie Européenne* 2, 16 (2001).

[223] F. Solano, S. Briganti, M. Picardo, and G. Ghanem, *Pigment Cell Res.* 19, 550 (2006).

In: Phytochemicals
Editor: Hai-Xue Kuang

ISBN: 978-1-62417-354-7


*Chapter XI*

# Mineral, Flavonoid, and Fatty Acid Concentrations in Ten Diverse *Lablab Purpureus* (L.) Sweet Accessions

***J. Bradley Morris*[1*]*, Michael A. Grusak*[2]*,***
***Ming Li Wang*[1]* and Brandon David Tonnis*[1]***
[1]USDA, ARS, PGRCU, Griffin, GA, US
[2]USDA, ARS,
Children's Nutrition Research Center,
Baylor College of Medicine, Houston, TX, US

## Abstract

Seeds of Hyacinth bean (*Lablab purpureus* [L.]) Sweet containing high concentrations of minerals, flavonoids and fatty acids may provide government agencies with a nutrient-dense and health-beneficial food for use in hunger stricken and nutrient deprived people. Seeds from ten hyacinth bean accessions were analyzed for mineral, flavonoid, and fatty acid concentrations using inductively coupled plasma-optical emission spectroscopy, reverse-phase high performance liquid chromatography, and gas chromatography, respectively. Mean concentrations of macrominerals (Ca, Mg, K, P, and S) and microminerals (Cu, Fe, Mn, Ni, and Zn) ranged from 0.78 - 15.1 mg/g DW, and 0.60 - 95.90 µg/g DW, respectively among accessions. Seed concentrations of myricetin, quercetin and kaempferol ranged from 0 - 19.46 µg/g DW, 0 - 28.49 µg/g DW, and 0 - 7.42 µg/g DW, respectively. Linoleic, linolenic, and oleic acid ranged from 50 - 55%, 9 - 14%, and 5 - 14%, respectively, as percent of total fatty acids measured (total fat ranged from 0.437 – 0.663%). The information presented here will be useful for the development of new hyacinth bean cultivars enriched in essential nutrients and health-beneficial phytochemicals.

* Email: Brad.Morris@ars.usda.gov.

# Experiment

The hyacinth bean (*Lablab purpureus* L. Sweet; $2n = 22$) is closely related to common bean [1], and is a marginal food crop in southern Asia and eastern Africa [2]. Hyacinth bean seeds contain useful phytochemicals which can contribute to human health. These include the flavonoids myricetin, which has been shown to contribute to pancreatic cancer cell death *in vitro* [3]; quercetin, which has been shown to inhibit the growth of pancreatic cancer stem cells [4]; and kaempferol, which has been shown to prevent skin cancer [5].

The land race accessions used in this study originated from Australia (PI 388017), China (PI 419086, PI 639277, PI 639279), India (PI 164772, PI 183451), the former Soviet Union (PI 392369) and Zambia (PI 338341). However, an elite line from the U.S.A. (PI 596358) and an Indian cultivar (PI 509114) were included also. Even though, more than 90 hyacinth bean accessions are curated at the USDA, Agricultural Research Service (ARS), Plant Genetic Resources Conservation Unit (PGRCU) facility in Griffin, GA, these 10 accessions were chosen because they typically produce more seed than most of the other accessions in our local environment.

Characterization information regarding these accessions can be found at http://www.ars-grin.gov/npgs/. Little is known about mineral, flavonoid, and fatty acid concentrations in the USDA, ARS, PGRCU hyacinth bean collection. Our objectives were to determine the concentrations of:

1) minerals using inductively coupled plasma-optical emission spectroscopy (ICP-OES), 2) flavonoids using reverse-phase high performance liquid chromatography (HPLC), and 3) fatty acids using a gas chromatograph, in ten hyacinth bean accessions from the USDA, ARS, (PGRCU) collection in Griffin, GA, USA. One month old seedlings from ten of the highest seed producing accessions were transplanted to 6-m long field plots in a randomized complete block design with 2 replications at Griffin, Georgia during 2009. Six meter long field plots and 2 replications per hyacinth bean accession were used because of limited field space. Dry, mature seeds were harvested from each accession at the end of the growing season.

For mineral analysis, seeds were oven-dried (60°C) and ground to a fine powder. Samples (~0.25 g DW) of each ground accession were wet digested with ultra-pure nitric acid and hydrogen peroxide (at 125 - 200°C) and then dried at 200° C. After cooling, digestates were re-suspended in 2% ultra-pure nitric acid overnight, then vortexed and transferred to plastic storage tubes until analysis for Ca, Mg, K, P, S, Fe, Zn, Mn, Cu, and Ni concentrations. Elemental analysis was performed using an ICP-OES (CIROS ICP Model FCE 12, Spectro, Kleve, Germany). The instrument was calibrated daily with certified standards. Tomato leaf standards (SRM 1573 A, National Institute of Standards and Technology, Gaithersburg, MD) were digested and analyzed along with the samples to ensure accuracy of the instrument calibration. For flavonoid analysis, 0.5 g of ground hyacinth bean seed tissue and 5 ml of 60% methanol and 1.2 M HCl was mixed and incubated at 80ºC for 2 h. The samples were centrifuged, and separations were performed by reverse phase HPLC (Zorbax Eclipse 3.0 X 150 mm, 5 µm, C18 column, Agilent Technologies) at 40ºC on an Agilent 1100 with a binary pump and autosampler. The injection volume was 5 µl, and analytes were monitored with a diode-array detector at 370 nm. Myricetin, quercetin and kaempferol (Sigma-Aldrich, St. Louis, MO) standards were dissolved in a 5 DMSO:3 methanol:2 water mix and diluted with 60% methanol to generate standard curves for analysis. The mobile phase consisted of HPLC-grade acetonitrile (B) and 0.1% formic acid in filtered, sterile water (A). The flow rate was 0.8 ml/min consisting of 15% B at time zero to 35% B at 20 min. The column was washed with 95% B for 5 min and equilibrated for 7 min between injections. Fatty acids were determined using an Agilent 7890 A gas chromatograph. Oil from ground seeds (~250 mg) was extracted in 3-ml heptanes and converted to fatty acid methyl esters with 500 µl 0.5 N sodium methoxide in methanol. An aliquot from the heptane layer was injected. Peak separations were performed on a DB-225 capillary column (15 m X 0.25 mm internal diameter with a 0.25 µm film). One µl of sample was injected at a 15:1 ratio into the column using a thermal gradient of: 195°C for 1 h, 195 to 200°C at 2.5°C/min and 200 to 230°C at 5°C/min. The carrier gas was helium, at 12 psi (~39 cm/sec at 195°C) and peaks were identified. The oven was equilibrated for 3.5 min between injections. Two extractions per replicated accession were used for flavonoid and fatty acid analysis. Two-way analysis of variance was conducted using the statistical analysis system (SAS, 2008) to analyze the mineral, flavonoid, and fatty acid data. Duncan's multiple range test was used to separate means.

## Discussion

Significant differences among ten hyacinth bean accessions occurred for all of the macro-minerals measured (Table 1). Potassium demonstrated the highest concentration of all the seed macrominerals, averaging 15.1 mg/g DW, while all other macrominerals, including Ca, Mg, P and S had average concentrations ranging from 0.78 mg/g DW (Ca) to 5.12 mg/g DW (P). In total, the concentrations noted in this study's ten accessions were higher (Ca), similar (Mg, P) or lower (K) than those reported in two previous papers with fewer lines studied [6, 7]. However, Alabi and Alausa reported lower Ca, P, K, and Mg concentrations than what we found in our study. Significant differences for most of the microminerals was observed in these accessions. The average seed Fe concentration was 95.90 µg/g DW, which was similar to the five cultivars measured by Kamatchi Kali *et al.* Zinc concentrations averaged 38.41 µg/g DW, with accession PI 509114 exhibiting the highest concentration (50.22 µg/g DW). Manganese concentrations averaged 31.12 µg/g DW among the hyacinth bean accessions with PI 388017 (36.25 µg/g DW) exceeding all other accessions. Kamatchi Kali *et al.* reported seed zinc concentrations averaging 27.41 µg/g DW, 30% lower than in this study; however, their average Mn concentrations (59.72 µg/g DW) exceeded ours by almost 2-fold. There is limited information in the literature about variability for mineral, flavonoid, and fatty acid concentrations in hyacinth bean. However, genotype and environment have been found to influence mineral concentrations in common bean seeds [9] which are a close relative of hyacinth bean. With respect to seed flavonoids, quercetin concentration among the hyacinth bean accessions averaged 8.43 µg/g DW. However, the Chinese accession, PI 639279 contained a significantly higher concentration of myricetin (19.46 µg/g DW). The Zambian accessions, PI 338341 and PI 509114 exhibited the highest concentrations of quercetin (28.49 and 25.97 µg/g DW, respectively). The accession PI 509114 also exhibited the highest concentration of kaempferol (7.42 µg/g DW), which was higher than that reported by Yang *et al.* Average quercetin concentrations were similar to those found by Cao *et al.*, however their kaempferol concentrations were superior to ours, except for PI 509114 (7.42 µg/g).

**Table 1. Macromineral, micromineral, and flavonoid concentrations in ten *Lablab purpureus* accessions**

| | Macrominerals (mg $g^{-1}$ DW) | | | | | Microminerals (μg $g^{-1}$ DW) | | | | | Flavonoids[a] (μg $g^{-1}$ DW) | | |
|---|---|---|---|---|---|---|---|---|---|---|---|---|---|
| | Ca | K | Mg | P | S | Cu | Fe | Mn | Ni | Zn | M | Q | K |
| Accession | | | | | | | | | | | | | |
| 596358 | 0.99a | 15.1bc | 2.09ab | 5.53ab | 2.22ab | 11.87bc | 97.36a | 33.46ab | 0.78ab | 40.34b | 0.00b | 19.97b | 1.83c |
| 639279 | 0.97a | 13.7d | 2.15a | 4.55c | 1.91d | 8.95d | 78.04a | 27.49cd | 0.50ab | 33.68c | 19.46a | 0.00c | 0.00e |
| 639277 | 0.96a | 15.4abc | 2.14a | 5.07bc | 2.13abcd | 12.82ab | 94.32a | 34.03ab | 0.60ab | 39.96b | 1.06b | 1.21c | 0.00e |
| 392369 | 0.94ab | 14.5cd | 1.87c | 4.97bc | 1.94cd | 13.36ab | 99.70a | 33.95ab | 0.53ab | 38.48bc | 2.61b | 1.16c | 0.00e |
| 388017 | 0.85abc | 15.5ab | 1.96bc | 4.59c | 1.97cd | 10.29cd | 105.79a | 36.25a | 0.40b | 35.77bc | 2.21b | 1.09c | 0.00e |
| 338341 | 0.83abc | 16.3a | 1.89c | 5.68ab | 2.15abc | 10.39cd | 96.91a | 31.11abcd | 0.54ab | 36.49bc | 0.00b | 28.49a | 2.56b |
| 509114 | 0.68bcd | 15.4abc | 2.12a | 6.07a | 2.27a | 14.15a | 103.47a | 32.12abc | 0.56ab | 50.22a | 0.00b | 25.97a | 7.42a |
| 419086 | 0.63cd | 15.4abc | 2.07ab | 5.19bc | 2.14abc | 10.31cd | 102.22a | 27.55cd | 0.52ab | 38.41bc | 0.00b | 1.00c | 0.00e |
| 183451 | 0.55d | 14.7bc | 1.82c | 4.57c | 2.05abcd | 7.10e | 78.86a | 29.21bcd | 0.91a | 36.44bc | 0.00b | 2.26c | 1.12d |
| 164772 | 0.42d | 15.5ab | 1.89c | 5.00bc | 2.01bcd | 9.01d | 102.32a | 25.98d | 0.68ab | 34.26c | 0.00b | 3.14c | 1.10d |
| Mean | 0.78 | 15.1 | 2.00 | 5.12 | 2.08 | 10.83 | 95.90 | 31.12 | 0.60 | 38.41 | 2.53 | 8.43 | 1.40 |

[a]The flavonoids are represented as myricetin (M), quercetin (Q), and kaempferol (K).

**Table 2. Total fat and percentages of individual fatty acids in ten *Lablab purpureus* accessions**

| | | Fatty acid % (of total fatty acids measured) | | | | | | | | | |
|---|---|---|---|---|---|---|---|---|---|---|---|
| Accession | Total fat % (DW basis) | 16:0 | 17:0 | 18:0 | 18:1 | 18:2 | 18:3 | 20:0 | 20:1 | 22:0 | 24:0 |
| 596358 | 0.437c | 21.07a | 0.25b | 2.81ef | 11.39abc | 49.94d | 11.34bcd | 0.47ef | 0.26ef | 0.55cd | 1.18b |
| 639279 | 0.540bc | 20.26a | 0.21c | 2.58f | 5.28f | 54.82a | 13.57a | 0.44f | 0.22g | 0.52de | 1.38a |
| 419086 | 0.487bc | 20.11ab | 0.26ab | 4.59a | 12.24abc | 50.06d | 9.68de | 0.58bc | 0.34a | 0.48ef | 1.33a |
| 388017 | 0.483bc | 20.02ab | 0.26ab | 3.59cd | 12.68ab | 50.01d | 10.56bcde | 0.50e | 0.31bc | 0.55cd | 0.76d |
| 509114 | 0.491bc | 19.98ab | 0.25b | 2.58f | 9.85cde | 52.64bc | 11.59bcd | 0.37g | 0.24fg | 0.43f | 0.84d |
| 338341 | 0.530bc | 19.90ab | 0.26ab | 3.23de | 10.27bcd | 51.17cd | 11.80ab | 0.57bcd | 0.29cd | 0.66a | 1.00c |
| 164772 | 0.568ab | 19.89ab | 0.26ab | 3.57cd | 7.39ef | 54.02ab | 11.65bc | 0.56cd | 0.26ef | 0.61abc | 0.98c |
| 639277 | 0.513bc | 19.87ab | 0.27ab | 4.04bc | 11.49abc | 51.45cd | 9.78cde | 0.61ab | 0.27de | 0.49def | 1.35a |
| 183451 | 0.663a | 18.87bc | 0.25ab | 3.23de | 8.57de | 54.69a | 11.18bcd | 0.52de | 0.28de | 0.59bc | 0.97c |
| 392369 | 0.580ab | 18.01c | 0.28a | 4.28ab | 13.59a | 51.58cd | 9.19e | 0.64a | 0.32b | 0.63ab | 1.04c |
| Mean | 0.529 | 19.80 | 0.26 | 3.45 | 10.28 | 52.04 | 11.03 | 0.53 | 0.28 | 0.55 | 1.10 |

The accessions, PI 338341 (28.49 μg/g), PI 509114 (25.97 μg/g) and PI 596358 (19.97 μg/g) exhibited higher concentrations of quercetin than those reported by Cao *et al.* Significant differences for seed fatty acids also occurred among these hyacinth bean accessions (Table 2). Total fat (per g DW) ranged from 0.437 % to 0.663%. The proportion of different fatty acids (i.e., percent of total fatty acids) also differed amongst the accessions (Table 2), with the human essential fatty acids, linoleic [18:2] and linolenic [18:3] acids, averaging 52.04% and 11.03%, respectively. The accessions, PI 639279 and PI 183451 exhibited the highest linoleic fatty acid percentages (54.82% and 54.69%, respectively). These linoleic values are about 11 percentage points higher than those found by (Kamatchi Kali *et al.*). The accession, PI 639279 exhibited the highest percentage of linolenic acid (13.57%), which is in the range of what Kamatchi Kali *et al.* reported. The second leading fatty acid exhibited amongst the hyacinth bean accessions was the saturated fatty acid, palmitic acid [16:0] averaging 19.80 %, with accessions ranging from 18.01% (PI 392369; former Soviet Union) to 21.07% (PI 596358; United States) of total fatty acids. Oleic acid [18:1] concentrations averaged 10.28% of total fatty acids, with PI 392369 exhibiting the highest level (13.59%). Our palmitic and oleic acid averages were about 9 and 3 percentage points lower, respectively, than those reported by Kamatchi Kali *et al.* Significant differences were observed in the seed concentrations of minerals, flavonoids, and fatty acids among ten hyacinth bean accessions. These phytochemicals are very important for human nutrition and as potential inhibitors of pancreatic cancer [3]; [4] and skin cancer [5].

These hyacinth bean accessions in the USDA, ARS, PGRCU collection contain little information regarding mineral, flavonoid, and fatty acid content. Several of these hyacinth bean accessions could be produced by farmers and marketed as a functional health food. This information will also help advance the development of new cultivars with enhanced mineral, flavonoid or fatty acid profiles for human health. Several of the hyacinth bean accessions contain significant levels of myricetin, quercetin, oleic, gadoleic [20:1], linoleic, and linolenic acids, which have potential use as functional food components. Seed mineral, flavonoid, and fatty acid concentrations differed across the accessions in this study. These differences indicate variability of these accessions originating from several countries of origin. Additional studies concerning potential toxic components and genotype by environmental effects of these compounds in hyacinth bean will need to be verified in future evaluations. However, this preliminary report provides evidence that these 10 hyacinth bean accessions in the USDA, ARS, PGRCU germplasm collection contain different concentrations of minerals, fatty acids, and flavonoids.

# References


[1] H. Valenzuela and J. Smith, *Sustainable agriculture green manure crops*, Manoa, HI: Cooperative Extension Service, College of Tropical Agriculture and Human Resources, University of Hawaii (2002).

[2] B.G. Cook, B.C. Pengelly, S.D. Brown, J.L. Donnelly, D.A. Eagles, M.A. Franco, J. Hanson, B.F. Mullen, I.J. Partridge, M. Peters and R. Schultze-Kraft, Tropical Forages: an interactive selection tool. [CD-ROM], CSIRO, DPIandF (Qld), CIAT and ILRI, Brisbane, Australia (2005).

[3] P.A. Phillips, V. Sangwan, D. Borja-Cacho, V. Dudeja, S.M. Vickers and A.K. Saluja, *Cancer Letters* 308, 181 (2011).

[4] S.N. Tang, J. Fu, D. Nall, M. Rodova, S. Shankar and R.K. Srivastava RK, (2011) *International Journal of Cancer* (2011) at doi: 10.1002/ijc.26323.

[5] K.M. Lee, K.W. Lee, S.K. Jung, E.J. Lee, Y.S. Heo, A.M. Bode, R.A. Lubet, H.J. Lee and Z. Dong Z, *Biochemical Pharmacology* 80, 2042 (2010).

[6] R.K. Deka and C.R. Sarkar, *Food Chemistry* 38, 239 (1990).

[7] Kali B. Kamatchi, Soris P. Tresina, V.R. Mohan and V. Vadivel, *Advances In Bioresearch* 1, 44 (2010).

[8] D.A. Alabi and A.A. Alausa, *World Journal of Agricultural Sciences* 2, 115 (2006)

[9] John T. Moraghan and Kenneth Grafton, *Journal of the Science of Food and Agriculture* 81, 404 (2001).

[10] Ray-Yu Yang, Lin Shou and George Kuo, *Asia Pacific Journal of Clinical Nutrition* 17, 275 (2008).

[11] Jie Cao, Wei Chen, Ying Zhang, Yuqiu Zhang and Xiujuan Zhao, *Food Science and Technology* 16, 395 (2010).

In: Phytochemicals
Editor: Hai-Xue Kuang

ISBN: 978-1-62417-354-7


*Chapter XII*

# Research Progress on Medicinal Plants of Urtica Linn. Genus

***Zhang Yanxiu, Wang Ying, Ju Zhihe and Guan Feng****

Key Laboratory of Chinese Materia Medica
(Heilongjiang University of Chinese Medicine), Ministry of Education, Harbin, China

## Abstract

As types of traditional folk herbs and ethnic drugs, medicinal plants of the Urtica Linn genus, with a long history in China,have been widely regarded by people and research in the field of medicine were gradually carried on. Due to the unique curative effect, widespread distribution and abundant resources, broad development prospects are present with these medicinal plants. In the other countries, *U. dioica*L. subsp. dioica, *U. urens*L. and *U. cannabina* L., etc. can still be widely used for the treatment of prostate disease, psoriasis, allergic rhinitis, leukemia, diabetes, anemia and eczema in addition to rheumatism arthritis. In recent decades, a great number of studies on the phytology, chemical constituent, bioactivity and clinical applications on the Urtica Linn genus have been done worldwide. This review summarizes and analyzes the research results in these fields, mostlyin China. It will provide the basics for further development and utilization of these important medicinal plants, not only for the pharmaceutical industry, but also forfood, cosmetics and so on.

## 1. Introduction

There are 47 generas, about 1300 species in Urticaceae, and ordinarily they distribute in tropical and temperate areas. There are 25 generas, 352 species, 26 subspecies and 63 varieties in China, and most of them grow in the subtropical areas and tropical areas south of

* E-mail address: traciiy-guan@hotmail.com.

the Changjiang River basin. They particularly like growing in a wet environment [1]. Stems and skins of many kinds plants of Urticaceae are rich in fiber, such as BoehmeriaJacq.,DebregeasiaGaudich., OreocnideMiq., Urtica Linn., GirardiniaGaudich, and LaporteaGaudich are all important fiber plants and most of them are used for disease treatment.

# 2. BotanicalStudies

## 2.1. Species and Biological Characteristics

Medicinal plants of Urtica Linn are annual or perennial shrubs (Figure 1). They all have sting-hairs and quadrangular stems, leaves opposite the margin that have teeth or split, monoecy or dioecy. Their florescenceis in July to August and they bear fruit in September.

U. laetevirens U. cannabina L. U. angusitifolia U. Fissa.

U. laetevirens U. Hyperborean.

U. macrorrhiza U. Angustifolia.

U. dioica U. fissa.

Figure 1. Several common species of Urtica Linn.in China.

**Table 1. Important medicinal plant species and distribution of Urtica Linn.in China**

| Plant Species | Medicinal part | Distribution |
|---|---|---|
| *Urticafissa* E. Pritz. | whole grass | Anhui, Zhejiang, Hunan, Hubei, Henan |
| *U. Angusitifolia*FischexHornem | whole grass | Dongbei, Neimenggu, Hebei |
| *U. artichocaulis*(Hand. -Mazz. )C. J. Chen | whole grass | Guizhou, Sichuan, Yunnan |
| *U. cannabina*L. | whole grass | Huabei, Xibei, Dongbei, Sichuan |
| *U. dioica* L. subsp. *dioica* | whole grass or root | Qinghai, Xinjiang, Yunnan, Xizang |
| *U. fissa* | whole grass | Shanxi, Gansu, Sichuan, Guizhou |
| *U. laetevirens*Maxim. subsp. laetevirens | whole grass | Xibei, Dongbei, Huabei, Xinan |
| *U. laetvirens* axim. subsp. dentata(Hand. - Mazz. )C. J. Chen | whole grass | Xibei, Xinan |
| *U. macrorrhiza* Hand. -Mazz. | whole grass | Yunnan |
| *U. mairei* | whole grass | Sihcuan、guizhou, yunnan, xizang |
| *U. thunbergiana* Sieb. etZucc. | leave | Taiwan |
| *U. triangularis*(Hand. -Mazz. )subsp. triangularis | whole grass | Sichuan, yunnan, Xizang, qinghai |
| *U. laetevirens* Maxim. subsp. *Lacyanec -ens* (Kom. )C. J. Chen | whole grass | Neimenggu, dongbei |
| *U. maireiLevl.* var. mairei | whole grass | Xizang, sichuan, yunnan |
| *U. dioica* L. subsp. gansuensis C. J. Chen | whole grass | Xizang, yunnan |
| *U. triangulanis* Hand-Mazz. subsp. trichocarpa C. J. Chen | whole grass | Gansu, qinghai, sichuan |
| *U. triangularis* Hand-Mass. subsp. Pinnatifida(Hand. -Mazz. )C. J. Chen | whole grass | Xizang, Qinghai, Yunnan, gansu |
| *U. urens*L. | whole grass or root | Xizang, Qinghai, Xinjiang, Liaoning |
| *U. silvatica* H. M. | whole grass | Guizhou |

There are about 50 species of plants and 23 of them are in China, mainly distributed in the temperate areas. They often grow onwet hills, grassland and valleys. *U. fissa*, *U. laetevirens* and *U. angusitifolia*are widely distributed in China and *U. laetvirens*Maxim., *U. dioica* L. subsp. Xinjianggensis C. J. Chen, *U. dioica L. subsp*gansuensis and *U. macrorrhiza* are endemic species of China [2]. Various plants of Urtica Linn (Table1) can be used as medicine and can relieve rheumatic pains, protect the liver and calm fear, remove metabolin from the intestines and relax the bowels and detoxify, and are used widely among the people. Most of them can be used with whole grass and some of them with radix and rhizomes, which are called Nettle Root [3-5].

The name of Urtica Linn came from Latin, which means painful urticant hair [6-7]. There are sting-hairs on the stems and leaves of these plants and people will feel unbearably pain if theyare bitten by these sting-hairs.

At the same time, allergic symptoms will fester on the skin that was hurt [8]. This kind of sting–hair is a kind of glandular hair. There is much venom in sting–hair and its ingredients are very complex including a special enzyme, formic acid and acetic acid, et al. It is said that daubing the wound with alkaline solution can relieve the pain [1, 9].

### 2.2. Identification of Crude Drug

Corsigabrielia and some other researchers [2]studied the organization of *U membrdnacae*Poiret's Sting hairs and made a comparative study of the microstructure of *U. dioica* L. subsp. dioica and *U. urens* L. Some Chinese scholars made a microscopic identification of *U. fissa* E. Pritz. and*U. laetevirens* Maxim. subsp. laetevirens [10-11].

At the same time they made a classification and identification of Urtica Linn. genus plants, found that this genus has inflorescence characteristics, the insertion way of stipule and the feature of fruits can be made the basis of classification [1]. By paraffin section and systematic observation, Liu Yueqiu [12] cleared the anatomical structure of the root, stem, leaves of U. dioica L. subsp. dioica. By making physical and chemical identification of *U. fissa*, *U.laetevirens*, *U. mairei*Levl., *U. artichocaulis*(Hand. -Mazz.) C. J. Chen, *U. dioica* L. subsp. gansuensis C. J. Chen, and *G. suborbiculata* C. J. Chen subsp. triloba (C. J. Chen) C. J. Chen.

Wang Mengyue [13-14] found that these six kinds of nettle herbs can be distinguished by the shape of the stem and leaves in traits and for the first time found a large branch of sting-hair. *U. cannabina* morphology and histology were studied by Li Chengyi and others [15], and its anatomical structure of roots, stems and leaves have been cleared.

## 3. Studies on Chemical Constituents

At present the types of compound which isoloted from Urtica Linn. plants mainly include flavonoids, organic acids, lignans, phenols, sterols, terpenoids, protein and polysaccharide, etc.

## 3.1. Flavonoids

There are about twenty kinds of flavonoids compounds [16-25] have been isolated from Urtica Linnplants, most belong to flavonoids and flavonols. The main basic nucleus of them for the following several types and compounds are shown in the Table 2.

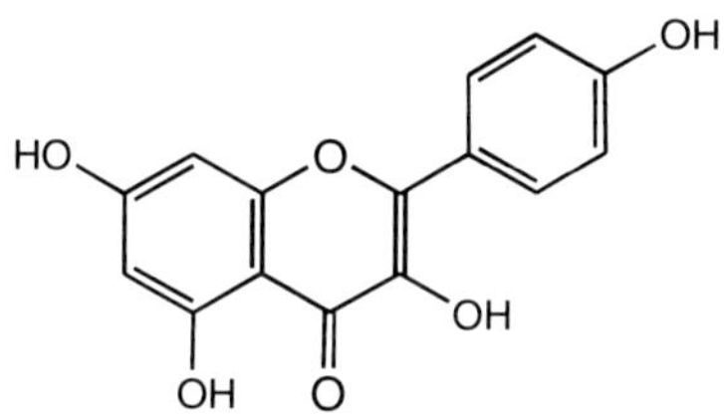


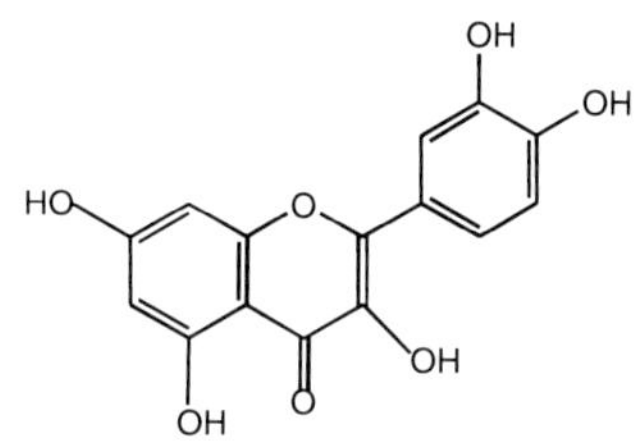


KaempferolQuercetin

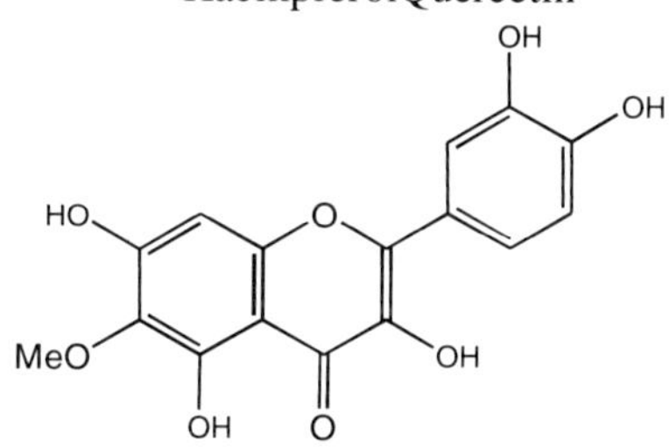


IsorhamnetinPatuletin

**Table 2. Important flavonoids constituents of Urticalinn. Plants**

| Compounds | Sources of plants | References |
|---|---|---|
| Kaempferol-3-O-glucoside | *U. dioica*L. | 16 |
| Kaempferol -3-O-rutinoside | *U. dioica*L. | 16 |
| Isorhamnetin-3-O-glucoside | *U. dioica*L. | 16 |
| Isorhamnetin-3-O-rutinoside | *U. dioica* L.; *U. fissa* E. Pritz. | 16, 20, 25 |
| Isorhamnetin-3-O-neohesperidin | *U. dioica*L. | 16 |
| 5, 2', 4'-trihydroxyl-7, 8-dimethoxyl-flavone | *U. dioica*L. | 16 |
| Quercitin-3-O-galactoside | *U. dioica*L. ; *U. urens*L. | 16 18 |
| Quercitin-7-O-glucoside | *U. dioica* L. ; *U. fissa* E. Pritz. | 16 20, 25 |
| Quercitin -3-O-glucoside | *U. dioica*L. | 16 |
| Quercitin -3-O-rutinoside | *U. dioica* L. ; *U. fissa* E. Pritz. | 16, 20, 22 |
| Quercitin | *U. dioica*L. | 17, 22, 23 |
| Kaempferol | *U. dioica*L. | 1723 |
| Isorhamnetin | *U. dioica* L. ; *U. fissa* E. Pritz. | 17, 20, 25 |
| Patuletin | *U. urens*L. | 18 |
| Patuletin-3-O-rutinoside | *U. urens*L. | 18 |
| Patuletin-3-O-glucoside | *U. urens*L. | 18 |
| Kampherol-7-O-rutinoside | *U. urens*L. | 18 |
| Isorhamnetin-3-O-galactoside | *U. urens*L. | 18 |
| Scutellarein-7-O-α-L-rhamnoside | *U. cannabina* | 19 |
| Apigenin-6, 8-di-c-glucoside | *U. cannabina* | 19 |
| Kaempferol-7-O-glucoside | *U. fissa* E. Pritz. | 20, 25 |
| Isoquercetin | *U. dioica*L. | 21 |
| 4'-methoxy isoquercetin | *U. dioica*L. | 21 |
| Apigenin | *U. sp.* | 23 |
| Luteolin | *U. sp.* | 23 |
| Quercitrin | *U. dioica*L. | 24 |

## 3.2. Lignans and Phenols

Analysis of *U. dioica* L. root extract were made by Kraus and others [22, 26, 30]through the method of GC-MS, they identified eighteen compounds of phenol and eight lignan. Four new compounds of lignanoids were separated from *U. mairei* Levi. by Wei wang [31-32]. The important compounds of lignans and phenolics are shown in the Table 3 and Figure 2.

**Table 3. Important lignansconstituents of Urtica Linn. Plants**

| Compounds | Sources of plants | References |
|---|---|---|
| didehydroisolariciresinol | U. dioicaL. | 26 |
| (+)-isolariciresinol | U. dioicaL. | 26 |
| (-)-secoisolariciresinol | U. dioicaL. | 26 |
| 5-methoxyl-8-hydroxyl-ringcleavage-isolariciresinol | U. dioicaL. | 26 |
| 5-methoxyl-ringcleavage-isolariciresinol | U. dioicaL. | 26 |
| olivil | U. dioicaL. | 26 |
| 2-(4-hydroxyl-3-methoxyl-phenyl)-3-methylol-4-(4-hydroxyl-3- methoxyl-phenyl) methylol-tetrahydrofuran | U. dioicaL. | 26 |
| (+)-neo-olivil | U. dioicaL. | 27 |
| urticol | U. mairei Levi. | 31 |
| urticene | U. mairei Levi. | 31 |
| urticaside A | UrticatriangularisHand. - Mazz. | 32 |
| urticaside B | UrticatriangularisHand. - Mazz. | 32 |

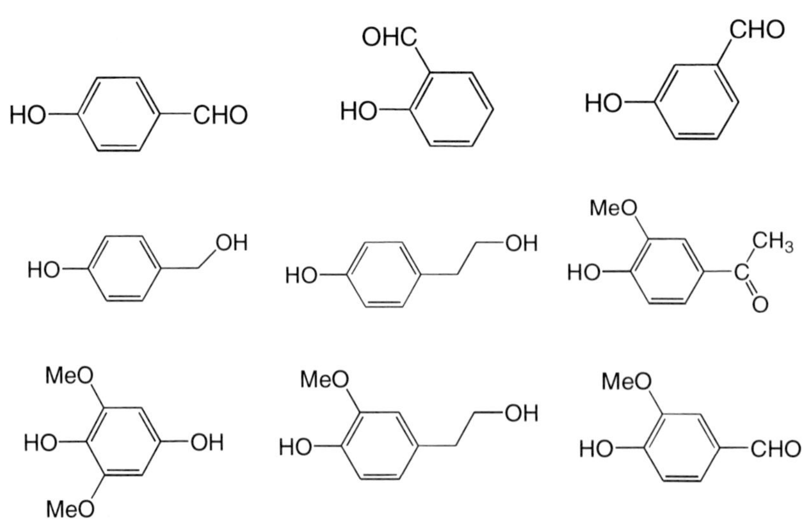

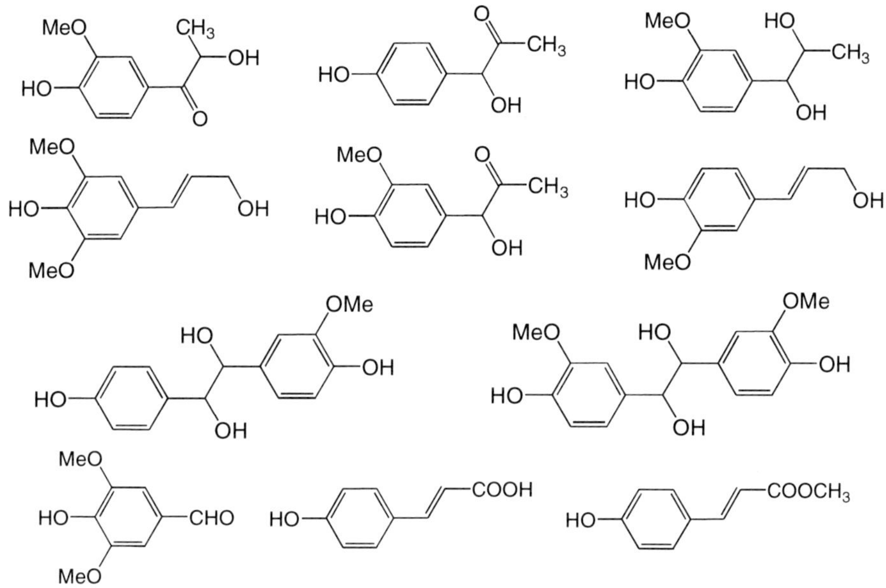

Figure 2. Important phenols constituents of Urtica Linn.plants(reference 28-30).

## 3.3. Organic Acids

More than twenty organic acids compounds which were separated from Urtica Linn. plants [22, 24-25, 30, 33-40]are shown in Table 4.

**Table 4. Important organic acids constituents of Urtica Linn. Plants**

| Compounds | Sources of plants | References |
|---|---|---|
| Linoleic Acid | *U. dioica*L. | 30 |
| 9-Octadecenoic Acid | *U. dioica*L. | 30 |
| Linoleic Acid | *U. dioica*L. | 30 |
| Erucidic Acid | *U. dioica*L. | 22, 38 |
| Vanillic Acid | *U. dioica*L. | 24 |
| 4-OH-Cinamic Acid | *U. cannabina* | 33 |
| Syringic Acid | *U. dioica*L. | 34 |
| Homovanillic Acid | *U. dioica*L. | 34 |
| P-Hydroxy Benzoic Acid | *U. dioica*L. | 34 |
| Tran-Cinamic Acid | *U. dioica*L. | 34 |
| 2-OH-Cinamic Acid | *U. dioica*L. | 34 |
| Apple Acid | *U. dioica*L. | 35 |
| Ethanedioic Acid | *U. dioica*L. | 35 |
| Orthophosphoric Acid | *U. dioica*L. | 35 |
| Kinic Acid | *U. dioica*L. | 35 |
| Amber Acid | *U. dioica*L. | 35 |
| Fumaric Acid | *U. dioica*L. | 35 |
| Threose-1, 4-Lactone | *U. dioica*L. | 35 |
| Citric Acid | *U. dioica*L. | 35 |
| Threonic Acid | *U. dioica*L. | 35 |
| Caffeic Acid | *U. dioica*L. | 35 |
| Glyceric Acid | *U. dioica*L. | 35 |

**Table 4. (Continued)**

| Compounds | Sources of plants | References |
|---|---|---|
| Caffeaoyl-Apple Acid | *U. dioica*L. | 36 |
| Ferulic Acid | *U. dioica*L. | 22, 36 |
| Caffeotannic Acid | *U. dioica*L. | 37 |
| E, Z-12-Hydroxyl-10, 12-Octadecadienoic Acid | *U. dioica*L. | 37 |
| Oleanolic Acid | *U. dioica*L. | 38 |
| Ursolic Acid | *U. dioica*L. | 22, 38 |
| Z, E-13-Hydroxyl-9, 11-Octadecadienoic Acid | *U. dioica*L. | 25 |
| Benzoic Acid | *U. fissa* | 39 |
| Fumarate | *U. fissa* | 39 |
| p-Coumaric Acid | *U. cannabina* | 39 |
| 3, 5-*Di*-Caffeaoyl-Kinic Acid | *U. fissa* E. Pritz. | 40 |

## 3.4. Proteins and Polysaccharides

There high amount of protein in Urtica Linn. plants and include all amino acids required for human beings. Fresh plants include protein 5%-6%, the dry plants have 23%-24% and 70% of them are digested proteins. A kind of plant protein was separated from the roots of *U. dioica* L. by Peumans in 1984 [41]which has the special red blood cells adhesion function and called Urticadioicaagglutinin(UDA).

In fact, UDA is a kind of mixture, includes 11 different exogenous lectin compositions, each kind of them consists of 80-90 amino acids, molecular weight is at between 8300-9500Da, this is single polypeptide chain [42]. It is usually combined with chitin and has agglutinate activity, including two combination areas of cysteine and chitin; these areas have highly homology with hevein [43-44].

Kazuaki and the others [45-46] researched the structure of UDA and expounded parts of the architectural feature.

Polysaccharide is one of the main active ingredients of Urtica Linn plants, rhamnose, mannose, arabinose, galactose, glucose and xylose and so on are included. Wagner and the others [47-48] had researched the ingredients of polysaccharide in *U. dioica*L. roots. Among this, RP1 and RP2 were pectin and molecular weight were respectively 15kD and 50kD. RP3 and RP4 were rhamnose, molecular weight were respectively 210kD and 18kD, RP5 is arabinogalactan (Type II). Song Xuewei [49] studied the polysaccharides extract technology in *U. laetevirens* by the spherically symmetrical design method and showed the polysaccharide content was 2.76%. Yang Aigang and others [50-51] optimized extraction technology of polysaccharide in *U. fissa* and determined the polysaccharide content of the roots of it and *U. maireiLevl.* var. mairei at the same time. HuoWenlan [52] analyzed and determined in detail the polysaccharide content of different parts of *Ucannabina*L. The results showed that these parts of *U. cannabina*L. all had a certain amount of polysaccharide. The order of content from high to low is that seeds(5. 76%)>leaves(3. 48%)>stems(1. 71%)>roots(1. 25%). Li Yan [53]made a study ofoptimized extraction technology of polysaccharide from *U. cannabina*L. in Guizhou.

The crude polysaccharide was separated and depurated after eliminating protein, DEAE-fiber column chromatograph and gel column chromatograph. They separated and successfully got five kinds of polysaccharide and analyzed their structures respectively. Guan Feng [54] got high purity polysaccharide from*U. Angusitifolia*Fisch ex Hornem in Heilongjiang and determined the polysaccharide content. The results showed it's 64.83% and 3.28% in crude drugs.

### 3.5. Terpenoids

Kraus and the other scientists [26]separated and got terpin and their glycosides from the root of *U. dioica*L. subsp. Dioica and the structures of basic nucleus are as follow:

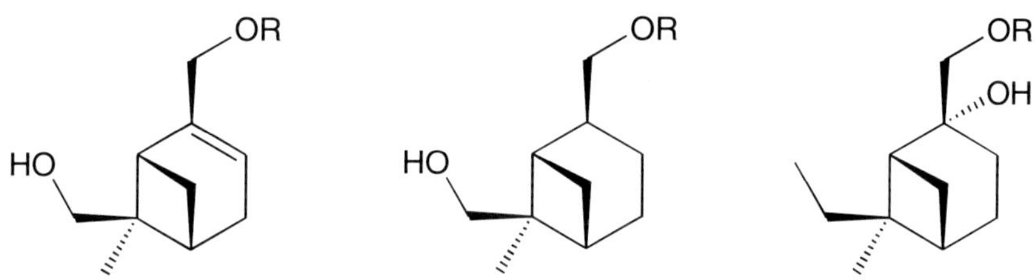


### 3.6. Sterols

Schomer and the other scholars [55] reported that they got β-sitosterol-O-β-D- pyranoid-glucose(1→4)-O-β-D-pyranoid-arabinoside, 7-β-hydroxy-sitosterol-3-O-β-D-glucoside, 7-α-hydroxy-sitosterol-3-O-β-D-glucoside and 24R-ethyl-5-α-cholestane-3β, 6α-diol, triglyceride, sterolesters, fatty acid and its methyl-esters, glyceryl-ether, diglyceride, galactosyldiglyceride, phospholipid, stigmasterol-4-ene-3-ketone, stigmasterol and so on from *Urticadioica.* There were still sitosterol and daucosterol in *Urticadioica* from the other reports [22, 38, 56-59].

### 3.7. Others

Guil-Guerrero [60] determined fatty acid and carotinoid in *U. dioica*L. subsp. dioica by the method of GC and RP-HPLC. H. Sowowa [61] extracted pigment and the oilness resin from *U. dioica*L. subsp. Dioicaby SFE. QianHua [62-63] analysed the chemistry composition of ethanol extraction of the roots, stems, leaves of *U. dioica L. subsp. Dioica.* by GC-MS. Most were aliphatic acids and their derivatives, borane complexes, heterocycle and hydrocarbon compounds. Meanwhile they compared chemistry composition of ethanol extraction of two kinds of *U. dioica L. subsp. Dioica,* growing in different environments.

Mehmet [64] and the other scientists reported the inorganic element content of *U. dioica*L. subsp. dioica. Qin Yuanman and others [65-66] determined eight microelements in *Urticafissa* E. Pritz. by flame atomic absorption spectrometry, they were Fe, Zn, Mn, Cu, Cr, Ni, Co and Cd. And meanwhile they determined ten inorganic elements in *U. Angusitifolia*Fisch ex H, which were at different picking times and analyzed the dynamic content trend [67]. Han Zhiping [68] studied the inorganic element of wild *Urticafissa* E. Pritz. fromYuLin, the experiment result showed that overground part of *Urticafissa* E. Pritz. wasthe richest in Ca among the macroelement and then were Ka, Na, Mg, P. successively. And it is richest in Fe among the trace elements; the other elements of the sequence were Zn, Ni, Mn, Cu, Co, Cr, Pb, Ge. The leaves of *U. dioica*L. subsp. *dioica* were rich in histamine, acetylcholine, formic acid, gallic acid, 5-hydroxytryptamine, vitamin A, vitamin C and other compositions [69]. They have got some other types of compounds yet, they were dotriacontane, nonacosane, 14-octacosanol, diosmin and sugarcane sterol and so on [22, 26, 37, 70-71].

# 4. Studies on Bioactivity

In recent years, researches in the biological activities of the Urtica Linn medicinal plants have been made significant progress.

## 4.1. Anti-Inflammatory and Analgesic Activity

Obertreis reported [72] that the extract of *U. dioica*L. subsp. dioica leaves possess activity of treating rheumatism. UDA from *U. dioica*L. subsp. dioica was a kind of anti-inflammatory. Rheuma-HeK capsule was the preparation of extracts of leaves from *U. dioica*L. subsp. dioica and *U. Angusitifolia*Fisch ex Hornem and it contained plenty of flavonoids, phenolic acids and theirs derivatives. Among them the major component is caffeoylmalic acid. The structure of caffeoylmalic acid was similar with salicylic acid and other non-steroidal anti-inflammatory drugs (NSAR), so extracts of leaves from *U. dioica*L. subsp. dioica and *U. Angusitifolia*Fisch ex Hornem may have the similar action principle.

Anti-inflammatory effects *of U. dioica*L. subsp. Dioica polysaccharide mixture (POR1)ofRPl, RP2, RP3 and RP4 can be compared with indomethacin. Five hours after given POR1 40mg/kg, inhibition rate of carrageenan induced swollen mice feet,which could reach 36.80%, and the indomethacin(10mg/kg) was 42.1%. Twenty-two hours later, activity of POR1 is obviously superior to indomethacin, the inhibition rate of two drugs were 63.6% and 27.3% respectively [73]. The results, which were made by Wang Mengyue and his workmates [74-76] showed that decoctions of three kinds of Urtica Linn. plants in Sichuan have significant anti-inflammatory and analgesic effects, and could obviously restrain rats auricle tumefaction induced by xylene, carrageenan induced mice feet swollen and pain of rats caused by hot plate. They also observed from weather the water extract and polysaccharide taken from six kinds of Urtica Linn. plants could influence the response of rats to pain by hot plate method. The results showed that these extracts and polysaccharides obviously have the analgesic action, in 0. 5 hours the force was strongest. Polysaccharides showed significant anti-inflammatory effects and could enhance the clearance ability of rat macrophages. Zhao Yongna and other researchers [77-79] found that the rough water extract of *U. macrorrhiza* Hand. -Mazz. roots could restrain mice feet swollen induced by carrageenan and obviously reduced the numbers of body torsion which is caused by acetic acid and meanwhile had certain therapeutic action of adjuvant arthritis. These results showed that the water extract of

*U. macrorrhiza* Hand. Mazz. Roots possessed certain anti-inflammatory and analgesic action. Wang Yali and other researchers reported [80]that 70% ethanol extracts of the roots, stems, leaves and roots water extract of *U. Angusitifolia*Fisch ex Hornem could obviously restrain rats auricle tumefaction induced by xylene, and significantly restrained the number of body torsions which were caused by 0. 5% acetic acid, and had the obvious anti-inflammation and analgesic activity. Other reports showed that [81] extract of *U. dioica*L. subsp. dioica had the resistant action to adrenaline, similar to dihydroergotamine, and could inhibit the secretion of TNF-α, IL-1β and IL-6 in the full blood of man which induced by LPS. UDA (*U. dioica*L. subsp. dioicalectin)could not restrain NF-kB(proinflammatory factor) to combine with DNA, but could prevent degradation of I-kB and so could restrain transcription of NF-kB to express anti-inflammatory effects. Perhaps this was the important mechanism of the treatment for rheumatoid arthritis [82]. Randal and other scholars carried out randomized and double blind experiments with 27 cases of arthritis of thumb and index finger. These patients were applied with extract of *U. dioica*L. subsp. dioica on the joint pain parts every day. After one week, it showed that the curative effect was obviously superior to placebo [83]. Flavonoid glycosides in the extract of *U. dioica*L. subsp. dioica had good therapeutic effects on white blood cells function shortage and chronic granulomatous disease [84]. Now it was sure that the exogenous lectin extracted from *U. dioica*L. subsp. dioica roots was the main active component on the anti-inflammatory effect and remedying prostatic disorders [85]. Zhao Yongna and her workmates [78]found that *Urticafissa* E. Pritz. could relieve the pain and had antipyretic activity through the pharmacological experiments. Another report showed that *Urticamembranacea* had the ability to relieve a cough, and as an anti-inflammatory it could be used for the treatment of tonsillitis [86].

## 4.2. Anticoagulant and Anti-Platelet Aggregation Activity

Qin Yuanman and others [87-88] determined the effect of the leavesof *U. Angusitifolia*Fisch ex Hornem and *U. cannabina*L. by the method of slide-slip and capillary. The results showed that blood coagulation time of all large, medium and small dose groups were significantly longer than normal saline groups, with the increase of the drug, and the results were more obvious. In mice,using the broken tail method to determine the bleeding time, they found that nettle can restrain the platelet aggregation of mice. It showed that nettle had the ability of promoting blood circulation, relieving pain and antithrombotic function and had the obvious anticoagulant effect. Other reports [89] showed that the nettle extract from water had the role of anti-platelet aggregation. In Germany, people had more experiences to use nettle to treat and prevent the cerebral thrombus.

## 4.3. Diuresis and Anti-Hypertension Activity

Abdelhafid and the others [90]gave the experimental rats continuous injections within 1. 25h with water extract of overground part of nettle, Doses were for 4mg/kg/h(Low dose group, LDG), 24mg/kg/h(high dose group, HDG) and furosemide(control group, CG) 2mg/kg/h. Experimental rats artery blood pressure of LDG and HDG were down by 15% and

38%, urine volume increased by 11% and 84% respectively, promoted the urine $Na^+$ excreting, 28% and 143% respectively. In the experimental rats of CG there was a 28% decrease in arterial blood pressure, urine volume and urine $Na^+$ excreting increased 85% and 155% respectively. The hypotensive effect of LDG and CG can be kept within 1h, and then gradually disappeared. And HDG could continue to hold antihypertensive action. The water extract of the overground part of nettle can be used to cure children with nocturnal enuresis,the water extract of roots of nettle and seeds(crushed into crackers food) can remedy urinary retention [91].

## 4.4. Inhibitory Activity of $Na^+$, $K^+$-ATP Enzyme Activity

Different extracts of the roots from *U. dioica* L. subsp. Dioica(0. 1mg/ml) could inhibit $Na^+$, $K^+$-ATP enzymeactivity(which inhypertrophy tissue of the prostate), inhibition rate was 27. 6%-81. 5%. Some steroids (such as stigmasterol and campesterol)with $10^{-3}$-$10^{-6}$M could inhibit $Na^+$, $K^+$-ATP enzymeactivity also and inhibition rate was 23%-67%. And this inhibitory action can affect the metabolism and growth of the prostate cells [92].

## 4.5. Inhibitory Activity of Aromatase Activity

Aromatase is the key enzyme of the metabolism of steroid hormones, and can adjust the transformation of androgens to estrogen [93-94]. The inhibition of aromatic enzyme system can cause the estrogen levels to lower, so as to improve the prostate disorders [95-96]. Methanol extract of *U. dioica*L. subsp. *dioica* root can inhibit aromatase activity, and it contained 14-octacosanol, oleanolic acid, ursolic acid and 9Z, 11E-13-hydroxy-octadecadienoic acid and so on [97].

## 4.6. Inhibitory Activity of Cell Proliferation

Johannes [98]built BPH experimental animal model through implanting UGS to the adult rats UGS at ventral area to study the inhibitory activity of water, methanol, ethyl acetate and butanol extracts of *U. dioica*L. subsp. *dioica* root cyclohexane on BPH.

The results showed that 20% methanol (ME-20) and water extracts possess significant inhibitory activity on BPH. And the inhibitory rates were 51.4% and 76.5% respectively. Of suramin and these extracts, the latter seemed to be the superior. At the same time, they analyzed the components of these group extracts, and it showed that protein, polysaccharides and other polarity compositions maybe the main active ingredients, however β-sitosterol has the opposite effect and can promote prostate hyperplasia. They did further research on UDA and found an inhibition rate of polysaccharide to BPH was 33.8% [99].

Results of the external experiment showed that ME-20 compared with controls, had a specific inhibitive effect on LNCaP proliferation and related with the time and concentration, but had no effect on proliferation of stroma cells (hPCPs) [100].

UDA still can inhibit proliferation of tumor cells and Hela cell besides antitumor activity and cytotoxicity as the general plant lectin [101]. Moreover, UDA can inhibit the combination of epithelial cell growth factor (EGF) and tumor cell receptor and the inhibition rate was 50%. Compared with wheat germ agglutinin and ConA, it was more active [102-103].

## 4.7. Affinity Activity with Sex Hormone Binding Globulin

Sex hormone binding globulin (SHBG) is a kind of transportprotein in the plasma with a high appetency of androgen and estrogen, which can adjust the plasma concentration of sex hormones. SHBG was related to physiological function of prostate and benign prostatic hyperplasia. There has been reported [104] that dilute ethanol extractof *Urticafissa* E. Pritz. root could interfere with the combination of SHBG and Dihydrotestosterone (DHT). In the external experiments, it showed that except (-)-rosin spirit, the rest lignans were great affined with SHBC and 3, 4-di-vanillyl-tetrahydrofuranwas the best but its relative contents was very low.

## 4.8. Immune Adjustment Activity

Polysaccharides from *U. dioica*L. subsp. *dioica* in vitro can stimulate lymphocyte proliferation, of which RP2, RP3 and RP4 in concentration is 10 to 100μg/ml, regardless of whether there was any ConA stimulation before, the exciting rate is always higher, and the activity of RP3 is the highest. In the preparation "Bazoton", polysaccharides content is 1. 7%, in the concentration for 100 μg/ml the stimulates rate is 100% [105-106].

UDA can stimulate the rat thymocyte and T-lymphocyte proliferation specifically too, the exciting activity is similar to ConA and dose-related [107-108]. Given the lack of MHC-I and MHC-II family mutant mice UDA, found that UDA could combine with glycoprotein receptors of cell membrane surface and T cell surface receptors, induction T lymphocyte proliferation [109].

The water-soluble part of the roots of hybrid *Urticafissa* E. Pritz.mainly contain polysaccharides, which can regulate the immune function of T lymphocytes, prevent epithelial cancer cell differentiation and proliferation [110]. Li Zhongkun and other researchers [111] determinedthe IL-2, IFN-γ, CD3, CD4 and CD8 level in mice serum to study immune adjustmentactivity *of U. macrorrhiza* Hand. -Mazz. extract, and the results showed that the extract can improve the mice IL-2, IFN-γ, CD3, CD4 and CD8 level, and can strengthen the immune function of the organism.

## 4.9. Antibacterial Activity

Cheng Zhen and other researchers [112-113]reported that part of the antibacterial activity of *U. cannabina*L. seeds was petroleum ether extract and ethyl acetate extract, and the petroleum ether extract was the strongest part and the minimum inhibition concentration (mic) to colun bacillus was 2. 5mg/ml. M. I. A ksu and other researchers [114] also reported the corresponding antibacterial activity of *U. dioica*L. subsp. dioica.

### 4.10. Liver Protective Activity

Liver cell toxicity experiment of the rats induced by carbon tetrachloride showed that the water extract of *U. dioica*L. subsp. dioica could protect the liver [115]. The researchers analyzed that UDA could stimulate Th and Tc lymphocyte activation, the production of the cell factors, reduced TNF-α and IL-1β levels and thus had liver protective activity.

### 4.11. Cerebrovascular Pharmacological Activity

The extract from the root of *U. dioica* is the vasodilator agents that can depress blood pressure by making vascular endothelial cell release NO to relax vessels, opening the potassium channel, and weakening the vascular contraction force [116]. In the internal and external experiments, the water extract from the *U. dioica*root can induce the mice heart to beat slowly, and increase cardiac contractility. It acted with similar effect to carbachol chloride and it was not affected by atropin [117].

Urtica Linn. plants can decrease the aggregation of oxygen radicals, increase the distribution of DNA and binding to AP-1 and NF-kB, decrease the oxidant injury and cerebral cell apoptosis, promote the secretion of nerve growth factor (NGF) and brain derived neurotrophic factor (BDNF), enhance the viability of brain cells. It is very important to reduce the brain cell death after cerebral injury [118].

Costantine [119]had reported the effect of reducing blood lipid of extract from *U. dioica.* After 30 days of oral drug supplementation, the total cholesterol, LDL, LDL/HDL ratio of rats were significant decreased, and there was no hepatic injury symptom. Marjan investigated the effects of extract of *U. dioica* and the hepatic histopathological results reflected the correlation of *U. dioica* extract with both liver weight and the levels of plasma TC and LDL-C. These results indicated that *Urticadioica* extract had hypocholesterolemic effects in the animal model [120].

### 4.12. Antimutagen Activity

After the extract of *U. dioica*L. subsp. *dioica*was diluted 20 times, it was treated in a water bath for 15 min and stored at room temperature for 45 min. The experiment animals were fed with the mutagen 1h before the 2ml/kg oral supplement of the treated extract. Then the damage of chromosome in bone was detected. The result showed that the extractdecreased the ratio of mutation chromosome, and revealed significant antimutagen activity in both pharmacologic and clinical experiments [121].

### 4.13. Antiallergic Activity

An allergic reaction is the abnormal or pathologic immunoreaction, the body's way of responding to an "invader ". When the body senses a foreign substance, called an antigen, the immune system is triggered. The *U. fissa* E. Pritz.extract has the function of antagonist histamine and treating allergic complications. The components in the *U. fissa* E. Pritz.extract contain many carboxyls, and antagonize the $H_1$ receptor or the $H_2$ receptor competing with antihistamine in biochemical reactions and the histamine from the reaction system [122].

### 4.14. Stimulate Bone Formation Activity

Liying Shi [123]studied the effect of eight extracts of four Urtica plants on bone formation and found that EtOH extract from the root of *U. triangu*laris, EtOH extract and 40% EtOH-$H_2O$ fraction with D101 macroporous resin from the $H_2O$ partition of EtOH extract from the whole plant of *U. laetevirens* Maxim. Subsplaetevirens at the concentration of 50 to 200g/ml significantly stimulated MC3T3-E1 cell proliferation. The three extracts also somewhat increased ALP activity, but had no significant difference.

### 4.15. Antilithic Activity

Xiang Ming [124] examined the potential antilithic effects of the traditional Chinese medicine Urticadentata Hand (UDH)in experimental rats and screened the optimal extract of UDH as a possible therapeutic agent for kidney stones. The rats in different experimental groups were then intragastrically given petroleum ether extract (PEE), N-butanol extract (NBE), aqueous extract (AqE) of UDH, Jieshitong (positive control drug), and saline, respectively. Treatment with NBE significantly reduced the elevated levels of urinary calcium, uric acid, phosphate, as well as increased urinary output. Accordingly, the increased calcium, oxalate levels and the number of calcium oxalate crystals deposits were remarkably reverted in the renal tissue of NBE-treated rats. In addition, NBE also prevented the impairment of renal function to decrease the contents of blood urea nitrogen (BUN) and creatinine. Taken together, these data suggest that NBE of UDH has a beneficial effect on calcium oxalate urinary stones in rats by flushing the stones out and protecting renal.

### 4.16. Other Activities

UDA can inhibit the HIV-1 and HIV-2, cytomegalovirus and respiratory syncytial virus in vivo [125]. The extract with ethyl acetate of *U. dioica*L. subsp. dioica can induce the differentiation of HL-60 with $ED_{50}$ as 4. 0μg/ml.

The compound can induce cell differentiation from cancer cell to normal cell and it may be an effective anticancer drug [126]. Abuharfeil [127] also reported that extract of *U. dioica*L. subsp. *dioica*with water could damage the tumor cell and inhibit tumor cell growth. Ilhami [128] reported the bioactivity as antioxidant and microbial resistance of *U. dioica*L.

subsp. *dioica* water extractive. And R. E. Uncini [129] had a relative report. JiBaoquan [130] studied the 5α-reductase inhibitory activity of the aqueous constituents in *Urticafissa* E. Pritzs' extract and the results showed that this kind of extract can inhibit the activity of the 5α-redutase in vitro.

## 5. Preparation Research

In German the *U. dioica*L. subsp. *dioica* was used in the treatment of the early BPH, andmany preparations with dozens of patents were explored, such as Urticur Capsule of BIOCUR Co., Rheuma-Hek Capsule of Strathmann Co. and Salus Teabag of Salushaus Co. The Cletan capsule of Intersan Co. (one capsule contains 192 mg stinging nettle dry extract) is used to treat benign prostatic hypertrophy. The Bazostad coating tablet of Stada Co. (one tablet contains 125 mg stinging nettle dry extract) can effectively improve the micturition status of patients with benign prostatic hypertrophy [131].

However, in China, there are rarely reports about the medicine preparations of Urtica Linn. plants. Li Zhongkun [132-133] investigated the microcapsule preparation of the extract of *Urticafissa* E. Pritz. and*Urticafissa* with orthogonal design. They found that the high output can obtain the ratio of capsule core and capsule material as 0. 5:1. 25 and 0.75:0. 5, temperature at 70°C, and stirring rate 800r $min^{-1}$ and 600r $min^{-1}$. Li Xinfang [134] prepared the dry extract of the roots of *U. dioica using* the method of extraction with water and sedimentation and then made it into a suppository for treatment of benign prostatic hypertrophy and had satisfactory results.

## Conclusion

The medicinal plants of Urtica Linn.aredistributed widely in northwest, north, northeast and southwest of China. It has the rich resource and popular usage but with no system to exploit and utilize at present. Both the extensive folk use in China and a large number of modern research reports in other countriesdemonstrated the good development prospect of Urtica Linn. plants [135-137]. It cannot only be used as medicine but also for cosmetics, health care products,nourishment and so on.

There was a long history of eating *U. dioica*and was edible as charlotte in Europe. There are plentiful Vitamin A and C, different minerals, top-quality plant protein, inorganic salt (such as Fe and Ca), abundant carotene and a higher content of chlorophyll than other vegetables in stems and leaves.

As reported [138], there are 77.88g water, 4.66g crude protein, 4.34g coarse fiber, 0.62g fat and 9.64g carbohydrate in every 100g tender stem and leafof *U. dioica*L. subsp. Dioica.In shoots, there are 140~300mg renieratene, 1000~2000mg Vitamin C, 20mg Vitamin E, 25 mg Vitamin K and a large number of Fe, Mg, Cu and Zn [139]. Aotergen [140] detected the nutrition constituent in the edible part of *U. cannabina*in Inner Mongolia, and found that there are crude protein, crude fat, crude fiber, $VB_1$, $VB_2$, $VB_6$, K, Na, Ca, Mg, Fe, Mn, Zn, Cu, palmitic acid, stearic acid, oleic acid, linoleic acid, linolenic acid and 17 amino acid in the leaves and tender stems. Among them, the values of crude protein, carotene, essential trace elements, unsaturated fatty acids, and essential amino acids were much higher, which suggested that the nutritional composition and edible contents of the edible partfrom *U. cannabina* have the characteristics of highgrade vegetables. Liu Cuihua [141]analyzed the nutrition composition of the wild *Urticafissa* E. Pritz. in Tibet, and found that it contains many kinds of essential amino acids with total quantity as 28 94 g/100g, especially lysine and tryptophan.

In many countries, the exploitation and utility of Urtica Linn. genus, especially *U. dioica*, has been achieved, which lays a good foundation to take full advantage of these plant resources. Relevant researchers fully explored the potential sources of this genus, so that its resources can be fully utilized, its germ to be utilized sufficiently and its germ plasmprotected. It guarantees further researchonUrtica Linn. plants.

# References

[1] Wang WC, Chen JR, Flora of China (Vol23, Fas2), *Science and Technology Press*, (1995).
[2] Wang MY, Wei YF, Huang XY, *Chinese herbal medicine*, 1, 58-60(2002).
[3] Li R, Qin Mj, Yu Gd, *Chinese wild plant resources*, 1, 24-26(2002).
[4] Wang MY, Wei YF, *Chin J. EthnomedEthnopharm*, 6, 345-346(2001).
[5] Wang MY, Wei YF, Xing F, et al, *Chinese herbal medicine*, 12, 857-859(2001).
[6] Madaus G, Lehebuch der biologischenHeilmittel Band, 2nded, OlmsVerlag, Hildesheim-New York, 2746-2754(1979).
[7] Hou KZ, Chinese Spermatophyte Genera Dictionary, Science Press, Beijing, 1982.
[8] Yang ST, Yang BM, Qin MJ, *Plants*, 1, 5(2001).
[9] Chen JS, Zheng S, Science Press, Beijing, 605-707(1987).
[10] China institute of pharmaceutical and biological products, et al, Chinese National Medicine Records (Vol1), Beijing, People's Medical Publishing House, 378-382(1984).
[11] The People's Republic of China Pharmacopoeia Committee, Ministry of Health of China, *Drug Standard* (Tibetan, Vol1), 69(1995).
[12] Liu YQ, Doctor Degree Dissertation of Beijing Forestry University, 2007.
[13] Wang MY, *Master Degree Dissertation of Chengdu University of Traditional Chinese Medicine*, 2000.
[14] Wang MY, Wei YF, Shu S, et al, *Sciences*, 3, 171-174(2002).
[15] Li CY, Yang FD, Lin L, Chinese Society Fifth Annual Academic Conference on Identification of Traditional Chinese Medicine-Chinese Medicine Identification. Beijing, Science Press, 144-146.
[16] Chaurasia N, Wichtl M., *Planta Med.*, 53, 432-434(1987).
[17] Kavtaradze N Sh, Alaniya M. D., Aneli J N., *Chemistry of Natural Compounds*, 3, 287-289(2001).
[18] SaeedAtea, El-Eraqy, Waffa, et al. *J. Pham. Sci*, 6, 415(1995).
[19] Zhang ML, Li ZP, Jia XM., *Natural product research and development*, 2, 175~176(2005).
[20] Wang MY, Lee, Li XB, *China Journal of Pharmaceutical Sciences*, 24, 1853-1855 (2005).

[21] VassolikiExarchou, Yiannis C. Fiamegos, et al, *J. Chromatographic A*, 1112, 293-302(2006).

[22] Ji TF, Liu CH, Wang AG, et al, *Chinese herbal medicine*, 6, 662-664(2007).

[23] SibelKarakaya, SedefNehir E., *Food Chemistry*, 66, 289-292(1999).

[24] Cao YP, Yang XL, *Food science*, 1, 288-290(2007).

[25] Tan L, Kang SH, Yuan AX, et al, *Chinese herbal medicine*, 6, 9(1989).

[26] Kraus R, Spiteller G., *Liebigs Ann. Chem.*, 12, 1205-1213(1990).

[27] Matthias Schottner, Josef Reiner, Francis S. K., *PhytoChemistry*, 6, 1107-1109(1997).

[28] Kraus R, Spiteller G., *PhytoChemistry,* 29, 1653(1990).

[29] Ma XM, Beijing university of Chinese medicine Ph. D. Thesis (2004).

[30] Quan YS.,*Chinese herbal medicine*, 1997, 20 (1): 8-10.

[31] Wei W, Xing GY, Li XD, et al, *Chinese Chemical Letters*, 19, 1212-1214(2008).

[32] Feng B, Yan X, Wang H, et al., *Fitoterapia*, 6, 607-609(2010).

[33] Yiannis C. Fiamegos, Christos G. Nanos, Jacques Vervoort, *J. of Chromatographya.*, 1041, 11-18(2004).

[34] Franck Bakke I. L, Thorsen E, Nordal A. *Med. Norsk. Farm. Selsk.*, 40, 181(1987).

[35] Beschia M., Leonte A. Bul. Univ. Calati, Fasc, 6, 59(1982).

[36] Budzianowski., *Planta Med.*, 57, 507(1991).

[37] KrausR.,SpitellerC., Barstsch W., *Liebigs Ann. Chem.*, 335(1991).

[38] Can D., Spiteller G., *Planta Med.*, 2, 138(1995).

[39] Ma XM, Guo YJ, Wang LS, *China Journal of Chinese MateriaMedica*, 5, 472(2004).

[40] Wang MY, Li B, Li XB, *Chinese Pharmaceutical Journal*, 24, 1853-1855(2005).

[41] Peumans W., De Ley M, Broekaert W. E., *FEBS Lett.*, 177, 99-103(1984).

[42] Van Damme EJM, Peumans W., *Physiol Plant*, 71, 328-334(1987).

[43] Beintema J., Peumans WJ.,*FEBS Lett.*, 299, 131-134(1992).

[44] Does MP, Ng DK, Dekker HL.,*Plant MolBiol*, 39, 335-3471999.

[45] KazuakiHarata, MichiroMuraki, *J. Mol. Biol.*, 297, 673-681(2000).

[46] Frederick A Saul, Paula Rovira, GinetteBoulot.,*Structure*, 6, 593-603(2000).

[47] Wagner H., Willer F., Kreher B., *Planta Med.*, 55. 452 (1989).

[48] Wagner H., Willer F., *Samtlegen R. Phytomedicine*, 1, 213 (1994).

[49] Song XW, Pan L, Han YP, et al, *Am. J. pathol.*, 9, 1344-134(2006).

[50] Yang Ag, Ji BQ, Tang L, et al, *J. zhongnanmedicine*, 1, 12-14(2007).

[51] Zhu YN, Yang AG, Liu JY et al, *Am. J. pathol.*, 8, 1195-1196(2007).

[52] Huo WL, Cao YP, *Food science and technology*, 12, 99-100, 104(2007).

[53] Li Y, *Master Dissertation of Military Medical Science Academy of the PLA*, 2005.

[54] Guan F, Wang BQ, Wang Y, *China pharmacy*, 9, 869-871(2011).

[55] Schotmer M, Gansser D, Spiteller G., *Planta M*ed. 6, 529-532(1997).

[56] Li F, Zhou BH, Yan XG, et al, *Chinese Journal of hospital pharmacy,* 28, 873-874(2008).

[57] Yan XG, Ji BQ, Zhou Y, et al, *Journal of Shenyang Pharmaceutical University*, 11, 856-58(2008).

[58] Ji BQ, Feng BM, Shi LY, et al, *China Journal of Pharmaceutical Sciences*, 9, 1372-1374(2009).

[59] Wang W, Zhou Y, Tang L, et al, *Journal of Shenyang Pharmaceutical Universit*y, 9, 711~713(2008).

[60] J. L. Guil-Guerrero, M. M. Rebolloso-Fuentes, M. E. TorijaIsasa, J. *of food composition and analysis*, 16, 111-119(2003).

[61] H. Sowowa, M. Sajfrtova, M. Bartlova, et al, *J. of superitical fluids*, 30, 213-224(2004).

[62] Qian H, Gao ZH, Wang GY, et al, *Zhejiang forestry science and technology*, 3, 31-33(2006).

[63] Qian H, Gao ZH, Wang GY, et al, *China modern medicine*, 5, 10-12(2006).

[64] Mehmet Musa Özcan, AhmetÜnver, TolgaUcar, et al, *Food Chemistry*, 3, 1120-1127(2008).

[65] Qin YM, Sun LP, Ma KR, et al, *Chinese Journal of clinical rehabilitation*, 33, 7556(2004).

[66] Qin Y, Ma KR, *World element medicine*, 3, 71-76(2004).

[67] Qin YM, Wei EK, *World element medicine*, 2, 69-72(2005).

[68] Han ZP, *Science technology and engineering*, 14, 3523-3525(2007).

[69] Wang KZ, *Plant Journal*, 1, 23(2001).

[70] Vishnevskii O. V., Proshunina D. V., *Farm. Zh.*, 2, 50(1989).

[71] Chaurasia N, Wichtl M., *J. Nat. Prod*, 5, 881(1987).

[72] Zhuang GL, Foreign Medicine, *Botany Fascicule*, 2, 66-67(1996).

[73] Wagner H., Willer F., Samtlegen R., Boos G., *Phytomedicine*, 1, 213-214(1994).

[74] Wang MY, Liu Y, Wei YF, *China Journal of national folk medicine*, 3, 229-231(2001).

[75] Wang MY, Wei YF Huang XY.,*Lishizhen Medicine and MateriaMedica Research*, 8, 676-677(2001).

[76] Wang MY, Wei YF, Shi Y, *Chinese herbal medicine*, 9, 666-667(2001).

[77] Zhao YN, ReanmongkolWantana, BoukingPisit, et al, Natural product research and development, Suppl, 69-72(2005).

[78] Zhao YN, ReanmongkolWantana, BoukingPisit, et a1, *Fitoterapia*, 76, 91-95(2005).

[79] Li XH, Zhao YN, Li BL, et al, *Journal of Yunnan University ofTraditional Chinese Medicine*, 6, 21-23(2007).

[80] Wang YL, Qin MJ, Dai Y, et al, *Research and practice of Chinese medicine*, 3, 42-43(2005).

[81] Barbara O, Tina R, Thomas T, et a1, *Arzneimittelforschung*, 4, 389-394(1996).

[82] Kristina R, Bert B, Klaus S., *FEBS*, 442, 89-94(1999).

[83] Colin Randall, Hester Ran, Frank Dobbs, et a1, *JR Soc Med*, 93, 305-309(2000).

[84] Pinar Akbay, A. AhmetBasaran, UlkuUndeger, et a1, *Phytother*, 17, 34-37(2003).

[85] Johannes Josef Lichius, Carols Muth.,*Planta Med*, 4, 307-310(1997).

[86] N. G., P. M., G. De Fine.,*Fitoterapia*, 1, 52-68(2007).

[87] Qin YM, Fu HJ, *Herald of medicine*, 3, 65-66(2001).

[88] Qin YM, Wei EK, *Lishizhen Medicine and MateriaMedica Research*, 7, 1203-1204(2006).

[89] Mek H, HaoM, LegssyerA, et al.,*Ethnopharmacol*, 223, 317(2004).

[90] Abdelhafid T., Sabah, Abdelkhaleq L., *J. Ethnopharmacology*, 73, 95-100(2000).

[91] Guarrera PM. Traditional phytotherapy in Central Italy (Marche, Abruzzo, Latium). *Fitoterapia*, 1, 1(2005).

[92] Hirans T, Homma M, Kitato O., *Planta Med*, 60, 30(1994).

[93] Henderson D. J., *Steroid Biochem*, 4, 905(1987).

[94] EL Etreby M. F., J. Atamestane.,*Steroid Biochem. Molec. Biol.*, 4, 565(1993).

[95] Thompson E. A., Jr., Siiteri P. K., *BiolChem*, 249, 5364(1974).

[96] Thompson E. A., Jr., Siiteri P. K., *BiolChem*, 249, 5373(1974).
[97] Pinar Akbay, A.AhmetBasaran, UlkuUndeger, et a1.,*Phytother,* 17, 34-37(2003).
[98] Johannes J, Carola M., *Plarta Med.*, 63, 307(1997).
[99] JOhannes J. L, Heiner R, Wolfgang B., *Planta Med.*, 65, 666(1999).
[100] Lutz Konrad, Hans-Helge Muller, Corinna Lenz.,*PlantaMedica*, 66, 44-47(2000).
[101] Rosner W., Hryb D. J., Khan M. S., *Steroid Biochem. Molec. Biol.*, 4, 813(1991).
[102] Schilcher H., Effenberger S. ZurStandardisierung von Brennesselwur-zel, *Dtsch. Apoth. Ztg*, 126, 79(1986).
[103] Wagner H., Geiger W. N., *Phytomedicine*, 4, 287(1995).
[104] Matthias Schottmer, Dietmar Ganger, Gerhard Spiteller, *PlantaMedica*, 63, 599-522(1998).
[105] Obertreis B., Giller K., Teucher T., *Arzneim-Forsch/Drug Res.*, 1, 52(1996).
[106] Kristina, Klaus S, Bert B., *FEBS Letters*, 1, 97-108(1998).
[107] Le Moa1 M., Col1e J., Calelli A., *Res. Immunol*, 143, 691(1992).
[108] Rovira P., Buckle M. I., *J. Immunol*, 29, 1571(1999).
[109] Paula Rovira, Malcolm Buckle, Jean-Pierre. abastado, et a1. Majoy histocompatibility class-I molecules present Urticadioica agglutinin a superanfigen of vegetal origin to Tlymphocytes. *J. Immunol*, 1999, 29: 1571~1580.
[110] Sun SY, *Modern agriculture*, 9, 12(2002).
[111] Li ZK, Yin WM, Wang CJ, et al., *Chinese Traditional Patent Medicine*, 5, 882-884(2011).
[112] Cheng Zhen, *Biotechnology*, 5, 30-32(2005).
[113] Cheng Zhen, *Master Dissertation of Xinjiang University*, 6, 18-22(2005).
[114] M. I. Aksu, M. Kaya, *Food control,* 15, 591-595(2004).
[115] M. K. Turkdogan, H. Ozbek. Z. Yener, et a1., *Phytother*, 17, 942-946(2003).
[116] Lara Testai, Silvio Chericoni, Vincenzo Calderone, et a1., *Ethnopharmacology*, 81, 105-109(2002).
[117] AbdelkhaleqLegssyer, AbderrahimZiyyat, HassaneMekhfiH, et a1., *Phytother*, 6, 503-507(2002).
[118] Anna Toldy, Krisztianstadler, Maria Sasvari, et a1.,*Brain Research Bulletin*, 65, 487-493(2005).
[119] Costantine F., Daher, KarmenG., Gerorge M., Baroody, *Fitoterapia*, 77, 183-188(2006).
[120] Marjan NASSIRI-ASL, Farzaneh ZAMANSOLTANI, Esmail ABBASI, et al., *Journal of Chinese Integrative Medicine*, 5, 428-433(2009).
[121] CPI(B), 9411, 97(1994).
[122] Nie L, *Chinese Journal of Ethnomedicine and Ethnopharmacy*, 41, 338-339(1999).
[123] Shi Ly, Yan Xg, Zhou Y, et al., *Lishizhen Medicine and MateriaMedica Research*, 10, 2315-2317(2008).
[124] Xiang M, Zhang SS, Lu JL, et al., *J. Huazhong Univ. Sci. Technol.* [Med Sci], 5, 673~677(2011).
[125] Balzarini J, Neyts J, Schols D., *Antiviral Res.*, 18, 191(1992).
[126] Suh N, Lu YG, Fong HS, *Anticancer Res.*, 15, 233(1995).
[127] AbuharfeilNM, MaraqaA, Von Kleist S., *Ethnopharmacol.*, 122, 55(2000).
[128] IlhamiGulcin, O. Irfan, MunirOktay, et al, *J. of Ethnopharmacology*, 90, 205-215 (2004).

[129] R. E. Uncini, et al., *J. of Ethnopharmacology*, 98, 323-327(2005).
[130] Ji BQ, Yang AG, Zhu YN, et al, *Lishi Zhen Medicine and MateriaMedica Re Search*, 8, 1852-1854(2009).
[131] Li XF, AziGuli, Mai ED, *Journal of Medicine and Pharmacy of Chinese Minorities*, 2, 19(2004).
[132] Li ZK, Du LY, Wang H, et al, China pharmacist, 11, 683-684(2003).
[133] Li ZK, Du LY, Yin WM, et al, *Journal of Chinese hospital pharmacy*, 6, 781-783(2007).
[134] Li XF, Yang HP, *Journal of Medicine and Pharmacy of Chinese Minorities*, 2, 60(2009).
[135] Qiu X, Foreign medicine, *Botany Fascicule*, 1, 9-14(2000).
[136] Zhang ML, Li ZP, *Botany fascicule*, 1, 12-15(2004).
[137] Hu JF, Sun XY, Chen HY, et al, *NingBo agricultural science*, 2, 19-22(2006).
[138] Yao GQ, Song L, Wang SL, *Journal of Chengde Nationalities Teachers College*, 2, 65-67(2002).
[139] Li HP, *Livestock and poultry industry*, 8, 18(1998).
[140] AoTegen, Shi HP, A Rong, et al, *Food research and development*, 5, 143-147(2007).
[141] Liu CH, Zhang HF, Daci ZG, *Tibet science and technology*, 6, 16-18(2007).

In: Phytochemicals
Editor: Hai-Xue Kuang

ISBN: 978-1-62417-354-7


*Chapter XIII*

# Research and Development on Chemical Constituents and Pharmacology of *Acanthopanax* Genus

***Gaofeng Liu, Chao Zeng and Zhiping Lin***
Department of Pharmacy, The Second Affiliated Hospital of Harbin Medical University, Harbin, Heilongjiang, China

## Abstract

The *Acanthopanax* genus plants of *Araliaceae* include 37 species in the world. They are mainly distributed in China, Korea, Japan and the far-eastern region of Russia of northeastern Asia.

The root and stem bark of these plants, belonging to the liver meridian and kidney channel, are warm-natured, acrid flavor and bitter. They mainly contain triterpenoids, diterpenoids, lignans, flavonoids, polyacetylenes, phenylpropanoids and polysaccharides. *Acanthopanax* genus plants have been used for a long time as tonics and sedatives to nourishthe liver and kidney, strengthen bones and muscle, stimulate circulation to end stasis, eliminate hydroncus, as well as having anti-tumor, anti-inflammatory, anti-fatigue, anti-oxidant, anti-diabetic, hepatoprotective effects, and immunomodulatory effects. They have shown effects on the brain, heart and nervous system, etc., and have been clinically used for the treatment of tumor, rheumatism, diabetes, hypertension, chronic bronchitis, ischemic heart disease, waist and knee aching and gastric ulcer. In recent decades, a great number of chemical, pharmacological and clinical studies on *Acanthopanax* genus have been carried out worldwide. The present review is an up-to-date and comprehensive analysis of the ecology, biology, chemistry, pharmacology and clinical applications of *Acanthopanax* genus. It will provide scientific references for further exploiting the bioactivity ingredients from the *Acanthopanax* genus, and provide a basis for further development and utilization of these plants in the food, cosmetic and pharmaceutical industry.

# 1. Introduction

The *Acanthopanax (A.)* genus, belonging to the *Araliaceae* family, includes 37 species (excluding varieties) around the world, which are mainly distributed in China, Korea, Japan and the far-eastern region of Russia of northeastern Asia. Some of them are also found in other regions, such as India, Bhutan, Mongolia, Malaysia, Thailand, Philippines, Nepal, Vietnam, etc. [1, 2], and the most common of *Acanthopanax* genus is *A. gracilistylus*, *A. senticosus*, *A. koreanum and A. brachypus*. The first abundant resources of *Acanthopanax* genus in the world is in China, with about 26 species and 18 subspecies. The root bark from *A. gracilistylus* W. W. Smith has been listed officially in the China Pharmacopoeia named as "Cortex *Acanthopanacis*" (*Wujiapi*) [3]. There are 11 kinds, 3 subspecies and 3 forms in Korea, and the root bark of *A. sessiliflorum Seemann* is recorded in Korea Pharmacopeia. The main plants of *Acanthopanax* genus are listed in Figure 1.

The roots and stem bark of the *Acanthopanax* genus, belonging to the liver meridian and kidney channel, are warm-natured, acrid flavor and bitter. Several kinds of chemical compounds have been reported, including triterpenoids, diterpenoids, lignans, coumarins, flavones, polyacetylenes, phenylpropanoids, essential oils and polysaccharides etc.[4-9], among which triterpenoids and lignans were considered to be the most active components.

Acanthopanax gracilistylus W.W.Smith

Acanthopanax senticosus

Acanthopanax koreanum

Acanthopanx giraldii

Acanthopanax brachypus

Figure 1. (Continued)

Acanthopanax divaricatus var. albeofructus

Acanthopanax sessiliflorus

Acanthopanax trifoliatus

Figure 1. The main plants of *Acanthopanax* genus.

For a long time, the *Acanthopanax* genus has been used as tonics and sedatives to nourish the liver and kidney, strengthen bones and muscle [10], stimulate circulation to end stasis, eliminate hydroncus, as well as having anti-tumor [11], anti-inflammatory [12], anti-fatigue [13], hepatoprotective effects[14], anti-oxidant [15], anti-diabetic [16], immunomodulatory effects[17], effects on brain, heart [18] and nervous system [19], antipyretic and analgesic[20], and have been clinically used for the treatment of rheumatism, diabetes, hypertension, chronic bronchitis, tumor, ischemic heart disease, waist and knee aching and gastric ulcers [4, 21, 22]. *A. gracilistylus* W. W. Smith, classified into the family of *Araliaceae*, is widely distributed in China. They have been used as a traditional oriental medicine for many ailments, including anti-rheumatic, anti-tumor, anti-stress, longitudinal bone growth, prophylactic functions for chronic bronchitis hypertension and ischemic heart disease [23-25]. What's more, the roots and bark of *A. senticosus* (also called *Eleutherococcus senticosus* or Siberian ginseng) have been used extensively in China, Russia, Japan and Korea as adaptogens to fight against fatigue, hypertension, ischemic heart disease, hepatitis and stress [26-28]. In addition, *A. koreanum* is a deciduous scrub of the *Araliaceae* family and an endemic species in Korea, and is often used in traditional Korean, Chinese and Japanese medicine. The roots and stem bark of *A. koreanum* have been traditionally administered as a tonic and sedative [29]. As an endangered perennial deciduous shrub in the wild, due to overharvesting and loss of habitats through deforestation, *A. brachypusHarm* is mostly distributed in a narrow geographical area of the loess plateau of the northwest of China [30]. Nowadays, different parts of this plant have been employed for various therapeutic purposes in China and Korea. For instance, the alcohol extracts of the bark have anti-inflammatory action, and also inhibit I, II and III type allergic inflammation, as well as adjuvant arthritis [31, 32]. The *Acanthopanax* genus plants have extensive pharmacological effects and clinical applications, with great development value. In recent years, with the growing concern on natural medicine, more and more attention has been paid to the *Acanthopanax* species. This article reviewed the wide research progress of *Acanthopanax* genus, including the ecology, biology, chemistry, pharmacology and clinical applications in the last decades. Based on this, the present situation and exploiting prospects of *Acanthopanax* genus were analyzed, with a view towards a follow-up study and to provide the foundation for the further development and utilization of them. It's also valuable for the reference of further exploitation of bioactivity ingredients from *Acanthopanax* genus.

## 2. Chemical Constituents of *Acanthopanax*Genus

Since the 1950s, researchers have done many studieson the chemical components and pharmacological activities of the *Acanthopanax* genus, and have made great achievements. The *Acanthopanax* genus mainly includes *A. gracilistylus*, *A. senticosus*, *A. koreanum*, *A. giraldii*, *A. chiisanensis*, *A. divaricatus*, *A. sessiliflorus*, *A. trifoliatus*, *A. nipponicus*, etc.. Lots of bioactive componentsand compounds from them have been previously investigated or isolated, for example, triterpenoids, lignans, flavonoids, diterpenoids, coumarins, essential oils, polysaccharides and fatty acids [4, 5, 7]. Specifically, *A. senticosus* mainly contain lignans, phenolic (eleutheroside B, eleutheroside E), flavonoids, triterpenoids, coumarins, alkaloids [33-35]. The leaves of *A. gracilistylus* contain a series of lupane-type triterpenoids, such as acankoreoside A, acankoreoside D, wujiapioside A, wujiapioside B, acankoreoside C, acantrifoside A, β-sitosterol, daucosterol, acankoreanogenin [36-38]. Many compounds from the fruits of *A.gracilistylus* are acankoreoside D, acantrifoside A, acankoreoside A, (-)-kaur-16-en-19-oic acid, protocatechuic acid, β-sitosterol, acanthopanax cerebroside A, acanthopanax cerebroside B, momor-cerebroside, acanthopanax cerebroside C, etc.[8, 39]. The leaves of *A. koreanum* contain triterpenoids, diterpenoids, lignans, polyacetylenes, phenylpropanoids, flavonoids and polysaccharides.

**Table 1.The chemical constituents from *Acanthopanax* genus**

| Species (distribution) | Part of plant | Classification | Chemical components | Reference |
|---|---|---|---|---|
| *A.gracilistylus* (China) | Root | Diterpenoids | acanthoic acid, kaurenoic acid, β-sitosterol, β-sitosterol glucoside | [40-43] |
| | | Other | ent-16a,17-dihydroxy-kauran-19-oic acid, eleutheroside B, linoleic acid,isofraxidinglycosides, D-sesamin, stearic acid, palmitic acid, carotene, tannins | |
| | Stem/ Bark | Diterpenoids | β-sitosterol, daucosterol, 16α-hydroxy-ent-kauran-19-ocid, 16αH,17- isovaleryloxy- ent-kauran-19-oic acid | [44] |
| | | Lignans | syringin, eleutheroside D, stigmasterol | |
| | | Others | coniferin,pentacosanoic acid,<br>(2S,3S, 4R, 8E)-2-[(2'R)-2'-hydroxy-pentadecanoylamino]- heptacosane-1,3,4-triol-8-ene,<br>(2S,3S,4R,8E)-2-[(2'R)-2'-hydroxy-octadecanoylamino]-lignocer-ane-1,3, 4-triol- 8-ene,<br>(2S, 3S,4R,8E)-2-[(2'R)-2'-hydroxy-heneicosanoylamino]-heneicosane-1,3,4-triol-8-ene,<br>(2S, 3S,4R,8E)-2-[(2'R)-2'-hydroxy-docosanoylamino]- eicosane-1,3,4-triol-8-ene,<br>(2S, 3S,4R,8E)-2-[(2'R)-2'- hydroxy-trico-sanoylamino]-nonadecane-1,3,4-triol-8-ene,<br>(2S,3S,4R,8E)-2-[(2'R)-2'-hydroxy-lignocera-noylamino]-cctadecane-1,3,4-tri-ol-8-ene,<br>1-O-β-D-glucopyranosyl-(2S, 3S, R,8E)-2-[(2'R)-2'-hydroxy-pentadecanoylamino]-nonadecane-1,3,4-triol-8-ene | |
| | Leaf | Triterpenoids | acankoreoside A, acankoreoside D, wujiapioside A, wujiapioside B, acankoreoside C, acantrifoside A,<br>3-epi-betulinic acid28-O-α-L-rhamnopyranosyl-(1→4)-β-D-glucopyranosyl-(1→6)-β-D-glucopyranosyl ester,<br>acankoreanogenin, β-sitosterol, daucosterol,<br>3α,11α- dihydroxy-20(29)-lupene-23, 28-dioic acid, acanthopanaxgric acid,<br>stigmast-5,22-dien-3-O-α-D-glucopyranoside, 3,11-dihydroxy-23-oxo-20(29)-lupen-28-oic acid,<br>3-hydroxy-23-oxo-20(29)-lupen-28-oic acid | [36-38, 45] |
| | | Flavonoids | rutin, quercetin | |
| | | Other | (-)-kaur-16-en-19-oic acid, kaempferol, myristin, protocatechuic acid, palmitic acid, 28-dioic acid | |
| | Fruit | Triterpenoids | acankoreoside A, acankoreoside D, acantrifoside A, β-sitosterol,<br>3α,11α-dihydroxylup-20(29)-en-28-oic acid, 3/3-([O-β-D-glucopyranuronosyl]oxy)-olean-12-ene-28-olc acid,<br>3β-([O-β-D-glucopyranuronosyl]oxy)-28-O-P3-D-glucopyranosyl-olean-12-ene-28-olc acid, oleanolic acid,<br>3-O-6'-O-methyl-β-D-(-)- kaur-16-en-19-oic acid | [8, 39] |
| | | Cerebrosides | acanthopanax cerebroside A, momor-cerebroside, acanthopanax cerebroside B, acanthopanax cerebroside C,<br>1-*O*-*β*-D-glucopyranose-(2*S*, 3*S*, 4*R*,8*E*)- 2-[(2′*R*)-2′- hydroxydocosanosylamino]-8(*E*)-heptadecene-1, 3, 4-triol | |
| | | Others | glucuronopyranoside, protocatechuic acid | |

| Species (distribution) | Part of plant | Classification | Chemical components | Reference |
|---|---|---|---|---|
| *A.senticosus*(China, Siberia, Korea, Japan, etc.) | Root | Lignans | syringin, eleutheroside E, eleutheroside $E_2$, chlorogenic acid, caffeic acid | [13, 26, 35, 46, 47] |
| | | Others | fatty acid, diolefine, Acanthopanax senticosides B, antimony, protocatechuic acid, liriodendrin, isofraxidin | |
| | Stem/ Bark | Lignans | (+)-Syringaresinol-di-O-β-D-glucoside, eleutheroside E, syringing(eleutheroside B), (+)-syringaresinol, (-)-sesamin | [48-51] |
| | | Essential oils | Isovaleraldehyde, butyl alcohol, valerldehyde, 2-methyl-1- butanol, 1-pentanol, n-hexanal, furfural, n-hexanol, 2-heptanone, 2-heptanone, 2-butylfuran, n-heptaldehyde, α-pinene, α-thujene, (Z)-2-heptenal, β-pinene, 1-octen-3-ol, 6-methyl-5-heptene-2-one, 2-n-pentylfuran, p-cymene, n-caprylaldehyde, 2,4,-heptadienal, limonene, melonal, pinocarveol, (cis)-β- ocimene, benzeneacetaldehyde, (E)-2-octenal, p-tolualdehyde, linalool, n-nonaldeyde, 4-acety-1-methyl- cyclohexene, (E)-2-nonenal, pinocarvone, p-menth-1-en-4-ol, n-decaldehyde, (E,E)2,4-nonadienal, 6,7-methyl-1,2,3,5,8,8α-hexa- hydronaphthalene, anisole, phellandral, isocaryophyllene, 2,4-decadienal, eugeol, β-famesene, humulene, dihydro-5-penyl-2(3H)-furanone, 2,6-methyl-6-(4- methyl-3-pen-tenyl)-2- norpinene, geranylacetone, 2-heptene,2- methyl- p-tolyl, spathulenol, 3,7,11-trimenthy-1,3,6,10-dodecatetraene, β-bilopanone, β-caryophyllene, caryophyllene oxide, tetradecanal, n-pentadecanol, 10,12-hexadecadienal, manoyl oxide, 10,13-hexadecadien-1-ol, 9,17-octadecadienal, 9,12-octadecadien-1-ol, n-octadecanol | |
| | | Others | isovanillin, 5-hydroxymethylfurfural, isofraxidin | |
| | Leaf | Triterpenoids | acanthopanaxosides A, ciwujianosides $A_2$, ciwujianosides $C_2$, ciwujianosides B, ciwujianosides $D_2$, Ciwujianoside $C_1$, acanthopanaxosides B, ciwujianosides $C_4$, hederasaponin B, ciwujianosides $C_3$, tauroside $H_1$, 3-O-α-rhamnopyranosyl-(1→2)-α-rabinopyranosyl mesembryanthemoidigenic acid, acanthopanaxosides C, sessiloside, chiisanoside,inermoside, 1-deoxychiisanoside, 24-ydroxychiisanoside, 11-deoxyisochiisanoside, chiisanogenin, ursolic acid, oleanolic acid, β-sitosterol, daucosterol | [33, 34, 47, 52-54] |
| | | Flavonoids | hyperin, quercetin, quercitrin, rutin | |
| | | Others | arsenic, mercury, selenium | |
| | Fruit | Triterpenoids | acanthopanaxoside E, silphioside F, copteroside B, hederagenin 3-*O*-β-D-glucuronopyranoside 6'-*O*- methyl ester, gypsogenin 3-*O*-β-D-glucuronide, silphioside G, ilexoside XLVIII, hederagenin 3-*O*-β-D-glucuronopyranosyl methyl ester-28-*O*-β-D-glucopyranoside, 3-*O*-β-D-glucuronopyranosyl gypsogenin 28-*O*-β-D-glucopyranoside, hederagenin 28-*O*-β-D-glucopyranoside, hemsgiganoside B, anhuienside C, ciwujianoside $C_3$, hederasaponin B, eleutheroside L, ciwujianoside $A_1$, 3,4-seco-lupane-20(29)-ene-3,28-dioic acid | [16, 53, 55] |
| | | Lignans | (-)-schisandrin B, (-)-sesamin, (-)-syringaresinol | |
| | | Flavonoids | hyperin, | |

**Table 1. (Continued)**

| Species (distribution) | Part of plant | Classification | Chemical components | Reference |
|---|---|---|---|---|
| | Flower | Flavonoids | hyperin | [53] |
| | Aerial parts | Diketopiperazines | eleutherazines A, eleutherazines B | [56, 57] |
| | | Others | oplopanone B, eleutheroside $B_2$ | |
| *A.koreanum* (Korean) | Root | Triterpenoids | impressic acid, acankoreoside A, ursolic acid, 3-epi-betulinic acid 28-O-[α-L-rhamnopyranosyl(1→4)-β-D-glucopyranosyl (1→6)]-β-D-glucopyranosyl] ester | [58-62] |
| | | Diterpenoids | 7β-hydroxy-ent-pimara-8(14),15-dien-19-oic acid, acanthoic acid, sumogaside, paniculosides ent-kaur-16-en-19-oic acid, 16α-hydroxy-ent-kauran-19-oic acid, 16αH,17-isovaleryloxy-ent-kauran-19-oic acid, 16α-hydroxy-17-isovaleryloxy-ent-kauran-19-oic acid, acanthokoreoic acid A, acanthol, kaurenoic acid | |
| | | Lignans | eleutheroside E, tortoside A, (+)-syringaresinol, hinokinin, hemiariensin, (-)-yatein | |
| | | Others | polysaccharide | |
| | Leaf | Triterpenoids | 3α-hydroxylup-20(29)-ene-23,28-dioic acid 28-O-[β-D- glucopyranosyl-(1→6)-β-D-glucopyranosyl] ester, 3α-hydroxylup-20(29)-ene-23,28-dioic acid, acankoreoside A, 3α,11α-dihydroxylup-23-oxo-lup- 20(29)-en-28-oic acid, acankoreoside D, 3α,11α,23-trihydroxy-lup-20(29)- en-28-oic acid, impressic acid, 3-O-β-D- glucopyranosyl 3α,11α-dihydroxylup-20(29)-en-28-oic acid, acantrifoside A, acankoreoside M, acankoreoside N, acankoreoside I, acankoreoside F, acankoreoside G, acankoreoside H, acankoreoside E, 3α,11α-dihydroxy-20,23-dioxo-30-norlupane-28-oic acid,acankoreoside O, acankoreoside J, acankoreoside K, acankoreoside L, (20R)-3α-hydroxy-29-dimethoxylupan-23,28-dioic acid | [6, 9, 10, 15, 17, 63, 64] |
| | | Diterpenoids | ent-kaur-16-en-19-oic acid, 16α-hydroxy-ent-kauran-19-oic acid, 17-hydroxy-ent-kaur-15-en-19-oic acid | |
| | | Phenylpropanoids | acanthopanic acid | |
| | | Others | betulabuside B, rutin,(4S)-α-terpineol 8-O-β-D- glucopyranoside,1,2-O-dicaffeoylcyclopenta-3-ol | |
| | Fruit | Triterpenoids | ursolic acid | [65] |
| *A.brachypus* (China) | Root | Flavonoids | (3R)-5,7-dihydroxy-8-(2''-O-veratroyl-β-D-glucopyranosyl)-3-(4'-hydroxyphenyl)-6-methylchroman-4-one, 4'-hydroxyisoflavone-7-O-[α-L-arabinopyranosyl-(1→6)-β-D-glucopyranoside], 7,8,4'-trihydroxyisoflavone, pachypodol, wogonin, kaempferol-7-*O*-β-D-glucopyranoside, baicalin, vitexin, orientin, hyperoside, quercitrin, kaempferol-3-*O*- rutinoside, 3'-*O*-β-D-glucopyranosylokanin 4-methyl ether | [5, 7] |

| Species (distribution) | Part of plant | Classification | Chemical components | Reference |
|---|---|---|---|---|
| | | Essential oils | n-hexanal, trans-2-hexenal, cis-3-hexenol, n-hexanol, α-thujene, α-pinene, camphene, benzaldehyde,verbenene, sabinene, (-)-1-octen-3-ol, β-pinene, octanal, limonene,α-phellandrene, 1,8-cineole, γ-terpinene, linalool, menthol, menthone, endo-fenchol, octan-3-ol acetate, β-terpineol, ipsdienol, isomenthone, endo-borneol, terpinen-4-ol, cryptone, verbenone, nerol, geraniol, geranial, nonanoic acid, isobornyl acetate, carvacrol, δ-elemene, trans-carvyl acetate, α-terpinyl acetate, citronellyl acetate, eugenol, neryl acetate, α-copaene, geranyl acetate, β-elemene, α-gurjunene, β-caryophyllene, linalyl butyrate, α-ionone, α-guaiene, cis-β-farnesene, cinnamyl acetate, β-humulene, geranyl acetone, valencene, trans-β-farnesene, zonarene,germacrene-D, β-selinene,α-muurolene, β-bisabolene, β-cadinene, δ-cadinene, elemol,γ-bisabolene, α-agarofuran, germacrene-B, geranyl butyrate, Z-isoeugenol acetate, globulol,β-calacorene, E-nerolidol, germacrene-D-4-ol, spathulenol, caryophyllene oxide, β-copaen-4-α-ol, guaiol,viridiflorol, α-acorenol, cis-isolongifolanone, cubenol, torreyol, β-eudesmol, α-cadinol, bulnesol, α-bisabolol, juniper camphor, α-oplopanone, octadecane, myristic acid, (E,E)-farnesyl acetate, methyl palmitate, palmitoleic acid, palmitic acid, kaurene, eicosane, methyl stearate, stearic acid, tetracosane, pentacosane, hexacosane, fatty acids/esters | |
| | Stem/ Bark | Triterpenoids | brachyposide A, brachyposide B, tabguticoside A, nipponoside D, palmatoside E, ciwujianoside $A_1$ | [5, 66, 67] |
| | | Essential oils | n-hexanal, cis-3-hexenol, n-hexanol,α-thujene, α-pinene, camphene, benzaldehyde,verbenene, sabinene, octanal, (-)-1-octen-3-ol, β-pinene, β-myrcene,α-phellandrene, δ-3-carene, p-cymene, limonene, 1,8-cineole, cis-linalool trans-β-ocimene, γ-terpinene, acetophenone, oxide,linalool,α-pinene oxide, endo-fenchol, β-thujone, E-pinene hydrate, camphor, β-terpineol, ipsdienol, menthone, endo-borneol, terpinen-4-ol, p-methylacetophenone, p-cymene-8-ol, cryptone, myrtenal, verbenone, trans-piperitol, citronellol, nerol, pulegone, myrtanol, piperitone, geraniol, geranial, thymol, carvacrol, δ-elemene, trans-carvyl cetate, eugenol, α-terpinyl acetate, α-cubebene, cyclosativene, carvacryl acetate, α-ylangene, α-copaene, geranyl acetate, β-elemene, β-bourbonene, β-cubebene, β-caryophyllene, linalyl butyrate, α-ionone, α-guaiene, epi-α-cedrene, β-humulene, geranyl acetone, γ-humulene, germacrene-D, valencene, bicyclogermacrene, α-muurolene, α-(E,E)-farnesene, β-bisabolene, β-cadinene, zonarene, γ-bisabolene, elemol, α-agarofuran, germacrene-B, Z-isoeugenol acetate, myristic acid,β-calacorene, E-nerolidol, germacrene-D-4-ol, spathulenol, globulol, β-copaen-4-α-ol, viridiflorol, bulnesol, γ-eudesmol, τ-cadinol, β-eudesmol, α-cadinol, α-bisabolol, juniper camphor, β-sinensal, (E,E)-farnesol, (E,E)-farnesyl acetate, methyl palmitate, 1-eicosene, manoyl oxide, palmitoleic acid, palmitic acid, eicosane, tetracosane, pentacosane, fatty acids/esters, heptanoic acid, vanillin, trans-Linalool oxide, α-phellandrene,1-methyl-2-(-methylethyl)-benaene, n-hexadecanoic acid, β-myrcene, benzaldehyde, 4-Acetyl-1-methylcy clohexene | |

**Table 1. (Continued)**

| Species (distribution) | Part of plant | Classification | Chemical components | Reference |
|---|---|---|---|---|
| | Leaf | Essential oils | n-hexanal, trans-2-hexenal, cis-3-hexenol, benzaldehyde, n-hexanol, α-thujene, α-pinene, camphene, verbenene, sabinene, (-)-1-octen-3-ol, β-pinene, β-myrcene, octanal, δ-3-carene, α-terpinene, p-cymene, limonene, 1,8-cineole, trans-β-ocimene, γ-terpinene, acetophenone, cis-linalool oxide, α-terpinolene, α-pinene oxide, linalool,α-camphenal, E-pinene hydrate, camphor, β-terpineol, menthone, thymol, endo-borneol, terpinen-4-ol, p-cymene-8-ol, myrtenal, trans-piperitol, verbenone, thymyl methyl ether, cuminal, myrtanol, E-cinnamaldehyde, isobornyl acetate, tridecane, δ-elemene, trans-carvyl acetate, piperitenone, α-terpinyl acetate, α-longipinene, eugenol, cyclosativene, α-copaene, β-elemene, α-gurjunene, β-caryophyllene, α-ionone, β-copaene, α-guaiene, epi-α-cedrene, β-humulene, geranyl acetone, alloaromadendrene, germacrene-D, β-selinene, valencene, bicyclogermacrene, α-muurolene, α-(E,E)-farnesene, β-bisabolene, β-cadinene, δ-cadinene, cadina-1,4-diene, γ-bisabolene, α-agarofuran, spathulenol, germacrene-B, Z-isoeugenol acetate, caryophyllene oxide, globulol, β-copaen-4-α-ol, viridiflorol, γ-eudesmol, methyl palmitate, τ-cadinol, cubenol, torreyol, α-cadinol, eicosane, β-sinensal, (E,E)-farnesol, octadecane, myristic acid, (E,E)-farnesyl acetate, nonadecane,1-eicosene, palmitoleic acid, palmitic acid, stearic acid, tricosane, tetracosane, hexacosane, fatty acids/esters | [5] |
| | Fruit | Essential oils | trans-2-hexenal, cis-3-hexenol, n-hexanol,α-thujene, α-pinene, camphene, sabinene, (-)-1-octen-3-ol, β-myrcene, octanal, α-phellandrene, limonene, 1,8-cineole,γ-terpinene, linalool, endo-fenchol, 1-octen-3-ol acetate, octan-3-ol acetate, menthone, verbenone, nerol, phenylethyl acetate, nonanoic acid, isobornyl acetate, thymol, trans-carvyl acetate,citronellyl acetate, thymyl acetate,eugenol, neryl acetate, α-copaene,geranyl acetate, β-bourbonene,guaiol,α-gurjunene, β-caryophyllene, linalyl butyrate, β-humulene, trans-α-bergamotene, geranyl acetone, trans-β- farnesene, γ-humulene, germacrene-D, β-selinene, α-muurolene, β-bisabolene, β-cadinene,δ-cadinene, zonarene, cubenol,γ-bisabolene, germacrene-B, geranyl butyrate, E-nerolidol, germacrene-D-4-ol, spathulenol, caryophyllene oxide, globulol, 2-phenylethyl tiglate, viridiflorol, α-acorenol, cis-isolongifolanone, γ-eudesmol, torreyol, β-eudesmol, α-cadinol, bulnesol, α-oplopanone, octadecane, myristic acid,(E,E)-farnesyl acetate, methyl palmitate,manoyl oxide, palmitoleic acid, palmitic acid, kaurene, methyl oleate, methyl stearate, stearic acid, fatty acids/esters | [5] |
| | Flower | Essential oils | n-hexanal, trans-2-hexenal, cis-3-hexenol, benzaldehyde, n-hexanol, α-thujene, α-pinene, camphene, verbenene, sabinene, (-)-1-octen-3-ol, β-pinene, β-myrcene, octanal, p-cymene, limonene, 1,8-cineole, trans-β-ocimene, thymol,γ-terpinene, α-terpinolene, linalool, β-thujone, camphor,α-camphenal, β-terpineol, menthone, endo-borneol, geranyl acetone, terpinen-4-ol, p-cymene-8-ol, α-terpinyl acetate,verbenone, cuminal, isobornyl acetate, carvacrol, tridecane, δ-elemene, trans-carvyl acetate, α-longipinene, eugenol, α-ylangene, β-elemene, α-gurjunene, β-caryophyllene, β-copaene, γ-elemene, α-guaiene, cis-β-farnesene, globulol, β-humulene, germacrene-D, β-selinene, α-muurolene, β-bisabolene, β-cadinene, cis-calamenene, | [5] |

| Species (distribution) | Part of plant | Classification | Chemical components | Reference |
|---|---|---|---|---|
| | | | δ-cadinene, cadina-1,4-diene, γ-bisabolene, α-agarofuran, spathulenol, germacrene-B, caryophyllene oxide, γ-eudesmol, cubenol, torreyol, α-bisabolol, myristic acid,nonadecane,eicosane,1-eicosene,tricosane, pentacosane, hexacosane, fatty acids/esters | |
| *A.chiisanensis* (Korean) | Root | Lignans | 3-(3,4-dimethoxybenzyl)-2-(3,4-methylenedioxybenzyl)butyrolactone, l-sesamin, savinin, helioxanthin, taiwanin C | [68-70] |
| | | Flavonoids | hyperin, (-)-sesamin | |
| | Leaf | Triterpenoids | chiisanoside, isochiisanoside, 22-hydroxychiisanoside, chiisanogenin(the aglycone of chiisanoside) | [71] |
| *A.sessiliflorus* (Korea, China, Japan, etc.) | Stem/ Bark | Biopolymers | 2,4,6-tri-O-methyl-D-glucitol, 2,3,6-tri-O-methyl-D-glucitol, 2,3,6-tri-O-methyl-D-galacitol, 3,4,6-tri-O-methyl-galacitol, 2,4-di-O- methyl-D-mannitol, 2,3,4-tri-O-methyl-D-galacitol, 2,3,4-tri-O-methyl-D-glucitol, 2-O-methyl-arabinitol | [72, 73] |
| | Leaf | Triterpenoids | chiisanoside, 11-deoxyisochiisanoside, isochiisanoside, sessiloside | [74] |
| | Fruit | Triterpenoids | (1R,11α,22α)-1,4-epoxy-11,22-hydroxy-3,4-secolupane-20(30)-ene-3,28-dioic acid, (1R,11α)-1,4-epoxy-11- hydroxyl-3,4-secolupane-20(30)-ene-3,28-dioic acid, 3-O-[(α-L-arabinopyranosyl)-(1→2)]-[β-D- glucuronopyranosyl-6-O-methylester]-olean-12-ene-28-olic acid, chiisanogenin, chiisanoside, ursolic acid, oleanolic acid, β-sitosterol, 22α-hydroxychiisanogenin, daucosterol, oleanolic acid-3-O-6'-O-methyl-β-D-glucuronopyranoside, oleanolic acid-3-O-β-D-glucuronopyranoside, oleanolic acid-3-O-β-D-glucopyranoside | [33, 75-77] |
| | | Others | sessiline, scoparone, (-)-sesamin, protocatechuic acid, hyperin, giraldii, 5-hydroxy methyl furfural | |
| *A.giraldii* (China,Japan, Siberia, etc.) | Stem/ Bark | Essential oils | hexanal, hexanoic acid, octanal, caprylic acid, thymol, E,E-2,4-decadienal, borneol, terpinen-4-ol, α-calacorene, calamenene, α-curcumene, α-cubebene, β-bourbonene, β-elemene, β-cadinene, α-bergamotene, trans-β-farnesene, γ-muurolene, β-selinene, α-selinene, aromadendrene, α-muurolene, β-bisabolene, γ-cadinene, δ-cadinene, dendrolasin, xanthomicrol, spathulenol, cargohpyllene oxide, tetradecanoic acid, pentadecanoic acid, palmitic acid, n-octadecane, hexabydrofarnesylacetone, mathyl palmitate, heptadecanioc, dibutyl phthalate, linolenic acid, oleic acid, octadecanoic acid, 11-tricosene, 11-tetracosane, n-tricosene, n-tetracosane, n-heptacosane | [78-81] |
| | | Lignans | eleutheroside E, eleutheroside B, chlorogenic acid, syringaresinol | |
| | | Triterpenoids | hederagenin 3-O-β-D-glucopyranosyl-(1→2)-α-L- arabinopyranoside, daucosterol oleanolic acid 3-O-β-D-glucopyranosyl-(1→2)-α-L-arabinopyranoside | |

**Table 1. (Continued)**

| Species (distribution) | Part of plant | Classification | Chemical components | Reference |
|---|---|---|---|---|
| | Leaf | Essential oils | caprylic acid, thymol, linalool, terpinen-4-o1, α-terpineol, α-Calacorene, dodecanoic acid, E-1-(2,6,6-trimathy1-1,3-cyclohexadien-gls)-2-buten-1-on, calamenene, α-curcumene, α-cubebene, copaene, β-bourbonene, β-elemene, humulene, β-cadinene, cis-caryophyllene, trans-caryophyllene, mathyl palmitate, phytol, trans-β-farnesene, β-selinene, α-selinene, β-bisabolene, γ-cadinene, δ-cadinene, dendrolasin, linolenic acid, anthracene, xanthomicrol, spathulenol, cargohpyllene oxide, tetradecanoic acid, pentadecanoic acid, dibutyl phthalate, (E,E)-farnesylacetone, octadecanoic acid, hexabydrofarnesylacetone, heptadecanioc, linoleic acid, 11-tricosene, n-tricosene, 11-tetracosane, n-tetracosane | [79] |
| | Fruit | Essential oils | p-cymene, thymol, E,E-2,4-decadienal, linalool, copaene, terpinen-4-o1, α-terpineol, α-calacorene, E-1-(2,6,6-trimathy1-1,3- cyclohexadien-gls)-2-buten-1-one, dodecanoic acid, calamenene, α-curcumene, α-cubebene, β-bourbonene, β-elemene, β-cadinene, trans-caryophyllene, trans-β-farnesene, γ-muurolene, β-selinene, α-selinene, α-muurolene, β-bisabolene, γ-cadinene, δ-cadinene, dibutyl phthalate, dendrolasin, xanthomicrol, spathulenol, linoleic acid, phytol, cargohpyllene oxide, tetradecanoic acid, pentadecanoic acid, (E,E)-farnesylacetone, heptadecanioc, hexabydrofarnesylacetone, mathyl palmitate, octadecanoic acid, oleic acid, 11-tricosene, n-tricosene, 11-tetracosane, n-tetracosane, n-heptacosane | [79] |
| *A.divaricatus*(Japan,etc.) | Leaf | Triterpenoids | 24-hydroxychiisanogenin, 22α-hydroxychiisanogenin, chiisanoside | [82, 83] |
| *A.trifoliatus* (Southeast of Asia:Vietnam, Thailand, etc.) | Stem/ Bark | Diterpenoids | acantrifoside D, ent-kaur-16-en-19-oic acid, 16αH,17-isovalerate- ent-kauran-19-oic acid, ent-pimara-8(14),15-dien-19-oic acid | [84, 85] |
| | | Phenylpropanoids | 1-β-D-glucopyranosyl-2,6-dimethoxy-4-propenylphenol, 1-[β-D-glucopyranosyl-(1→6)-β-D-glucopyranosyl]-2,6- dimethoxy-4-propenylphenol | |
| | Leaf | Triterpenoids | acantrifoic acid A, acantrifoside C, acanthodiolglycoside, 24-nor-11α-hydroxy-3-oxo-lup-20(29)-en-28-oic acid, 28-O-α-L-rhamnopyranosyl-(1→4)- β-D-glucopyranosyl-(1→ 6)- β-D-glucopyranosyl ester | [86-88] |
| | | Others | nevadensin, kaur-16-en-19-oic acid, taraxerol, taraxerol-acetate | |
| *A.nipponicus/A.japonicus* (Japan) | Leaf | Triterpenoids | acanjaposide D, acanjaposide E, acanjaposide F, acanjaposide G, acanjaposide H, acanjaposide I, acanjaposide A, acanjaposide B, acanjaposide C, nipponosides A, nipponosides B, nipponosides C, nipponosides E, nipponosides D, kalopanaxsaponin G | [89-91] |

The essential oils from flowers, leaves, bark and roots of *A. brachypus*, have been identified for 160 different constituents, which mainly consistof monoterpenes and sesquiterpene hydrocarbons, as well as oxygenated monoterpenes and sesquiterpenes, while the fruit oil mainly consists of fatty acids and esters [5]. Different parts of the *Acanthopanax* genus plants contain different chemical compositions, and Table 1 shows the detailed information. According to the Table 1, the main active components of *Acanthopanax* genus are terpenoids (Figure 2-5). Until now, lots of studies on the triterpenoids have been carried out, with more than 120 triterpenoids being reported. On the basis of the type of aglycon, these triterpenoid saponins can be classified into the oleanane, lupane and 3,4-seco-lupane classes. The triterpenoids-containing fractions from different parts of *Acanthopanax* genushave been reported to have several important pharmacological effects, which are exemplified by the reduction of acute myocardial infarction and so on.

## 2.1. Oleanane-Type Triterpenoids

All the oleanene-type triterpenoids possess the oleanene triterpene skeleton as the aglycone. Many triterpene glycosides were isolated from the leaves of *A. japonicus* FRANCH *et.* SAVART (=*A. nipponicus*Makino) and elucidated structurally by mass, 1D, and 2D NMR spectroscopy. The structures of these oleanene glycosides were established as acanjaposide A(1), acanjaposide B(2), acanjaposide C(3), acanjaposide D(4), acanjaposide E(5), acanjaposide F(6), acanjaposide G(7), acanjaposide H(8), acanjaposide I(9), nipponosides A(10), kalopanaxsaponin G(11), nipponosides B(12), nipponosides C(13), nipponosides D(14) and nipponosides E(15), respectively[89-91]. What's more, three novel 3α-hydroxy-oleanane-type triterpene oligoglycosyl esters, spinoside $C_1$(16), spinoside $C_4$(17) and spinoside $C_5$(18), were isolated from the leaves of *A. spinosus* [92]. Additionally, lots of triterpenoid saponins were isolated from *A. senticosus*, such as acanthopanaxosides A(19), ciwujianosides $A_2$(20), ciwujianosides $C_2$(21), ciwujianosides B(22), ciwujianosides $D_2$(23), ciwujianosides $C_1$(24), acanthopanaxosides B(25), ciwujianosides $C_4$(26), hederasaponin B(27), ciwujianosides $C_3$(28), tauroside $H_1$(29), 3-O-α-rhamnopyranosyl-(1→2)-α-arabinopyranosyl mesembryanthemoidigenic acid(30), acanthopanaxosides C(31), acanthopanaxoside E(32), ciwujianoside $C_3$(33), hederasaponin B(34), eleutheroside L(35), ciwujianoside $A_1$(36)[16, 52]. (Figure 2).

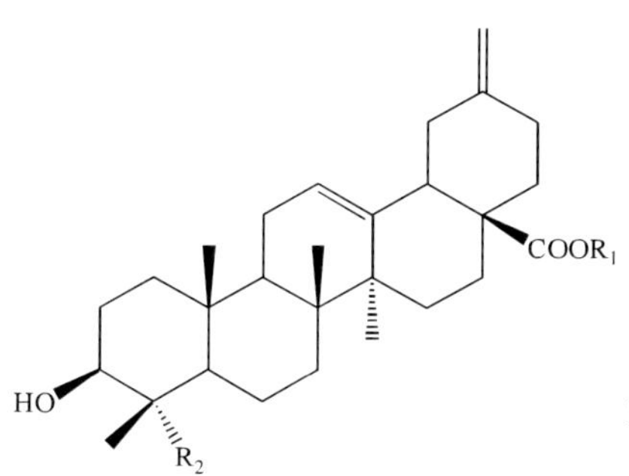


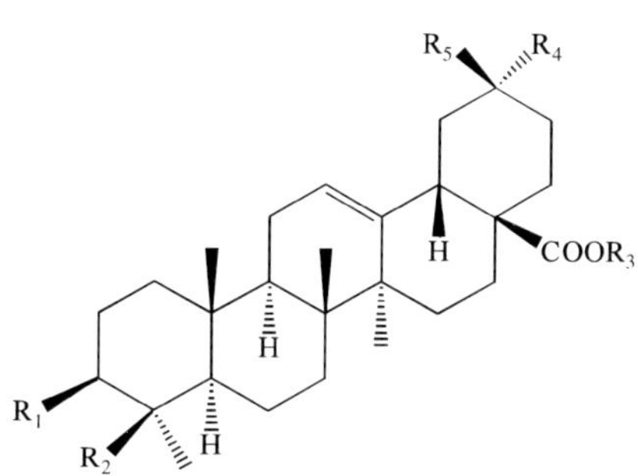


1: $R_1$=S $R_2$=COOH 2: $R_1$=S $R_2$=CHO compound 3-18

Figure 2. (Continued)

| | $R_1$ | $R_2$ | $R_3$ | $R_4$ | $R_5$ |
|---|---|---|---|---|---|
| 3 | OH | OHO | $Glc^6$-$Glc^4$-Rha | OH | $CH_3$ |
| 4 | OH | $CH_2OH$ | $Glc^6$-$Glc^4$-Rha | COOH | $CH_3$ |
| 5 | OH | COOH | $Glc^6$-$Glc^4$-Rha | $CH_3$ | $CH_2OH$ |
| 6 | OH | COOH | $Glc^6$-$Glc^4$-Rha | $CH_2OH$ | $CH_3$ |
| 7 | OH | COOH | $Glc^6$-$Glc^4$-Rha | $CH_3$ | $CH_3$ |
| 8 | OH | CHO | $Glc^6$-$Glc^4$-Rha | $CH_2OH$ | $CH_3$ |
| 9 | OH | CHO | $Glc^6$-$Glc^4$-Rha | COOH | $CH_3$ |
| 10 | =O | $CH_2OH$ | $Glc^6$-$Glc^4$-Rha | $CH_3$ | $CH_3$ |
| 11 | βOH | $CH_2OH$ | $Glc^6$-$Glc^4$-Rha | $CH_3$ | $CH_3$ |
| 12 | βO-Glc | $CH_3$ | $Glc^6$-$Glc^4$-Rha | $CH_3$ | $CH_3$ |
| 13 | βO-Glc | CHO | $Glc^6$-$Glc^4$-Rha | $CH_3$ | $CH_3$ |
| 14 | βOH | $CH_2OH$ | $Glc^6$-$Glc^4$-Rha | $CH_2OH$ | $CH_3$ |
| 15 | βOH | $CH_2OH$ | $Glc^6$-$Glc^4$-Rha | OH | $CH_3$ |
| 16 | αOH | $CH_3$ | $Glc^6$-$Glc^4$-Rha | $CH_2OH$ | $CH_3$ |
| 17 | αOH | CHO | $Glc^6$-$Glc^4$-Rha | $CH_2OH$ | $CH_3$ |
| 18 | αOH | $CH_2OH$ | $Glc^6$-$Glc^4$-Rha | $CH_2OH$ | $CH_3$ |

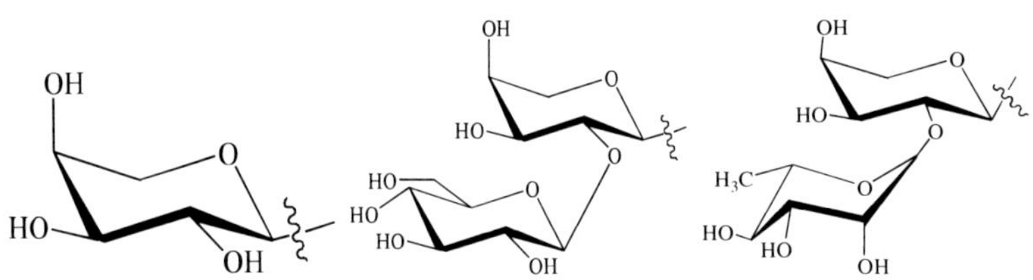

$S_1$ $S_2$ $S_3$

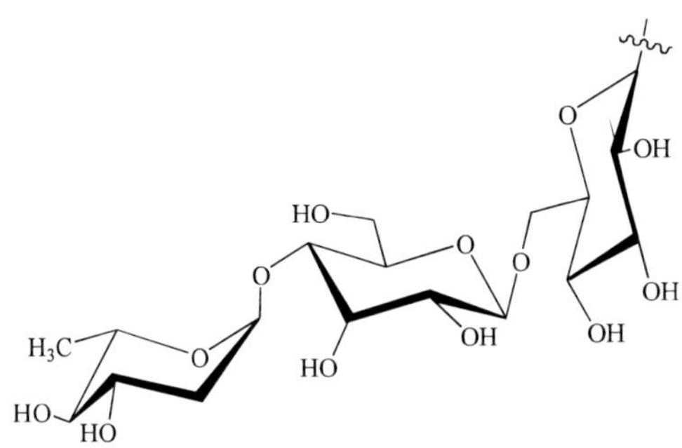

$S_4$

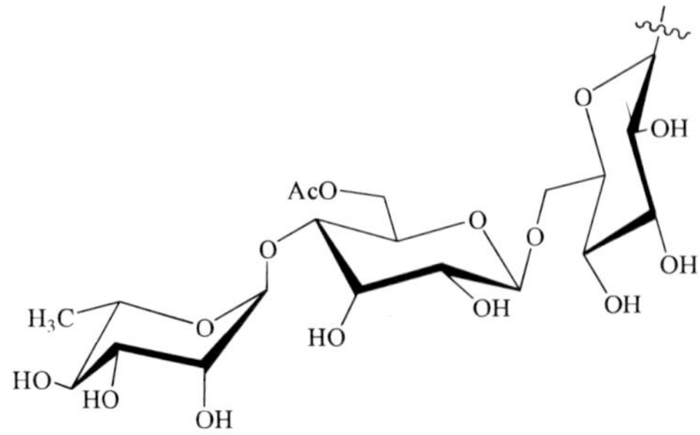

$S_5$

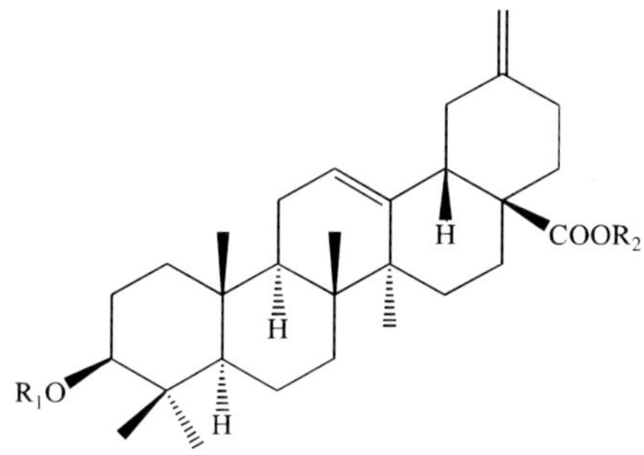

compound 19-24

19: $R_1$=$S_2R_2$=$S_5$ 20: $R_1$=$S_2R_2$=$S_4$ 21: $R_1$=$S_3R_2$=$S_5$
22: $R_1$=$S_3R_2$=$S_4$ 23: $R_1$=$S_1R_2$=$S_5$ 24: $R_1$=$S_1R_2$=$S_4$

| | $R_1$ | $R_2$ | $R_3$ | $R_4$ | $R_5$ |
|---|---|---|---|---|---|
| 25 | $S_2$ | $CH_3$ | H | $S_5$ | $CH_3$ |
| 26 | $S_3$ | $CH_3$ | H | $S_5$ | $CH_3$ |
| 27 | $S_3$ | $CH_3$ | H | $S_4$ | $CH_3$ |
| 28 | $S_1$ | $CH_3$ | H | $S_4$ | $CH_3$ |
| 29 | $S_3$ | $CH_3$ | OH | $S_4$ | $CH_3$ |
| 30 | $S_3$ | $CH_3$ | H | H | $CH_2OH$ |
| 31 | $S_3$ | CHO | H | H | COOH |
| 32 | GlcA | $CH_3$ | OH | Glc | $CH_3$ |
| 33 | Ara | $CH_3$ | H | S | $CH_3$ |
| 34 | $Ara^2$-Rha | $CH_3$ | H | S | $CH_3$ |
| 35 | $Ara^4$-Rha | $CH_3$ | H | S | $CH_3$ |
| 36 | $Ara^2$-Glc | $CH_3$ | H | S | $CH_3$ |

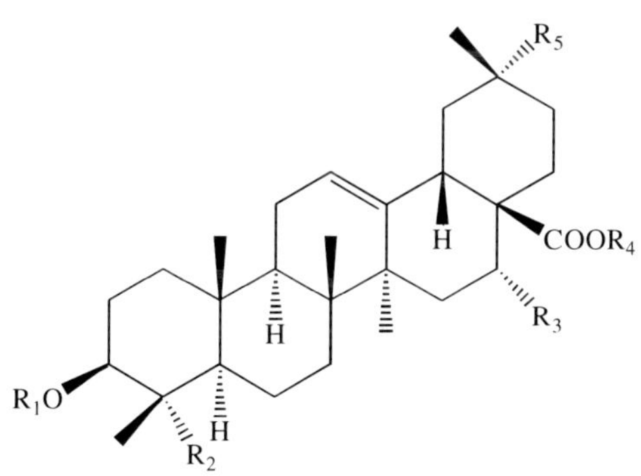

compound 25-36

(Glc: β-D-glucopyranosyl and Rha: α-L-rhamnopyranosyl, S: $Glc^6$-$Glc^4$-Rha).

Figure 2. Structures ofoleanane type triterpenoids isolated from *Acanthopanax* genus.

## 2.2. Lupane-Type Triterpenoids

Twenty-two lupane-type triterpenoids(compound1-22) were isolated from the steamed leaves of *A. koreanum*. Their chemical structures were determined as 3α-hydroxylup-20(29)-

ene-23,28-dioic acid 28-O-[β-D-glucopyranosyl-(1→6)-β-D-glucopyranosyl]ester(1),3α-hydroxylup-20(29)-ene-23,28-dioic acid(2), acankoreoside A(3), 3α,11α-dihydroxylup-23-oxo-lup-20(29)-en-28-oic acid(4), acankoreoside D(5), 3α,11α,23-trihydroxy-lup-20(29)-en-28-oic acid(6), impressic acid(7), 3-O-β-D-glucopyranosyl 3α,11α-dihydroxylup-20(29)-en-28-oic acid(8), acantrifoside A(9), acankoreoside M(10), acankoreoside N(11), acankoreoside I(12), acankoreoside F(13), acankoreoside G(14), acankoreoside H(15), acankoreoside E(16), 3α,11α-dihydroxy-20,23-dioxo-30-norlupane-28-oic acid(17), acankoreoside O(18), acankoreoside J(19), acankoreoside K(20), acankoreoside L(21), (20R)-3α-hydroxy-29-dimethoxylupan-23,28-dioic acid(22) [6, 10, 63, 64]. Moreover, lupane-triterpene glycosides were isolated from the leaves of *A. gracilistylus W. W. Smith.*

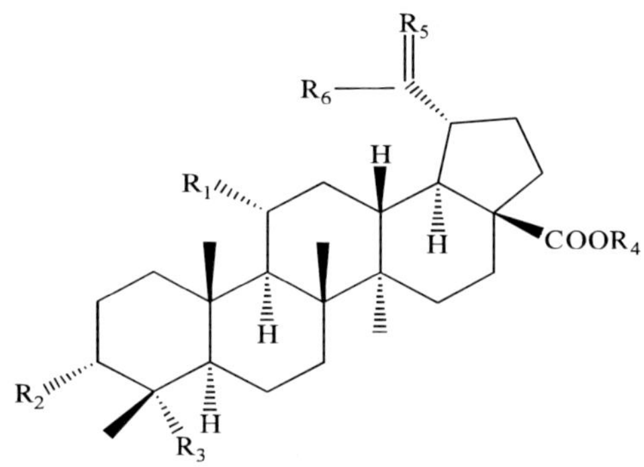

Compound 1-17, 23-29

| | $R_1$ | $R_2$ | $R_3$ | $R_4$ | $R_5$ | $R_6$ |
|---|---|---|---|---|---|---|
| 1 | H | OH | COOH | $Glc^6$-Glc | $CH_2$ | $CH_3$ |
| 2 | H | OH | COOH | H | $CH_2$ | $CH_3$ |
| 3 | OH | OH | COOH | $Glc^6$-$Glc^4$-Rha | $CH_2$ | $CH_3$ |
| 4 | OH | OH | CHO | H | $CH_2$ | $CH_3$ |
| 5 | OH | OH | CHO | $Glc^6$-$Glc^4$-Rha | $CH_2$ | $CH_3$ |
| 6 | OH | OH | $CH_2OH$ | H | $CH_2$ | $CH_3$ |
| 7 | OH | OH | $CH_3$ | H | $CH_2$ | $CH_3$ |
| 8 | OH | O-Glc | $CH_3$ | H | $CH_2$ | $CH_3$ |
| 9 | OH | OH | $CH_3$ | $Glc^6$-$Glc^4$-Rha | $CH_2$ | $CH_3$ |
| 10 | OH | OH | COOH | $Glc^6$-$Glc^4$-Rha | $CH_2$ | $CH_2OH$ |
| 11 | OH | O-Glc | $CH_3$ | $Glc^6$-$Glc^4$-Rha | $CH_2$ | $CH_2OH$ |
| 12 | OH | OH | CHO | $Glc^6$-$Glc^4$-Rha | $CH_2$ | $CH_2OH$ |
| 13 | H | OH | COOH | $Glc^6$-$Glc^4$-Rha | $CH_2$ | $CH_2OH$ |
| 14 | H | OH | CHO | $Glc^6$-$Glc^4$-Rha | $CH_2$ | $CH_2OH$ |
| 15 | H | α-OH,β-H | CHO | $Glc^6$-$Glc^4$-Rha | H,COOH | H |
| 16 | H | α-OH,β-H | COOH | $Glc^6$-$Glc^4$-Rha | H,CHO | H |
| 17 | OH | OH | CHO | H | O | $CH_3$ |
| 23 | H | H | CHO | $Glc^6$-$Glc^4$-Rha | $CH_2$ | $CH_3$ |
| 24 | H | H | $CH_2OH$ | $Glc^6$-$Glc^4$-Rha | $CH_2$ | $CH_3$ |
| 25 | OH | β-D-Glc | $CH_3$ | $Glc^6$-$Glc^4$-Rha | $CH_2$ | $CH_3$ |
| 26 | H | H | $CH_3$ | $Glc^6$-$Glc^4$-Rha | $CH_2$ | $CH_3$ |
| 27 | H | H | COOH | H | $CH_2$ | $CH_3$ |
| 28 | H | $CO_2$ $CH_3$ | COOH | H | $CH_2$ | $CH_2OH$ |
| 29 | H | $CO_2$ $CH_3$ | COOH | $Glc^6$-$Glc^4$-Rha | $CH_2$ | $CH_2OH$ |

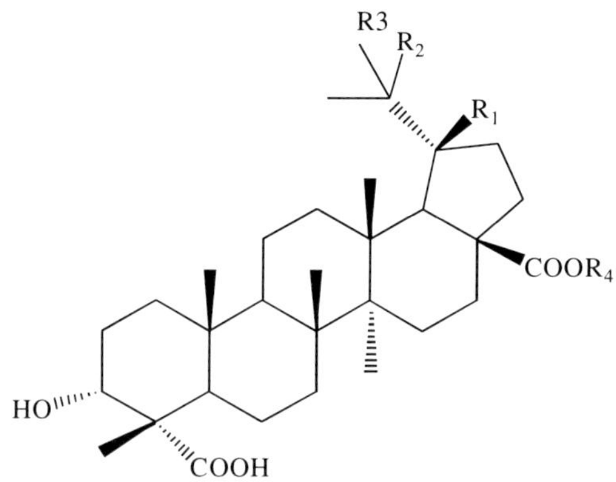

$R_1$ $R_2$ $R_3$ $R_4$

18 OH H $CH_3$ $Glc^6$-$Glc^4$-Rha

19 H =O - $Glc^6$-$Glc^4$-Rha

20 H OH $CH_2OH$ $Glc^6$-$Glc^4$-Rha

21 H H CH $(OH_3)_2$ $Glc^6$-$Glc^4$-Rha

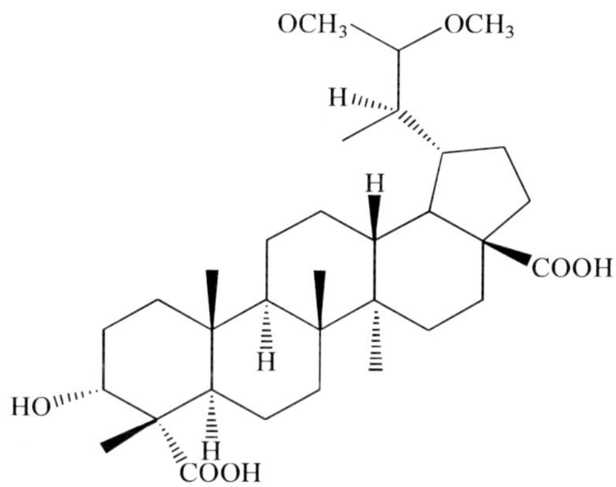

22

(Glc: β-D-glucopyranosyl and Rha: α-L-rhamnopyranosyl).

Figure 3. Structures oflupane-type triterpenoids isolated from *Acanthopanax* genus.

They were acankoreoside A, acankoreoside D, wujiapioside A(23), wujiapioside B(24), acankoreoside C(25), 3-epi-betulinic acid 28-O-α-L-rhamnopyranosyl-(1→4)-β-D-glucopyranosyl-(1→6)-β-D-glucopyranosyl ester(26), acantrifoside A, acankoreagenin(27) [36-38]. Additionally, two lupane-triterpene carboxylic acids, called acantrifoic acid A(28) and acantrifoside C(29) have been isolated from the leaves of *A. trifoliatus* [86].(Figure 3).

## 2.3. 3,4-Seco-Lupane Triterpenoids

The 3,4-*seco*-lupane triterpene glycosides are characteristic of the *Acanthopanax* genus. Since chiisanoside(1) was isolated from *A. chiisanensis* as the first 3,4-*seco*- lupane triterpene glycoside in 1985, several *Acanthopanax* species(*A. senticosus, A. divaricatus,* etc.) have been found containing 3,4-*seco*-lupane triterpenes.

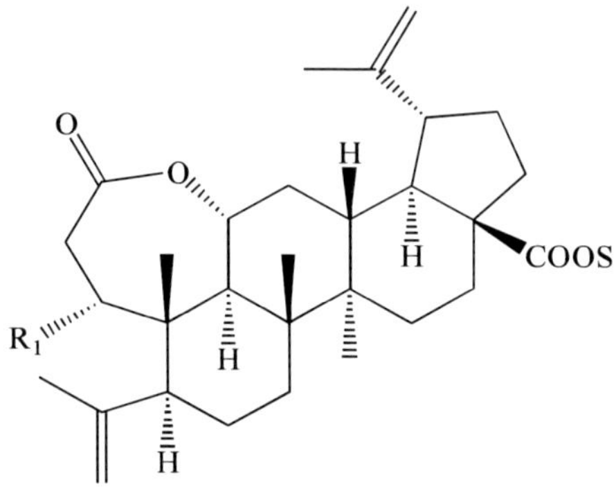

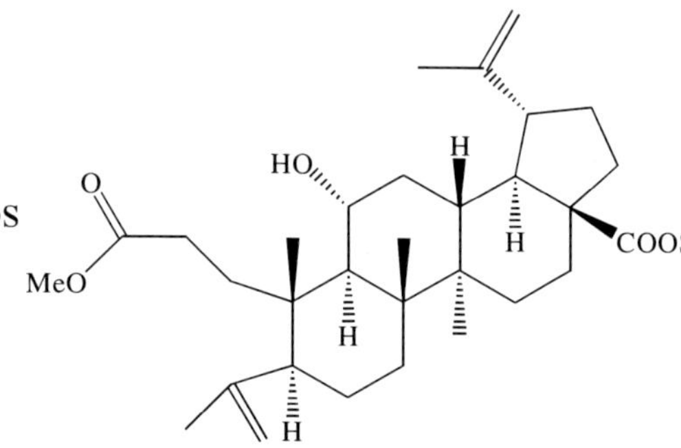

1: $R_{1}$= OH 2: $R_{1}$=H 3

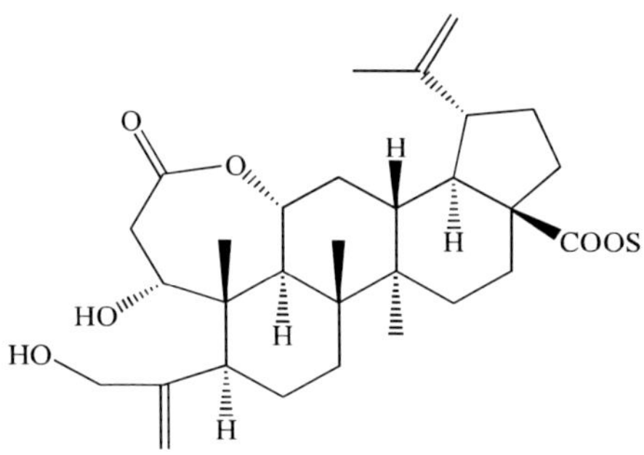

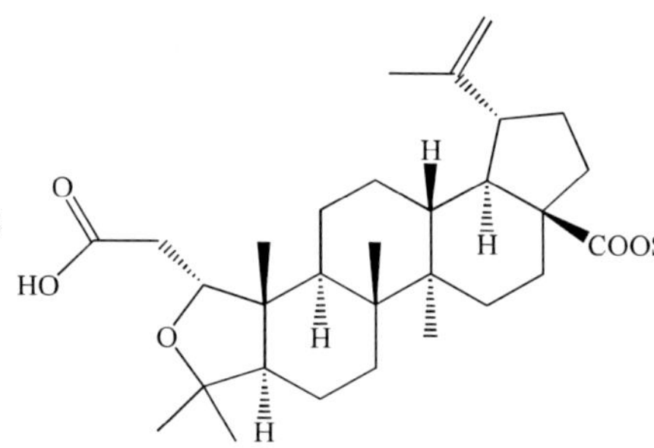

4 5

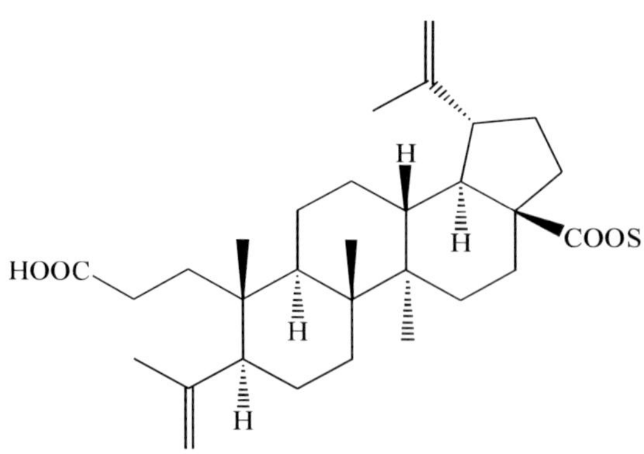

6

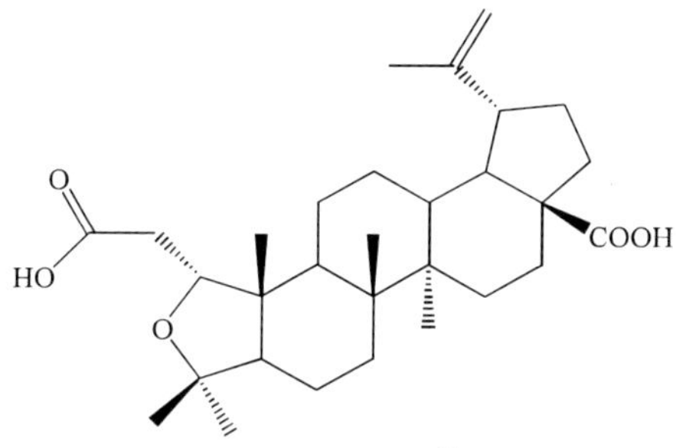

7

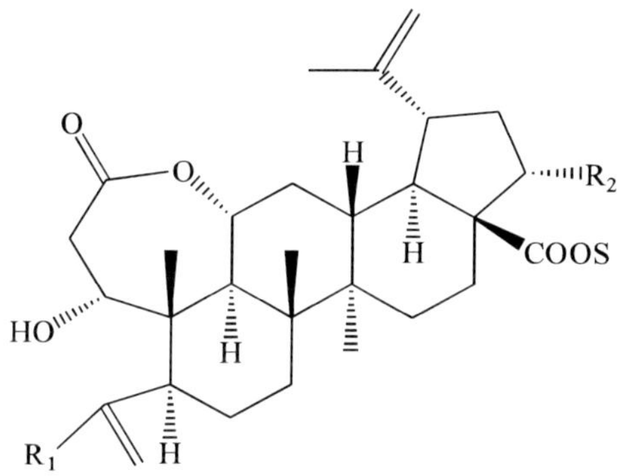

8: $R_1$= $CH_2OH$ $R_2$=H 9: $R_1$= $CH_3$ $R_2$=OH 10: $R_1$= $CH_3$ $R_2$=H
(Glc: β-D-glucopyranosyl and Rha:α-L-rhamnopyranosyl, S: $Glc^6$-$Glc^4$-Rha).

Figure 4. Structures of 3,4-seco-lupane-type triterpenoids isolated from *Acanthopanax* genus.

Five 3,4-seco-lupane-type triterpenoids(2-6) were isolated from the leaves of *A. senticosus*, and one(7) was isolated from *A. senticosus*(Rupr. et Maxim) Harms. The structures of 2-7 were established as1-deoxychiisanoside(2), inermoside(3), 24-hydroxychiisanoside(4), 11-deoxyisochiisanosideand(5), sessiloside(6) and3,4-seco-lupane-20(29)-ene-3,28-dioic acid(7) [52, 54, 55]. In addition, four 3,4-seco-lupane triterpenoids [24-hydroxychiisanogenin(8) and 22α-hydroxychiisanogenin(9), chiisanogenin(10), chiisanoside] were found on the leaves of *A. divaricatus* var. albeofructus [82,83].These data may be useful for the chemotaxonomy of the *Acanthopanax* species. (Figure 4).

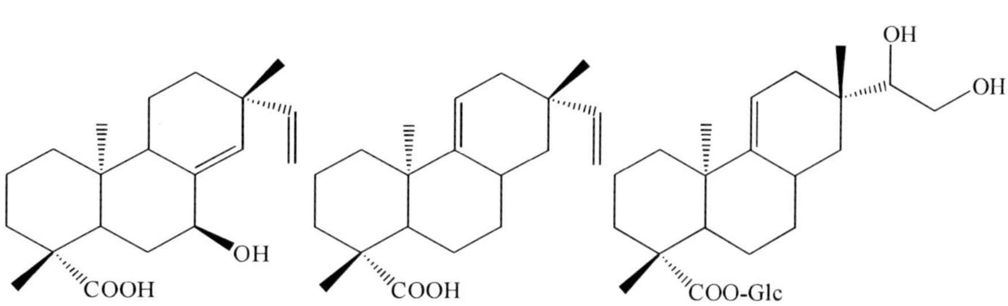

1 2 3

Figure 5. (Continued)

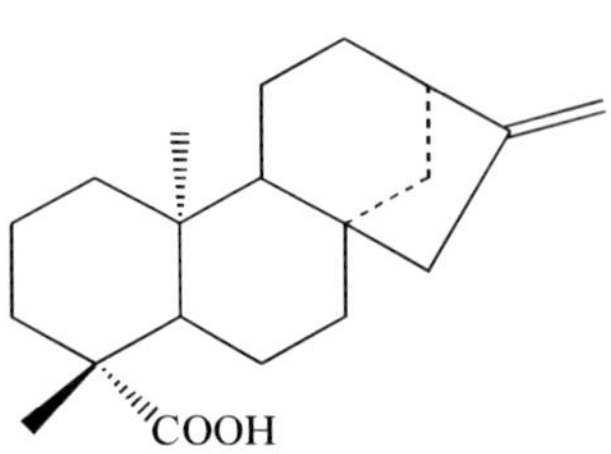
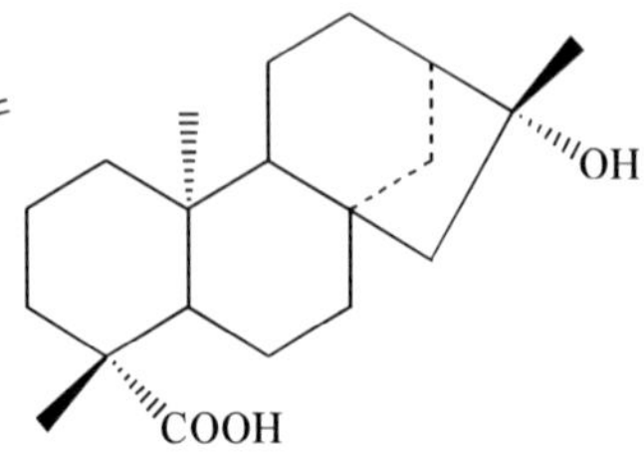
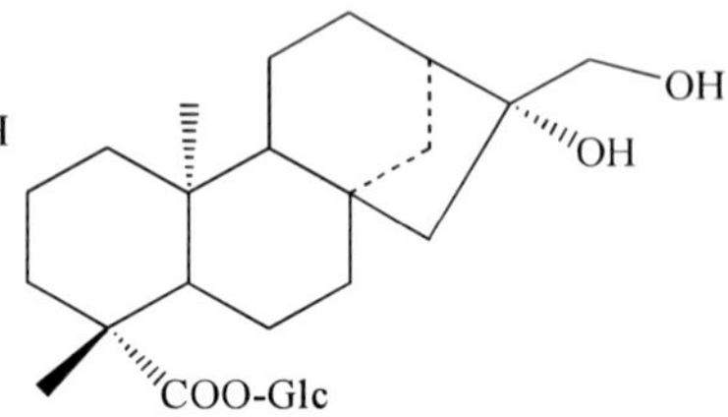

4 5 6

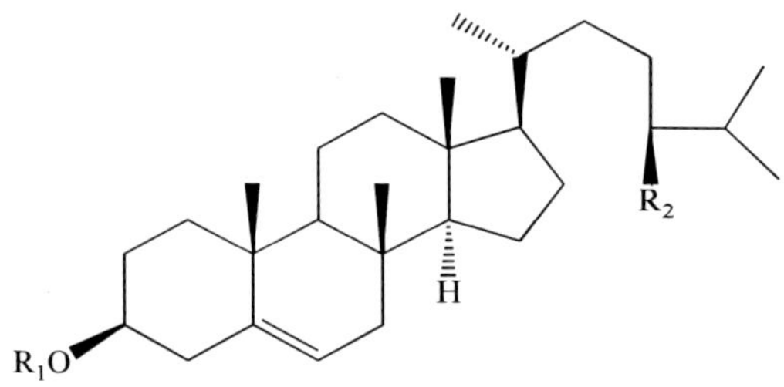

7: $R_{1=}HR_{2=}CH_2CH_3$ 8: $R_{1=}GlcR_{2=}CH_2CH_3$

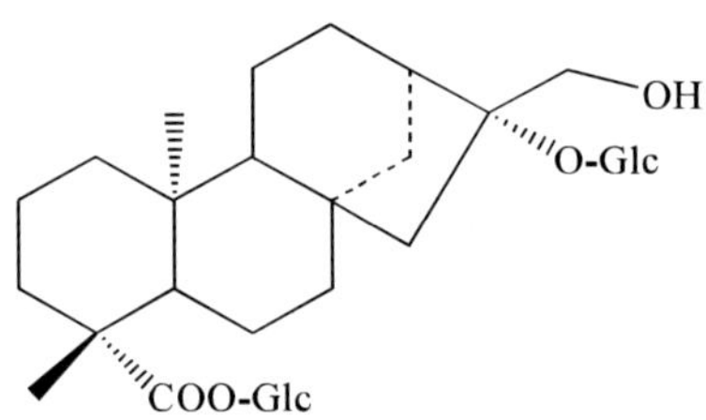

9

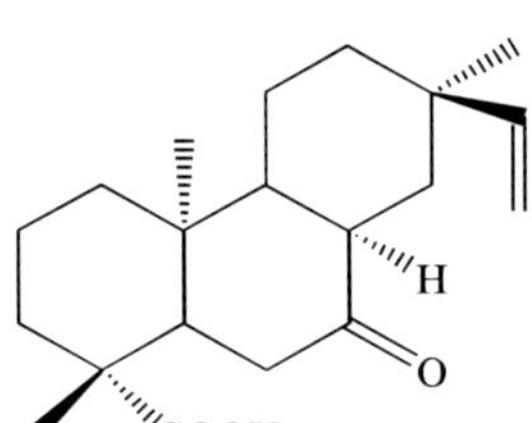
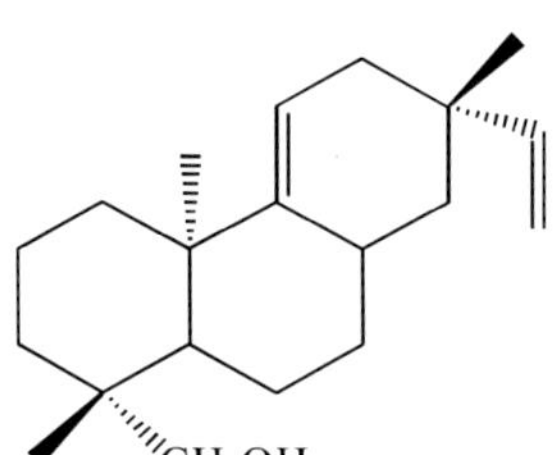

10 11

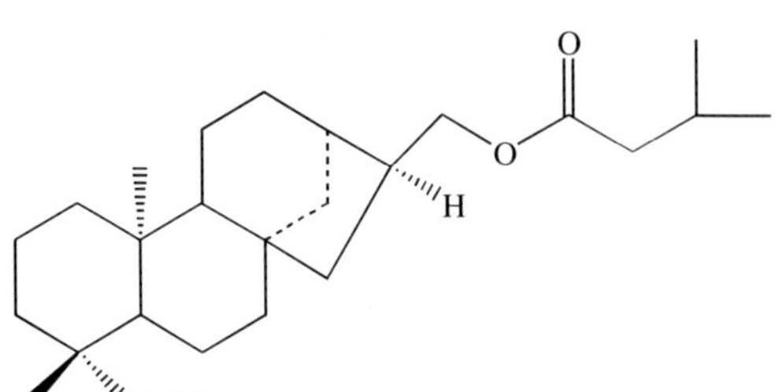

12

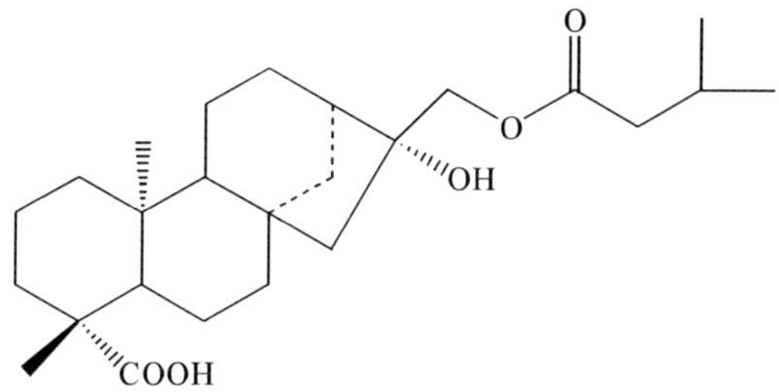

13

(ent-pimara-8(14),15-dien-19-oic acid(1), acanthoic acid(2), sumogaside(3), kaurenoic acid(4), 16α-hydroxy-ent-kauran-19-oic acid(5), paniculosides IV(6), β-sitosterol(7), daucosterol(8), acantrifoside(9)), acanthokoreoic acid A(10), acanthol(11), 16αH,17-isovaleryloxy-ent-kauran-19-oic acid(12), 16α-hydroxy-17-isovaleryloxy-ent-kauran-19-oic acid(13)).

Figure 5. Structures of diterpenoids isolated from *Acanthopanax* genus.

# 3. Pharmacological Effects and Clinical Applications

## 3.1. Anti-Inflammatory and Analgesic Effects

Nowadays, the different parts of *Acanthopanax* genus such as the roots, leaves, fruits, flowers and their active ingredients, are employed for various therapeutic purposes. There have been many reports on the anti-inflammatory effectsof terpenoids isolated from *A. koreanum*. For example, acanthoic acid, a pimarane-type diterpene from *A. koreanum*, exhibited potent inhibitory activity on the interleukin-8(IL-8) secretion of the tumor necrosis factor(TNF)-α-stimulated human colon adenocarcinoma cells (HT-29) and on the TNF-α secretion of the trypsin-stimulated human leukemic mast cells (HMC-1) [61]. Further studies showed thatacanthoic acid might inhibit TNF-α-mediated IL-8 production by blocking in both the mitogen-activated protein kinases(MAPKs) and nuclear factor(NF)-κB pathways in TNF-α-stimulated human colon epithelial cells [93]. What's more, acanthoic acid inhibited gene expression and biological function of TNF-α [94], and exhibited beneficial properties in experimental studies involving colitis [95]. In another study, it has supported the evidence that acanthoic acid may be used to block the development of inflammation caused by mast cells [96]. In addition, lupane-type triterpenoids(20R)-3α-hydroxy-29-dimethoxylupan-23,28-dioic acid and3α,11α-dihydroxy-20,23-dioxo-30-norlupane-28-oic acid, isolated from the *A. koreanum* leaves, inhibited the lipopolysaccharide(LPS)-stimulated production of pro-inflammatoy cytokine(including TNF-α, IL-6, and IL-12) in bone marrow-derived dendritic cells(BMDCs). Thus, these lupane-triterpenoids may ameliorate both the symptom of mortality and endotoxic shock, andmay possess potent anti-inflammatory action [6, 12]. Besides, acankoreoside F,H,G(lupane-type saponins) from *A. koreanum* leaves were assessed for anti-inflammatory activity by examining their effects on nitric oxide(NO) and prostaglandin E2(PGE2) by LPS-induced in RAW 264.7 cells. As a result, acankoreoside F most potently inhibited PGE2 and NO production [63]. Moreover, *A. koreanum* fruit waste extracts have similar inhibitory effects in LPS-stimulated macrophages, and they were considered as anti-inflammatory candidates for topical application [65]. Furthermore, impressic acid(triterpenoid), 16αH,17-isovaleryloxy-ent-kauran-19-oic acid (diterpenoid) and hemiariensin (lignan) from *A. koreanum*were found to exhibit the inhibitory activity against NFAT transcription factor [58-60, 97]. 16αH,17-isovaleryloxy-ent-kauran-19-oic acid also inhibitedthe activity of TNF-α secretion in trypsin-stimulated HMC-1 [98].

Excess production of reactive oxygen species and NO are the characteristics of inflammation. *A. senticosus* extract significantly suppresses NO production and iNOS gene expression *in vitro* and *in vivo* through the inhibition of intracellular peroxides production, which has been implicated in the activation of NF-κB. Exposure to *A. senticosus* extract had no effect on cell viability and systemic toxicity [99, 100]. In addition, *A. senticosus* extract significantly suppressed reactive oxygen species production by mouse peritoneal macrophages *in vitro* and *in vivo*, and did not affect the cell viability or systemic toxicity [101]. These may be partly responsible for the anti-inflammatory function. On the other hand, syringaresinol is the hydrolysate of liriodendrin which is isolated by activity-guided fractionation from the ethyl acetate extracts of the stem bark of *A. senticosus*. The anti-inflammatory and antinociceptive effects of liriodendrin after oral administration were attributable to the transformation of syringaresinol *in vivo*, which may be the active constituent [102]. Additionally,(+)-Syringaresinol-di-O-β-D-glucoside, a phenolic compound from *A. senticosus* Harms, suppressed proinflammatory mediators in SW982 human synovial sarcoma cells by inhibiting activating protein-1 and/or NF-κB activities [49].

3-(3,4-dimethoxybenzyl)-2-(3,4-methylenedioxybenzyl) butyrolactone, TaiwaninC, hyperin, helioxanthin and savininfrom the roots of *A. chiisanensis* showed the potent inhibitory activity on PGE2 production [68, 69]. It was also found that hyperin inhibited NO production, and thus participated in the anti-inflammatory activity of the extract of *A. chiisanensis* [103, 104]. Chiisanoside isolated from the leaves of *A. chiisanensis,* have anti-inflammatory properties due to the inhibition of iNOS, COX-2, TNF-αand IL-1β expression through the down-regulation of NF-κB binding activity [71]. Moreover, total glycosides of *A. Giraldii* for inhibiting the production of NO and PGE2inhibited COX-2 mRNA expression without TNF-αchange, and that's their anti-inflammatory mechanisms [105]. In addition, methanol extract of *A. giraldii* has pronounced analgesic action and reduced not only high temperature,induced by proteinic jelly, but also normal body temperature in rabbits [20].

## 3.2. Anti-Tumor Activities

An *A. senticosus* injection (ASI) can regulate the cellular immunity and treatpatients with lung cancerby such methods as increasing the levels of IgA, IgG, IgM and the activity of TNF-β and natural killer cell(NKC) [11]. Isofraxidin, a major coumarin component isolated from the stem bark of *A. senticosus*, inhibited expression of TPA (12-O-tetradecanoylphorbol-13-acetate)-induced MMP-7(matrix metalloproteinase-7), and *in vitro* cell invasion at a non-toxic level through inhibiting the phosphorylation of ERK1/2 (extracellular signal-regulated kinase 1/2) in hepatoma cells. These results suggested that isofraxidin might be an effective agent for suppressing hepatoma cell invasion [48]. What's more, sesamin, another component of *A. senticosus*, suppressed the growth and induced apoptosis in the cells. These findings suggested that growth inhibition by *A. senticosus* resulted from the apoptosis induced by sesamin [106].

The water-extracted *A. senticosus* exhibited anti-metastatic activity by stimulating the immune system. By using gel filtration, EN-SP, an active glycoprotein fraction, was isolated from the soluble protein layer(GF-AS) of *A. senticosus*. While both GF-AS and EN-SP stimulated the proliferatation of splenocytes of normal mice, EN-SP showed higher anti-metastatic activity and more potently stimulated the proliferation of splenocytes compared with GF-AS. These results suggestedthat the EN-SP could be used as a therapeutic agent to prevent or inhibit tumor metastasis [107]. Additionally, *A. senticosus* aqueous extract (GF100) had antitumor activity to inhibit tumor metastasis prophylactically as well as therapeutically, and its antitumor effect was associated with activation of macrophages and NK cells [108].

*A. gracilistylus* extract(AGE) markedly inhibited the proliferation of tumor cells (such as MT-2, Raji, HL-60, TMK-1 and HSC-2) by way of affecting the activity of phosphorylated retinoblastoma(Rb) proteins and cyclin-dependent kinases(Cdks) to suspend the proliferation of tumor cells [109, 110]. Moreover, *A. giraldii* polysaccharides(AGP) selectively inhibited the colony forming ability, the proliferation, the viability and function of human gastric cancer cells through the low protein expression of c-myc, bcl-2 and the high protein expression of p53, fas, fas-L and the cell factor TGF β. The different inhibitory characteristics on the cancer cells and normal cells were possibly caused by gene and cell factor expressions [111]. What's more, AGP inhibited the growth of solid Sarcoma 180 and prolonged the survival time significantly. In tumor-bearing mice, AGP enhanced the phagocytosis and chemiluminescence of macrophages [112]. In addition, on the human gastric cancer cell, apoptotic peaks in cell cycle analysis and the DNA ladder on the agarose gel were observed after treatment with AGP-I [113]. These results demonstrated that AGP may inhibit cancer by enhancing immune responses or inducing cancer cell apoptosis.

### 3.3. Anti-Fatigue Activities

The liposoluble fraction of *A.senticosus*, whose main components are saturated fatty acid, unsaturated fatty acid, unsaturated alcohol and diolefine, delays the accumulation of lactic acid in muscle by enhancing the burden of swimming capacity of mice by increasing the lactate dehydrogenase(LDH) and then protects the muscle.These were effectivein preventing the increment of blood urea nitrogen (BUN) concentration and increasing fat utilization [13]. In the same way, the eleutherosides, which are the exact active constituents of *A.senticosus*, possess the potent abilities to alleviate fatigue both, physical and mental [26], especially for eleutheroside B [114]. Moreover, *A. senticosus* stem bark extracts (ASSE) could extend the swimming time before exhaustion of the mice, as well as increase the tissue glycogen contents, and then decrease the blood lactate and serum urea nitrogen contents. This indicates that ASSE had anti-fatigue activity and could elevate the exercise tolerance [115]. Besides, the extract of *A. senticosus* could increase not only the stress resistance, but also the mean lifespan of the nematode *C. elegans* in a dose-dependent way [116]. Further study on *A. senticosus* showed that they relieved fatigue during exercise by the inhibition of exercise-induced 5-hydroxy tryptamine(5-HT) synthesis and the tryptophan hydroxylase (TPH) expression in the dorsal raphe [117].

### 3.4. Antioxidant Activities

Compoundsisolated from the methanol extract of *A. koreanum* leaves have high antioxidant activity, and their activities were evaluated by the intracellular reactive oxygen species (ROS) radical scavenging 2',7'-dichlorofluorescein diacetate assay [15]. Moreover, the leaf aqueous extracts of *A. trifoliatus* showed a high level of antioxidative activity due to high levels of phenolic and flavonoid compounds [118]. Different doses of *A.senticosus* aqueous extracts were evaluated for the antioxidant activity against oxidative stress in mice induced by t-BHP(tert-butyl hydroperoxide) through detecting antioxidant enzyme activity, histopathology of the liver, concentration of antioxidant, and related gene and protein expression. As a result, *A.senticosus* aqueous extracts protected against oxidative stress via the induction of Nrf2(NF-E2-related factor-2) and related antioxidant enzymes [119]. In addition, the water extract of the stems exhibited significant increases in antioxidant enzyme activities such as hepatic cytosolic superoxide dismutase, catalase and glutathione peroxidase. Eleutheroside B isolated from *A. senticosus* was also found to cause a moderate free radical scavenging effect [120]. In another study, *A. senticosus* extract was found to be excellent for strengthening the antioxidative defense system, reducing the generation of ROS, damaging oxidative substances and maintaining membrane fluidity(MF) in the liver of a type 2 diabetes mouse [121].What's more, by the way of examining the effects of *A. senticosus* extract on the microarray-based expression profiled using an Affymatrix Gene Chip Rat genome 230 2.0 Array, it's found that *A.senticosus* extract acted as a strong antioxidant in addition to exerting anti-HES (heat environmental stress) effects [122]. Furthermore, *A. senticosus* saponins (such as *Acanthopanax* senticosides B) protected cardiomyocytes against oxidative-stress injury induced by $H_2O_2$ through reduction of lipid peroxidation and enhancement of the activity of antioxidant defense [46]. Besides, *A. senticosus* supplementations have beneficial effects against oxidative stress and improve serum lipid profiles without subsequent side effects in postmenopausal women [123].

### 3.5. Immunomodulatory Effects

*A. gracilistylus* extract (AGE) markedly suppressed the proliferative responses of human peripheral blood lymphocytes stimulated with mitogens Con A (concanavalin A) and SAC (Staphylococcus aureus Cowan I).B cell and T cell, production of interferon(IFN)-γ and immunoglobulin, all were suppressed by AGE. They also suppressed the alloantigen-specific cytotoxic T lymphocyte response, and enhanced monocyte function to produce cytokines. These results suggested that AGE has an immunomodulating activity on human lymphocytes and their properties could be clinically applied in the treatment of several diseases such as autoimmune and allergic diseases [124]. The immunomodulatory effects of *A. koreanum* were found to be due to the lupane-triterpene glycosides in the leaves and stems of this plant. Moreover, the genuine immune stimulating effects were assessed by IL-2 and IFN-γ release in splenocytes treated with eleven lupane-triterpene glycosides,amongwhich acankoreoside A, acankoreoside D, acantrifoside A and acankoreoside L significantly increased IL-2 and IFN-γ release in spleen cells [17]. Furthermore, another study demonstrated that *A. koreanum* directly activates B cells, but not T cells, and that *A. koreanum* has a broader receptor profile than LPS in B cells [62].

On the other hand, polysaccharide, isolated from a cell culture of *A. senticosus*, activates B cells and macrophages by interacting with toll-like receptors (TLRs) and leading to the subsequent activation of mitogen-activated protein kinases and NF-κB [125].

Further studies on biopolymers showed thatthe AS-3 and AS-5, biopolymers from A. sessiliflorus, possessed immunomodulatory activities due to the presences of glucose (1→3), (1→4) linkages, arabinogalactan and rhamnogalactan [73].

## 3.6. Effects on Diabetes and Obesity

The addition of *A. senticosus* extract(ASE) inhibited α-glucosidase activity and 2'-deoxy-D-glucose(DG) uptake in human intestinal Caco-2 cells. Further kinetic analysis indicated that ASE inhibited intracellular DG uptake chiefly by inhibiting transport via a glucose transporter. In addition, ASE oral administration significantly inhibited α-glucosidase activity in the small intestine mucosa extirpated from the mice [126]. Moreover, eight compounds that might bind toα-glucosidase inhibitory activities were screened from the leaves of *A. senticosus* Harms, and seven of them were successfully identified [127].

These findings manifested that ASE may be useful as an ingredient of functional foods to improve postprandial hyperglycemia and prevent type II diabetes mellitus, and the leaves of *A. senticosus Harms* may have potential anti-diabetic application. Furthermore, oral administration of ASE seemed to lower the weight gain, serum LDL-cholesterol concentration, and liver triglycerides accumulation in mice with obesity induced by high fat diets [128].

Additionally, oral administration of the aqueous extract from *A. senticosus* root had the ability to improve insulin sensitivity and delay the development of insulin resistance in rats [129]. Thus, ASE may be used as an adjuvant therapy for patients with insulin resistance and may have a role in the prevention ofdiabetes and obesity. Furthermore, *A. senticosus* stem bark ethanol extract was demonstrated to be an effective treatment for insulin resistance, non-alcoholic fatty liver and hepatic steatosis in ob/ob mice by decreasing hepatic lipid synthesis [130].

On the other hand, by using a pancreatic lipase-inhibiting assay system, the crude saponin fraction isolated from the fruits of *A.senticosus* showed inhibitory activity on pancreatic lipase, which is a key enzyme in lipid digestion.

Among the isolated compounds, copteroside B, silphioside F, gypsogenin 3-O-β-D-glucuronopyranoside and hederagenin 3-O-β-D-glucuronopyranoside 6'-O-methyl ester showed inhibitory activity toward pancreatic lipase, and the free carboxylic acid groups in position 28 within their chemical structures were required for enhancement of pancreatic lipase inhibition [16]. In another study, fifteen triterpenoid saponins isolated from the leaves of *A. senticosus* were examined against pancreatic lipase. As a result, Ciwujianoside $C_1$, sessiloside, tauroside $H_1$, acanthopanaxoside C, chiisanoside and 3-O-α-rhamnopyranosyl-(1→2)-α-arabinopyranosyl mesembryanthemoidigenic acid also inhibited pancreatic lipase activity *in vitro*. In turn, ciwujianosides $C_2$, $C_3$, $C_4$, $D_2$ and hederasaponin B enhanced this enzyme [52].The $CH_2Cl_2$-soluble extractions from the roots of *A. koreanum*and their active fractions (such as acanthoic acid, ent-kaur-16-en-19-oic acid and 6αH,17-isovaleryloxy-ent-kauran-19-oic acid) were found to inhibit protein tyrosine phosphatase 1B (PTP1B) activity, and the inhibition of PTP1B has been proposed to treat type 2 diabetes and obesity [131].

Chiisanoside and sessiloside, the main components of *A. sessiliflorus* leaves, had the potential to treat patients with obesity as a lipase inhibitor which suppresses fat absorption *in vivo* [74, 132].

## 3.7. Hepatoprotective Effects

The release of pro-inflammatory cytokines in both chronic (such as high mobility group box 1 protein (HMGB1)) and acute (such as IL-1β and TNF-α) phases, is thought to play critical roles in the development of fulminant hepatitis. It is generally believed that compounds display hepatoprotective effects by acting as an anti-inflammatory agent to inhibit the production of cytokines, such as TNF-α, IFN-γ, IL-2, etc.. The *A. koreanum* extract was reported to protect against D-galactosamine- or LPS-induced fulminant hepatitis by inhibiting the systemic release of TNF-α [28, 133]. Moreover, acankoreanogenin A(AA) which was extracted from the leaves of the *A. gracilistylus* showed its inhibiting effects on TNF-β, IL-1α and HMGB1 release *in vitro*.

On the other hand, AA inhibited the systemic release of HMGB1 and dose-dependently, rescued the mice from lethal GalN/LPS-induced fulminant hepatitis *in vivo*, which indicates this component can be used as a candidate therapy for fulminant hepatitis [14]. In addition, the diterpenoid acanthoic acid, a major constituent in Araliaceae, was reported to suppress liver fibrosis and nodular formation through reducing TNF-αand IL-1production [134], and had a protective effect against hepatotoxicity induced by t-BHP or carbon tetrachloride *in vitro* and *in vivo* [135]. What's more, it's found that acanthoic acid protected liver tissue from the oxidative stress caused by APAP-induced liver damage, which indicates that the hepatic protection mechanism of acanthoic acid would relate to antioxidation and hypoxia factor on APAP-induced hepatotoxicity [136]. Furthermore, *A. giraldii* polysaccharides significantly inhibited the elevation of serum ALT in $CCl_4$-, D-galactosamine- and thioacetamide-intoxicated rats and mice [137]. Glycoprotein(GF-AS) isolated from the stem bark of *A. senticosus* could alleviate alcohol-induced hepatotoxicity through increasing ethanol and lipid metabolism, as well as antioxidant defense systems in the treatment of livers injured by acute- and chronic-alcoholism [138].

## 3.8. Brain and Nerve Protection

Treatment with *A. senticosus* resulted in prophylactic effects on 1-methyl-4- phenyl-1,2,3,6-tetrahydropyridine (MPTP)-induced Parkinsonian bradykinesia and catalepsy and cytoprotective effects against MPTP-induced loss of dopamine (DA) cells. Further study demonstrated that sesamin from *A. senticosus* have the same effects. These results suggested that it may be possible to use *A. senticosus* and sesamin for the prevention of nigral degenerative disorders, e.g., Parkinson's disease with depression, caused by exposure to toxic substances [139, 140]. Another study found that *A. senticosus* may act by regulating noradrenaline (NA) and DA levels in specific brain regions related to stress response and Parkinson's disease [141]. Besides, after peritoneal injection of *A. senticosus* water extracts, the brain infarct volume of rats was markedly reduced compared with the control, and *A. senticosus* water extracts significantly inhibited both cyclooxygenase-2 and OX-42 expressions in the penumbral regionafter middle cerebral artery occlusion. These results suggested that *A. senticosus* has a neuroprotective effect by inhibiting inflammation and microglial activation in brain ischaemia [142].

*A. senticosus* could prevent the reduction of rats learning and memory ability induced by sleep deprivation, and the mechanism may involve regulating the sleep deprivation caused by disorders of the monoamine neurotransmitter, accelerating the metabolic conversion of 5-HT, regulating norepinephrine(NE) and acetylcholinesterase (AchE) activity in the hippocampus, and improvement of the synaptic plasticity [143,144]. Based on enhancement of GABA (Gamma-amino butyric acid) and nigra-striatum TH (tyrosine hydroxylase) expression derived from the effective parts of *A. senticosus*, Yang *et al.* speculated that the effective parts could protect brain injuries in micesuffering from Parkinson's disease induced by MPTP [145]. In addition, an *A. senticosus* pill could increase the expression of TH and TPH (tryptophan hydroxylase) in the hippocampus of depressed rats [146]. What's more, *A. senticousus* saponins reversed LTP depression in epileptic rats induced by penicillin, and promoted remarkably LTP in hippocampal CA1 in rats, which indicates that it improves cerebral learning and memory [147,148].

## 3.9. Cardioprotective Effects

An *A. senticosus* injection (CASI) has protective effects on myocardial ischemia-reperfusion arrhythmia, which may be related to scavenging the oxygen free radicals and $Ca^{2+}$ overload which formed during reperfusion [27]. Pretreatment with *A. senticosus* saponins could induce tolerance againstischemia in PC12 cell lines. The elevated expression and increased DNA binding activity of hypoxia-inducing factor (HIF)-1α, and the overexpression of its downstream target erythropoietin (EPO), may be the molecular mechanism [18]. Moreover, *A. senticosus* saponins have protective effects on myocardial ischemia-reperfusion injury due to the function of improving free radicals and myocardial metabolism, decreasing plasma ET(endothelin), Ang II (angiotensin II) and TXA2(thromboxane A2) levels and increasing plasma PGI2 (prostacycline) level and PGI2/TXA2ratio etc. [149].

## 3.10. Effects on Bone Metabolism

The compounds isolated from the methanol extract of the leaves of *A. koreanum* significantly increased the growth of osteoblastic MC3T3-E1 cells and caused a significant elevation of osteoblastic cell differentiation. These results suggested that the compounds have a direct stimulatory effect on bone formation *in vitro* and may contribute to the prevention of osteoporosis [10]. What's more, *A. senticosus* extract supplementation may have beneficial effects on bone remodeling in Korean postmenopausal women and has no significant adverse events [150].

### 3.11. Antiplatelet Aggregation Activities

*A. gracilistylus* saponins inhibited human platelet aggregation, platelet factor 4(PF4) liberation, and thrombosis *in vitro*, suggesting its possible antithrombotic action in man [151]. Moreover, chiisanogenin, isolated from the leaves of *A.senticosus* and the fruits of *A. sessiliflorus*, showed about a 50-fold higher potency than acetylsalicylic acid(ASA) on U46619-induced platelet aggregation and 10~20-fold higher effects than ASA on epinephrine- and arachidonic acid(AA)-induced aggregation, respectively [33]. In addition, three triterpenes from *A. sessiliflorus* fruits also showed antiplatelet aggregation activity on ADP-induced platelet aggregation [76].

### 3.12. Anti-Cytotoxic Activities

Lupane-triterpene glycosides, acankoreoside A, acankoreoside D, acankoreoside I, acankoreoside F, acankoreosides J-L, and acantrifoside A, were isolated from the leaves of *A.koreanum*. The cytotoxic activities of these compounds were evaluated with four cancer cells such as A-549 (lung), HL-60 (leukemia), MCF-7 (breast), and U-937 (leukemia). Among them, acankoreoside I showed a growth inhibitory effect in A-549, HL-60, and MCF-7 cells, whereas it was less active in the U937 cells, and acankoreosides J-L showed moderate activity against four cells [152, 153]. In addition, the *in vitro* cytotoxicity of chiisanoside, isolated from leaves of *A. divaricatus* var. albeofructus, was inferior to that of its metabolite chiisanogenin. However, the anti-rotaviral activity of chiisanoside was more potent than that of chiisanogenin [82]. Besides, *A. senticosus* extract protected mice against LPS/D-GalN-induced endotoxic shock involving inhibition of NF-κB activation, which caused down-regulation of TNF-α and involved up-regulation of IL-10[154]. Moreover, *A. senticosus* injection antagonized the gentamicin ototoxicity by inhibiting the expression of caspases-3 [149].

### 3.13. Antibacterial Activities

The antimicrobial activities of the essential oils of *A. brachypus* were evaluated against 11 microorganisms using agar disc diffusion and broth microdilution methods. As a result, gram-positive bacteria and gram-negative bacteriawere more sensitive to oils than yeasts, and the result will provide some scientific references for the further pharmacological research of *A. brachypus* [5, 67]. Additionally, chiisanogenin isolated from the leaves of *A. senticosus* revealed broad but moderate antibacterial activities against gram-positive and gram-negative bacteria [156].

### 3.14. Anti-Mutagenic Activities

*A. divaricatus* var. albeofructus(ADA) extracts were not mutagenic for Salmonella typhimurium strains TA 98, TA 100, TA1535, TA1537, and Escherichia coli by the Ames assay, but they showed a strong anti-mutagenic activity against 2-(2-furyl)-3-(5-nitro-2-furyl) acrylamide(AF-2), 2-Aminoanthrancene (2-AA)and sodium azide-induced mutagenesis[157]. The anti-mutagenic property of ADA provides additional health supplemental value to the other claimed therapeutic properties of the plant.

### 3.15. Anti-Anaphylactic Activities

Mast cells synthesize and secrete chemical mediators play a central role in anaphylaxis. *A. senticosus* stem inhibited passive cutaneous anaphylaxis reaction and histamine release from mast cells in a dose-dependent manner. Moreover, they had an inhibitory effect on anti-dinitrophenyl IgE-induced TNF-α production from mast cells [158]. It's also been discovered that *A. senticosus* root may possess effective anti-anaphylactic activity [159].

### 3.16. Other Activities

Yoo *et al.* collectively demonstrated that savinin, helioxanthin and 3-(3",4"-dimethoxybenzyl)-2-(3',4'-methylenedioxybenzyl)butyrolactone from *A. chiisanensis* inactivated the CYP3A4 enzyme in a mechanism-based mode [160]. In addition, chiisanoside and chiisanogenin from *A. chiisanensis* leaves significantly reduced rheumatoid arthritis(RA) and C-reactive protein factors in the Freund's complete adjuvant reagent-induced rat, and they also inhibited xanthine oxidase activity and increased superoxide dismutase(SOD), glutathione peroxidase and catalase, indicating that both compounds scavenged reactive oxygen species(ROS) [161]. On the other hand, savinin isolated from the root of *A. chiisanensis* exhibited anti-estrogenic activity in Ishikawa cells [70].

Pharmacological experiments have shown that nevadensin, which is isolated from the leaves of *A. trifoliatus,* possess expectorant and antitussive actions [162].

*A. giraldii* has pronounced inhibitory effects on the central nervous system(CNS). It prolonged the hypnotic time of pentobarbital sodium, reduced the spontaneous activity and antagonized amphetamine-induced hyperactivity [163].

To explore the mechanism of *A. senticosus* on diabetic nephropathy(DN) patients, Ni *et al.* observed the effect of *A. senticosus* injection on urinary albumin excretion (UAE) and renal endothelin(ET) metabolism. As a result, the protective effect on DN is probably correlated with its inhibition on ET synthesis in the kidney [164]. Besides, *A. senticosus* extract suppressed intestinal drug transporter(P-gp) and peptide transporter activities in a non-competitive manner [165], and exerted a protective effect against ethanol (EtOH)-induced apoptosis of human neuroblastoma cells [166].

## Conclusion

In summary, scientists have conducted abundant research on the biology, chemistry, pharmacologyand clinical applications of the *Acanthopanax* genus in the last decades, and have made great achievements. Phytochemical investigations on the *Acanthopanax* genus have resulted in the isolation of terpenoids, lignans, and flavonoids, etc. Extensive pharmacological investigations of these plants reported anti-inflammatory, antioxidant, immunostimulatory, and anti-diabetic activities. The root and stem bark have effects that calm the heart and nerves, replenish qi to invigorate the spleen, activate blood, as well as dispel melancholy, and so on. Due to rich resourcesand extensive applications in the field of medicine, the *Acanthopanax* genus could be widely used as an alternative medicine. Thus, it can be further researched as to the following aspects: (1)The genetic relationship and biological characteristics of the *Acanthopanax* species should be determined by molecular biology technology in order to define their relationship and further develop these resources; (2)In order to expand the other medicinal parts, the chemical compositions and pharmacological effects of the plants commonly applied such as *A. gracilistylus* and *A. senticosus,* should be researched; (3)Based onthe findings of the *Acanthopanax* species, researchers can clarify the structure-activity relationship and material foundation to develop more and more over-the-counter drugs and functional foods, such as health supplements; (4)The plants(*A. gracilistylus, A. senticosus*, etc.) with clear pharmacological effects should be artificially cultivated to ensure the sustainable utilization of resources. Recently, the *Acanthopanax* genus has attracted significant interest amongscientists, who have not only isolated chemical components but have also evaluated pharmacological activities and clinical applications. However, the mechanisms of some of these effects are not yet clear. In future studies, more detailed works about these plants with regard to their bioactive constituents and their mechanisms of action are still needed.

## References


[1] Yook, C.S., *Medicinal Herbs of Acanthopanax in Asia*, Kyungwon Media, Seoul (2001).

[2] Kim, C.H. and B.Y. Sun, *New Taxa and Combinations in Eleutherococcus (Araliaceae) from Eastern Asia*, Kyungwon Press, Seoul (2000).

[3] Pharmacopoeia Commission of People's Republic of China, *Pharmacopoeia of the People's Republic of China*, Medical Science Press of China, Beijing (2010).

[4] Huang L, Zhao H, Huang B, Zheng C, Peng W, Qin L, *Pharmazie*. 66, 83-97 (2010).

[5] Hu HB, Zheng XD, Hu HS, Li Y, *Arch. Pharm. Res*. 32, 699-710(2009).

[6] Kim JA, Yang SY, Koo JE, Koh YS, Kim YH, *Bioorg. Med. Chem. Lett*. 20, 6703-6707, (2010).

[7] Hu H.B. and Zhu J.H., *Chem. Pharm. Bull*. 59, 135-139 (2011).

[8] Zhang JY, Pu SB, Qian SH, Liu D, Wang KC, *Chin. Med. Mat*. 34, 226-229 (2011).

[9] Kim JA, Song SB, Yang SY, Kim YH, *Nat. Prod. Commun*. 6, 1233-1236 (2011).

[10] Park SH, Nhiem NX, Kiem PV, Choi EM, Kim JA, Kim YH, *Arch. Pharm. Res*. 33, 75-80 (2010).

[11] Huang D.B., Ran R.Z., and Yu Z.F., *Chin. J. Chin. Med. Mat*. 30, 621-624 (2005).

[12] Nhiem NX, Kiem PV, Minh CV, Tai BH, Quang TH, Soung KS, Koo JE, Koh YS, Kim YH, *Arch. Pharm. Res*. 34, 1593-1598 (2011).

[13] Huang LZ, Huang BK, Liang J, Zheng CJ, Han T, Zhang QY, Qin LP, *Phytother. Res*. 25, 940-943 (2011).

[14] Zhang BX, Li N, Zhang ZP, Liu HB, Zhou RR, Zhong BY, Zou MX, Dai XH, Xiao MF, Liu XQ, Fan XG, *Int. Immunopharmacol.* 11, 1018-1023 (2011).

[15] Nhiem NX, Kim KC, Kim AD, Hyun JW, Kang HK, Van Kiem P, Van Minh C, Thu VK, Tai BH, Kim JA, Kim YH, *J. Asian Nat. Prod. Res.* 13, 56-61 (2011).

[16] Li F, Li W, Fu H, Zhang Q, Koike K, *Chem. Pharm. Bull.* 55, 1087-1089 (2007).

[17] Nhiem NX, Kiem PV, Minh CV, Tai BH, Tung NH, Hado T, Soung KS, Kim JH, Ahn JY, Lee YM, Kim YH, *Bioorg. Med. Chem. Lett.* 20, 4927-4931 (2010).

[18] Chen J., Zhu L. and Pan Y.J., *Chin. J. Pediatr.* 45, 138-142 (2007).

[19] Bai Y, Tohda C, Zhu S, Hattori M, Komatsu K, *J. Nat. Med.*65, 417-423 (2011).

[20] Deng H.Z., S.Y. Sun, and S.Y. Li, *Chin. J. Chin. Mater. Med.* 19, 38-40, 63 (1994).

[21] Li W, Liu M, Feng S, Wu B, Zhang S, Yang W and Liu GJ, *Cochrane Db. Syst. Rev.* 2009(3): CD007032.

[22] Spelman K, Burns J, Nichols D, Winters N, Ottersberg S, Tenborg M, *Altern. Med. Rev.* 11, 128-150 (2006).

[23] Park S.Y., *Studies on RAPD Analysis and Triterpenoidal Constituents of Acanthopanax Species,* Kumamoto University Press, Japan (2002).

[24] Kim H.J., "Effect of *A.senticosus* on longitudinal bone growth in adolescent male rats", Organized by PSK, Seoul (2002).

[25] Yi J.M., M.S. Kim, and S.W. Seo, *Clin. Chim. Acta.* 312, 163-168 (2001).

[26] Huang LZ, Huang BK, Ye Q, Qin LP, *J Ethnopharmacol.* 133, 213-219(2011).

[27] Wang L, Yu XF, Qu SC, Xu HL, Sui DY, *Chin. J. Chin. Mate. Med.* 32, 2174-2177 (2007).

[28] Nan JX, Park EJ, Nam JB, Zhao YZ, Cai XF, Kim YH, Sohn DH, Lee JJ. J. *Ethnopharmacol.* 92, 71-77 (2004).

[29] Bensky D. and Gamble A., *Materia Medica*, Eastland press, Washington (1986).

[30] Zheng WJ, *Record Chinese Woody Plants,*Chinese Forestry Press, Beijing (1985).

[31] Guo X.Q. Gansu, *Agric. Sci. Technol.* 44, 2004 (2004).

[32] Liu L.P., *Ziwuling Woody Flora*, Lanzhou University Press, 215-216 (1997).

[33] Jin JL, Lee S, Lee YY, Kim JM, Heo JE, Yun-Choi HS, *Planta. Med.* 70, 564-566 (2004).

[34] Chen M, Song F, Guo M, Liu Z, Liu S. *Rapid Commun.Mass Spectrom.* 16, 264-271 (2002).

[35] Li Q, Jia Y, Xu L, Wang X, Shen Z, Liu Y, Bi K, *Biol.Pharm. Bull.* 29, 532-534 (2006).

[36] Liu X.H., S.Y. Chang, and C.S. Yook, *J. Lanzhou Univ.(Nat. Sci.)*, 42, 86-91 (2006).

[37] Liu XQ, Chang SY, Park SY, Nohara T, Yook CS, *Arch. Pharm. Res.* 25, 831-836 (2002).

[38] Yook CS, Liu XQ, Chang SY, Park SY, Nohara T, *Chem. Pharm. Bull.* 50, 1383-1385 (2002).

[39] Zhang JY, Pu SB, Qian SH and Liu D. *Chin. J. Nat. Med.* 9, 105-107 (2011).

[40] Liu X.Q., Yook C.S., and Chang S., *Chin. Tradit. Herbal Drugs,* 35, 250-252 (2004).

[41] Liu X., Kyung Hee University, Korea (2003).

[42] Ni L., Research on the active ingredients from the leaves of *Aeanthopanax gracilistylus* W. W. Smith and other *Acanthopanax Miq.* Plants, Central South University, Changsha (2008).

[43] Tang X., Ma Y., and Li P., *Chin. J. Chin. Mater. Med.* 20, 231, 253 (1995).

[44] Xian, L.N., S.H. Qian, and Z.L. Li, *Chin. Med. Mat.* 33,538-542 (2010).
[45] An SY, Qian SH, Jiang JQ, Wang KC, *Chin. Tradit. Herbal Drugs,* 40, 1529- 1534 (2009).
[46] Liang Q, Yu X, Qu S, Xu H, Sui D, *Eur. J. Pharmacol.* 627, 209-215 (2010).
[47] Liu CT, Hou HG, Fan NY, Li HH, *Spectrosc. Spect. Anal.* 30, 1123-1125 (2010).
[48] Yamazaki T. and Tokiwa T., *Biol. Pharm. Bull.* 33, 1716-1722 (2010).
[49] Yamazaki T, Shimosaka S, Sasaki H, Matsumura T, Tukiyama T, Tokiwa T, *Toxicol. In Vitro.* 21, 1530-1537 (2007).
[50] Ryu J, Son D, Kang J, Kim HS, Kim BK, Lee S, *Arch. Pharm. Res.*27, 912-914 (2004).
[51] Yu W, Zhang H, Huang W, Chen J, Liang X. *Chin. J. Chromatog.* 23, 196-201 (2005).
[52] Jiang W, Li W, Han L, Liu L, Zhang Q, Zhang S, Nikaido T, Koike K. *J. Nat. Prod.* 69, 1577-1581 (2006).
[53] Cao JG, YG Zu, FJ Yang, ZH Zhao. *Chin. J. Appl. Ecol.* 16, 1007-1010 (2005).
[54] Park SY, Chang SY, Yook CS, Nohara T, *J. Nat. Prod.* 63, 1630-1633 (2000).
[55] Yan ZW, Liu JP, Lu D, Groundwater P, He CY, Li PY, *Nat. Prod. Res.* 24, 1523-1527 (2010).
[56] Li ZF, Xu N, Feng BM, Zhang QH, Pei YH, *J. Asian Nat. Prod. Res.* 12, 51-55 (2010).
[57] Li ZF, Wu ZH, Chen G, Zhang QH, Pei YH, *J. Asian Nat. Prod. Res.* 11, 716-719 (2009).
[58] Cai XF, Lee IS, Shen G, Dat NT, Lee JJ, Kim YH, *Arch. Pharm. Res.* 27, 825-828 (2004).
[59] Cai XF, Lee IS, Dat NT, Shen G, Kim YH., *Phytother. Res.* 18, 677-680 (2004).
[60] Cai XF, Lee IS, Dat NT, Shen G, Kang JS, Kim DH, Kim YH, *Arch. Pharm. Res.* 27, 738-741 (2004).
[61] Cai XF, Shen G, Dat NT, Kang OH, Kim JA, Lee YM, Lee JJ, Kim YH, *Chem. Pharm. Bull.* 51, 605-607 (2003).
[62] Han SB, Park SK, Ahn HJ, Yoon YD, Kim YH, Lee JJ, Lee KH, Moon JS, Kim HC, Kim HM, *Int. Immunopharmacol.* 3, 683-691 (2003).
[63] Choi HS, Kim HJ, Nam SG, Kim IS, Lee KT, Yook CS, Lee YS, *Chem. Pharm. Bull.* 56, 1613-1616 (2008).
[64] Park SY, Choi HS, Yook CS, Nohara T. *Chem. Pharm. Bull.* 53, 97-99 (2005).
[65] Yang EJ, Moon JY, Lee JS, Koh J, Lee NH, Hyun CG, *J. Biomed. Biotechno.* 2010, 715-739 (2010).
[66] Hu HB, Zheng XD, Cao H, Guo XQ, Hu HS, *Chem. Pharm. Bull.* 57, 1000-1003 (2009).
[67] Hu, H.S., Hu H.B., and Zheng X.D., *Chin. Med. Ma. 32*, 67-70 (2009).
[68] Hong J, Shin KH, Lim SS, Kwak JH, Zee O, Ishihara K, Hirasawa N, Seyama T, Ohuchi K, *Inflamm. Allergy Drug Targ.* 7, 195-202 (2008).
[69] Lee S, Ban HS, Kim YP, Kim BK, Cho SH, Ohuchi K, Shin KH, *Phytother. Res.* 19, 103-106 (2005).
[70] Lee S, Yoo HH, Piao XL, Kim JS, Kang SS, Shin KH, *Arch. Pharm. Res.* 28, 186-189 (2005).
[71] Won JH, Park SY, Nam SG, Park HJ, Choi JW, Lee KT, *Biol. Pharm. Bull.* 28, 1919-1924 (2005).
[72] Jeong SC, Yang BK, Jeong YT, Rao KS, Song CH, *J. Microbiol. Biotechnol.* 17, 21-28 (2007).

[73] Jeong SC, Jeong YT, Yang BK, Song CH, *J. Biochem. Mol. Biol.* 39, 84-90 (2006).

[74] Yoshizumi K, Hirano K, Ando H, Hirai Y, Ida Y, Tsuji T, Tanaka T, Satouchi K, Terao J, *J. Agric. Food. Chem.* 54, 335-341(2006).

[75] Yang C, An Q, Song Y, Xiong Z, Li F, *Chin. J. Chin. Mat. Med.* 34, 715-717 (2009).

[76] C Yang, Q An, Z Xiong, Y Song, K Yu, F Li, *Planta Med.* 75, 656-659 (2009).

[77] Lee S, Ji J, Shin KH, Kim BK, *Planta Med.* 68, 39-941 (2002).

[78] Wang Z., Zhang L., and Sun Y., *J. Chromatogr. Sci.* 43, 249-252 (2005).

[79] Zhang LX and Liu H, *Chin. J. Basic Med.* 7, 24-27(2001).

[80] kong, D.Y. and S.Q. LUO, *Chin. Pharm. J.* 21, 202-204 (1989).

[81] Wang ZW, Zhang QH, Zhang LX, Sun YQ, *Chin. J. Pharm. Anal.* 25, 765-768 (2005).

[82] Bae EA, Yook CS, Oh OJ, Chang SY, Nohara T, Kim DH, *Biol. Pharm. Bull.* 24, 582-585 (2001).

[83] Oh OJ, Chang SY, Yook CS, Yang KS, Park SY, Nohara T, *Chem. Pharm. Bull.* 48, 879-881 (2000).

[84] Kiem PV, Cai XF, Minh CV, Lee JJ, Kim YH, *Planta. Med.* 70, 282-284 (2004).

[85] Kiem PV, Chau VM, Dat NT, Cai XF, Lee JJ and Kim YH, *Arch. Pharm. Res.* 26, 1014- 1017 ( 2003).

[86] Kiem PV, Cai XF, Minh CV, Lee JJ, Kim YH, *Chem. Pharm. Bull* (Tokyo). 51, 1432-1435 (2003).

[87] Kiem PV, Minh CV, Cai XF, Lee JJ, Kim YH, *Arch. Pharm. Res.* 26,706-708(2003).

[88] Yook CS, Chang SY, Lai JH, Ko SK, Jeong JH, Nohara T, *Arch. Pharm. Res.* 22, 629-632 (1999).

[89] Park S.Y., C.S. Yook, and T. Nohara, *Chem. Pharm. Bull.* 53, 1147-1151 (2005).

[90] Park SY, Chang SY, Oh OJ, Yook CS, Nohara T, *Phytochemistry.* 59, 379-384 (2002).

[91] Miyakoshi M, Shirasuna K, Hirai Y, Shingu K, Isoda S, Shoji J, Ida Y, Shimizu T. *J. Nat. Prod.* 62, 445-448 (1999).

[92] Miyakoshi M, Ida Y, Isoda S, Shoji J, *Phytochemistr*y, 34, 1599-1602 (1993).

[93] Kim JA, Kim DK, Jin Tae, Kang OH, Choi YA, Choi SC, Kim TH, Nah YH, Choi SJ, KimYH, Bae KH, Lee YM. *Clin. Chim. Acta.* 342, 193-202 (2004).

[94] Kang HS, Song HK, Lee JJ, Pyun KH and Choi I, *Mediators Inflamm. 7,* 257-259 (1998).

[95] Kang OH, Kim DK, Cai XF, Kim YH, Lee YM, *Arch. Pharm. Res.* 33, 87-93 (2010).

[96] Kang OH, Choi YA, Park HJ, Kang CS, Song BS, Choi SC, Nah YH, Yun KJ, Cai XF, Kim YH, Bae KH, Lee YM, *J. Ethnopharmacol.* 105, 326-331 (2006).

[97] Kim JA,Yang SY, Song SB, Kim YH. *Arch. Pharm. Res.* 34, 1347-1351 (2011).

[98] Cai XF, Shen G, Dat NT, Kang OH, Lee YM, Lee JJ, Kim YH, *Arch. Pharm.* Res. 26, 731-734 (2003).

[99] Lin QY, Jin LJ, Cao ZH, Xu YP, *J. Ethnopharmacol.* 118, 231-236 (2008).

[100] Lin QY, Jin LJ, Ma YS, Shi M, Xu YP, *Phytother. Res.* 21, 879-883 (2007).

[101] QY Lin, LJ Jin, ZH Cao, YN Lu, HY Xue, YP Xu, *Phytother. Res.* 22, 740-745 (2008).

[102] Jung HJ, Park HJ, Kim RG, Shin KM, Ha J, Choi JW, Kim HJ, Lee YS, Lee KT, *Planta Med.* 69, 610-616 (2003).

[103] Lee S, Park HS, Notsu Y, Ban HS, Kim YP, Ishihara K, Hirasawa N, Jung SH, Lee YS, Lim SS, Park EH, Shin KH, Seyama T, Hong J, Ohuchi K. *Phytother. Res.* 22, 1552-1556 (2008).

[104] Lee S, Jung SH, Lee YS, Yamada M, Kim BK, Ohuchi K, Shin KH, *Arch. Pharm. Res.* 27, 628-632 (2004).

[105] Yuan F, Chen J, Xu GM, Zheng JJ, Long QC, *Chin. J. Integr. Med.* 15, 210-215(2009).

[106] Hibasami H, Fujikawa T, Takeda H, Nishibe S, Satoh T, Fujisawa T, Nakashima K, *Oncol. Rep.* 7, 1213-1216 (2000).

[107] Ha ES, Hwang SH, Shin KS, Yu KW, Lee KH, Choi JS, Park WM, Yoon TJ, *Arch. Pharm. Res.* 27, 217-224 (2004).

[108] Yoon TJ, Yoo YC, Lee SW, Shin KS, Choi WH, Hwang SH, Ha ES, Jo SK, Kim SH, Park WM. *J. Ethnopharmacol.* 93, 247-253 (2004).

[109] Shan BE, Zeki K, Sugiura T, Yoshida Y and Yamashita U, *Jpn. J. Cancer Res.* 91, 383-389 (2000).

[110] Shan BE, Fu XM, Hua ZX, Li Q, Liang W, Liu J, Zhang H, Liu G, *Chin. J. Inte. Trad. West Med.* 25, 825-828 (2005).

[111] Lu X, Su M, Li Y, Zeng L, Liu X, Li J, Zheng B, Wang S, *Chin. Med. J (Engl).* 115, 716-721 (2002).

[112] Wang JZ, Tsumura H, Shimura K, Ito H, *Cancer.Lett.* 65, 79-84 (1992).

[113] Lü X, Zeng L, Cai Y, Li Y, Liu X, *J. West Chin. Univ. Med. Sci.* 33, 265-266 (2002).

[114] Li C, Wang XY, Hu XW, Fang HT, Qiao SY, *Chin. J. Chin. Mater. Med.* 33, 2800-2802 (2008).

[115] Zhang XL, Ren F, Huang W, Ding RT, Zhou QS and Liu XW, *Molecules.* 16, 28-37 (2010).

[116] Wiegant FA, Surinova S, Ytsma E, Wikman G, Post JA, *Biogerontology.* 10, 27-42 (2009).

[117] Rhim YT, Kim H, Yoon SJ, Kim SS, Chang HK, Lee TH, Lee HH, Shin MC, Shin MS, Kim CJ, *J. Ethnopharmacol.* 114, 38-43 (2007).

[118] Sithisarn P. and Jarikasem S. *Med. Princ. Pract.* 18,393-398 (2009).

[119] Wang X, Hai CX, Liang XS, Yu X, Zhang W, Li YL, *J. Ethnopharmacol.* 127, 424-432 (2010).

[120] Lee S, Son D, Ryu J, Lee YS, Jung SH, Kang J, Lee SY, Kim HS, Shin KH, *Arch. Pharm. Res.* 27, 106-110 (2004).

[121] Hong, J.H., Y.S. Cha, and S.J. Rhee, *J. Clin. Biochem. Nutr.* 45, 101-109 (2009).

[122] Kim KJ, Hong HD, Lee OH, Lee BY, *Toxicology.* 278, 217-223 (2010).

[123] Lee YJ, Chung HY, Kwak HK, Yoon S, *Biochem. Biophy. Res.Commun.* 375, 44-48 (2008).

[124] Shan BE, Yoshita Y, Sugiura T, Yamashita U, *Clin. Exp. Immunol.* 118, 41-48 (1999).

[125] Han SB, Yoon YD, Ahn HJ, Lee HS, Lee CW, Yoon WK, Park SK, *Int. Immunopharmacol.* 3, 1301-1312 (2003).

[126] Watanabe K, Kamata K, Sato J, Takahashi T, *J. Ethnopharmacol.* 132, 193-199 (2010).

[127] H Zhou, JP Xing, S Liu, FR Song, ZW Cai, ZF Pi, ZQ Liu, SY Liu. *Phytochem. Anal.*(2011).

[128] Cha Y.S., Rhee S.J., and Heo Y.R., *J. Med. Food.* 7, 422-429 (2004).

[129] Liu TP, Lee CS, Liou SS, Liu IM, Cheng JT, *Clin. Exp. Pharmacol. Physiol.* 32, 649-654 (2005).

[130] Park SH, Lee SG, Kang SK, Chung SH, *Arch. Pharm. Res.* 29, 768-776 (2006).

[131] Na M, Oh WK, Kim YH, Cai XF, Kim S, Kim BY, Ahn JS, *Bioorg. Med.Chem. Lett.* 16, 3061-3164 (2006).

[132] Yoshizumi K, Murota K, Watanabe S, Tomi H, Tsuji T, Terao J, *Biosci. Biotechnol. Biochem.* 72, 1126-1129 (2008).

[133] JX Nan, XJ Jin, LH Lian, XF Cai, YZ Jiang, HR Jin, Jung Joon Lee, *Biol. Pharm. Bull.*31,738-742 (2008).

[134] Kang HS, Kim YH, Lee CS, Lee JJ, Choi I, Pyun KH, *Cell Immunol.* 170, 212-221 (1996).

[135] Park EJ, Zhao YZ, Kim YH, Lee JJ, Sohn DH, *Planta Med.* 70, 321-327 (2004).

[136] Wu YL, Jiang YZ, Jin XJ, Lian LH, Piao JY,Wan Y, Jin HR, Lee J,Nan JX, *Phytomedicine.* 17, 475-479 (2010).

[137] Dang Y, *Chin.J.Chin. Mater. Med.*22, 176-178 (1997).

[138] Choi JS, Yoon TJ, Kang KR, Lee KH, Kim WH, Suh YH, Song J, Jung MH, *Biol. Pharm. Bull.* 29, 306-314 (2006).

[139] Fujikawa T, Miguchi S, Kanada N, Nakai N, Ogata M, Suzuki I, Nakashima K, *J. Ethnopharmacol.* 97, 375-381 (2005).

[140] Fujikawa T, Kanada N, Shimada A, Ogata M, Suzuki I, Hayashi I, Nakashima K, *Biol. Pharm. Bull.* 28, 169-172 (2005).

[141] Fujikawa T, Soya H, Hibasami H, Kawashima H, Takeda H, Nishibe S, Nakashima K, *Phytother. Res.* 16, 474-478 (2002).

[142] Bu Y, Jin ZH, Park SY, Baek S, Rho S, Ha N, Park SK, Kim H; Sun Yeo Kim, *Phytother. Res.* 19, 167-169 (2005).

[143] Zhu L., Zhang R., Li T.L., *Chin. J. Exp. Tradit. Med. Form.* 18, 219-223 (2012).

[144] Zhang R., Li T.L., Liu, X.Y., Zhu, L., *Pharmacol. Clin. Chin. Mater. Med.* 27, 63-66 (2011).

[145] Yang T.T., Liu S.M., Cui L.R., Liu, X.W., *Chin. Tradit. Patent Med.* 31, 1662-1666 (2009).

[146] Shu W., Wu W.B., Zhao Y.Q, Liu L., *Progress in Modern Biomedicine.* 11, 4247-4249 (2011).

[147] Liu P., Yang T. and Ding Y.Q., *Chin. J. Inte. Tradit.West Med.* 3, 390-392 (2008).

[148] Yang X., Cai X.J. and Liu L., *Chin. J. of Clin Drug Use.* 2, 4-6 (2009).

[149] Sui DY, Qu SC, XF Yu, YP Chen, XY Ma, *Chin. J. Chin. Mater. Med.* 29, 71-74 (2004).

[150] Hwang YC, Jeong IK, Ahn KJ, Chung HY, *J. Bone. Miner. Metab.* 27, 584-590 (2009).

[151] Chen XC, Xia L, Hu S, Huang G, *Acta. Pharmacol. Sin.*17, 523-526 (1996).

[152] Nguyen XN, Phan VK, Chau VM, Do TH, Buu HT, Pham HY, Nguyen HT, Hyun JH, Kang HK, Kim YH, *Planta Med.* 76, 189-194 (2010).

[153] Nhiem NX, Tung NH, Kiem PV, Minh CV, Ding Y, Hyun JH, Kang HK, Kim YH, *Chem. Pharm. Bull.* 57, 986-989 (2009).

[154] Lin QY, Jin LJ, Cao ZH, Li HQ, Xu YP, *J. Ethnopharmacol.* 118, 495-502 (2008).

[155] Jia L.L. and Tang H, *J. Appl. Physiol.* 22, 246-249 (2006).

[156] Lee S, Shin DS, Oh KB, Shin KH, *Arch. Pharm. Res.* 26, 40-42 (2003).

[157] Hong CE, Cho MC, Jang HA, Lyu SY, *J. Toxicol. Sci.* 36, 661-668 (2011).

[158] Yi JM, Hong SH, Kim JH, Kim HK, Song HJ, Kim HM, *J. Ethnopharmacol.* 793, 347-352 (2002).

[159] Yi JM, Kim MS, Seo SW, Lee KN, Yook CS, Kim HM, *Clin. Chim. Acta,* 312, 163-168 (2001).
[160] Yoo HH, Lee SH, Jin C, Kim DH, *Planta Med.* 74, 822-827 (2008).
[161] Jung HJ, Nam JH, Choi J, Lee KT, Park HJ, *J. Ethnopharmacol.* 97, 359-367 (2005).
[162] Du J. and Gao L., *Chin. J. Chin. Mater. Med.* 17, 356-357 (1992).
[163] Deng HZ, Hou LB, Sun SY, Li SY, *Chin. J. Chin. Mater. Med.* 19, 171-174 (1994).
[164] Ni HX, Luo SS, and Shao GM, *Chin. J. Inte. Trad. West Med,* 21, 105-107 (2001).
[165] Takahashi T, Kaku T, Sato T, Watanabe K, Sato J, *J. Nat. Med.* 64, 55-62 (2010).
[166] Jang MH, Shin MC, Kim YJ, Kim CJ, Chung JH, Seo JC, Kim EH, Kim KY, Lee CY, Kim KM, *Am. J. Chin. Med*, 31, 379-388 (2003).

In: Phytochemicals
Editor: Hai-Xue Kuang

ISBN: 978-1-62417-354-7


*Chapter XIV*

# Simultaneous Determination of Five Triterpenoid Saponins in the Root of *Caulophyllum Robustum* Maxim by HPLC-ELSD

***Guo-Yu Li*[1], *Yong-Gang Xia*[2], *Jun Liang*[2], *Yan-Hai Zhang*[2] *and Hai-Xue Kuang*[2]***

[1]Pharmaceutical College, Harbin Medical University, Harbin, China
[2]Key Laboratory of Chinese Materia Medica
(Heilongjiang University of Chinese Medicine),
Ministry of Education, Harbin, China

## Abstract

A high-performance liquid chromatographic (HPLC) method coupled with evaporative light scattering detector was developed for the simultaneous quantification of five major bioactive constituents in the root of *Caulophyllum robustum* Maxim. The five constituents were separated on a Kromasil ODS column by gradient elution using. The validation of the method included recovery, linearity, accuracy and precision (intra- and inter-day variation). The range of recoveries of this method was 95.75–99.79%, with all the constituents showing good linearity ($r2 > 0.99$). The accuracy and precision were satisfactory, with the overall intra- and inter-day variation being less than 4%. The present method has been successfully applied for the determination of all five constituents in *C.robustum*. The study may be considered helpful to the quality control of *Caulophyllum robustum* Maxim.

**Keywords**: *Caulophyllum robustum* Maxim; triterpenoid saponin; Determination; HPLC-ELSD

# Introduction

*Caulophyllum robustum* Maxim is the roots and rhizomes of Berberidaceae family plant. The herb which was recorded in "Chinese traditional medicine dictionary" is a folk medicine that has already been utilized widely to cure many diseases such as rheumatoid arthritis, stomachache and bellyache, fracture injuries and menstruation disorder due to its strong and wide biological activities [1]. It contains alkaloids and saponins mainly, according to literature. Among them, triterpend saponins are regarded as the active constituents. Thus it is essential to establish an analytical method for the triterpenoid saponins to control the quality of *Caulophyllum Robustum* Maxim.

As a quasi-universal detector, the evaporative light scattering detector (ELSD) response does not depend on the analytes' optical characteristics, which eliminates the common problems associated with other HPLC detectors. It has been successfully applied to the analysis of non-volatile compounds such as saccharides, saponins and so on. Although, to the best of our knowledge, saponins occur at high concentrations in *Caulophyllum* [5,9,11], there were no reports regarding the simultaneous determination of saponins. In the present study, an HPLC-ELSD method was developed and validated for the simultaneous determination of five major bioactive components of Leonticin D, Ranunoside 1 [5], Cauloside G, Cauloside D, and Cauloside B.

# Experimental

## Samples, Chemicals and Reagents

The dry roots of *C. robustum* were collected in March 2007 from Suiling of Heilongjiang Province, China and identified by Prof. Zhenyue Wang of Heilongjiang University of Chinese Medicine.

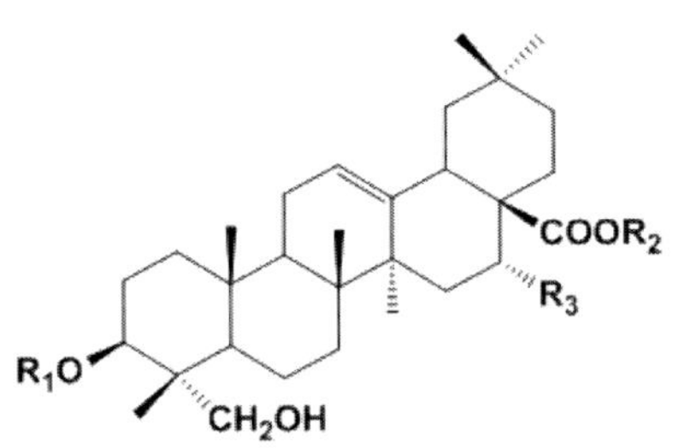


| | $R_1$ | $R_2$ | $R_3$ |
|---|---|---|---|
| **1** | Ara | Glc(6-1)Glc(4-1)Rha | OH |
| **2** | Ara(3-1)Glc | Glc(6-1)Glc(4-1)Rha | H |
| **3** | Ara(2-1)Glc | Glc(6-1)Glc(4-1)Rha | H |
| **4** | Ara | Glc(6-1)Glc(4-1)Rha | H |
| **5** | Ara | H | OH |

Figure 1. Structures of five triterpenoid saponins.

The voucher specimen (20070093) was deposited at Herbarium of Heilongjiang University of Chinese Medicine, Harbin, P. R. China.

HPLC-grade methanol was purchased from Merck (Darmstadt, Germany). Ultrapure water was prepared from Milli-Q system (Millipore, Bedford, MA, USA). Acetic acid was purchased from Chengdu Chemical Factory (China); other chemicals were of analytical grade. All solvents and samples were filtered through 0.45 μm membrane fi lters before injecting into HPLC.

Standards, namely Leonticin D, Ranunoside 1, Cauloside G, Cauloside D, and Cauloside B, were isolated by the author from the root of *C. robustum*.Their purities were over 98% by HPLC analysis. Structures of the standards are shown in Figure 1. Details of NMR and MS data of standards were below [2-7].

Compound 1 (Leonticin D): white powder, negative ESI-MS: *m/z* 1089.3 $[M-H]^-$, $^1$H-NMR (pyridine-$d_5$, 400 MHz): δ0.92, 0.95, 1.00, 1.02, 1.14, 1.75(each 3H, s, Me), 1.69(3H, d, *J*=6.0Hz, H-6 of *α*-rha), 4.96(1H, d, *J*=7.6Hz, H-1 of *β*-glc`), 4.97(1H, d, *J*=6.8Hz, H-1 of *α*-ara), 5.26(1H, br.s, H-16), 5.59(1H, like-t, H-12),5.84(1H, s, H-1 of *α*-rha), 6.23 (1H, d, *J*=8.1Hz, H-1 of *β*-glc). $^{13}$C-NMR(pyridine-$d_5$, 100 MHz): *δ*38.9(C-1), 26.2(C-2), 82.0(C-3), 43.5(C-4), 47.7(C-5), 18.2(C-6), 33.2(C-7), 40.1(C-8), 47.3(C-9), 37.0(C-10), 23.9(C-11), 122.8(C-12), 144.5(C-13), 42.1(C-14), 36.2(C-15), 74.3(C-16), 49.1(C-17), 41.2(C-18), 47.2(C-19), 30.8(C-20), 35.9(C-21), 32.2(C-22), 64.5(C-23), 13.7(C-24), 16.4(C-25), 17.7(C-26), 27.2(C-27), 176.0(C-28), 33.2(C-29), 24.7(C-30),106.7, 73.1, 74.7, 69.7, 67.0(3-O-araC1-5),95.8, 73.9, 78.6, 70.7, 78.0, 69.2(28-O-glc C1-6),104.9, 75.3, 76.5, 78.2, 77.1, 61.3(28-O-glc`C1-6),102.7, 72.6, 72.7, 73.8, 70.3, 18.5(28-O-rhaC1-6). Liebermann-Burchard reaction purple.

Compound 2 (Ranunoside 1): white powder,positive ESI-MS: *m/z*1259.2$[M+Na]^+$, negative ESI-MS: *m/z* 1235.0 $[M-H]^-$, $^1$H-NMR (pyridine-$d_5$, 400 MHz): δ0.85, 0.86, 0.92, 0.95, 1.09, 1.17(each 3H, s, Me), 1.68(3H, d, *J*=6.0Hz, H-6 of *α*-rha), 4.96(1H, d, *J*=6.5Hz, H-1 of *α*-ara), 4.98(1H, d, *J*=7.6Hz, H-1 of *β*-glc``), 5.30(1H, d, *J*=7.9Hz, H-1 of *β*-glc),5.39(1H, like-t, H-12),5.84(1H, s, H-1 of *α*-rha), 6.22 (1H, d, *J*=8.1Hz, H-1 of *β*-glc`). $^{13}$C-NMR(pyridine-$d_5$, 100 MHz): *δ*40.3(C-1), 27.7(C-2), 83.3(C-3), 45.0(C-4), 49.0(C-5), 19.6(C-6), 34.0(C-7), 41.4(C-8), 49.7(C-9), 38.4(C-10), 25.4(C-11), 124.4(C-12), 145.6(C-13), 43.6(C-14), 29.8(C-15), 24.8(C-16), 48.5(C-17), 43.1(C-18), 47.7(C-19), 32.2(C-20), 35.4(C-21), 34.3(C-22), 65.7(C-23), 15.2(C-24), 17.7(C-25), 19.1(C-26), 27.6(C-27), 178.1(C-28), 34.6(C-29), 25.2(C-30),108.0, 73.0, 85.7, 70.8, 68.5(3-O-araC1-5),107.7, 77.2, 79.7, 73.5, 80.2, 64.2(glc at ara, C1-6), 97.1, 75.3, 80.2, 72.3, 79.5, 70.6(28-O-glc` C1-6),106.3, 76.8, 78.0, 79.8, 78.6, 62.7(28-O-glc``C1-6),104.2, 74.1, 74.2, 75.5, 71.8, 20.0(28-O-rhaC1-6). Liebermann-Burchard reaction purple.

Compound 3 (Cauloside G): white powder, positive ESI-MS: *m/z*1259.3$[M+Na]^+$, negative ESI-MS: *m/z* 1235.0 $[M-H]^-$, $^1$H-NMR (pyridine-$d_5$, 400 MHz): δ0.85, 0.86, 0.95, 1.01, 1.09, 1.14(each 3H, s, Me), 1.68(3H, d, *J*=6.1Hz, H-6 of *α*-rha), 4.97(1H, d, *J*=7.8Hz, H-1 of *β*-glc``), 5.15(1H, d, *J*=5.6Hz, H-1 of *α*-ara), 5.17(1H, d, *J*=7.4Hz, H-1 of *β*-glc),5.39(1H, like-t, H-12),5.83(1H, s, H-1 of *α*-rha), 6.22 (1H, d, *J*=8.1Hz, H-1 of *β*-glc`). $^{13}$C-NMR(pyridine-$d_5$, 100 MHz): *δ*39.9(C-1), 27.1(C-2), 83.3(C-3), 44.6(C-4), 48.9(C-5), 19.3(C-6), 33.9(C-7), 41.0(C-8), 49.3(C-9), 38.0(C-10), 24.9(C-11), 124.0(C-12), 145.2(C-13), 43.2(C-14), 29.4(C-15), 24.4(C-16), 48.1(C-17), 42.8(C-18), 47.3(C-19), 31.9(C-20), 35.1(C-21), 33.6(C-22), 65.9(C-23), 14.6(C-24), 17.3(C-25), 18.7(C-26), 27.2(C-27),

177.1(C-28), 34.2(C-29), 24.8(C-30),105.0, 82.4, 74.8, 69.4, 66.1(3-O-araC1-5),106.9, 77.4, 79.3, 72.5, 79.3, 63.6(glc at ara, C1-6), 96.7, 74.9, 79.8, 71.9, 79.1, 70.3(28-O-glc` C1-6),105.9, 76.4, 77.6, 79.4, 78.3, 62.4(28-O-glc``C1-6),103.8, 73.7, 73.9, 75.1, 71.4, 19.7(28-O-rhaC1-6). Liebermann-Burchard reaction purple.

Compound 4 (Cauloside D): white powder, negative ESI-MS: $m/z$ 1073.3 $[M-H]^-$, $^1H$-NMR (pyridine-$d_5$, 400 MHz): δ0.85, 0.86, 0.92, 0.96, 1.11, 1.16(each 3H, s, Me), 1.69(3H, d, $J$=6.4Hz, H-6 of $\alpha$-rha), 4.97(1H, d, $J$=6.8Hz, H-1 of $\alpha$-ara), 4.98(1H, d, $J$=8.0Hz, H-1 of $\beta$-glc`), 5.39(1H, like-t, H-12),5.85(1H, s, H-1 of $\alpha$-rha), 6.23 (1H, d, $J$=8.0Hz, H-1 of $\beta$-glc). $^{13}C$-NMR(pyridine-$d_5$, 100 MHz): $\delta$38.8(C-1), 26.1(C-2), 81.9(C-3), 43.5(C-4), 47.6(C-5), 18.2(C-6), 32.8(C-7), 39.9(C-8), 48.2(C-9), 36.9(C-10), 23.9(C-11), 122.9(C-12), 144.1(C-13), 42.1(C-14), 28.3(C-15), 23.3(C-16), 47.0(C-17), 41.6(C-18), 46.2(C-19), 30.7(C-20), 33.9(C-21), 32.5(C-22), 64.4(C-23), 13.6(C-24), 16.2(C-25), 17.6(C-26), 26.1(C-27), 176.5(C-28), 33.1(C-29), 23.7(C-30),106.7, 73.1, 74.7, 69.7, 67.0(3-O-araC1-5),95.6, 74.0, 78.7, 70.8, 78.0, 69.1(28-O-glc C1-6),104.8, 75.3, 76.5, 78.2, 77.2, 61.3(28-O-glc`C1-6),102.7, 72.6, 72.8, 73.9, 70.3, 18.5(28-O-rhaC1-6). Liebermann-Burchard reaction purple.

Compound 5 (Cauloside B): white powder, positive ESI-MS: $m/z$1263.8.3$[2M+Na]^+$. $^1H$-NMR (pyridine-$d_5$, 400 MHz): δ0.93, 0.99, 1.05, 1.07, 1.18, 1.80(each 3H, s, Me), 4.99(1H, d, $J$=6.8Hz, H-1 of $\alpha$-ara), 4.43(1H, dd, $J$=7.6,8.4Hz, H-2 of $\alpha$-ara), 4.07(1H, dd, $J$=3.4,9.0 Hz, H-3 of $\alpha$-ara), 4.27(1H, m, H-4 of $\alpha$-ara), 3.72(1H, t, $J$=12.2 Hz, H-5a of $\alpha$-ara), 4.29(1H, m, H-5b of $\alpha$-ara), 5.24(1H, br.s, H-16), 5.65(1H, like-t, H-12). $^{13}C$-NMR(pyridine-$d_5$, 100 MHz): $\delta$38.8(C-1), 26.1(C-2), 81.8(C-3), 43.4(C-4), 47.6(C-5), 18.2(C-6), 33.1(C-7), 39.9(C-8), 47.2(C-9), 36.9(C-10), 23.9(C-11), 122.1(C-12), 144.7(C-13), 42.0(C-14), 36.1(C-15), 74.5(C-16), 48.7(C-17), 41.3(C-18), 47.1(C-19), 30.9(C-20), 36.1(C-21), 32.8(C-22), 64.4(C-23), 13.6(C-24), 16.2(C-25), 17.5(C-26), 27.1(C-27), 179.5(C-28), 33.2(C-29), 24.7(C-30), 106.4, 72.9, 74.5, 69.4, 66.8(araC1-5). Liebermann-Burchard reaction purple.

## HPLC Apparatus and Conditions

The HPLC system consisted of waters 2695 pump, waters 2695 controller, waters2996PAD, ELSD2000(Alltech, America) and air compressor(Japan). An Kromasil ODS column(200×4.6mm, 5μm) with a $C_{18}$ guard column was used at a column temperature of 25⁰C. The chromatographic data were recorded and processed by Empower chromatographic Work station software. The mobile phase which consisted of methanol-water was applied in the gradient elution at a flow rate of 1.0 ml/min$^{-1}$ (Table 1). Temperature for the drift tube of ELSD was set at 90⁰C.The gas flow rate was 2.1ml.min$^{-1}$ with the pressure of air compressor 0.4 MPa.

**Table 1. Methanol- water gradient elutiong program**

| Time | 0 | 10 | 20 | 21 | 35 | 50 |
|---|---|---|---|---|---|---|
| Methanol | 60 | 63 | 63 | 80 | 80 | 100 |
| Water | 40 | 37 | 37 | 20 | 20 | 0 |
| Curve | | 6 | 6 | 11 | 6 | 6 |

## Preparation of Standard Solutions

The mix sapoins solution containing 1 (0.65 mg/ml), 2 (0.70 mg/ml), 3 (0.30 mg/ml) 4 (0.60 mg/ml) and 5 (0.1 mg/ml) were prepared with deionized water.

Working standard solutions were further obtained by appropriate dilution of the stock mix monosaccharide sample solutions with deionized water. All sample solutions were filtered through a 0.45 μm syringe filter. All the solutions prepared were stored in a refrigerator at +4 °C.

## Preparation of Sample Solutions

The roots of the plant (1.5 kg) were extracted twice with 70 % ethanol (6L, each time) in Soxhlet's extractor for 2 hour.

The extracts were homogenized in $H_2O$ and was poured into AB-8 Macroporousadsorption resin column and eluated with $H_2O$, 70% EtOH and 95% EtOH, respectively. The 70% ETOH eluate was concentrated to dryness which was for determination.

## Method Validation [8-13]

The regression equations were calculated in the form of $y = a\,x + b$, where $y$ was the peak areas and $x$ was the concentration of analytes. The signal-to-noise of 3:1 and 10:1 were used to establish LOD and LOQ, respectively. The measurement of intra- and inter-day variability was utilized to determine the precision of this newly developed method. The intraday variation was determined by analyzing the same mixed standard water solution for five times within 1 day.

While for inter-day variability test, the solution was examined in triplicate for consecutive 3 days. Stability of sample solution was tested at 0, 4, 8, 12, 24 and 48 h within 2 days. All solutions were kept at 4 °C before analysis. The analytes showed very stable in water solution (RSD$<$3.42%) at 4 °C during the tested period.

# Results and Discussions

## Optimization of Chromatographic Conditions

The chromatographic conditions were optimized to obtain chromatograms with a good resolution of adjacent peaks within a short analysis time.

Different types of chromatographic column from three were tested. Kromasil ODS column, Hypersil ODS II column and Dima Diamand $C_{18}$ column are suitable to different kinds of chemical constituents. The sample showed different retention behaviors on these columns.

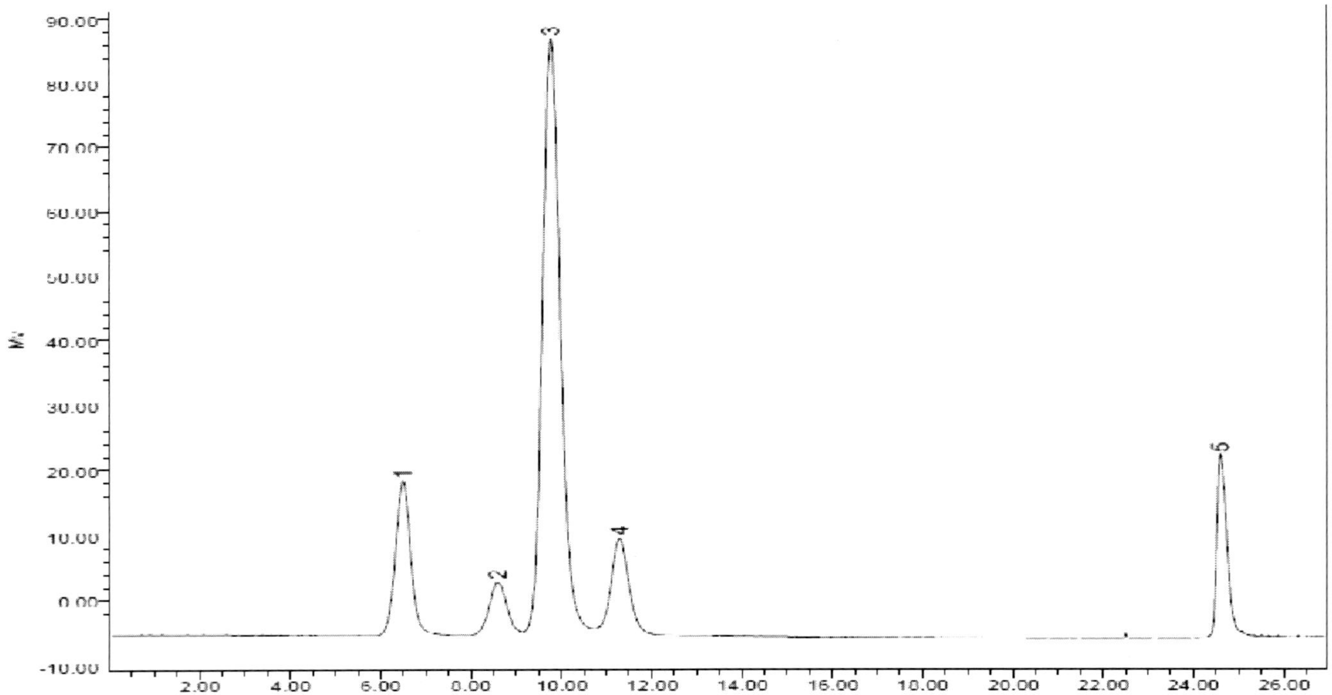

1
2
3
4
5
90.00
80.00
70.00
60.00
40.00
30.00
20.00
10.00
0.00
-10.00
2.00
4.00
6.00
10.00
12.00
14.00
16.00
20.00
22.00
24.00
26.00

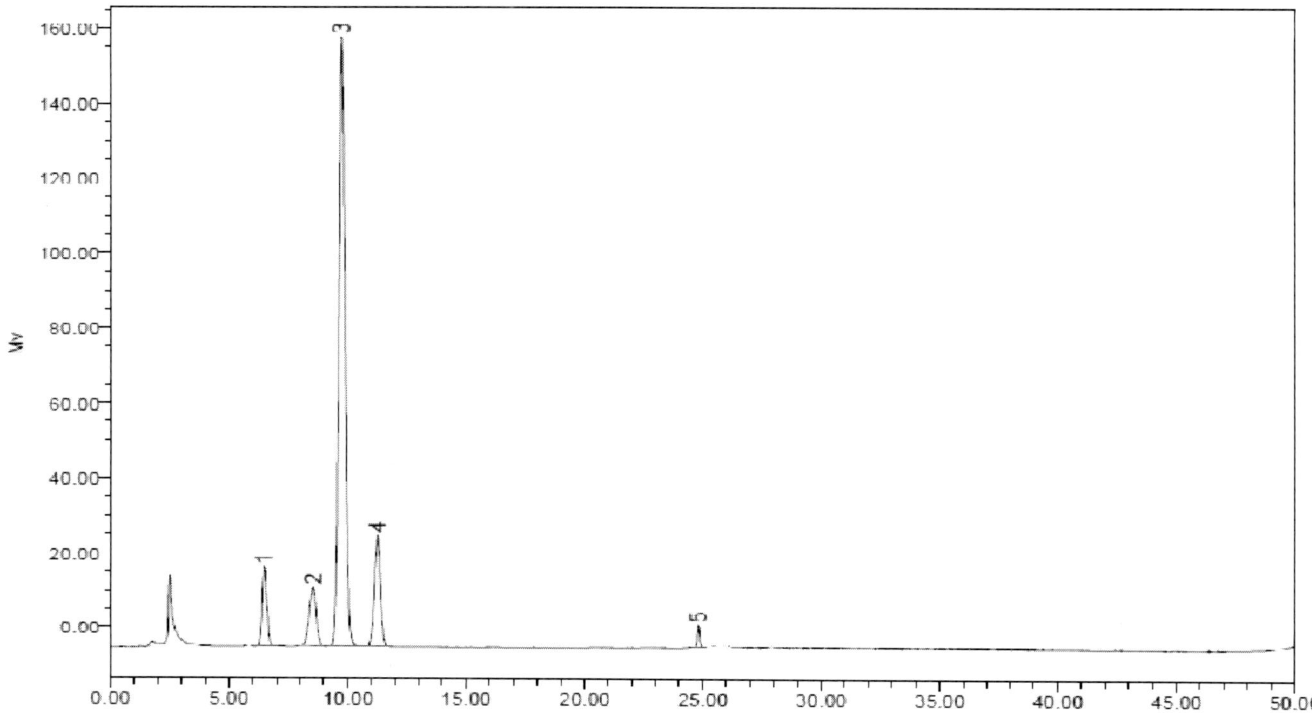


Figure 2. HPLC-ELSD chromatograph of standards (A) and sample (B).

The analysis time did not vary significantly on three columns, while the resolution of Kromasil ODS column was better than the rest. Different mobile phase compositions were also optimized. ACN had the effect of accelerating the elution of the early compounds, which were not well separated from each other. Methanol-water system showed more powerful separation ability for all active components than ACN-water system. In order to keep good repeatability.

There was no sharp effect upon changing the column temperature (25-30 °C) on either the peak symmetry or the resolution of the eluted peaks, but only a marked increasing in the retention time and so the column temperature of 30 °C was used. The effect of different flow rates (0.5-1.0 mL/min) has been also examined.

Tailing and poor peaks symmetry appeared at lower values of the flow rate, so 1.0 mL/min was selected as the optimum flow rate.

Under these optimum conditions (Table 1), all the studied chemical constituents were well separated from each other (Figure 2). In addition, Temperature for the drift tube of ELSD was $90^{0}$C.The gas flow rate was 2.1 ml/min-1 with the pressure of air compressor 0.4 MPa.

The chromatographic peaks of samples were identified by injecting and comparing with the retention time of each target compound (Figure 2).

## Optimization of Sample Extraction Conditions

In order to obtain quantitative extraction of plant samples, variables involved in the procedure such as solvent, extraction methods and extraction time were optimized. Various extraction solvents were compared, including different concentrations of methanol, ethanol and the mobile phase. The results demonstrated that 70% ethanol is the most efficient extraction solvent. After comparing several extraction methods, Soxhlet's extraction was found to be the most suitable extraction method. The optimized extraction time was found to be 2 hour.

According to these, the exhausted extraction of sample powder could be attained by means of Soxhlet's extraction with 70% ethanol 6L for 2hour each time.

## Method Validation

As shown in Table 2, all calibration curves showed good linear regression ($r \geq 0.9779$) within the test ranges. The LOD (S/N = 3) and the LOQ (S/N = 10) were less than 0.0082 and 0.163 μM for all analytes, respectively. The measurement of intra- and inter-day variability was utilized to determine the precision of this newly developed method, which was investigated by the relative standard deviations (RSDs) of the migration time and peak area rations of analytes. The overall intra- and inter-day variability of peak area rations were less than 3.42% (Table 3).

**Table 2. Linearity of HPLC-ELSD method of different analytes**

| Com. | Regression equation [a] | r | Linear range (μmol/L) | LOQ (μmol/L) | LOD (μmol/L) |
|---|---|---|---|---|---|
| 1 | $y = 1.762\,x + 10.804$ | 0.9996 | 0.32 – 6.38 | 0.163 | 0.0087 |
| 2 | $y = 1.792\,x + 10.516$ | 0.9988 | 0.35 – 6.99 | 0.165 | 0.0086 |
| 3 | $y = 1.457\,x + 11.266$ | 0.9991 | 1.65 – 19.80 | 0.168 | 0.0089 |
| 4 | $y = 1.804\,x + 10.351$ | 0.9996 | 0.32– 6.40 | 0.165 | 0.0086 |
| 5 | $y = 1.776\,x + 10.310$ | 0.9989 | 0.30 – 2.40 | 0.166 | 0.0082 |

[a] $y$ and $x$ stand for the peak area and the contentration (μmol/L) of compounds, respectively.

**Table 3. Precision of the retention time and peak area of analytes in the present method**

| Com. | Intra-day precision (RSD%, n = 5) | | | Inter-day precision (RSD%, n =3) | | |
|---|---|---|---|---|---|---|
| | Retention time | Peak area | RSD(%) | Retention time | Peak area | RSD(%) |
| 1 | 6.49 | 218700 | 3.30 | 6.50 | 236159 | 3.42 |
| 2 | 8.59 | 291980 | 0.94 | 8.60 | 324189 | 3.06 |
| 3 | 9.77 | 1297262 | 3.01 | 9.77 | 2610742 | 2.21 |
| 4 | 11.28 | 361341 | 2.93 | 11.29 | 303506 | 2.02 |
| 5 | 24.60 | 17251 | 3.01 | 24.63 | 18203 | 2.32 |

### *Application and Recovery*

Samples were injected directly and separated under the optimum condition described above. Figure 2 shows typical chromatograms of *C. robustum* samples.

As can be seen, five saponins from samples could be still perfectly baseline separated from each other. Peaks were identified through a comparison of migration times with those of the standard mixtures and by spiking the sample with individual monosaccharides.

The results show that the sapoins in *C. robustum* were composed of Leonticin D, Ranunoside 1, Cauloside G, Cauloside D, and Cauloside B in the contents of 5.35±0.15,6.53±0.11 21.53±0.35, 8.07±0.13 and 1.66±0.01 mM, respectively.

Furthermore, recovery experiments were performed in order to investigate the accuracy of the method. Known amounts of saponins solutes were added to the sample detected, and the resulting spiked sample was subjected to the entire analytical sequence.

Each solute was spiked at a close concentration with the sample and recoveries were calculated based on the difference between the total amount determined in the spiked samples and the amount observed in the non-spiked samples.

All analyses were carried out in triplicate. The results show that the mean ± SD percent recoveries of Leonticin D, Ranunoside 1, Cauloside G, Cauloside D, and Cauloside B were 95.75 ± 0.0898, 98.23± 0.0094, 98.39± 0.1970, 98.73 ± 0.0580and 99.79 ± 0.0049, respectively, and their RSD values were both less than 3.5% (Table 4). Such results further demonstrate that this method is precise and practical for the analysis of triterpenoid saponins in *C. robustum*.

**Table4. Recoveries of 1-5 in sample analysis (n = 6)**

| Com. | added amount (mM) | found amount (mM) | recovery(%)[a] means ± SD | RSD(%)[b] |
|---|---|---|---|---|
| 1 | 5.36 | 5.13 | 95.75 ± 0.0898 | 1.75 |
| 2 | 6.53 | 6.41 | 98.23± 0.0094 | 0.15 |
| 3 | 27.55 | 27.11 | 98.39± 0.1970 | 0.73 |
| 4 | 8.09 | 7.99 | 98.73 ± 0.0580 | 0.73 |
| 5 | 1.67 | 1.67 | 99.79 ± 0.0049 | 2.93 |

[a] Recovery (%) = [(mean of measured concentration−spiked concentration)/spiked concentration]×100.
[b] RSD (%) = (SD/mean)×100.

# Conclusion

It is evident from a number of reports that saponins are responsible for the biological activities of herb medicine. Techniques for the simultaneous qualitative and quantitative analysis are therefore of great importance. The developed HPLC-ELSD method, offering good linearity, sensitivity and accuracy, makes it possible to evaluate the quality of *C. robustum* through the simultaneous determination of multi-components in a single run.

# References

[1] Jiangsu New Medicinal College, 1979. The Dictionary of Chinese herbal medicine.Vol. 1. Shanghai People's Publishing Press, Shanghai, pp.998-999.

[2] Jhoo JW, Sang S, He K, Cheng X, Zhu N, Stark RE, Zheng QY, Rosen RT, Ho CT. Characterization of the triterpene saponins of the roots and rhizomes of blue cohosh (Caulophyllum thalictroides). *J. Agric Food Chem.* 2001,49 (12):5969-5974.

[3] M.Ganzera, H.R.W.Dharmaratne, N.P.D. Nanayakkara, et al. Determination of Saponins and Alkaloids in Caulophyllum thalictroides (Blue Cohosh) by High-performance Liquid C-hromatography and Evaporative Light Scattering Detection. *Phytochemical analysis*.2003, 14:1-7.

[4] Chen M, Wu WW, Nanz D, Sticher O. Leonticins D-H, five triterpene saponins from Leontice kiangnanensis. *Phytochemistry*. 1997, 4(3):497-504.

[5] Christian Weger and Matthias Hamburger.Tensioactive Compounds from the Aquatic plant Ranunculus fluitans L(Ranunculaceae).*Helvetica Chimica Acta*.2000,83:1454-1464.

[6] V.I.Grishkovets, E.A.Sobolev, A.S.Shashkov et al. Triterpenid Glycosides of Fatsia japonica II.Isolation and Structure of Glycosides from the Leaves.*Chemistry of Natural compounds*.2000, 36(5):501-505.

[7] Haruhisa Kizu, Shingo Kitayama,Fukio Nakatani et al.Studies on Nepalese Crude Drugs. III On the saponins of Hedera nepalensis K.Koch. *Chem Pharm Bull*.1985,33(8):3324-3329.

[8] Yo sh ikawa M, M urakam i T, O ruk i K, et al. Bioactive saponins and glycosides. X III1 Horse chest-nut. (3): Quantitative analysis of escins Ia, Ib, IIa and IIb by means of high performance liquid chromatography. [J]. *YakugakuZasshi*, 1999, 119 (1): 81-87.

[9] Kuwajima H, Taneda Y, Cnen W Z, et al. Variation of chemical constituents in processed licorice root: quantitative determination of saponin and flavoniod constituents in bark removed and roasted licorice roots. [J]. *YakugakuZasshi*, 1999, 119 (12): 945-955.

[10] Samukawa K I, Hideyuk i Y A, HideakiM A, et al. Simultaneous analysis of saponins in ginseng radix by high performance liquid chromatography. [J]. *Chem. Pharm. Bull*, 1995,43 (1) : 1371.

[11] He X G. Online identification of phytochemical constituents in botanical extanical by combined high performance liquid chromatograph icdiode array detection masspectrometric technique. [J]. *J. Ch. Romatogr. A*, 2000, 880 (2) : 203-232.

[12] LIW, Fitzloff J F. Determination of 24 (R) -pseudoginsenoside F (11) in North American ginseng using high performance liquid chromatophy with evaporative light scattering detection. [J]. *J. Pharm. Biomed. Anal.*, 2001, 25 (2): 257-265.

[13] Tbebe Z. Woldemariam,Joseph M. Betz, Peter J. Houghton. Analysis of aporphine and quinolizidine alkaloids from Caulophyllum thalictroides by densitometry and HPLC. *Journal of Pharmaceutical and Biomedical Analysis*. 1997,15:839-843.

In: Phytochemicals
Editor: Hai-Xue Kuang

ISBN: 978-1-62417-354-7


*Chapter XV*

# Chemical Study and Medical Application of Saponins from the Genus Aralia

***Hai-Xue Kuang*[*], *Yong-Gang Xia and Guang-Rui Mu***
Key Laboratory of Chinese Materia Medica
Heilongjiang University of Chinese Medicine,
Ministry of Education, Harbin, China

## Abstract

The genus of Aralia (Araliaceae), composing ca. 40 species, is mostly native to Asia, Oceania and North America. In these species, half are used mainly for medicinal purposes such as anti-tumor, adaptoge-like effect, effects on nervous system, effects on heart, anti-inflammatory activity, anti-diabetic activity and hepatoprotective effects. Saponins are a group of naturally occurring plant glycosides, characterized by their strong foam-forming properties in aqueous solution. Due to the great variability of their structures, saponins always display anti-tumorigenic effects through varieties of antitumor pathways. In addition, there are a large amount of saponins that still either remain to be trapped or be studied in detail by the medicinal chemists. In the past few years, a lot of triterpene saponins have been isolated from the leaves, bark, stem and tender shoot of Aralia genus. In recent years, with the growing concern on natural medicine, more and more attention was paid to Aralia species. In this article, we will summarize the progress on chemical structure of triterpene saponins and bioactivities of this genus from the literature until April, 2012.

[*] E-mail address: hxkuang@hotmail.com.

# Introduction

*Aralia* is a genus of shrubs or small trees in the family Araliaceae. There are about 40 species [1], mostly native to Asia, Oceania and North America. China has nearly 30 Aralia genus, they are distributed in Heilongjiang, Anhui, Zhejiang and other provinces. In these species, half are used mainly for medicinal purposes such as *A. elata*, *A. cordata*, *A. continentalis*, *et al* [2]. *A. elata* is widely distributed in Korea, Japan, and China. *A. elata*, classified as a shrub belonging to a species of Araliaceae, is one of the most popular edible mountain vegetables in East Asia. Its bark and root have been used in traditional medicine as a tonic, anti-arthritic, and anti-diabetic agent. *A. elata* has various documented anti-inflammatory effects. *A. cordata* has been used to treat arthritis and lower back pain. In an in vitro study, it has been reported that *A. cordata* inhibited COX-2 dependent PGE2 generation and showed effectiveness regarding analgesia, hypothermia, and duration of pentobarbital induced anesthesia. In a vitro study, *A. cordata* inhibited the cartilage and chondrocyte destruction through the down-regulation of MMP activities and the inhibition of proteoglycan and collagen degradation. Furthermore, studies indicates that the oral administration of *A. cordata* has therapeutic effects on an osteoarthritis rabbit model by prominent cartilage protection compared to celecoxib through the down-regulations of COX-2 expression, PGE2 production, csapase-3 activity, and the acceleration of chondrocyte proliferation; and has no adverse effects. *A. continentalis*, known as "Dokwhal" in Korea, has been used widely in traditional Korean medicine for dental diseases such as toothache, periodontal disease and gingivitis, and also has been used for neuralgia, analgesia, sweating, and as an antirheumatic [3]. There are several reports suggesting that *A. continentalis* has pharmacological properties as an analgesic and anti-inflammatory.

In the past few years, a lot of triterpene saponins have been isolated from the leaves, bark, stem and tender shoots of *Aralia.* Genus [4]. In recent years, with the growing concern on natural medicine, more and more attention was paid to the *Aralia.* species. In this article, we will summarize the progress on phytochemical studies and pharmacological activities of this genus from literature reported until July, 2011.

*A.elata* (Miq.) Seem. The leaves of *A.elata* (Miq.) Seem.

The leaves of *A. dasyphylla* Miq.

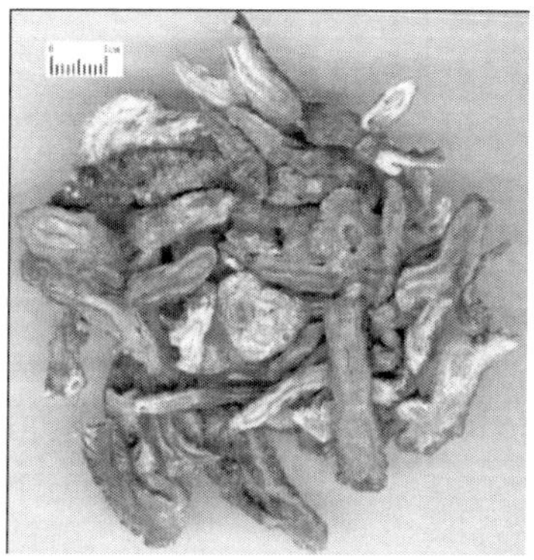

The herbs of *A.continentalis* Kitagawa The fruit of *A.continentalis* Kitagawa.

The leaves of *A.continentalis* Kitagawa.

*A. echinocaulis* Hand.-Mazz.

A. dasyphylla Miq.

## Extraction and Purification of the Saponin

The main chemical ingredient of Aralia is a saponin. According to the nature and structure of saponin, the extract can be used as an extraction solvent. Saponins has strong hydrophilic, so extraction solvent can be used in polar water or ethanol, methanol.

Because molecular saponin may contain a structure of carboxyl, acidic reaction with alkali and salt dissolved in water, it can also be used with lye as the extraction solvent. Chooseing a different extraction solvent should be based on the nature of the raw materials and the composition and structure of the saponins.

Huang Fushan [5] used ethanol as a solvent and determined the content of aralosides by ultrasonic extraction. Based on the single-factor test, the optimal extraction conditions of aralosides in *A. elata* (Miq.) Seem leaves were studied by the orthogonal experiment. The results showed that the optimal extraction conditions were 70% ethanol concentration, 40 multiples of solvent, an extraction temperature of 70°C, and an extraction time of 75 minutes. The extraction content of aralosides reached 92.77mg/g under the optimal extraction conditions. The optimized ultrasonic extraction process is simple, rapid, and efficient, and the results are stable. With the advances in technology, new extraction methods have been proposed. The conditions for the extraction of total saponin from *A.elata* Seem were optimized using a Herbal flash extractor. The contents of saponins from *A. elata* Seem were determined using a colourmetric method. Orthogonal experiments showed that the best extraction conditions were as follows: the alcohol concentration of 70%, extraction voltage of 180 volts and extraction time of 90 seconds. This method was simple and fast, with a high extracting rate [6]. About the purification of saponins, we usually use perforated resin. Liu Yan [7] used static and dynamic adsoption-desorption methods. Kuang HX [8] compared five kinds of macroporous resin adsorption performance. They all found that Macroporous resin AB-8 had the best separating efficiency when the content in *A. chinensis* liquid was 0.1 g/mL equivalent to raw material. The volume of drug is 6 BV (resin bed volume) with the adsorption-power 2.5 BV/h and the volume of 50% ethanol as eluant 4 BV with desorption-power 2 BV/h. After the treatment of AB-8 resin, the purity of araloside A could reach 30% and total araloside to 80%.

## Chemical Constituents of Saponins

Saponins are a vast group of glycosides, widely distributed in higher plants. Their surface-active properties are what distinguish these compounds from other glycosides. They dissolve in water to form colloidal solutions that foam upon shaking.

Currently, 146 kinds of saponins isolated from this genus can be divided into six types according to the differences of their parent nucleus (Chart 1.1, Table 1). The aglycons of these isolated and identified saponins were basically pentacyclic triterpenoid saponins including oleanolic acid, ivy, gill pouch acid and caulophyllogenin, which often had a Glc residue O-linked at C-3 or (and) C-28, forming mono- and bisdesmoside triterpene glycosides. Their glycosyls mainly contained D-glucuronic acid, D-glucose, L-arabinose, D-galactose, L-rhamnose and D-xylose and their derivatives.

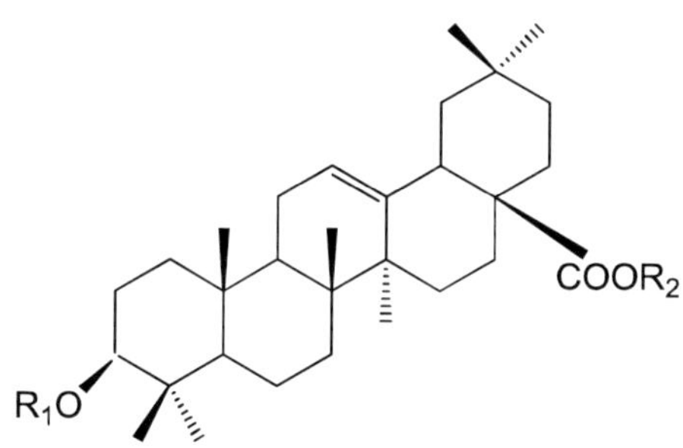


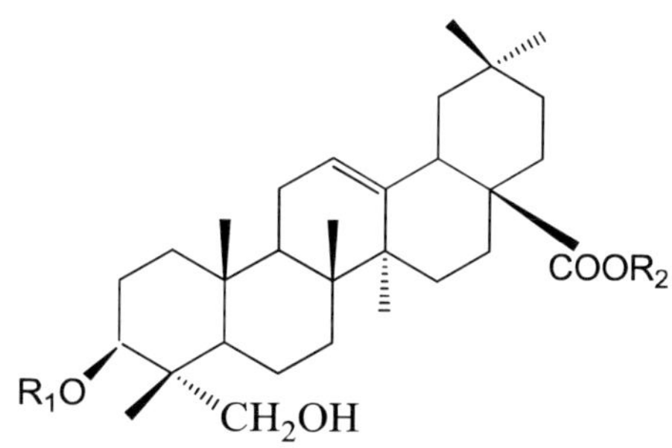


1

2

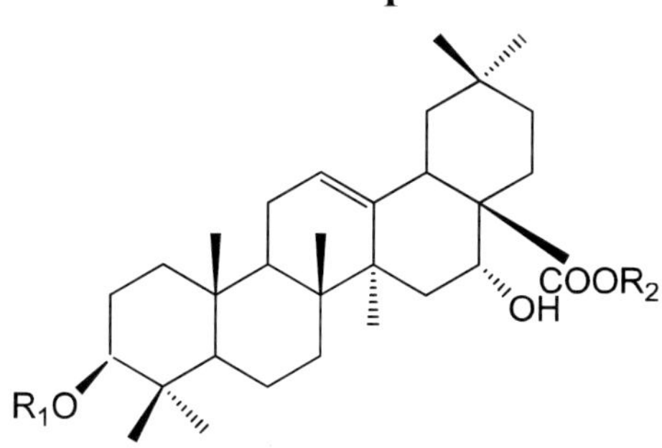


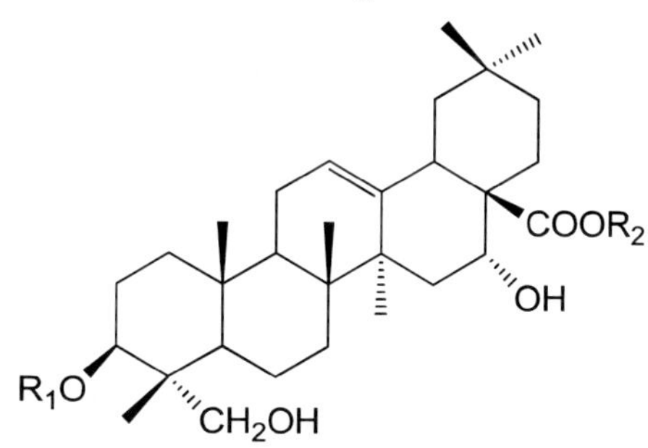


3

4

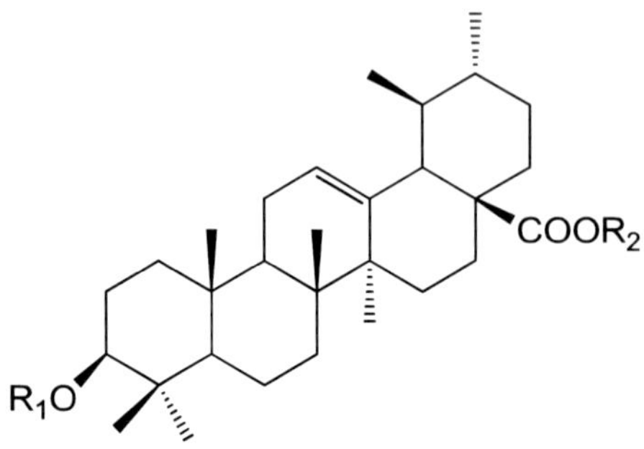

5

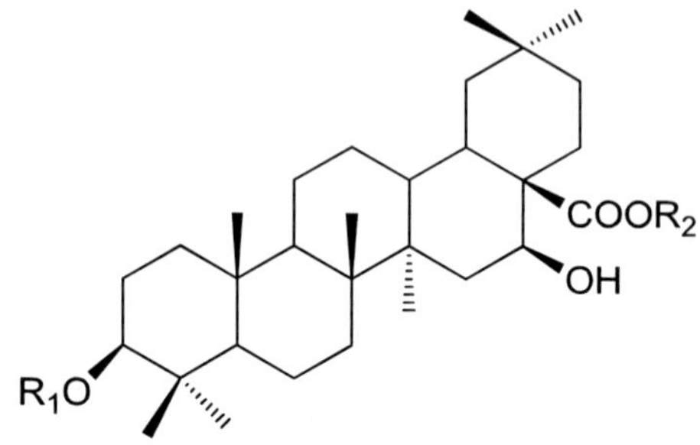

6

**Table 1. Saponins from Plant of Aralia**

| NO | type | $R_1$ | $R_2$ | Plant | Part | Refer |
|---|---|---|---|---|---|---|
| 1 | 1 | -Ara(p)2-Rha | -Glc6-Glc | *A.elata* | leaf | 9 |
| 2 | 3 | -Glc4-Glc | -H | *A.elata* | sprout | 39 |
| 3 | 1 | -GlcA4-Ara(f) | -Glc | *A. elata* | | 22 |
| | | | | *A.armate* | | |
| | | | | *A.echinocaullis* | | |
| | | | | *A.spinfolia* | | |
| | | | | *A.decaisneana* | | |
| | | | | *A.chinensis* | | |
| | | | | *A.taibaiensis* | | |
| 4 | 2 | -Glc2-Glc | -H | *A.elata* | sprout | 37 |
| 5 | 1 | -Ara(p)2-Xyl2- Glc | -H | *A.chinensis* | root | 20 |
| 6 | 1 | -GlcA$^{2}$-Xyl<br>Glc$^{3}$ | -H | *A.elata* | root | 10 |
| 7 | 1 | -GlcA$^{4}$-Glc<br>Glc$^{3}$ | -Glc | *A.elata* | root | 17 |
| 8 | 1 | -Glc3-Glc | -Glc | *A.elata* | sprout | 37 |
| 9 | 2 | -GlcA(Me) | -H | *A.armate* | root | 13 |
| 10 | 1 | -GlcA | -Glc | *A. elata* | root | 14, 16 |
| | | | | *A.armate* | | |
| | | | | *A.spinfolia* | | |
| | | | | *A.taibaiensis* | | |
| 11 | 1 | -Glc3-Glc | -H | *A. elata* | leaf | 19 |
| 12 | 1 | -Rha2-Ara(p) | -H | *A.elata* | leaf | 20 |
| 13 | 1 | -Ara(p)2-Rha | -Glc6-Xyl | *A.elata* | root | 9 |
| 14 | 2 | -GlcA(Me)4-Xyl | -Glc | *A.taibaiensis* | root | 15 |

| NO | type | $R_1$ | $R_2$ | Plant | Part | Refer |
|---|---|---|---|---|---|---|
| 15 | 1 | -GlcA(Me)$^2$-Xyl<br>\|<br>Glc$^3$ | -Glc | *A. elata* | root | 21 |
| 16 | 2 | -Ara(p)2-Rha | -Glc6-Xyl | *A. elata* | leaf | 11 |
| 17 | 4 | -GlcA | -H | *A. elata* | young root | 26 |
| 18 | 1 | -Glc3-Rha2-Ara(p) | -Glc6-Glc | *A.elata* | leaf | 20 |
| 19 | 1 | -Rha2-Ara(p) | -H | *A. elata* | leaf | 20 |
| 20 | 1 | -GlcA2-Glc | -Glc | *A.elata* | root | 11, 12 |
| | | | | *A.armate* | | |
| 21 | 1 | -GlcA(Me)$^2$-Xyl<br>\|<br>Gal$^3$ | -Glc | *A. elata* | root | 21 |
| 22 | 1 | -GlcA4-Ara(f) | -Gal | *A.spinfolia* | root | 16 |
| 23 | 1 | -GlcA(Me) | -Me | *A.taibaiensis* | root | 15 |
| 24 | 3 | -Ara3-Glc3-Glc | -Glc | *A. elata* | root | 36 |
| 25 | 1 | -GlcA2-Glc | -H | *A.armate* | root | 12 |
| 26 | 3 | -Glc3-Glc3-Glc | -Glc | *A.elata* | root | 36 |
| 27 | 1 | -GlcA$^4$-Ara(*f*)<br>\|<br>Glc$^3$ | -Glc6-Glc | *A. elata* | root | 37 |
| 28 | 1 | -GlcA3-Glc | -H | *A.armate* | root | 13 |
| 29 | 1 | -GlcA$^2$-Xyl<br>\|<br>Gal$^3$ | -Glc | *A. elata* | root | 11 |
| 30 | 1 | -GlcA3-Ara(p) | -H | *A.armate* | root | 13 |
| 31 | 3 | -Glc$^3$-Glc<br>\|<br>Glc$^2$ | -Glc | *A. elata* | root | 35 |
| 32 | 2 | -Glc$^3$-Glc<br>\|<br>Glc$^2$ | -Glc | *A.elata* | root | 35 |
| 33 | 4 | -Glc$^3$-Glc$^2$-Glc<br>\|<br>Glc$^2$ | -Glc | *A. elata* | root | 35 |
| 34 | 1 | -Gal$^3$-Glc<br>\|<br>Xyl$^2$ | -H | *A. decaisneana* | root | 27 |
| 35 | 5 | -Ara$^3$-Glc<br>\|<br>Xyl$^2$ | -Glc | *A.decaisneana* | root | 27 |
| 36 | 2 | -GlcA4-Xyl | -Me | *A.taibaiensis* | root | 15 |
| 37 | 1 | -Ara$^3$-Glc<br>\|<br>Glc$^6$-Glc | -Glc6-Glc | *A. elata* | root | 37 |

**Table 1. (Continued)**

| NO | type | $R_1$ | $R_2$ | Plant | Part | Refer |
|---|---|---|---|---|---|---|
| 38 | 1 | -Ara$^3$-Glc$^3$-Glc<br>\|<br>Glc$^6$-Glc | -Glc6-Glc | *A. elata* | root | 37 |
| 39 | 1 | -GlcA(Bu)$^3$-Gal<br>\|<br>Xyl$^2$ | -Glc | *A. elata* | root | 37 |
| 40 | 3 | -Glc$^3$-Glc$^3$-Glc<br>\|<br>Glc$^2$ | -- | *A.elata* | root | 37 |
| 41 | 1 | -GlcA$^4$-Ara(*f*)<br>\|<br>Glc$^3$ | -H | *A.taibaiensis* | root | 37 |
| 42 | 2 | -Rha2-Ara(p) | -Glc6-Glc | *A. elata* | leaf | 20 |
| 43 | 1 | -Ara(p)3-Glc | -Glc6-Glc | *A.decaisneana* | root | 27 |
| 44 | 5 | -Ara$^3$-Glc<br>\|<br>Xyl$^2$ | -H | *A.decaisneana* | root | 27 |
| 45 | 1 | -GlcA(Me)2-Gal | -Glc | *A.taibaiensis* | root | 15 |
| 46 | 2 | -Ara(p)3-Glc | -Glc | *A. elata* | root | 35 |
| 47 | 1 | -Ara(*p*)$^3$-Xyl<br>\|<br>Gal$^3$ | -H | *A.elata* | root | 11 |
| 48 | 2 | -Ara(p)2-Rha | -Glc$^6$-Glc<br>\|<br>Rha | *A.elata* | root | 11 |
| 49 | 1 | -GlcA$^2$-Gal<br>\|<br>Gal$^3$ | -H | *A. elata* | root | 11 |
| 50 | 1 | -GlcA(Bu) | -H | *A.armate* | root | 13 |
| 51 | 1 | -GlcA$^3$-Glc<br>\|<br>Ara$^4$ | -H | *A. elata* | root | 17 |
| | | | | *A.armate* | | |
| 52 | 2 | -Ara(p)2-Rha3-Glc | -Glc6-Glc | *A. elata* | leaf | 11 |
| 53 | 3 | -Ara3-Glc | -Glc | *A. elata* | root | 34 |
| 54 | 1 | -Glc$^3$-Glc<br>\|<br>Xyl$^2$ | -Glc6-Glc | *A.chinensis* | root | 27 |
| 55 | 3 | -GlcA3-Glc | -H | *A. elata* | young root | 26 |
| 56 | 1 | -Gal$^3$-Glc<br>\|<br>Xyl$^2$ | -Glc6-Glc | *A.chinensis* | root | 28 |
| 57 | 1 | -GlcA4-Ara(f) | -H | *A. elata* | root | 18 |

| NO | type | $R_1$ | $R_2$ | Plant | Part | Refer |
|---|---|---|---|---|---|---|
| | | | | *A.armate* | | |
| | | | | *A.echinocaullis* | | |
| | | | | *A.spinfolia* | | |
| | | | | *A.decaisneana* | | |
| | | | | *A.chinensis* | | |
| 58 | 4 | -Ara(p)3-Glc | -H | *A. elata* | sprout | 38 |
| 59 | 1 | -GlcA(Me) | -Glc | *A. elata* | root | 13, 17 |
| | | | | *A.armate* | | |
| 60 | 1 | -Glc4-Glc3-Glc | -H | *A. elata* | root | 17 |
| 61 | 1 | -Glc | -Glc | *A. elata* | leaf | 19 |
| 62 | 1 | -Gal | -Glc | *A. elata* | leaf | 19 |
| 63 | 1 | -GlcA4-Xyl | -Me | *A.taibaiensis* | root | 15 |
| 64 | 1 | -H | -Glc | *A. elata* | root | 13, 14 |
| | | | | *A.armate* | | |
| 65 | 1 | -GlcA-Gal | -Me | *A.taibaiensis* | root | 15 |
| 66 | 2 | -GlcA4-Gal | -Me | *A.taibaiensis* | root | 15 |
| 67 | 2 | Ara$^3$-Glc<br>\|<br>Glc$^2$ | -Glc | *A. elata* | leaf | 35 |
| 68 | 2 | -Ara(p)2-Rha | -H | *A. elata* | leaf | 9 |
| 69 | 1 | -GlcA4-Ara(f) | -H | *A. elata* | root | 17 |
| 70 | 1 | -GlcA(Me) | -H | *A.armate* | root | 13 |
| 71 | 1 | -GlcA3-Glc | -Glc | *A. elata* | root | 11 |
| 72 | 1 | -GlcA(Me)3-Gal | -H | *A.armate* | root | 13 |
| 73 | 1 | -GlcA(Bu)3-Gal | -H | *A.armate* | root | 13 |
| 74 | 1 | -GlcA(Bu) | -Glc | *A.armate* | root | 13 |
| 75 | 1 | -GlcA$^2$-Glc<br>\|<br>Ara(*f*)$^4$ | -Glc6-Glc | *A.chinensis* | root | 28 |
| 76 | 1 | -GlcA$^3$-Glc<br>\|<br>Ara(*f*)$^4$ | -Glc | *A. elata* | root | 23 |
| | | | | *A.echinocaullis* | | |
| 77 | 1 | -GlcA$^3$-Ara(*f*)<br>\|<br>Ara(*f*)$^4$ | -Glc | *A. elata* | root | 22 |
| 78 | 2 | -Rha2-Ara(p) | -Rha$^4$-Glc<br>\|<br>Glc | *A. elata* | leaf | 20 |
| 79 | 1 | -Ara(*p*)$^3$-Glc<br>\|<br>Xyl$^2$ | -Glc6-Glc | *A.decaisneana* | root | 27 |

**Table 1. (Continued)**

| NO | type | $R_1$ | $R_2$ | Plant | Part | Refer |
|---|---|---|---|---|---|---|
| 80 | 1 | -Ara(*p*)$^3$-Glc<br>Glc$^2$ | -Glc | *A.chinensis* | root | 28 |
| 81 | 1 | -GlcA(Me)4-Ara(f) | -Glc | *A. elata* | root | 23 |
| | | | | *A.armate* | | |
| | | | | *A.spinfolia* | | |
| 82 | 2 | -Glc3-Ara(f) | -Glc | *A. elata* | leaf | 19 |
| 83 | 1 | -Ara(*p*)$^3$-Xyl<br>Glc$^3$ | -H | *A. elata* | root | 10 |
| 84 | 1 | -Glc$^2$-Xyl<br>GlcA$^3$ | -H | *A.spinfolia* | root | 16 |
| 85 | 1 | -GlcA4-Xyl | -Glc | *A.taibaiensis* | root | 15 |
| 85 | 2 | -GlcA(Me)2-Xyl | -Glc | *A.taibaiensis* | root | 15 |
| 87 | 1 | -Glc$^3$-Glc$^2$-Glc<br>Glc$^2$ | -Glc | *A. elata* | root | 37 |
| 88 | 1 | -GlcA(ButyI) | -Glc | *A.taibaiensis* | | 29 |
| 89 | 2 | -Ara(p)$^2$-Rha$^3$-Glc | -H | *A. elata* | leaf | 9 |
| 90 | 1 | -GlcA$^2$-Glc$^2$-Ara | -Glc | *A.armate* | root | 14 |
| 91 | 1 | -GlcA$^4$-Xyl<br>Gal$^3$ | -Glc | *A. elata* | root | 17 |
| 92 | 1 | -GlcA(Me)$^2$-Glc<br>Ara(*f*)$^4$ | -Glc | *A. elata* | root | 21 |
| 93 | 1 | -Rha2-Ara(p) | -Glc6-Glc | *A. elata* | root | 20 |
| 94 | 1 | -Rha2-Ara(p) | -Xyl6-Glc | *A. elata* | leaf | 20 |
| 95 | 1 | -Ara(p)2-Rha | -Glc6-Glc | *A. elata* | leaf | 9 |
| 96 | 1 | -GlcA$^2$-Gal<br>Gal$^3$ | -Glc | *A. elata* | root | 11 |
| 97 | 1 | -Glc$^2$-Glc<br>Xyl$^3$ | -Glc6-Glc | *A.decaisneana* | root | 25 |
| 98 | 2 | -Glc3-Rha2-Ara(p) | -Glc6-Glc | *A. elata* | leaf | 20 |
| 99 | 1 | -Ara(p)2-Rha3-Glc | -Glc6-Glc | *A. elata* | leaf | 9 |
| 100 | 1 | -Glc$^3$-Glc<br>Glc$^2$ | -Glc | *A. elata* | root | 35 |
| 101 | 1 | -GlcA$^2$-Glc<br>Glc$^3$ | -Glc | *A. elata* | young root | 26 |

| NO | type | $R_1$ | $R_2$ | Plant | Part | Refer |
|---|---|---|---|---|---|---|
| 102 | 1 | -GlcA | -H | *A. elata* | root | 10-13 |
| 103 | 1 | -Ara(p)2-Rha3-Glc | -H | *A. elata* | leaf | 9 |
| 104 | 1 | -GlcA(Me)4-Ara(f) | -H | *A.armate* | root | 13 |
| 105 | 1 | -Gal$^3$-Glc<br>Xyl$^2$ | -Glc | *A.decaisneana* | root | 27 |
| 106 | 1 | -Ara(*p*)$^3$-Glc$^3$-Glc<br>Xyl$^2$ | -Glc | *A.decaisneana* | root | 27 |
| 107 | 5 | -Ara3-Glc | -Glc6-Glc | *A.decaisneana* | root | 27 |
| 108 | 5 | -Glc$^3$-Glc<br>Xyl$^2$ | -Glc | *A.decaisneana* | root | 27 |
| 109 | 1 | -Glc$^3$-Glc<br>Gal$^2$ | -Glc | *A.chinensis* | root | 28 |
| 110 | 1 | -GlcA$^2$-Glc<br>Ara(*f*)$^4$ | -H | *A.chinensis* | root | 28 |
| 111 | 1 | -GlcA($CH_2CH_3$)<br>Ara(*f*)$^4$ | -Glc | *A.taibaiensis* | root | 29 |
| 112 | 1 | -GlcA($CH_2CH_3$)$^3$-Glc<br>Ara(*f*)$^4$ | -Glc | *A.taibaiensis* | root | 29 |
| 113 | 1 | -GlcA(Decyl) | -Glc | *A.taibaiensis* | root | 30 |
| 114 | 1 | -GlcA($CH_2CH_3$)$^3$-Glc<br>Xyl$^2$ | -Glc | *A.taibaiensis* | root | 31 |
| 115 | 1 | -GlcA(Butyl)$^3$-Glc<br>Xyl$^2$ | -Glc | *A.taibaiensis* | root | 31 |
| 116 | 1 | -GlcA(Butyl)$^3$-Glc<br>Xyl$^2$ | -Glc | *A.taibaiensis* | root | 31 |
| 117 | 6 | -H | -Glc | *A.lasyhylla* | leaf | 32 |
| 118 | 2 | -Glc$^3$-Glc<br>Xyl$^2$ | -H | *A. elata* | leaf | 33 |
| 119 | 2 | -Glc$^3$-Glc$^3$-Glc<br>Gal$^2$ | -H | *A. elata* | leaf | 33 |
| 120 | 4 | -Ara3-Glc | -Glc | *A. elata* | root | 34 |
| 121 | 1 | -Ara(p)3-Glc | -H | *A. elata* | root | 21 |
| 122 | 2 | -Ara(p)2-Rha | -Glc6-Glc | *A. elata* | leaf | 9 |
| 123 | 1 | -GlcA(Me)2-Gal | -Glc | *A.armate* | root | 13 |

**Table 1. (Continued)**

| NO | type | $R_1$ | $R_2$ | Plant | Part | Refer |
|---|---|---|---|---|---|---|
| 124 | 1 | -Ara(p)2-Xyl | -Glc | *A. elata* | root | 10 |
| 125 | 1 | -GlcA2-Xyl | -H | *A. elata* | root | 37 |
| 126 | 2 | -Rha2-Ara(p) | -Xyl6-Glc | *A. elata* | leaf | 20 |
| 127 | 1 | -GlcA3-Gal4-Gal | -Glc | *A.armate* | root | 24 |
| | | | | *A.decaisneana* | | |
| 128 | 2 | -Glc3-Glc | -H | *A. elata* | leaf | 19 |
| 129 | 1 | -GlcA$^3$-Glc<br>\|<br>Ara(*p*)$^2$ | -- | *A. elata* | root | 37 |
| 130 | 3 | -Glc3-Glc | -H | *A. elata* | sprout | 38 |
| 131 | 1 | -GlcA3-Glc6-Glc | -Glc6-Glc | *A. elata* | root | 40 |
| | | | | *A. elata* | leaf | 41 |
| | | | | *A. elata* | leaf | 41 |
| | | | | *A. elata* | leaf | 41 |
| 132 | 1 | -GlcA$^2$-Glc<br>\|<br>Glc$^3$ | -H | *A. elata* | young root | 26 |
| 133 | 1 | -Glc$^2$-Glc<br>\|<br>Glc$^3$ | -H | *A. elata* | young root | 26 |
| 134 | 1 | -Glc$^3$-Glc<br>\|<br>Gal$^2$ | -Glc | *A.chinensis* | root | 28 |
| 135 | 1 | -GlcA$^3$-Glc$^6$-Glc<br>\|<br>Ara(*f*)$^4$ | -Glc6-Glc | *A. elata* | root | 37 |
| 136 | 3 | -GlcA-Glc | -H | A. elata | leaf | 42 |
| 137 | 1 | -GlcA3-Glc | -H | *A. elata* | root | 11 |

## Quality Control

In recent years, scholars have undertaken extensive research in saponins of *Aralia* species. They isolated and identified a variety of Aralia saponins from different parts of the plants of the genus *Aralia*. The most commonly used method is the use of the HPLC analysis. Araloside A was determined by HPLC in root bark of A. elata [43]. The calibration curve showed good linearity over the range of 6.325～25.30μg (r =0.9999) and average recovery was 96.63% with RSD of 1.23%. So we can get the conclusion that this method is sensitive, accurate, specific and reproducible, it can be used for quality control of *A. elata*. The HPLC method is also used to determine Araloside A in leaves and their extracts [44,45]. However, the saponins in *Aralia* are complex, a single detection method is difficult to satisfy the requirement of quality control of *Aralia* genus. So multi-component of quantitation could become a future trend. Guo [46], *et al* reported on an extensive investigation of a set of *Aralia* saponins firstly isolated from leaves of *A. elata* by ESI-MSn. Some diagnostic key fragment ions that are important for the structure elucidation of *Aralia* saponins were characterized. The spectra were acquired in both positive and negative ion modes for the selected fragmentation route up to MS5. Two isomeric saponins from leaves of *A. elata*, which cannot be differentiated by negative ESI-MSn, can be successfully identified by positive ESI-MSn. Because of the isomeric saponins existing in the leaves of *A. elata*, ESI-MSn cannot be used for the direct analysis of the isomeric saponins in mixtures from crude plant extracts. An efficient HPLC-MS method has been successfully developed for the direct characterization of saponin in crude extracts from leaves of *A. elata,* thus avoiding time-consuming and material-consuming efforts in phytochemistry separation. Furthermore, the differentiation of two isomeric saponins by positive ESI-MSn was confirmed by HPLC-MS using in-source collision-induced dissociation (CID).

# Bioactivity

## Anti-Tumor Effect

*A. elata* was used as therapeutic medicine in oriental courtiers a few hundred years ago, as one important constituent extracted from *A.elata*, *A. elata* total saponins (AETS) has been shown to treat tumors. Li Rong[47] researched the antitumor activity of AETSs in vivo and the effect on immunological function in tumor-bearing mice. The result showed that AETSs could elevate the level of IL-2, IL-6, IL-12 and TNF-α. AETSs can inhibit the growth of the tumor and increase the tumor-bearing cytokine level [48]. Also, it can regulate growth and differentiation of lymphocytes, activated macrophages. During research on examining the antitumor effects of Araloside A on human kidney cancer cell lines GRC-1 and 786-O, Liu Yu found that Araloside A caused a considerable reduction of cellular viability of GRC-11 and 786-O cells in a time-dependent manner. The number of Tunnel-positive cancer cells was also higher in cells treated with Araloside A than untreated cells. The real-time polymerase chain reaction (PCR) techniques showed that Araloside A was able to increase the expression of bax mRNA and inhibited the expression of bcl-2 mRNA. Conclusively, Araloside A has a remarkably antitumor effect on kidney tumor cells through regulating bax/bcl-2 ratio [49].

Many reports showed that AETSs was able to inhibit the growth of human lung cancer xenografts of nude mice in vitro, inhibit COX-2 expression, and reduce microvessel density [50]. Additionally, Kuang HX [51] found AETSs could also inhibit the expression of mutant p53 protein in the H22 tumor-bearing mice, reducing the expression of PCNA protein. The mechanism may be related to inhibit the fission and proliferation of tumor cells or induce the apoptosis of tumor cells [52] [53]. So AETSs plays an important role in the anti-tumor.

## "Adaptogen"-Like Effect

Both AETSs and ginseng saponins were found to have the similar "adaptogen"-like effect that can make the body enhance its non-specific defense capability[54]. However, it was more toxic than ginseng saponins and also possessed a potent anti-fatigue activity. Low doses of AETSs could significantly enhance the superoxide dismutase activity to scavenge excess free radicals, slow down the degree of lipid peroxidation and reduce free radical damage in liver tissue [55]. Chinese scholars also found that AETSs obviously improved the hypoxia tolerance ability in mice, reduced the oxygen *consumption* rate, and improved the toxic lack of oxygen and cerebral circulation barrier [56].

## Effects on Nervous System

AETSs showed a potent inhibition of central nervous system and a certain analgesic effect. To study the analgesic effects of saponins from the foot bark of *Aralia* echinocaulis [57], the results showed A. echinocaulis foot bark obviously inhibited the swelling of a mouse ear induced by xylene and edema of rat hind paw elicited by egg white, significantly depressed the increase of proliferation of granulom a led by implanting a cotton pellet and the torsion body of rats, and decreased pain threshold induced by the hot plate method. Additionally, a synergistic effect of AETSs and pentobarbital and chlorpromazine was found to have an effect of central inhibition. Additionally, AETSs could antagonize amphetamine-induced excitability, however, the report showed that they could not antagonize convulsions and caffeine toxicity [58]. AETSs could also significantly reduce the anesthetic effect of chloral hydrate and barbital sodium [59]. Moderate AETSs had the tendency to reduce the spontaneous activity of rats, and increase sleep time of rats. Besides, the AETSs could minimize the effects of cardiac contraction, increase amplitude and excite the intestine [60].

## Effects on Heart

AETSs could enhance the heart contraction force in addition to the amplitude. Large doses of AETSs had no significant effect on respiration and had only momentary restraint sometimes, but they could lead to cardiac spasm.

Research showed that *A. cordata* has anti-ischemic activity and its active component is oleanolic acid [61]. Chinese scholars found that AETSs protected myocardial ischemia injury in rats. Some reports indicated that AETSs could improve ECG changes caused by isoproterenol. In addition, AETSs could significantly decrease the release of creatine phosphokinase (CPK) and reduce the release of free fatty acid (FFA) in animal ischemic myocardial tissue and plasma. Besides, AETSs could protect the SOD activity in myocardial tissue [62]. Liu and Chang found that AETSs could improve relative myocardial hypoxia in over-training rats, and reduce the formation of free radicals to alleviate the oxidative damage of myocardial mitochondria [63]. Cui Xianjun did an investigation about the mechanism of aralosides in antagonism of myocardial ischemia and provided a theoretical base for it used for preventing and treating coronary heart disease. The result showed that the expression of CD40L decreased, and vigour of myocardial GSH-PX increased in the Aralosides group. The level of MDA decreased in the aralosides group. So Aralosides can decrease the expression of CD40L, improve the vigour of myocardial GSH-PX and reduce the level of MDA in rats with myocardial ischemia and reperfusion[64]. AETS could significantly decrease the mortality of rats with myocardial infarction.

## Anti-Inflammatory Activity

Zhou [65] reported that AETSs could obviously inhibit carrageenan in rats. In addition, AETSs could also significantly reduce PGE2 and histamine of nystatin in rats. Besides, AETSs showed a significant inhibition of granulation tissue hyperplasia and leukocyte migration. The results demonstrated that AETSs could significantly inhibit kinin formation and release of lysosomal. Kim TD [66] also found that the root of *A. cordata* is a traditional medicine for the treatment of inflammation, fever, pain, and spasm in the various diseases in Korea. These results indicated that OSA has analgesic and anti-inflammatory effects due to its inhibitory effects on capillary permeability, COX activities, and histamine release. AETSs were also found to significantly increase cAMP in mouse kidney and lung and reduce cGMP. In brief, AETSs could be used to provide an immune basis for the treatment of rheumatism, rheumatoid arthritis, nephritis, lupus and other diseases [67].

## Anti-Diabetic Activity

Saponins are the active components for anti-diabetes. A previous study has demonstrated that the anti-diabetic activity of the extract of root bark of *A.taibaiensis* was correlated with its combined anti-oxidant and anti-glycation properties.

Researchers investigated the aldose reductase inhibitory activity, antioxidant activity and anticataract capacity of *Aralia* extract using various experimental systems. They found *Aralia* extract inhibits aldose reductase and acts in vitro as an antioxidant, and suggested that these activities have a preventive effect on in vivo in streptozotocin induced diabetic rats [68].

To further confirm the constituents responsible, 12 triterpenoid saponins were isolated from AETSs and examined for their antioxidant and anti-glycation activities. The antioxidant activities of the pure compounds and AETSs were evaluated by studying the inhibition of lipid peroxidation in rat liver microsomes induced by ascorbate/Fe2+, cumine hydroperoxide (CHP) or CCl4/reduced form of nicotinamide-adenine dinucleotide phosphate (NADPH). The antioxidant capacities were also evaluated by studying the scavenging of 2,2-diphenyl-1-picrylhydrazyl (DPPH) free radical. The antiglycation activities of the pure compounds and AETSs were evaluated by hemoglobin–δ-gluconolactone (δ-Glu) assay, bovine serum albumin (BSA)-glucose assay and N-acetyl-glycyllysine methyl ester (GK peptide)-ribose assay. AETSs outperformed other compounds in all the assays. The compounds with best antioxidant (TA7, TA24 and TA35) and antiglycation (TA21, TA9 and TA24) activities in different assays were screened out. The results suggest that the anti-oxidant and anti-glycation properties of AETSs could be explained, at least in part, by the synergistic effect of pure compounds isolated from it [69].

Song *et al.* [70] reported that blood glucose in normal mice could be reduced when AETSs were continuously administered at doses of 120 and 60 mg·kg-1 orally. Also, Kim et al. [71] found that oral administration of AETSs at doses of 100mg kg-1 and subcutaneous administration of AETSs (50–100 mg kg-1) inhibited hyperglycemia in diabetic mice induced by urinary purine. Gong [72] proved that AETSs could significantly reduce blood glucose. In another research, Jiang investigated the inhibition effect of AETSs on LH-60 white blood cell growth [73 74]. Meanwhile, AETSs could significantly reduce consumption of high cholesterol foods in mice and reduce protein levels. Besides, AETSs could restore phospholipid levels. Therefore, AETSs possess a good hypoglycemic activity. The saponin elatoside rat oral glucose tolerance test confirmed that it inhibited blood plasma glucose levels, while the separation of several oleanolic acid glycosides have a hypoglycemic effect, and confirmed that the 3-O-mono-demoside was the main structure of hypoglycemic effect [75 76].

To determine the effect of total saponin extracted from *A. taibaiens* is (AT-saponin) on oxidative damage of experimental type 2 diabetes mellitus in rats [77]. The results showed AT- saponin significantly improved the oxidative damage of experimental type 2 diabetic rats and showed good dosage effect relationship. In the serum, liver tissue, muscle tissue and adipose tissue, MDA content was decreased except in the muscle tissue, SOD activity was increased and the AGEs level was decreased in the high-dose and middle-dose AT- saponin groups. The pathological symptoms of pancreas were relieved in the high-dose and middle-dose AT- saponin group. So we could say that AT- saponin relieves the oxidative damage of experimental type 2 diabetes mellitus and consequently protects against STZ- induced pancreatic injury in rats.

## Hepatoprotective Effects

Nishida examined the protective effects of AETSs on liver injury. The results indicated that AETSs could reduce transaminase, suggesting that AETSs had significant protective effects on carbon tetrachloride-induced acute liver injury in rats. Also, AETS could be used for the treatment of acute jaundice hepatitis. Japanese scholars also reported that AETSs could effectively prevent CCl4-induced liver injury, and showed more potent prevention than glycyrrhizins [78]. In Korea, the root of *Aralia* continentalis Kitagawa has been used in traditional Korean medicine to relieve pain and to treat inflammation. They wanted to investigate the protective effects of the extract of *A. continentalis* roots (AC) against hepatotoxicity induced by carbon tetrachloride (CCl4) and the mechanism of its hepatoprotective effect. And the results indicate that AC plays a critical protective role in CCl4-induced acute liver injury by promoting anti-oxidative protein expression [79]. Additionally, AETSs could inhibit the release of alanine aminotransferase [80]. Yi *et al.* and Zheng *et al.* found that AETSs could significantly inhibit the proliferation of NIH3T3 cells. AETSs could also reduce the number of cells to resist hepatic fibrosis [81] [82]. Some reports also showed that AETSs had some effects such as reducing enzyme and dropping turbidity [83]. It might be an ideal drug for the treatment of acute and chronic hepatitis. Wang et al. [84] administered AETSs from 50 to 300 mg·kg-1 orally for 5 days, and the results showed that AETSs significantly reduced ALT and GOT levels caused by acute liver injury. Thus, it proved that increasing nucleic acids of liver cells, protein synthesis and glycogen synthesis and energy synthesis might be the mechanism for protection of AETSs on liver injury. A study observed the effect of the *A. elata*(Miq.) They took the damaged organism of the overtraining rates as a model, and extracted to the liver tissue's lipid peroxidation, SOD and GPT. [85] They randomly divided male wistar rats into four groups: sedentary control group(A), training group(B), training and administered with the low dose of *A. elata*( Miq.) (D), and training and administered with the high dose of *A.elata*(Miq.) (E). After 6 weeks swimming training by the plan, and singly bewitch the groups of D and E by the low and the high dose of Extraction of the *A. elata*( Miq.), at last, the content of the MDA and the activity of GPT and SOD was determined. At last they found after overtraining, the liver tissue of rats', the content of the MDA and the activity of the serum GPT increased remarkably, and the liver tissue of the activity of SOD drops. Low dose extraction significantly inhibits the trend. The effect of high dose of *A. elata* (Miq.) extraction is not obvious. So they said low dose of Extraction of the *A. elata*(Miq.) has a preventative effect against the Oxidative Damage to the rat's liver during overtraining. All these experiments prove that the saponins have a hepatoprotective effect.

## Alcoholism Treatment

Japanese scholars reported that total saponins from *A. elata* Seem could inhibit alcohol absorption to treat alcoholism and monosaccharide oxygen glycoside in C-3 of oleanolic acid showed a strong inhibition of alcohol absorption. However, disaccharides oxygen glycosides in positions 3 and 28 of oleanolic acid showed weaker antibacterial activity, which indicated that monosaccharide oxygen glycoside in position 3 of oleanolic acid was a necessary structure for the treatment of alcoholism [86].

## Anti-Prostaglandin and Anti-Epoxy Nucleotide Activities

AETSs could induce synthesis of prostaglandins E2 and E20, and showed a significant increase of cAMP content of cyclooxygenase nucleotide and a significant decrease of cGMP content, but they had no effect on histamine release [87] [88]. Besides, AETSs could greatly increase prostate-seminal vesicle, levator ani muscle-sponge balls muscle and the weight of the adrenal gland in mice, and considerably reduce the content of VC in adrenal glands of rats [89].

## Other Activities

AETSs were reported to exhibit a marked inhibitory effect against lipid oxidation [90] and the production of MDA. In addition, AETSs could improve catalase CAT and superoxide dismutase SOD activities in rats [91]. Deng *et al.* reported that AETSs could protect the ischemic myocardial tissue SOD in mice from myocardial ischemia induced by isoproterenol [90]. In their subsequent work, the effects of AETSs on various gastric ulcer models in mice and rats had been also investigated. AETSs showed an obvious effect on scavenging free radicals, which provided a basis for anti-aging and myocardial ischemia. Besides, Wu *et al.* reported that AETSs could inhibit the mutant P53 expression ($p<0.05$), which indicated that AETSs could reverse mutational P53 gene [92].

Joon-Soo SIM investigated the intestinal absorption enhancing effect of the saponins from the root bark of *A. elata* in Caco-2 cell monolayers and rats. AETSs at concentrations of 0.04% and 0.08% (w/v) decreased the transepithelial electrical resistance (TEER) values and increased the paracellular uptake of chondroitin sulfates (CSs) having different molecular weights (MW 500, 4500, and 18000) in a dose-dependent manner. They also evaluated the cytotoxicity of SRBAE to determine its proper concentration as an absorption enhancer.

In conclusion, these results suggest that AETSs act as an efficient absorption enhancer and makes it easier for hydrophilic molecules to penetrate the intestinal epithelium [93]. So, AETSs could protect cartilage and inhibit apoptosis. Also, some studies showed that *Aralia* extract inhibits aldose reductase and acts in vitro as an antioxidant, and suggested that these activities have a preventive effect on cataractogenesis [94].

# Conclusion

Nowadays, Araliaceae genus has attracted more and more attention of scientists, who not only isolated chemical components but also evaluated pharmacological activities and clinical applications. Many research has been done on the chemistry, biology, and pharmacology in the last decades, and great achievements have been made. Pharmacological investigations of these plants reported anti-tumor effect, "adaptogen"-like effect effects on nervous system, effects on heart, anti-inflammatory activity, anti-diabetic activity, hepatoprotective effects, alcoholism treatment, anti-prostaglandin and anti-epoxy nucleotide activities and so on. Due to rich resources and extensive applications in the field of medicine, AETSs could be widely used as an alternative medicine. It was very common to study chemical compositions and pharmacological effects. However, the mechanisms of some effects are not clear yet. In the future study, more detailed works about these plants are still needed.

# References

[1] Tropicos. Missouri Botanical Garden. Available oneline: http://www.tropicos.org/NameSearch.aspx?name= Sambucus/ (accessed on 18 April 2011).

[2] S. G. Sparg, M. E. Light, J. van Staden. *Journal of Ethnopharmacology*, 94, 219. (2004).

[3] Yong Pil Hwang, Jae Ho Choi, Eun Hee Han, Hyung Kyun Kim, Shin Keon Kang, Young Chul Chung, Hye Gwang Jeong. *Food and Chemical Toxicology*, *46*, 3512. (2008).

[4] Da-Hong Lee, Bo-Ra Seo, Ha-Yeon Kim, Gi-Chun Gum, Hyeon Hee Yu, Hyung-Keun You, Tong Ho Kang, Yong-Ouk You. *Journal of Ethnopharmacology*, 137, 979. (2011).

[5] Huang FS, Zhao HT, *et al.* Research of Aralia leaves saponins ultrasonic extraction process optimization. (2011).

[6] Liu J, Xu J, *et al. Modern Food Science and Technology*, 622. (2010).

[7] Liu Y, Tian J, *et al. Chinese Traditional and Herbal Drugs*, 694. (2011).

[8] Lv SW, Liu D, Xia YG, Yang BY, Xiao HB, Kuang HX, *Joumal of Chinese Medicinal Materials*, 33,1959, (2010).

[9] Saito, S, Sumita, S, Tamura, N, *et al. Chem. Pharm. Bull*, 38, 411. (1990).

[10] Yoshikawa, M, Matsuda, H, Harada, E, Elatoside, E. *Chem. Pharm. Bull*, 42, 1354. (1994).

[11] Yoshikawa, M, Mastuda, H, Harada, E, *Chem. Pharm. Bull*, 41, 2069. (1993).
[12] Fang, ZP, Lei, JL, Zeng, XY. *Bulletin of Botany*, 37,74. (1995).
[13] Hu, M, Ogawa, K, Sashida, Y, *et al. Phytochemistry*,39,179. (1995).
[14] Jiang, YT, Xu, SX, Chen, YJ. *Journal of Shenyang Pharmaceutical University*, 8, 265.(1991).
[15] Kawai, H, Nishida, M, Tashiro, Y, *et al. Chem.Pharm.Bull*,37,2318. (1989).
[16] Yu, SS, Yu, DQ, Liang, XT. *J Nat Prod*, 57,978. (1994).
[17] Jiang, YT, Xu, SX, Chen, YJ. *Acta Pharmaceutica Sinica*, 27,528. (1992).
[18] Sun, WJ, Zhang, DK, Sha, ZF,*et al. Acta Pharmaceutica Sinica*, 26,197. (1991).
[19] Yang, GH, Zhou, J, Song, CC, *et al. Chinese Traditional and Herbal Drugs*, 26, 514. (1995).
[20] Saito, S, Ebashi, J, Sumita, S, Furumoto, T, Nagamura, Y, Nishida, K, Ishiguro, *Chem,Pharm,Bull*, 41, 1395. (1993).
[21] Satoh, Y, Sakai, S, Katsumata, M.*et al. Phytochemistry*, 36, 147. (1994).
[22] Kochetkov, NK, Khorlin, AJ, Vaskovsky, VE. *Dokl Akad Nauk*, 50, 1289. (1963).
[23] Jiang, YT, Xu, SX, Ren, L,*et al. Chinese Traditional and Herbal Drugs*, 23, 403. (1992).
[24] Fang, ZP, Zhou, UX, Zeng, XY. *Bulletin of Botany*, 34, 461. (1992).
[25] Zhu, ZY, Cui, LY, Hu, BH *et al. Acta Pharmaceutica Sinica*, 29, 840. (1994).
[26] Yoshikawa, M, Yoshizumi, S, Ueno, T *et al Chem.Pharm.Bull*, 43, 1878. (1995).
[27] Miyase, T, Sutoh, N, zhang, D, *Phytochemistry*, 41, 1411. (1996).
[28] Miyase, T, Sutoh, N, zhang, D, *Phytochemistry*, 42,1123. (1996).
[29] Tang, HF, Yi, YH, Wang, ZZ *et al. Acta Pharmaceutica Sinica*, 31, 517. (1996).
[30] Tang, HF, Yi, YH, Wang, ZZ *et al. Acta Pharmaceutica Sinica*, 32, 685 (1997).
[31] Tang, HF, Wang, ZZ, Yi, YH *et al. Chinese Pharmceutical Journal*, 6, 75. (1997).
[32] Yi, YH, Gu, JQ, Xiao, K *et al. Acta Pharmaceutica Sinica*, 32, 769. (1997).
[33] Kuang, HX, San, H, Zhang, N.*et al. Chem.Pharm.Bull*, 44, 2183. (1996).
[34] Song, SJ, Nakamula, N, Ma, CM *et al, Chem Pharm Bull*, 48, 838. (2000).
[35] Song, SJ, Nakamula, N, Ma, CM *et* al. *Phytochemistry*, 56, 491. (2001).
[36] Sakai, S, Katsumata, M *et al. Phytochemistry*, 35, 1319. (1994).
[37] Niu, XS, Tao, SL, Ma, ZQ, Song, SJ. *Acta Chinese Medicine and Pharmacology*, 37, 52. (2009).
[38] Ma, ZQ, Song, SJ, Chen, GT, Yang, YJ, Xu, SY. *Journal of Shenyang Pharmaceutical University*, 268. (2004).
[39] Ma, ZQ, Song, SJ, Zhang, GJ, Xu, SX. bud, 21, 230. (2004).
[40] Song, SJ, Peng, Y, Lu, LM, Xu, SY. *Journal of Shenyang Pharmaceutical University*. 18, 233. (2001).
[41] Wang, QH, Zhang, J, Ma, X, Ye, XY, Yang, BY, Xia, YG, Kuang, HX. *Chinese Journal of Natural Medicines*,9, 17 (2011).
[42] Wang, QH, Zhang, J, MA, X, YE, XY, Yang,BY, Xia,YG, Kuang,HX,*Chinese Journal of Natural Medicines*. 9, 17, (2011).
[43] Tian, J, Liu,Y, *et al. Journal of Luzhou Medical College*,141-143. (2010).
[44] Zhang, Y, Song, SJ, Ma, ZQ, *Journal of Shen yang Pharmaceutical University*, 810. (2008).
[45] Tian, J, *et al, Li Shi Zhen medicine and materia medica research*, 2426. (2009).

[46] Guo, MQ, Zhang, L, Liu, ZQ, *The Japan Society for Analytical Chemistry*, 25, 753, (2009).
[47] Li, R, *et al Shizhen medicine and materia medica research*, 2199. (2011).
[48] Makoto, T, *et al Cancer letters*, 20. (2003).
[49] Yin, LY, Kuang, HX, *et al. Acta Chinese Medicine and Pharmacology*. 18. (2010).
[50] Wang L. Y, Xiao HB, Zhang XP, Wang MH, Li LY, *Acta Chinese Medicine and Pharmacology*, 38,30. (2010).
[51] Yin, LY, Bian, XY, Xiao, HB, Kuang, HX, *Acta Chinese Medicine and Pharmacology*, 38, 18, (2010).
[52] Yin, LY, Bian, XY, Xiao, HB. *Acta Chinese Medicine and Pharmacology*,27,107. (2010).
[53] Yin, LY, Bian, XY, Xiao, HB, Kuang. HX. *Acta Chinese Medicine and Pharmacology*, 38, 18 (2010).
[54] Ren, Meiping, Liu, Minghua Chen, Yi, Xiao, Shunhun, LI, Hua, Zhang, Zhuo. *Lishizhen Medicine and Materia Medica Research*, 20, 2417. (2009).
[55] Wang, QF, Yi, XJ. *Journal of Shenyang Sport University*, 26, 63. (2007).
[56] Huang, LH, Jiang, P. *Northwest Pharmaceutical Journal,* 3, 13. (1988).
[57] Chen, GD, Wen, DJ, *Lishizhen medicine and materia medica research*, 997. (2008).
[58] Gao, B, Zhang, P, Fei, ZX. *Fourth Military Medical University Journal*, 9, 55. (1988).
[59] Xu, XD, Yang, JS, Zhu, ZY. *Acta Pharmaceutica Sinica*, 32, 711. (1997).
[60] Hou, JR, Yang, SH, Sun, XB, Xu, HB. *Special Wild Economic Animal and Plant Research,* 22, 53. (2000).
[61] Soon, OC, *et al. Archives of Pharmacal Research*, 923. (2009).
[62] Deng, HW, Li, YJ, Shen, N, Chen, X, Zhou, ZC. *Chinese Journal of Pharmacology and Toxicology*, 2, 20. (1988).
[63] Liu, JH, Chang, B. *Journal of Shenyang Sport University*, 26, 54. (2007).
[64] Cui, XJ, Li, M, Lu, WX, *Journal of Beijing University of Traditional Chinese Medicine*, 323. (2010).
[65] Zhou, CC. *Journal of Pharmacology and Toxicology*, 5, 30. (1991).
[66] Kim, TD, Lee, JY. *Arch Pharm Res*. (2010).
[67] Zhou, CC. *Journal of Pharmacology and Toxicology,* 5, 34. (1991).
[68] Young Shin, Chung, *et al. Journal of Ethnopharmacology*.49. (2005).
[69] Xi, MM, *et al.* Antioxidant and antiglycation properties of triterpenoid saponins from Aralia taibaiensis traditionally used for treating diabetes mellitus, 20. (2010).
[70] Song, SJ, Xu, SX, Cao, YL *Traditional Chinese Medicine Research and Information*, 7, 7. (2005).
[71] Kim, OK, Lee, EB, Kang, SS *Saengyak Hakhoechi*, 24, 219. (1993).
[72] Gong, XB. *Chinese Medicine and Pharmacology*, 2, 56. (1989).
[73] Jiang, YT. *Pharmaceutica Sinical*, 27, 528. (1992).
[74] Jiang, YT. *Journal of Shenyang Pharmaceutical University*, 8, 290. (1991).
[75] Yosh, ikaw aM, M urakam i T, *Chem Pharm Bull*(Tokyo), 44, 1923. (1996).
[76] Yosh ikaw aM, *et al. Chem Pharm Bull*(Tokyo), 43, 1878. (1995).
[77] Tang, HF, *et al. Central South Pharmacy*,1-5. (2009).
[78] Nishida, K. *Igaku to Seibutsugaku*, 123, 271. (1991).

[79] Yi, L, Ji, HW, Wei, XZ, *Chinese Journal of Integrated Traditional and Western Medicine on Liver Diseases,* 18, 196, (2008).

[80] Yong, PH, *et al. Food and Chemical Toxicology*,75. (2009).

[81] Zheng, LL, Pei, LP, *China National Journal of Medicine*, 6, 57. (2010).

[82] Song, GQ, Xu, YL, Wang, GJ, Pu, JB, Huo, H. *Chinese Journal of Experimental Traditional Medical Formula*, 7, 17. (2001).

[83] He, YZ. *International Journal of Chinese Medicine volumes*, 12, 63. (1990).

[84] Wang, LY, Xiao, HB, Wang, MH. *Chinese Journal of Information on Traditional Chinese Medicine*, 26, 29. (2009).

[85] Yoshikawa, M, Matsuda, H, Harada, E, Murakami, T, Wariishi, N, Yamahara, J, Murakami, N. Elatoside E, *Chemical Pharmaceutical Bulletin,* 42, 1354. (1994).

[86] Wang, QF, Yi, XJ, *Journal of Shenyang Sport University*, 64. (2007)

[87] Satoh, Y, Sakai, S, Katsumata, M, Nagasao, M, Miyakoshi, M, Ida, Y, Shoji, J, *Phytochemistry*, 36, 147. (1994).

[88] Ren, J. *Chinese Pharmacological Bulletin*, 4, 150. (1998).

[89] Sun, XB. *Chinese herbal medicine*, 22, 78. (1991).

[90] Jiang, YT. *Journal of Shen-yang College of Medicine*, 8, 290. (1991).

[91] Zhou, CC, *Chinese Journal of Pharmacology and Toxicology*, 5, 30. (1991).

[92] Deng, HW. *Chinese Journal of pharmacology and Toxicology*, 2, 20. (1998).

[93] Wu, BY, Zhu, SX, Wang, J. *Chinese Pharmaceutical Journal*, 31, 52. (2003).

[94] Joon-Soo, Sim, *et al. Pharm.Bull.* 1043. (2005).

In: Phytochemicals
Editor: Hai-Xue Kuang
ISBN: 978-1-62417-354-7


*Chapter XVI*

# Chemical Constituents and Bioactivities of Plants of the Genus *Hydrangea*

***Yang Chunjuan*[*] *and Xiao Yang***
College of Pharmacy, Harbin Medical University, Harbin, China

## Abstract

Hydrangea is one of the largest genera of Hydrangeaceae (Saxifragaceae), comprising about seventy-three species. They are distributed in eastern Asia, and eastern-north America, and extend southward to the tropic of both hemispheres. The plants of genus Hydrangea are always used as ornaments, flowers and a kind of natural medicine. They mainly contain isocoumarins, secoididoid glucosides, iridoid glucoside, sesquiterpene glucosides, triterpenes, flavonol glycosides, cyanogenic glycosides, alkaloids, coumarins, phenolic acids and their derivants. The isocoumarins are the principle constituents and their contents are higher than the other type constituents. The bioactivities of the crude extracts and the compounds of Hydrangea include antiallergic activity, immunomodulatory activity, inhibitory effect on histamine release, inhibitory effects on degranulation and release of cells, antimicrobial and antifungal activities, anti-diabetic activities, hepatoprotective and cholagoic activity, anti-malarial activities and antiulcer activities. The present review summarized the chemistry and pharmacological effects of the genus.

## 1. Introduction

*Hydrangea* is one of the largest genera of Hydrangeaceae (Saxifragaceae), comprising about seventy-three species. The genus is divided into five sections, i.e. sect. Petalanthea (Maxim). Wei, sect. Heteromallae (Rehd.) Wei, sect. Hydrangea, sect. Cornidia (Ruizrt Pav.) Engl. and sect. Calyptranthe Maxim.They are distributed in eastern Asia, eastern-north

[*] E-mail address: chunjuanyang@126.com.

America, and extend southward to the tropic of both hemispheres. The plants of genus *Hydrangea* are always used as ornaments flowers and a kind of natural medicine. In Japan, the processing Hydrangea Dulcis Folium (HDF), the fermented and dried leaves of *Hydrangea macrophylla* SERINGE var. *Thunbergii* MAKINO, even is listed in the Japanese Pharmacopoeia VIII, and extensively used in confectionary, drinks, and foods as an oral refrigerant and sweeting. Forty-six species and ten varieties are now recognized in China and mostly concentrated in the regions of south-west China, south China, central China, eastern China and Taiwan. The main plants of *Hydrangea* genus are listed in Figure 1.

Plants of the genus *Hydrangea* are known to contain isocoumarins, dihydroisocoumarins and their glycosides, iridoid glycosides, secoiridoid glycosides, which are the main constituents of the genus. Flavones and their glycosides, alkaloids, are isolated from the plants of this genus, and the cyanogenic glycoside compounds are the specified type constituents from the natural medicines HDF. The classical sweeting isolated from this genus, a dihydroisocoumarins, phyllodulcin is a hydrongenol derivative, which is 1000 times sweeter than sugar. The sepal color of the plants of this genus is famous for its color changes under different cultivation conditions. *H. macrophylla cv.* Hovaria is known as the chameleon hydrangea, because its sepals change color four times during maturation and senescence (Figure 2). It is clarified that these changes are relative with the anthocyanin and the co-pigments.

Except for as a sweetener, the plants of this genus are extensively used in confectionery, drinks, and an oral refrigerant. HDF is the crude drug mentioned in Japanese pharmacopoeia, and has found that the methanolic extract of this natural medicine exhibited potent antiallergic, antiulcer, antimicrobial and antifungal, antimalarial and cholagoic activities.

*H.* macrophylla Seringe var. Thunbergii Makino.

*H.* chinensis Maxim.

*H.* anomala D. Don.

*H.* macrophylla (Thunb.) Ser. Var. normalis Wils.

*H.* mangshanensis Wei. Sp. Nov.

Figure 1. (Continued)

*H.chungii* Rehd. In.

*H*. paniculata Sieb.

*H*. obovatifolia Hayata.

*H*. heteromalla D. Don.

*H*. sargentiana Rehder.

*H*. xanthoneura Diels var. setchuenensis Rehd.

Figure 1. (Continued)

*H*. sargentiana Rehd.

*H*. heteromolle D. Don.

*H*. obovatifolia Hayata.

*H. serrata* (Thunb.) Ser. f. acuminata (Sieb.et Zucc.) auct. non Wils.

*H.* strigosa Rehd.

Figure 1. The main plants of *Hydrangea* genus.

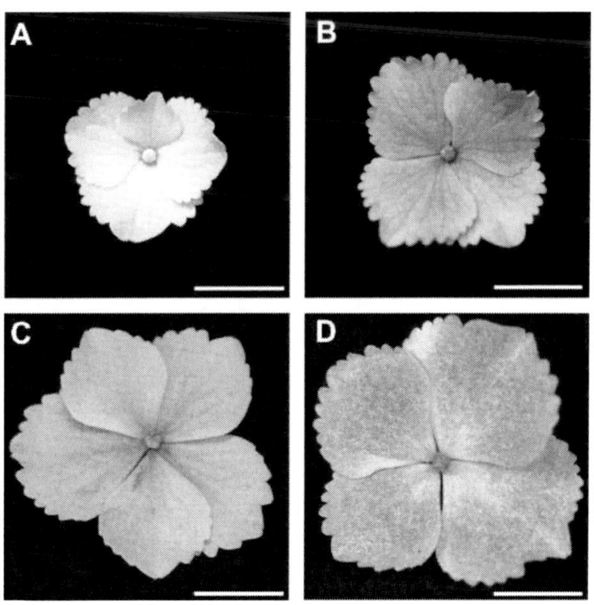


Figure 2. The sepal color changes of *H. macrophylla* cv. Hovaria during maturation and senescence.

The activities of chemical constituents are also investigated including differentiation inducing activities against M1 cells, immunomodulatory activities on Lymphocyte proliferation, inhibitory effect on histamine release, etc. In this review, we summarize the chemical constituents isolated from *Hydrangea* and the bioactivities of the crude extracts and the compounds of this genus.

# 2. Chemical Constituents

139 Compounds obtained from the plants of *Hydrangea* can be assigned to the following compound classes: isocoumarins, secoiridoid glycosides, iridoid glucoside, flavonoid glycosides, cyanogenic glycosides, alkaloids, triterpenes, norsesquiterpenoids, coumarins and phenolic acids [1-57]. The chemical constituents from *Hydrangea* genus are listed in Table 1.

**Table 1. The chemical constituents from *Hydrangea* genus**

| No. | Name | Source | References |
|---|---|---|---|
| 1 | thunberginol A | Hydrangeae Dulcis Folium | [1] [2] |
| 2 | thunberginol B | Hydrangeae Dulcis Folium | [1][2] |
| 3 | 3'-deoxythunberginol | Hydrangeae Dulcis Folium | [3] |
| 4 | 6-hydroxy-dehydrophyllodulcin | Hydrangeae Dulcis Folium | [3] |
| 5 | dehydrophyllodulcin | Hydrangeae Dulcis Folium | [4] |
| 6 | thunberginol C | Hydrangeae Dulcis Folium | [1] |
| 7 | thunberginol D | Hydrangeae Dulcis Folium | [1] |
| 8 | thunberginol E | Hydrangeae Dulcis Folium | [1] |
| 9 | thunberginol G | Hydrangeae Dulcis Folium | [1] |
| 10 | 3'- hydroxy- thunberginol D | *H. macrophylla* (THUNB.) Ser. | * |
| 11 | hydrangenol | *H. macrophylla* SERINGE var. *Thunbergii* MAKINO | [5][6] [46] [53] |
| | | *H. macrophylla* subsp. *Serrata* (THUNB.) MAKINO | [15] |
| | | Hydrangeae Dulcis Folium | [45] |
| 12 | 3*R*-phyllodulcin | *H. macrophylla* SERINGE var. *Thunbergii* MAKINO | [6][7][8] [44][46] 53] |
| | | *H. thunbergii* Sieb | [9] |
| | | *H. opuloides* Steud. var. *Thunbergii* | [11] [52] |
| | | *H. serrata* SERINGE var. | [51] |
| 13 | 3*S*-phyllodulcin | *H. macrophylla* SERINGE var. *Thunbergii* MAKINO | [6][7][8] [52] |
| 14 | phyllodulcin monomethyl ether | Hydrangeae Dulcis Folium | [10] |
| 15 | thunberginol A dimethyl ether | Hydrangeae Dulcis Folium | [10] |
| 16 | macrophyllol | *H. macrophylla* subsp. *Serrata* (THUNB.) MAKINO | [11] |

| No. | Name | Source | References |
|---|---|---|---|
| 17 | 8-acetoxy-3-(3-cetoxy-4-methoxyphenyl)-3,4-dihydroisocoumarins | *H. serrata* SERINGE var. *Thunbergii* | [5] |
| 18 | 3,4-dihydro-3-(3,4-dimethoxyphenyl)-8-hydroxy isocoumarin | *H. serrata* SERINGE var. *Thunbergii* | [5] |
| 19 | 8-acetoxy-3-(acetoxyphenyl)-3,4-dihydroisocoumarin | *H. serrata* SERINGE var. *Thunbergii* | [5] |
| 20 | 3,4-dihydro-8-hydroxy-3-(p-methoxyphenyl)isocoumarin | *H. serrata* SERINGE var. hunbergii | [5] |
| 21 | 3,4-dihydro-8-hydroxy-3-(p-hydroxyphenyl)isocarbostyryl | *H. serrata* SERINGE var. *Thunbergii* | [5] |
| 22 | 8-hydroxy-3-(acetoxyphenyl)-3,4-dihydroisocoumarin | *H. macrophylla* (THUNB.) Ser. | * |
| 23 | thunberginol F | Hydrangeae Dulcis Folium | [1][2] [48] |
| 24 | hydramacrophyllol A | Hydrangeae Dulcis Folium | [4][49] |
| 25 | hydramacrophyllol B | Hydrangeae Dulcis Folium | [4] |
| 26 | thunberginol G 3'-*O*-glucoside | Hydrangeae Dulcis Folium | [1] |
| 27 | 3*R*-thunberginol I 4'-*O*-glucoside | *H. macrophylla* SERINGE var. *Thunbergii* MAKINO | [7] |
| 28 | 3*S*-thunberginol I 4'-*O*-glucoside | *H. macrophylla* SERINGE var. *Thunbergii* MAKINO | [7] |
| 29 | thunberginol I 8-*O*-glucoside | *H. macrophylla* SERINGE var. *Thunbergii* MAKINO | [7] [13] |
| | | Hydrangeae Dulcis Folium | [12] |
| 30 | 3*S*-phyllodulcin 8-*O*-glucoside | *H. macrophylla* SERINGE var. *Thunbergii* MAKINO | [7][13] |
| | | Hydrangeae Dulcis Folium | [45] |
| 31 | hydrangenol 8-*O*-glucoside | *H. macrophylla* SERINGE var. *Thunbergii* MAKINO | [7][14] |
| | | *H. macrophylla* subsp. *serrata* | [43] |
| | | Hydrangeae Dulcis Folium | [45] |
| 32 | 3*R*-hydrangenol-4'-*O*-apiosylglucoside | *H. macrophylla* SERINGE var. *Thunbergii* MAKINO | [7] |
| 33 | 3*S*-Hydrangenol-4'-*O*-apiosylglucoside | *H. macrophylla* SERINGE var. *Thunbergii* MAKINO | [7] |
| 34 | 3*R*-hydrangenol 4'-*O*-glucoside | Hydrangeae Dulcis Folium | [1] |
| | | *H. macrophylla* SERINGE var. *Thunbergii* MAKINO | [7] |
| 35 | 3*S*-hydrangenol 4'-*O*-glucoside | *H. macrophylla* SERINGE var. *Thunbergii* MAKINO | [7][14] |
| 36 | 3*R*-phyllodulcin 8-*O*-glucoside | *H. macrophylla* SERINGE var. *Thunbergii* MAKINO | [7][8] [13] |
| 37 | 3*R*-phyllodulcin 3'-*O*-glucoside | *H. macrophylla* SERINGE var. *Thunbergii* MAKINO | [7] [8] |

**Table 1. (Continued)**

| No. | Name | Source | References |
|---|---|---|---|
| 38 | 3*S*-phyllodulcin 3'-*O*-glucoside | *H. macrophylla* SERINGE var. *Thunbergi*i MAKINO | [7] |
| 39 | 3*R*-hydrangenol 8, 4'-di-*O*-*β*-D-glucopyranoside | *H. macrophylla* subsp. *Serrata* (THUNB.) MAKINO | [14] [15] |
| 40 | 3*R*-thunberginol H 8-*O*-glucoside | *H. macrophylla* SERINGE var. *thunbergii* MAKINO | [8] |
| 41 | 3*S*-thunberginol H 8-*O*-glucoside | *H. macrophylla* SERINGE var. *thunbergii* MAKINO | [8] |
| 42 | hydramacroside A | *H. macrophylla* SERINGE var. *thunbergii* MAKINO | [16][17] |
| 43 | hydramacroside B | *H. macrophylla* SERINGE var. *thunbergii* MAKINO | [16][17] |
| 44 | hydrangenoside A | *H. macrophylla* Ser. var. *macrophylla* | [18] |
| | | *H. macrophylla* subsp. Serrata (THUNB.) MAKINO | [19] |
| | | *H. hortensis* | [56] |
| 45 | hydrachoside A | *H. chinensis* Shimizu and Kao | [20] |
| 46 | hydrangenoside E | *H. chinensis* Shimizu and Kao | [20] |
| | | *H. scandens*(LINN.fil.) SERINGE | [38] |
| 47 | hydrangenoside C | *H. macrophylla* (THUNB.) Ser. var. *macrophylla* | [21] |
| | | *H. scandens*(LINN.fil.) SERINGE | [38] |
| 48 | hydrangenoside A dimethyl acetal | *H. macrophylla* subsp. *Serrata* (THUNB.) MAKINO | [19] |
| 49 | hydrangenoside B | *H. macrophylla* (THUNB.) Ser. var. *macrophylla* | [21] |
| | | *H. scandens*(LINN.fil.) SERINGE | [38] |
| 50 | hydrangenoside D | *H. macrophylla* (THUNB.) Ser. var. *macrophylla* | [21] |
| | | *H. scandens*(LINN.fil.) SERINGE | [38] |
| 51 | hydrangenoside E | *H. scandens*(LINN.fil.) SERINGE | [22] [38] |
| 52 | hydrangenoside | *H. scandens*(LINN.fil.) SERINGE | [22] [38] |
| 53 | hydrangenoside G | *H. scandens*(LINN.fil.) SERINGE | [22] [38] |
| 54 | hydrangenoside H | *H. macrophylla* (THUNB.) Ser. var. *macrophylla* | [33] |
| 55 | macrophyllanoside A | *H. macrophylla* subsp. *serrata* | [23] |
| 56 | macrophyllanoside B | *H. macrophylla* subsp. *Serrata* | [23] |
| 57 | macrophyllanoside C | *H. macrophylla* subsp. *Serrata* | [23] |
| 58 | macrophyllanoside D | *H. macrophylla* subsp. *serrata* | [23] |
| 59 | vogeloside | *H. macrophylla* SERINGE var. *thunbergii* MAKINO | [17] |
| 60 | *epi*-vogeloside | *H. macrophylla* SERINGE var. *thunbergii* MAKINO | [17] |
| 61 | secologanin dimethyl acetal | *H. macrophylla* subsp. *Serrata* (THUNB.) MAKINO | [19] |
| 62 | n-butyl vogeloside | *H. macrophylla* subsp. *Serrata* (THUNB.) MAKINO | [19] |
| 63 | n-butyl *epi*-vogeloside | *H. macrophylla* subsp. *Serrata* (THUNB.) MAKINO | [19] |

| No. | Name | Source | References |
|---|---|---|---|
| 64 | (7*R*)-*n*-butyl morroniside | *H. macrophylla* subsp. *Serrata* (THUNB.) MAKINO | [19] |
| 65 | 7-deoxyloganic acid *β*-D-glucopyranosyl ester | *H. macrophylla* subsp. *Serrata* (THUNB.) MAKINO | [15] |
| 66 | asystasioside A | *H. macrophylla* subsp. *Serrata* (THUNB.) MAKINO | [15] |
| 67 | tecomoside | *H. macrophylla* (THUNB.) Ser. | * |
| 68 | ugandoside | *H. macrophylla* (THUNB.) Ser. | * |
| 69 | euphroside | *H. macrophylla* (THUNB.) Ser. | * |
| 70 | 7-deoxygardoside | *H. macrophylla* (THUNB.) Ser. | * |
| 71 | adoxosidic acid | *H. macrophylla* (THUNB.) Ser. | * |
| 72 | Mussaenosidic acid | *H. macrophylla* (THUNB.) Ser. | * |
| 73 | (6*R*,7*E*,9*R*)-megastigma 4,7-dien-3-one 9,13-di-*O*-*β*-D-glucopyranoside | *H. macrophylla* subsp. *Serrata* (THUNB.) MAKINO | [15] * |
| 74 | inamoside | *H. macrophylla* subsp. *Serrata* (THUNB.) MAKINO | [15] |
| 75 | citroside A | *H. macrophylla* SERINGE var. *thunbergii* MAKINO | [8] |
| 76 | blumenol C glucoside | *H. macrophylla* SERINGE var. *thunbergii* MAKINO | [8] |
| 77 | (6*R*,7*E*,9*R*)-9-hydroxy-megastigma-4,7-dien-3-one-9-*O*-*β*-D-glucoside | *H. macrophylla* SERINGE var. *thunbergii* MAKINO | [8] |
| 78 | ampelopsisionoside | *H. macrophylla* (THUNB.) Ser. | * |
| 79 | lauroside A | *H. macrophylla* (THUNB.) Ser. | * |
| 80 | 23-hydroxy-3*β*-[(*O*-*β*-D-glucopyranosyl)oxy]lup-20(29)-en-28-oic acid 28-*O*-*β*-D-glucopyranosyl ester | *H. macrophylla* (THUNB.) Ser. | * |
| 81 | 23-hydroxy-3b-[(*O*-*β*-D-glucopyranosyl)oxy]lup-20(29)-en-28-oic acid | *H. macrophylla* (THUNB.) Ser. | * |
| 82 | 23-hydroxy-lup-20(29)-en-28-oic acid 28-*O*-*β*-D-glucopyranosyl ester | *H. macrophylla* (THUNB.) Ser. | * |
| 83 | betulic acid 3',5'-dioxy cinnamic acid ester | *H. macrophylla* (THUNB.) Ser. | * |
| 84 | isoarborinol | Hydrangeae Dulcis Folium | [2] [10] |
| 85 | rubiarbonol B | *H. macrophylla* SERINGE var. *thunbergii* MAKINO | [8] |
| | | Hydrangeae Dulcis Folium | [2] [10] |
| 86 | prunin | *H. macrophylla* SERINGE var. *thunbergii* MAKINO | [8] |
| 87 | kaempferol 3-*O*-*β*-rutinoside | *H. macrophylla* SERINGE var. *thunbergii* MAKINO | [25] |
| 88 | kaempferol 3-*O*-*β*-D-xylopyranosyl-(1→2)-*β*-D-galactopyranoside | *H. macrophylla* SERINGE var. *thunbergii* MAKINO | [25] |

**Table 1. (Continued)**

| No. | Name | Source | References |
|---|---|---|---|
| 89 | quercetin 3-*O*-*β*-D-glucopyranosyl(1→2)-*β*-D-glucopyranoside | *H. macrophylla* SERINGE var. *thunbergii* MAKINO | [8] |
| 90 | kaempferol 3-*O*-*β*-D-glucopyranosyl(1→2)-*β*-D-glucopyranoside | *H. macrophylla* SERINGE var. *thunbergii* MAKINO | [8] |
| 91 | kaempferol 3-*O*-*α*-L-rhamnopyranosyl-(1→6)-[*β*-D-xylopyranosyl-(1→2)]-*β*-D-galactopyranoside | *H. macrophylla* SERINGE var. *thunbergii* MAKINO | [25] |
| 92 | kaempferol 3-*O*-*β*-D-glucopyranosyl(1→2)-[*α*-L-rhamnopyranosy(1→6)]*β*-D-glucopyranoside | *H. macrophylla* SERINGE var. *thunbergii* MAKINO | [8] |
| 93 | delphinidin 3-glucoside | *H. macrophylla* sepal | [24][35] [37] |
| 94 | 3-*O*-*β*-D-glucopyranosyl delphinidin | *H. macrophylla* cv. Hovaria | [32] |
| 95 | 3-*O*-*β*-D-glucopyranosyl cyanidin | *H. macrophylla* cv. Hovaria | [32] |
| | | *H. macrophylla* | [54] |
| 96 | 3-*O*-(2-*O*-*β*-D-xylopyranosyl)-*β*-D-glucopyranosyl delphinidin | *H. macrophylla* cv. Hovaria | [32] |
| 97 | 3-*O*-(2-*O*-*β*-D-xylopyranosyl)-*β*-D-glucopyranosyl cyanidin | *H. macrophylla* cv. Hovaria | [32] |
| 98 | hydracyanoside A | *H. macrophylla* (Thunb.) Ser. | [26] |
| 99 | hydracyanoside B | *H. macrophylla* (Thunb.) Ser. | [26] |
| 100 | hydracyanoside C | *H. macrophylla* (Thunb.) Ser. | [26] |
| 101 | hydranitriloside $A_1$ | *H. macrophylla* (Thunb.) Ser. | [27] |
| 102 | hydranitriloside $A_2$ | *H. macrophylla* (Thunb.) Ser. | [27] |
| 103 | hydranitriloside $B_1$ | *H. macrophylla* (Thunb.) Ser. | [27] |
| 104 | hydranitriloside $B_2$ | *H. macrophylla* (Thunb.) Ser. | [27] |
| 105 | isofebrifugine | *H. macrophylla* var. Otaksa | [47] |
| 106 | hydrachine A | *H. chinensis* Shimizu and Kao | [20] |
| | | *H. chinensis* | [29] |
| 107 | (-)-neodichroine | *H. chinensis* Shimizu and Kao | [20] |
| 108 | acetone adduct | *H. chinensis* | [29] |
| 109 | febrifugine | *H. chinensis* | [29] |
| | | *H. macrophylla* (THUNB.) Ser. var. *macrophylla* | [47][50] |
| 110 | 7-hydroxycoumarin(umbelliferone) | Hydrangeae Dulcis Folium | [10] |
| | | *H. macrophylla* (THUNB.) Ser. var. *macrophylla* | [41] [42] [55] |
| | | *H. hortensis* | [56] |
| 111 | umbelliferone 7-*O*-glucoside | *H. macrophylla* SERINGE var. *thunbergii* MAKINO | [8] |

| No. | Name | Source | References |
|---|---|---|---|
| 112 | 8-methoxy umbelliferone 7-*O*-glucoside | *H. macrophylla* (THUNB.) Ser. | * |
| 113 | 7- methoxycoumarin | *H. chinensis* | [29] |
| 114 | 7-hydroxy 8-methoxycoumarin | *H. chinensis* | [29] |
| 115 | 5-hydroxycoumarin | *H. chinensis* | [29] |
| 116 | dihydroresveratrol | Hydrangeae Dulcis Folium | [10] |
| 117 | hydrangeic acid | Hydrangeae Dulcis Folium | [4] [10][48] |
| | | *H. macrophylla* SERINGE var. *thunbergii* MAKINO | [34] |
| 118 | 4-hydroxy-hydrangeaic acid | Hydrangeae Dulcis Folium | [4] |
| 119 | 3'-hydroxy-hydrangeaic acid | Hydrangeae Dulcis Folium | [30] |
| 120 | 3,4'-dihydroxybibenzyl-2-carboxylic acid | *H. macrophylla* roots | [40] |
| 121 | 3,4'-dihydroxybibenzyl | *H. macrophylla* roots | [40] |
| 122 | 3,4'-dimethoxy-2-hydroxybibenzyl | *H. macrophylla* roots | [40] |
| 123 | hydrangea glucoside A | *H. serrata* SERINGE var. *thunbergii* SUGIMOTO | [57] |
| 124 | hydrangea glucoside B | *H. serrata* SERINGE var. *thunbergii* SUGIMOTO | [57] |
| 125 | 3-*O*-caffeoylquinic acid | *H. macrophylla* | [31] |
| | | *H. macrophylla* cv. Hovaria | [32] |
| 126 | 5-*O*-caffeoylquinic acid | *H. macrophylla* | [31] |
| | | *H. macrophylla* cv. Hovaria | [32] |
| 127 | 5-*O*-*p*-coumaroylquinic acid | *H. macrophylla* | [31] |
| | | *H. macrophylla* cv. Hovaria | [32] |
| 128 | 3-*p*-Coumaroylquinic acid | Hydrangea flower | [36] |
| | | *H. macrophylla* sepal | [37] |
| 129 | hydrangea glucoside C | *H. serrata* SERINGE var. *thunbergii* SUGIMOTO | [57] |
| 130 | *p*-aminophenyl-*O*-D-glucoside | *H. macrophylla* SERINGE var. *thunbergii* MAKINO | [39] |
| 131 | caffeic acid-4-*O*-*β*-D-glucopyranoside | *H. macrophylla* (THUNB.) Ser. | * |
| 132 | coumaric acid-4-*O*-*β*-D-glucopyranoside | *H. macrophylla* (THUNB.) Ser. | * |
| 133 | vanillic acid-4-*O*-*β*-D-glucopyranoside | *H. macrophylla* (THUNB.) Ser. | * |
| 134 | isovanillic acid-4-*O*-*β*-D-glucopyranoside | *H. macrophylla* (THUNB.) Ser. | * |
| 135 | syringaldehyde | *H. chinensis* | [29] |
| 136 | 4-hydroxy-cinnamic acid methyl ester | *H. chinensis* | [29] |
| 137 | *p*-hydroxybenzaldehyde | *H. chinensis* | [29] |
| 138 | protocatechuic acid | *H. chinensis* | [29] |
| 139 | gallic acid | *H. serrata* SERINGE var. *thunbergii* SUGIMOTO | [57] * |

Hydrangeae Dulcis Folium is the fermented and dried leaves of H. macrophylla SERINGE var. Thunbergii MAKINO.

* We obtained these compounds from H. macrophylla (THUNB.) Ser.. The dried leaves of H. macrophylla were collected from Shizhu, Sichuan Province, China in May.

## 2.1. Isocoumarins

Forty-one isocoumarin-type compounds were isolated from the plants of this genus. Among these compounds, five compounds, namely thunberginol A (1), thunberginol B (2) [1, 2], 3'-deoxythunberginol (3), 8-dehydrohydrangenol A (4) [3] and dehydrophyllodulcin (5) [4] presented double bond between C-3 and C-4 positions, and there no isocoumarins glycosides was obtained from this genus with double bond between C-3 and C-4 positions. Fifteen dihydroisocoumarins aglycone were obtained in the previous researches, and they are named thunberginol C~G (6~9) [1], 8-dehydrangenol (10), hydrangenol (11) [5, 6, 15, 45, 46, 53], 3*R*-phyllodulcin(12) [6~9, 44, 46, 51~53], 3*S*-phyllodulcin (13) [6~8, 52], phyllodulcin monomethyl ether (14), thunberginol A dimethyl ether (15) [10], macrophyllol (16) [11], 8-acetoxy-3-(3-cetoxy-4-methoxyphenyl)-3,4-dihydroisocoumarins (17), 3,4-dihydro-3-(3,4-dimethoxyphenyl)-8-hydroxyisocoumarin (18), 8-acetoxy-3-(acetoxyphenyl)-3,4-dihydro-isocoumarin (19), 3,4-dihydro-8-hydroxy-3-(*p*-methoxyphenyl)isocoumarin (20), 3,4-di-bhydro-8-hydroxy-3-(p-hydroxyphenyl)isocarbostyryl (21) [5], 8-hydroxy-3-(acetoxyphenyl)-3,4-dihydroisocoumarin (22), thunberginol F (23) [1, 2, 48], hydramacrophyllol A (24) [4, 9] and hydramacrophyllol B(25) [4]. A chiral carbon at the C-3 position is a special structural characteristic for the compounds of dihydroisocoumarins. The absolutely stereostructures of (±) phyllodulcin, the important constituent from *H. macrophylla*, was determined through the CD spectrum, and the compounds 3*R*-phyllodulcin (12) and 3*S*-phyllodulcin (13) showed the different characteristic CD curve ($[\theta]^{25}$ (nm): -8000 (240), +10000 (256), +2500 (314); $[\theta]^{25}$ (nm): +12700 (240), -10800 (256), -2900 (314); resp.). Meanwhile, the 3*R*- and 3*S*-phyllodulcin from enantiomeric mixture was separated by HPLC with chiral column, and the quality ratio of compound 12 versus 13 was 5:1 [8].

A novel dihydroisocoumarins 21 with the azote replacement in 1 position and forming intramolecular hydrogen bond was identified from *H. serrata* SERINGE var. *Thunbergii*. Compounds 23, 24 and 25 were the phthalide derivatives obtained from the HDF and belonged to the benzylidenephthalide type isocoumarins.

Sixteen isocoumarin glycosides compounds were isolated from this genus, and all the glycosides were the dihydroisocoumarin derivatives. Thunberginol G 3'-*O*-glucoside (26) [1], 3*R*-phyllodulcin 3'-*O*-glucoside (37) [7, 8] and 3*S*-phyllodulcin 3'-*O*-glucoside (38) [7] took place the glucoside replacement in C-3' position. 3*R*-thunberginol I 4'-*O*-glucoside (27), 3S-ThunberginolI 4'-*O*-glucoside (28), 3*R*-hydrangenol-4'-*O*-apiosylglucoside (32), 3*S*-Hydrangenol-4'-*O*-apiosylglucoside (33) [7], 3*R*-hydrangenol 4'-*O*-glucoside (34) [1, 7] and 3*S*-hydrangenol 4'-*O*-glucoside (35) [7, 14]took place the glucoside replacement in C-4' position. Thunberginol I 8-*O*-glucoside (29) [7, 12, 13], 3*S*-phyllodulcin 8-*O*-glucoside (30) [7, 13, 45], hydrangenol 8-*O*-glucoside (31) [7, 14, 43, 45], 3*R*-phyllodulcin 8-*O*-glucoside (36) [7, 8, 13], 3*R*-thunberginol H 8-*O*-glucoside (40) and 3*S*-thunberginol H 8-*O*-glucoside (41) [8] took place the glucoside replacement in C-8 position. (3*R*)-hydrangenol 8, 4'-di-*O*-*β*-D-glucopyranoside (39) [14, 15] was only dioxy-substitution glycosides. The absolutely stereostructures of dihydroisocoumarin glycosides were clarified by the circular dichroism spectrum and the chiral HPLC.

Isocoumarin-type compounds are characteristic constituent of the genus *Hydrangea*. Mehmet and Andreas reported on efficient synthetic route by which compounds 11, 16 and (±) phyllodulcin can be obtained easily and in high yield (Figure 3) [11].

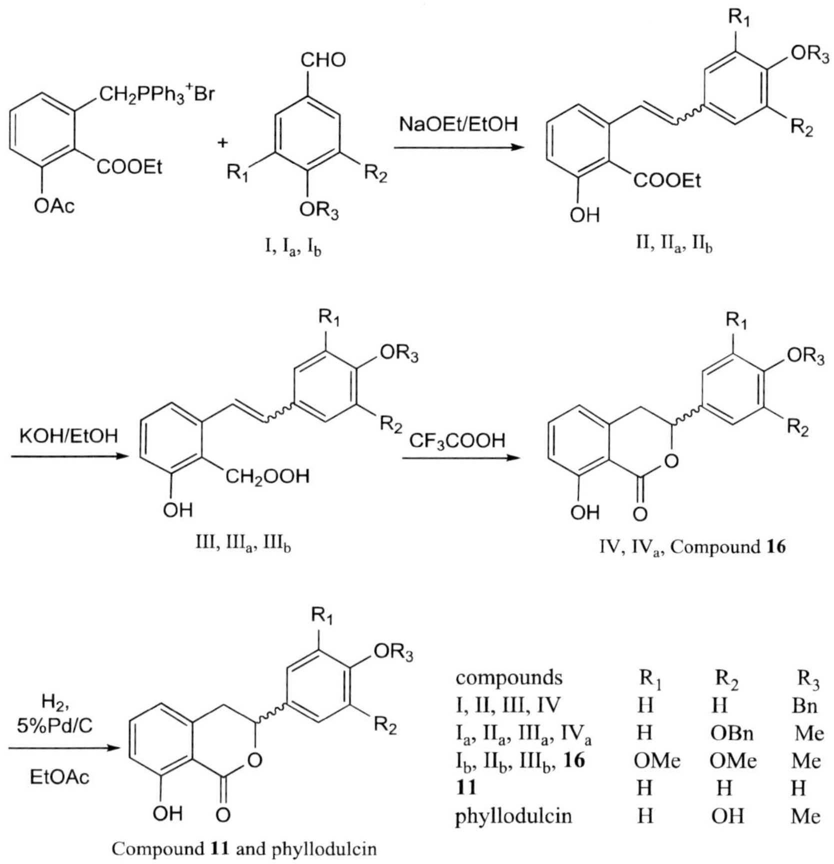


| compounds | $R_1$ | $R_2$ | $R_3$ |
|---|---|---|---|
| I, II, III, IV | H | H | Bn |
| $I_a$, $II_a$, $III_a$, $IV_a$ | H | OBn | Me |
| $I_b$, $II_b$, $III_b$, **16** | OMe | OMe | Me |
| **11** | H | H | H |
| phyllodulcin | H | OH | Me |

Figure 3. Synthesis of compounds 11, 16 and (±) phyllodulcin.

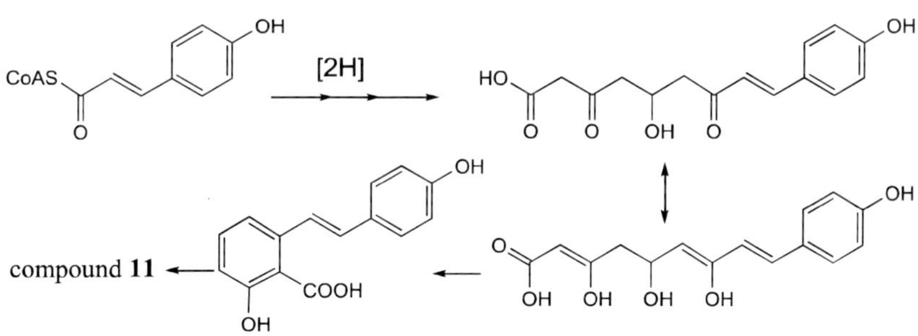


Figure 4. The biosynthesis of stilbenecarboxylates and derivatives in *H. macrophylla*.

Christian et al. suggest that stilbenecarboxylates (hydrangeic acid) can be synthesized by plant type III polyketide synthases and are derived from 4-coumaroyl starter residues (Figure 4).

The formation of the backbones can be explained by a stilbene-type reaction with 4-coumaroyl-CoA that retains the terminal carboxyl group (stilbenecarboxylate synthase) [28].

## 2.2. Secoididoid Glucosides

Secoididoid glucosides type structure is thought to be biosynthesized through an aldol-type condensation of secologanin with a unit formed by the shikimate- malonate route followed by decarboxylation. Twenty-three secoididoid glucosides were obtained from the plants of genus *Hydrangea*. Thirteen secoididoid glucosides with aromatic lateral chain isolated and structure elucidation from species of *H. macrophylla*, *H. chinensis* and *H. scandens* Seringe, named hydramacroside A (42), hydramacroside B (43) [16, 17], hydrangenoside A (44) [18, 19, 56], hydrachoside A (45) [20], hydrangenoside E (46) [20, 38], hydrangenoside C (47) [21, 38], hydrangenoside A dimethyl acetal (48) [19], hydrangenosides B (49), hydrangenosides D (50) [21, 38], hydrangenosides E (51), hydrangenosides F (52) hydrangenosides G (53) [22, 38] and hydrangenosides H (54) [H]. Macrophyllanosides A (55), macrophyllanosides B (56), macrophyllanosides C (57) and macrophyllanosides D (58), belonged to a novel type of secoididoid glucosides that connected secologanin and 1-*O*-glyceryl-*β*-D-galactopyranoside, shikimic acid or myo-inositol by acetal structure [23]. Vogeloside (59), *epi*-vogeloside (60) [17], secologanin dimethyl acetal (61), n-butyl vogeloside (62) and n-butyl *epi*-vogeloside (63) [19] were the secoididoid glucosides presented the vogeloside skeleton, and the glycon were all *β*-D-glucose. (7*R*)-n-butyl morroniside (64) [19], 62 and 63 were the secoididoid glucosides with the normal-butyl lateral chain. The side chain of secoididoid glucosides type structure possessed chirality carbon, and the absolutely stereostructures of these compounds were always elucidated on the basis of chemical and physicochemical evidence which compared with the similar known structure and used the modified Mosher's method [16, 17].

## 2.3. Iridoid Glucosides

Eight iridoid glucosides were isolated from *H. macrophylla* (THUNB.) Ser. and *H. macrophylla* subsp. *Serrata* (THUNB.) MAKINO. 7-deoxyloganic acid *β*-D-glucopyranosyl ester (65) and 7-deoxyloganic acid *α*-D-glucopyranosyl ester (66) [15] were the glucopyranosyl ester that connected 11-position hydroxyl group of iridoid acid and the glycon. In our research, tecomoside (67), ugandoside (68) euphroside (69), 7-deoxygardoside (70), adoxosidic acid (71) and mussaenosidic acid (72) were obtained from the plants of *H. macrophylla* (THUNB.) Ser.. Compounds 67, 68 and 69 were the iridoid glucosides with the aldehyde group at C-11 position and the compounds 70, 71 and 72 with the carboxyl group at C-11 position.

## 2.4. Norsesquiterpenoid Glucosides

Seven sesquiterpene glucosides were isolated from Hydrangeae Dulcis Folium [8, 15] and *H. chinensis*. (6*R*, 7*E*, 9*R*)-megastigma 4,7-dien-3-one 9,13-di-*O*-*β*-D-glucopyranoside (73), inamoside (74) [15], citroside A (75), blumenol C glucoside (76) and (6*R*,7*E*,9*R*)-9-hydroxy-megastigma-4,7-dien-3-one-9-*O*-*β*-D-glucoside (77) [8]. We obtained ampelopsisionoside (78) and lauroside A (79) from the EtOH extract of *Hydrangea chinensis*, compound 78, 79 were isomerides and the optical rotation value $[\alpha]_D^{25}$ were -4.8°, -30.6°, resp.

## 2.5. Triterpenes

Lupine- (*β*-amyrane) and fernaned-type triterpene were funded from *Hydrangea* genus, and 23-hydroxy-3b-[(*O*-*β*-D-glucopyranosyl)oxy]lup-20(29)-en-28-oic acid 28-*O*-*β*-D-glucopyranosyl ester (80), 23-hydroxy-3*β*-[(*O*-*β*-D-glucopyranosyl)oxy]lup-20(29)-en-28-oic acid (81), 23-hydroxy-lup-20(29)-en-28-oic acid 28-*O*-*β*-D-glucopyranosyl ester (82) and betulic acid 3', 5'-dioxy cinnamic acid ester (83) were isolated from *H. macrophylla*. Isoarborinol (84) and rubiarbonol (85) were obtained from *H. macrophylla* SERINGE var. *thunbergii* MAKINO [2, 8, 10].

## 2.6. Flavonol Glycosides

Twelve flavonol glycosides including flavonone, flavon and cyanidin were obtained from the genus *Hydrangea*. Prunin (86), kaempferol 3-*O*-*β*-rutinoside (87), kaempferol 3-*O*-*α*-L-rhamnopyranosyl-(1→6)-[*β*-D-xylopyranosyl-(1→2)]-*β*-D-galactopyranoside (88), quercetin3-*O*-*β*-D-glucopyranosyl(1→2)-*β*-D-glucopyranoside (89), kaempferol 3-*O*-*β*-D-glucopyranosyl(1→2)-*β*-D-glucopyranoside (90), kaempferol 3-*O*-α-L-rhamnopyranosyl-(1→6)-[*β*-D-xylopyranosyl-(1→2)]-*β*-D-galactopyranoside (91), kaempferol3-*O*-*β*-D-glucopyranosyl(1→2)-[*α*-L-rhamnopyranosy(1→6)]*β*-D-glucopyranoside (92), were obtained from the leaves of *H. macrophylla* SERINGE var. *thunbergii* MAKINO [8, 25].

Delphinidin 3-glucoside (93) [24, 35, 37], 3-*O*-*β*-D-glucopyranosyl delphinidin (94) [32], 3-*O*-*β*-D-gluc opyranosyl cyaniding (95) [32, 54], 3-*O*-(2-*O*-*β*-D-xylopyranosyl)-*β*-D-glucopyranosyl delphinidin (96), 3-*O*-(2-*O*-*β*-D-xylopyranosyl)-*β*-D-glucopyranosyl cyanidin (97) [32] were isolate from *H. macrophylla* sepal. These compounds were relative with the sepal colors of the pants of *Hydrangea* in different stages.

## 2.7. Cyanogenic Glycosides

In June 2008, the cases of food poisoning that present symptoms of sickness or disgust, etc. after eating the leaves of *H. macrophylla* were generated in Osaka and Ibaraki prefectures of Japan. The Ministry of Health, Labor, and Welfare of Japan advised the possibility that the food poisoning was caused by cyanogenic glycosides in the leaves of *H. macrophylla*. In April 2009, Three new cyanogenic glycosides named hydracyanosides A (98), B (99), and C (100) were isolated from the stems of *H. macrophylla* in China. The cyanogenic glycosides would generate the cyanhydric acid (HCN), which was a kind of poisonous material under the acid conditions (Figure 5). The absolute stereostructures of hydracyanoside were characterized on the basis of NMR data and crystal X-ray crystallographic analysis [26].

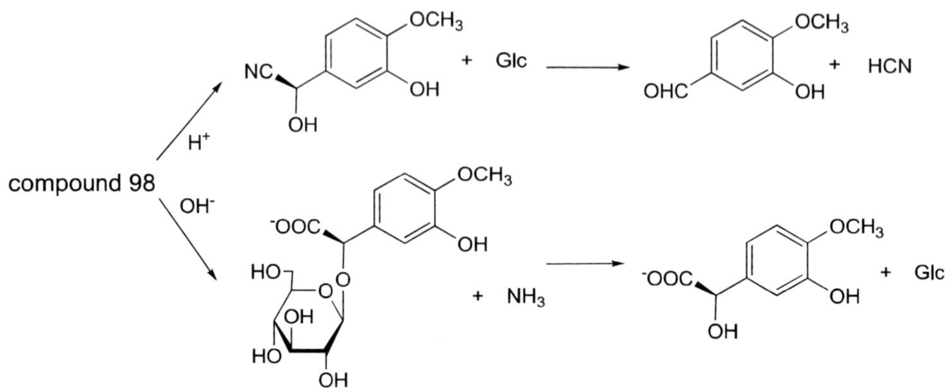


Figure 5. The degradation product of hydracyanosides A in acid and base condition.

Four non-cyanogenic cyanoglucosides including hydranitrilosides $A_1$ (101), $A_2$ (102), $B_1$ (103), and $B_2$ (104), Non-cyanogenic glucosides are less common than their cyanogenic counterparts [27]. They represent a small group of compounds comprising *ca.* 30 representatives with a scattered distribution in plants. While their function in plants is not fully understood, they possess some interesting bioactivities. For example, they were found to enhance the antibiotic activity of rifampicin, tetracycline, and ampicillin against the *Gram*-positive and *Gram*-negative bacteria, and also to enhance the absorption of drugs, vitamins, and nutrients through the gastrointestinal membrane and increase their bioavailability.

## 2.8. Alkaloids

Five alkaloid were obtained from *H. chinensis* in Taiwan of China. Isofebrifugine (105), hydrachine A (106), (-)-neodichroine (107), acetone adduct (108), febrifugine (109) [20, 29, 47, 50] have a same mother nucleus of quinazolone, while compounds 107~109 have a same side chain of quinolizine.

## 2.9. Coumarins

Six coumarins and their glucosides were isolated from the genus *Hydrangea.* 7-hydroxycoumarin (110), umbelliferone 7-*O*-glucoside (111), 8-methoxy umbelliferone 7-*O*-glucoside (112), 7- methoxycoumarin (113) and 7-hydroxy 8-methoxycoumarin (115) [8, 10, 41, 42, 55, 56] at 7-position of their structure have a hydroxyl or D-glucose, and 5-hydroxycoumarin (114) [29] at 5-position have a hydroxyl.

## 2.10. Phenolic Acids and Derivants

In the literature, many phenolic acids and their derivant were reported. Dihydroresveratrol (116) [10], hydrangeic acid (117) [4, 10, 34, 48], 4-hydroxy-hydrangeaic acid (118) [4], 3'-hydroxy-hydrangeaic acid (119) [30], 3,4'-dihydroxybibenzyl-2-carboxylic acid (120) [40], 3,4'-dihydroxybibenzyl (121) [40], 3,4'-dimethoxy-2-hydroxybibenzyl (122) [40], hydrangea glucoside A (123), hydrangea glucoside B (124) [57] have bibenzyl structure and important contribution in the biosynthesis course of the isocoumarin and secoididoid possessed special side chain (Figure 6).

The sepal color of *H. macrophylla* cv. Hovaria changes in four stage, from colorless to blue, then to green, and finally to red, during maturation and the senescence periods. At each stage the components of the sepals had been changes. The chemical constituents include compounds 94~97, 3-*O*-caffeoylquinic acid (125), 5-*O*-caffeoylquinic acid (126), 5-*O*-*p*-coumaroylquinic acid (127) [31, 32], and 3-*p*-Coumaroylquinic acid (128) [36, 37].

Hydrangea glucoside C (129), gallic acid (139) [57] were isolated from *H. serrata* SERINGE var. *thunbergii* SUGIMOTO. *p*-Aminophenyl-*O*-D-glucoside (130) [39] were isolated from *H. macrophylla* SERINGE var. *thunbergii* MAKINO. caffeic acid-4-*O*-*β*-D-glucopyranoside (131), coumaric acid-4-*O*-*β*-D-glucopyranoside (132), vanillic acid-4-*O*-*β*-D-glucopyranoside (133), isovanillic acid-4-*O*-*β*-D-glucopyranoside (134) were isolated from *H. macrophylla* (THUNB.) Ser.. syringaldehyde (135), 4-hydroxy-cinnamic acid methyl ester (136), *p*-hydroxybenzaldehyde (137), protocatechuic acid (138) [29]were isolated from *H. Chinensis.*

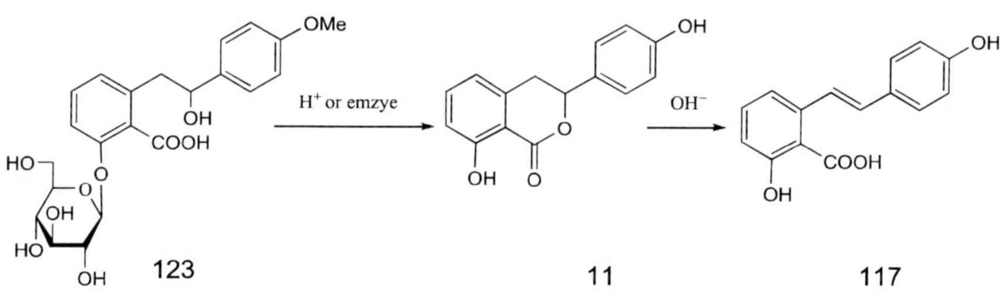


123 11 117

Figure 6. The product of hydrangea glucoside A (123) in acid (or emzye) and base condition.

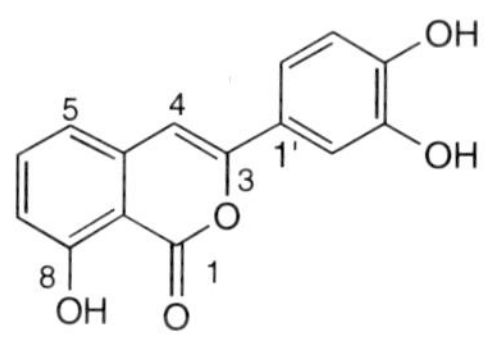


1

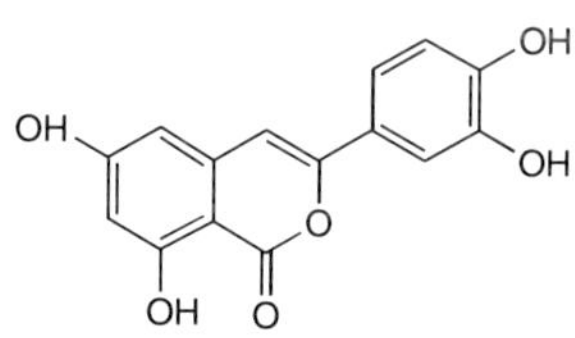


2

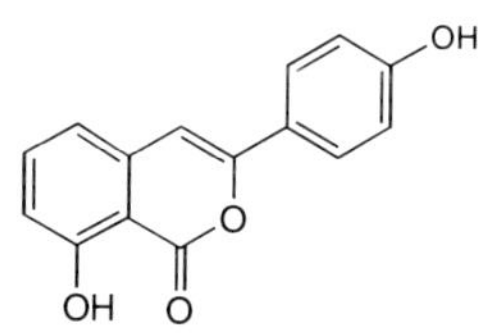


3

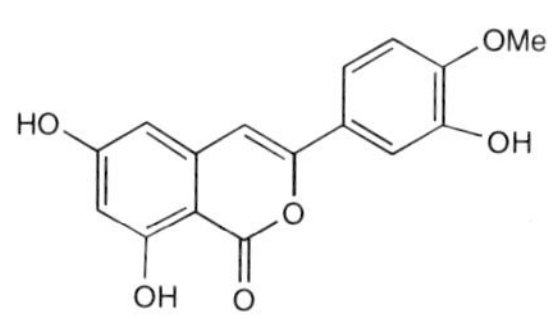


4

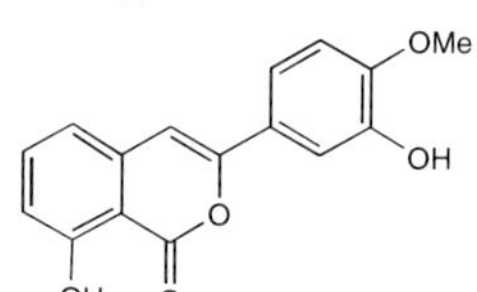


5

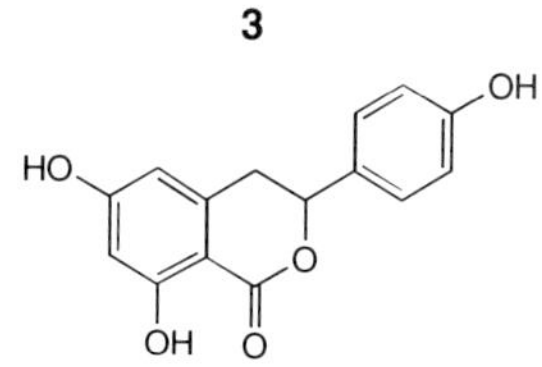


6

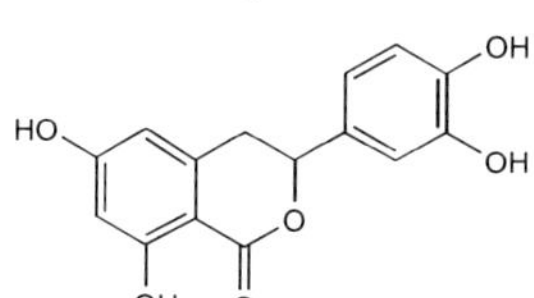


7

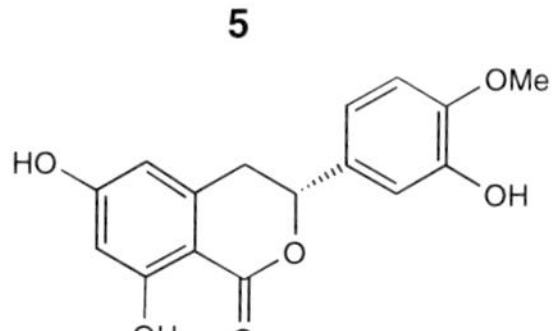


8

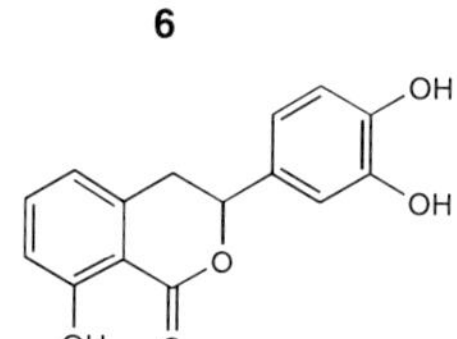


9

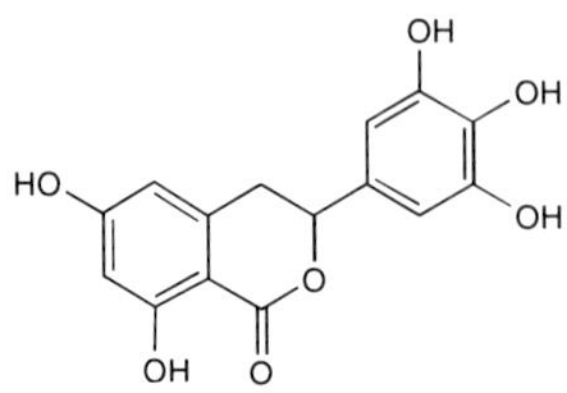

**10**

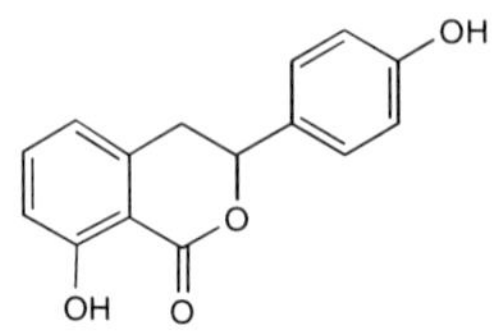

**11**

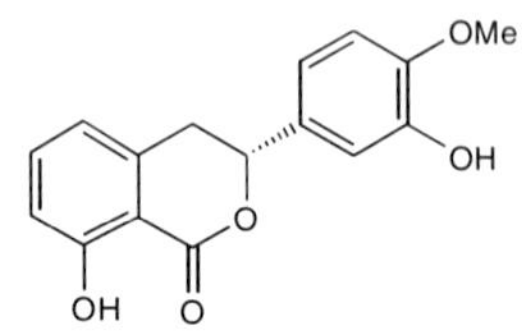

**12**

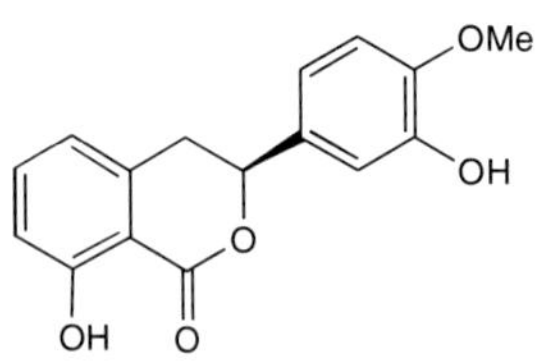

**13**

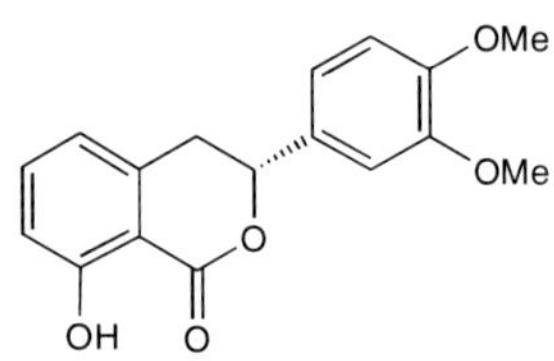

**14**

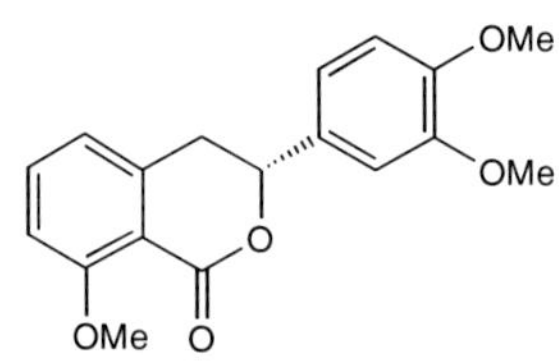

**15**

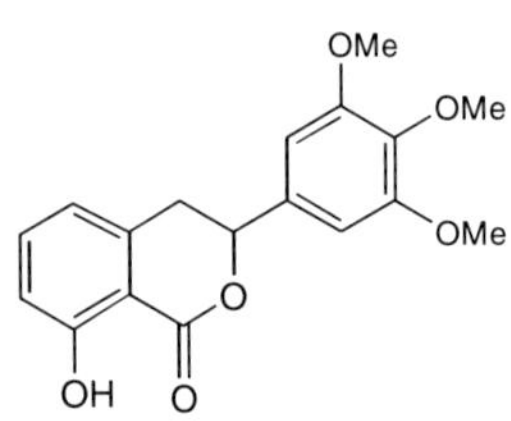

**16**

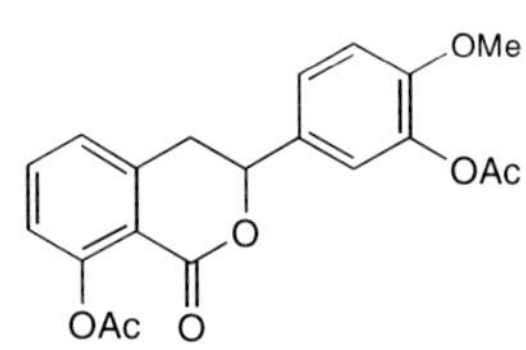

**17**

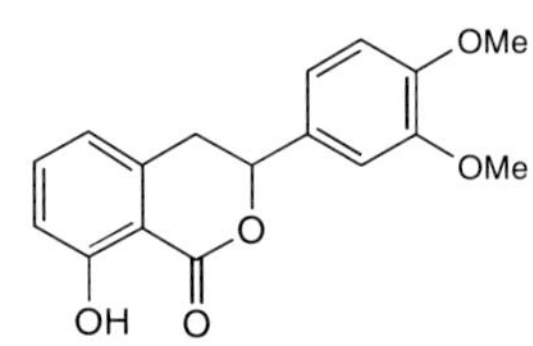

**18**

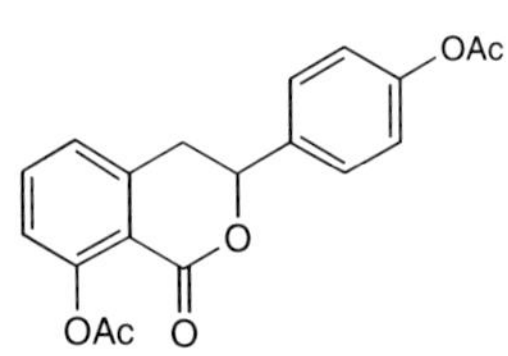

**19**

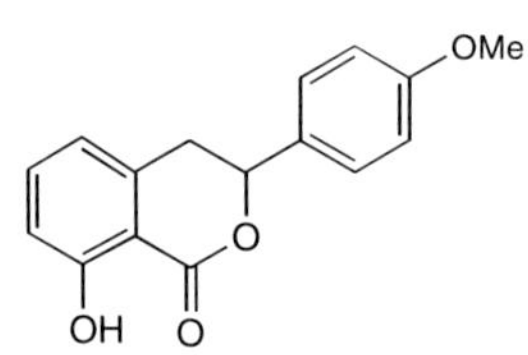

**20**

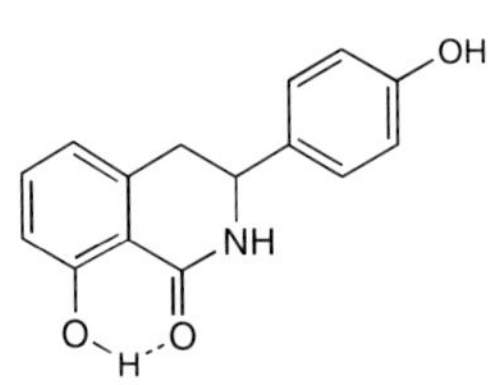

**21**

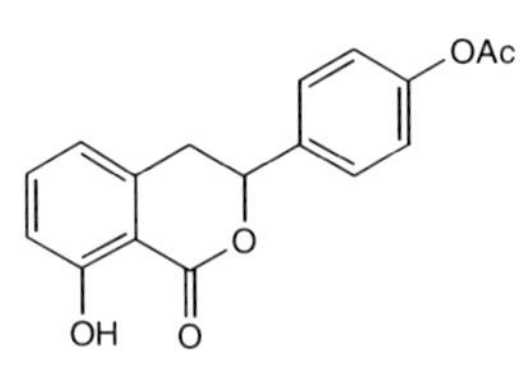

**22**

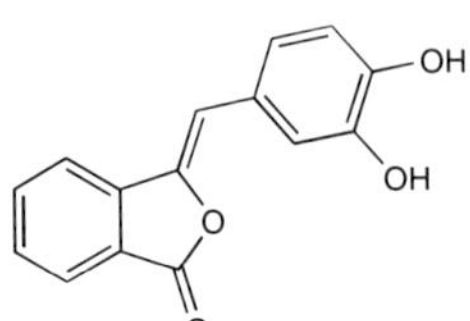

**23**

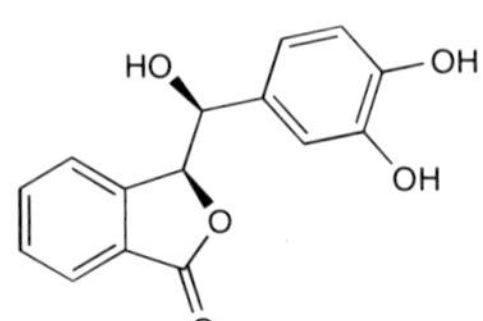

**24**

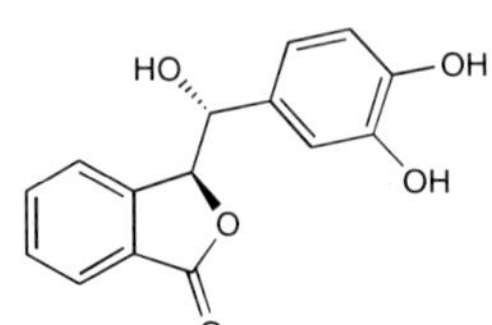

**25**

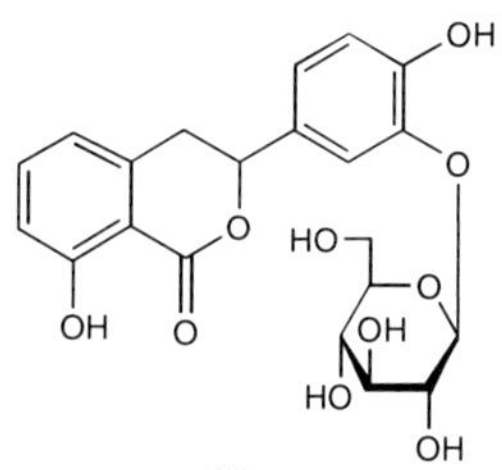

**26**

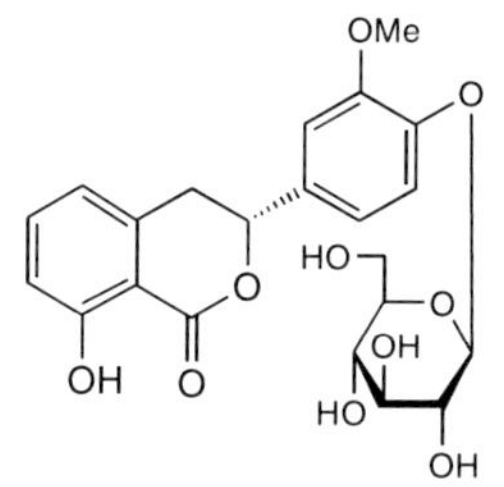

**27**

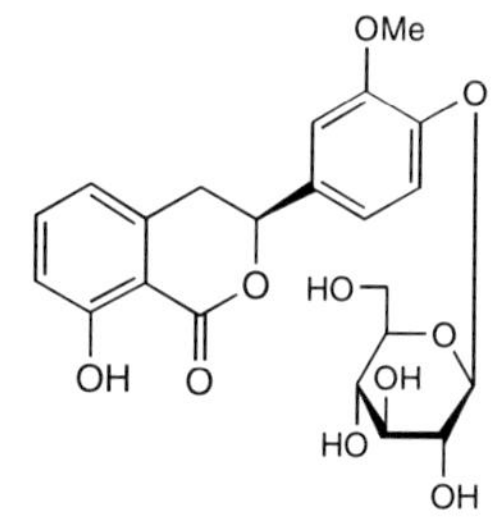

**28**

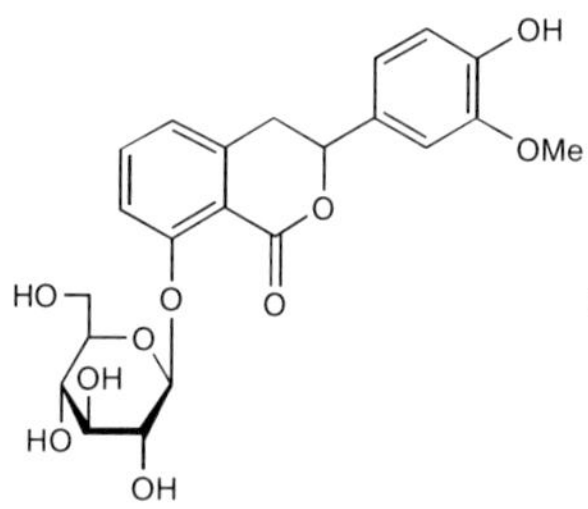

**29**

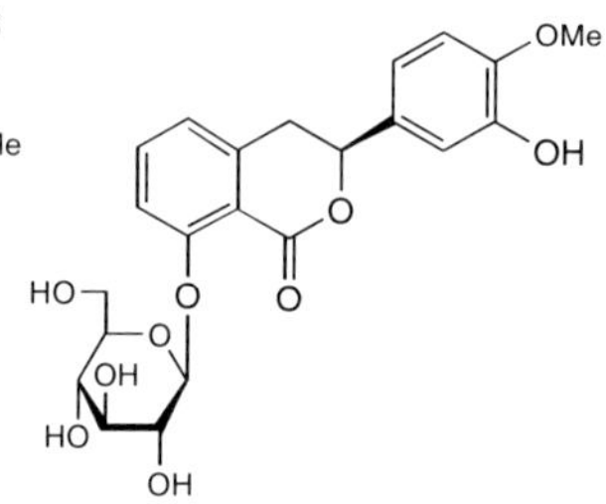

**30**

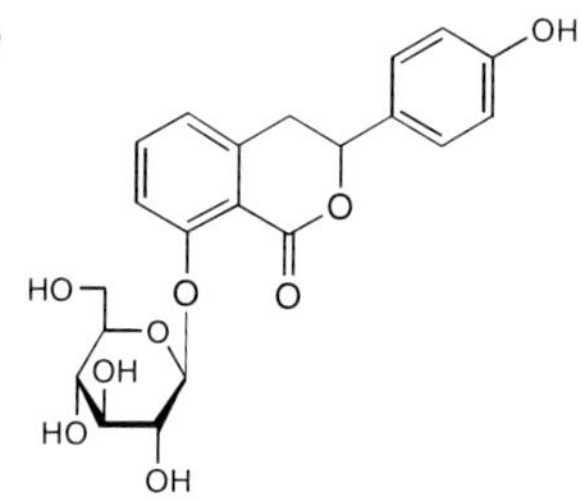

**31**

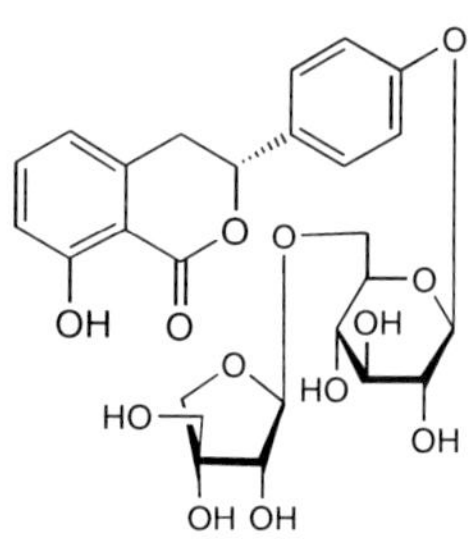

**32**

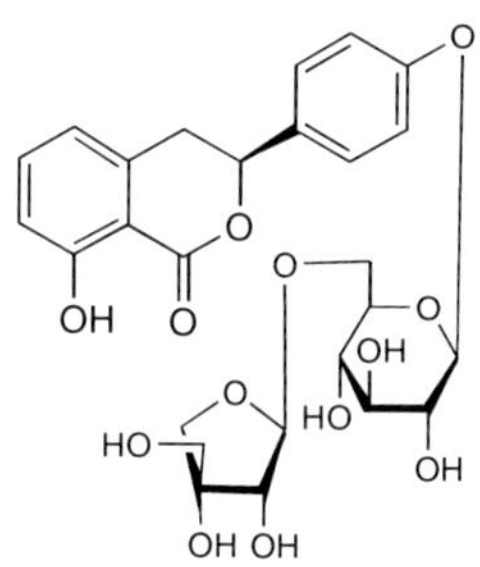

**33**

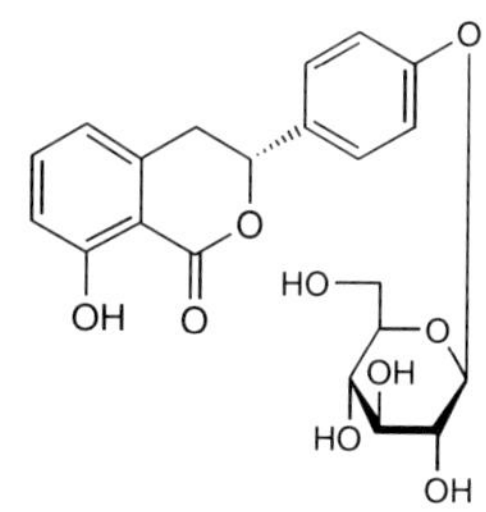

**34**

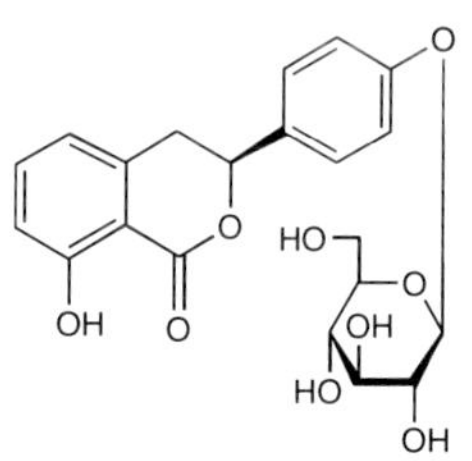

**35**

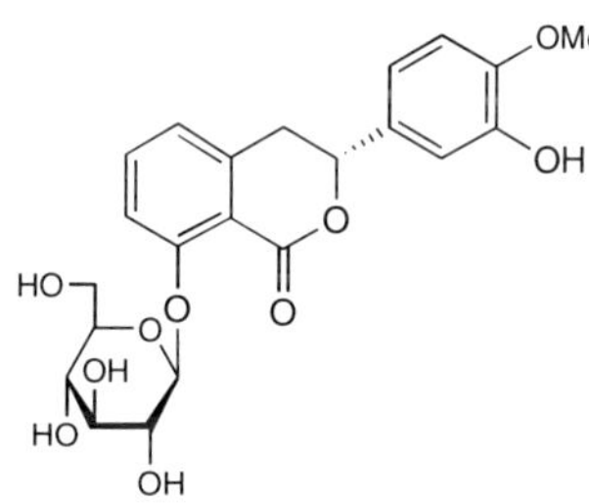

**36**

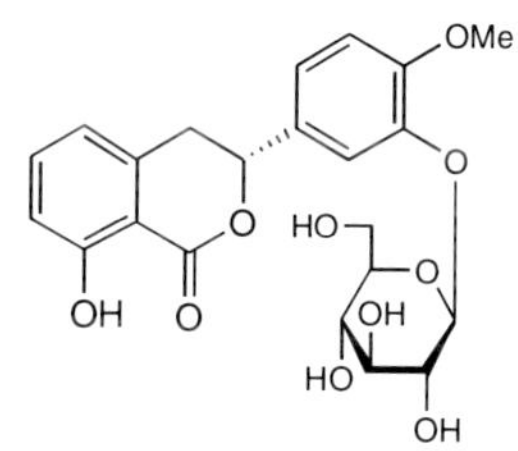

**37**

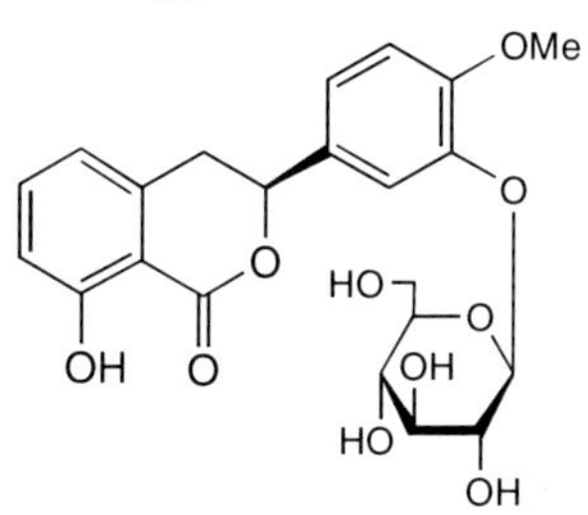

**38**

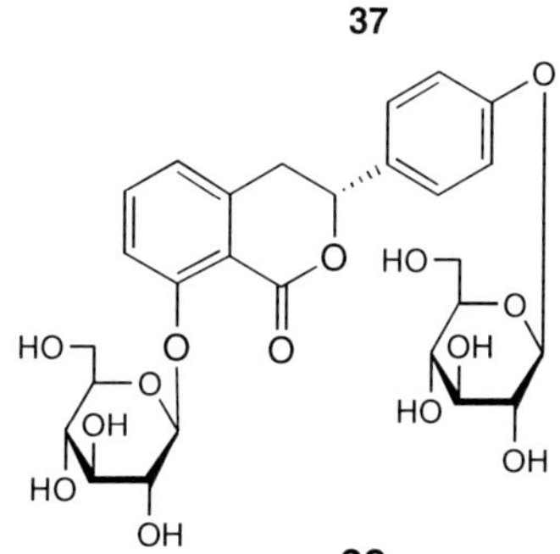

**39**

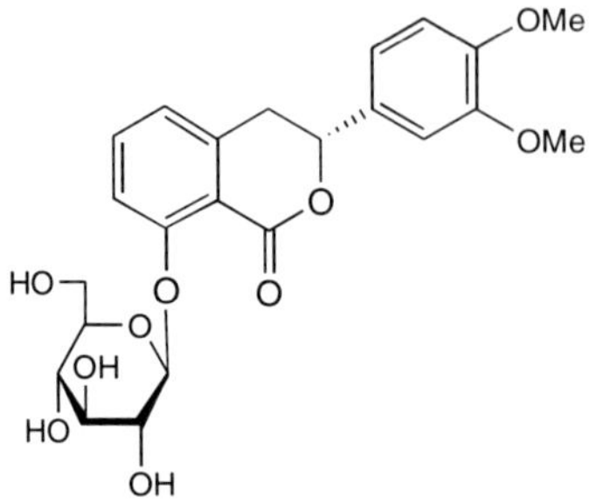

**40**

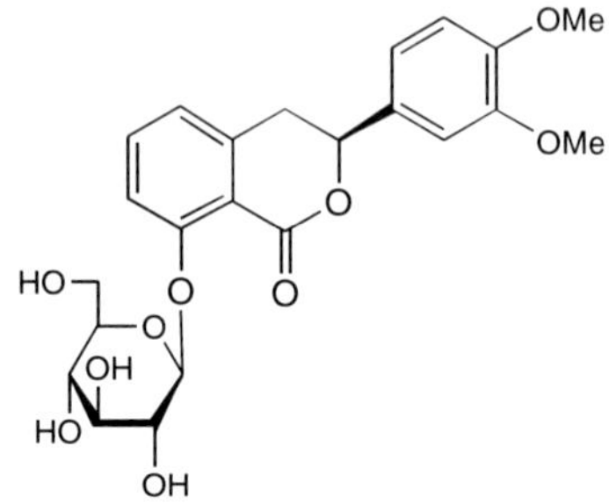

**41**

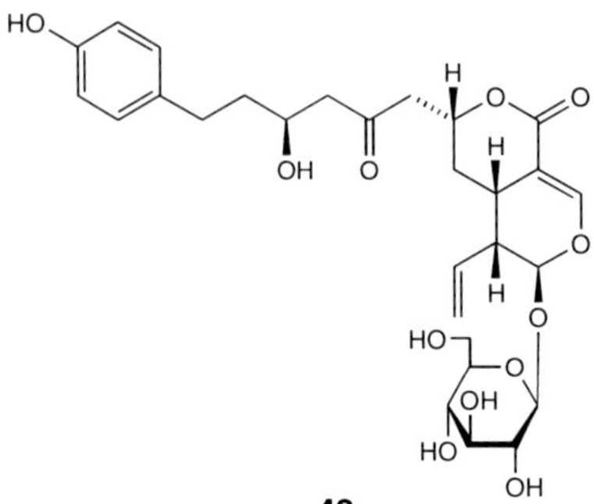

**42**

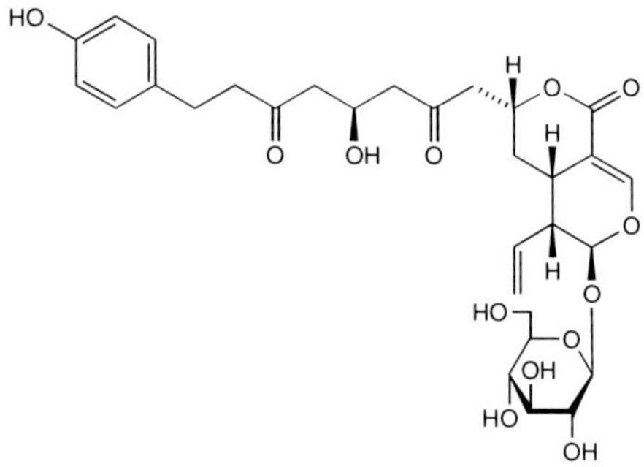

**43**

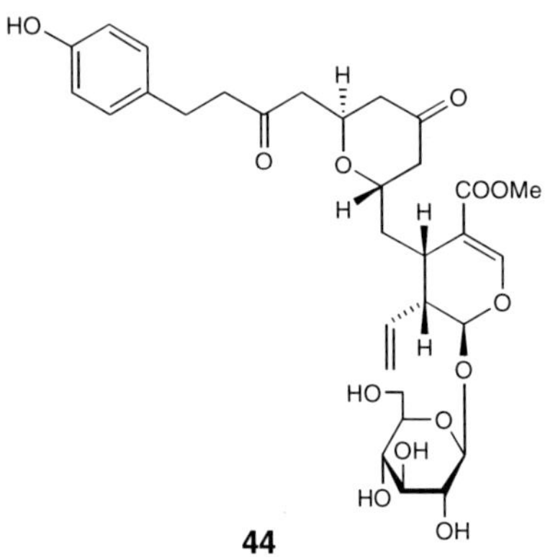

**44**

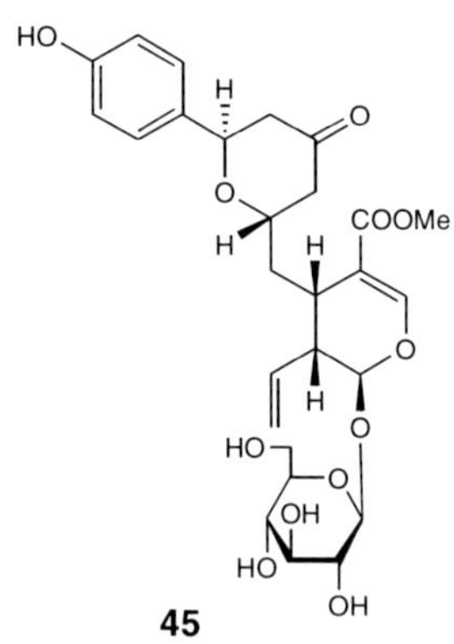

**45**

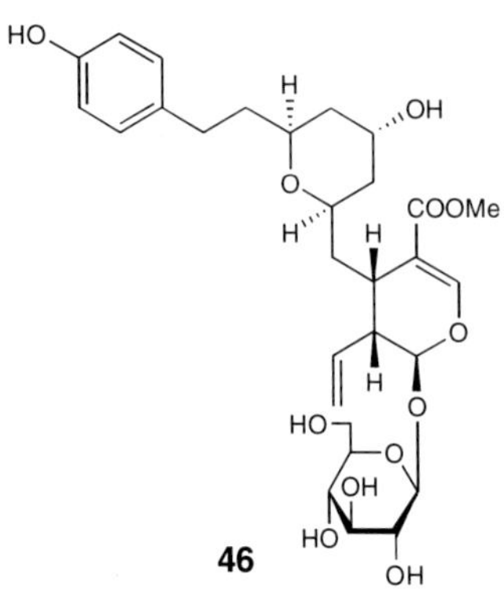

**46**

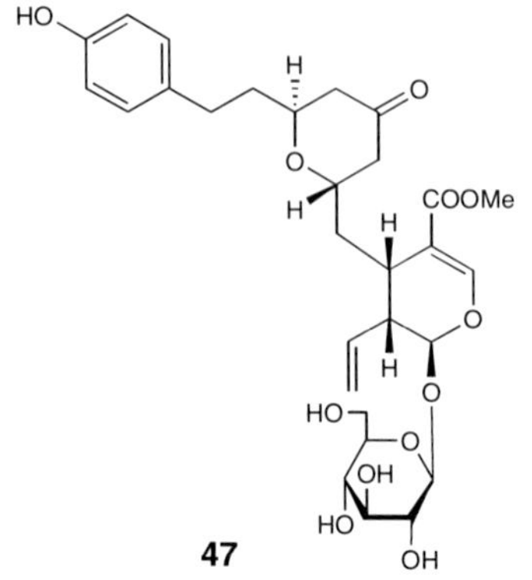

**47**

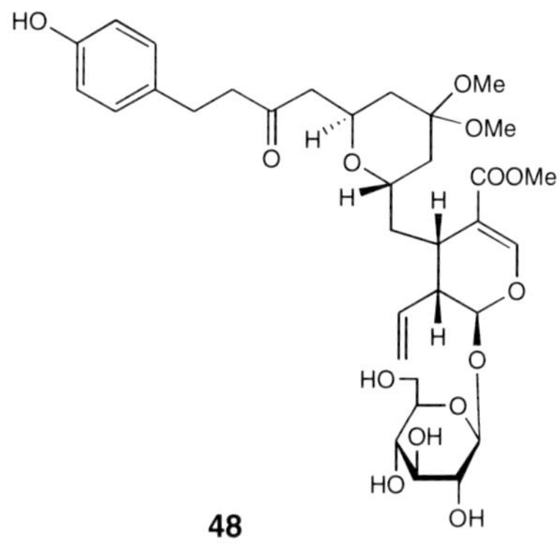

**48**

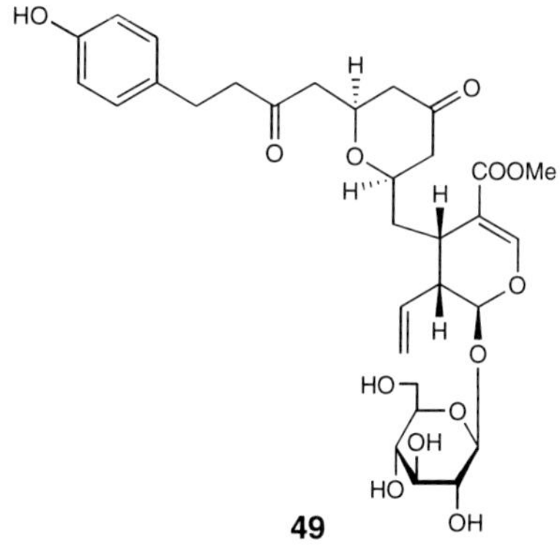

**49**

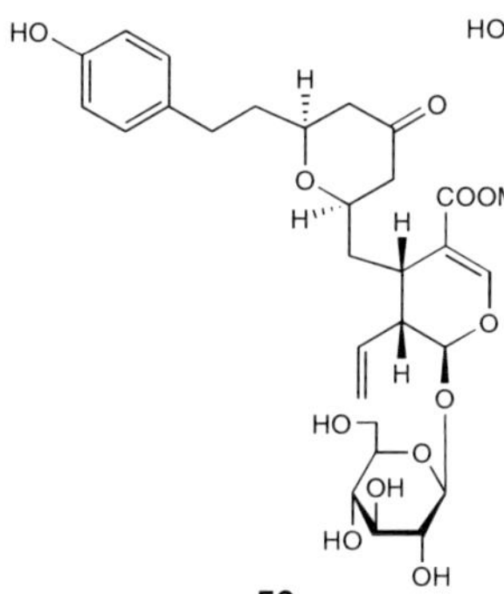

**50**

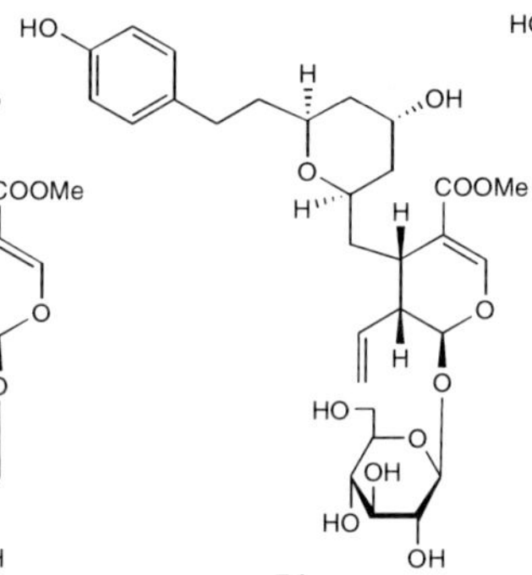

**51**

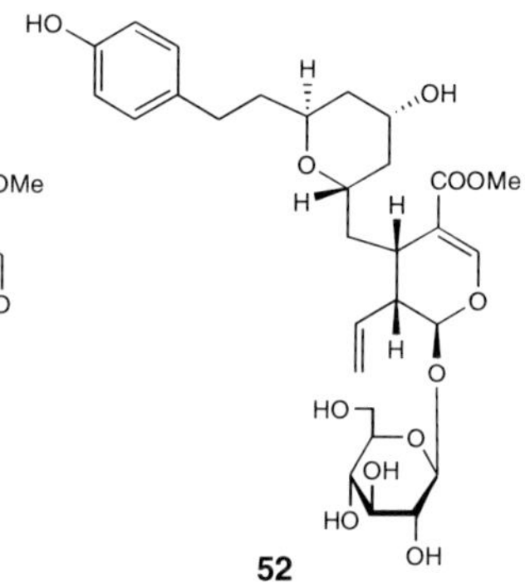

**52**

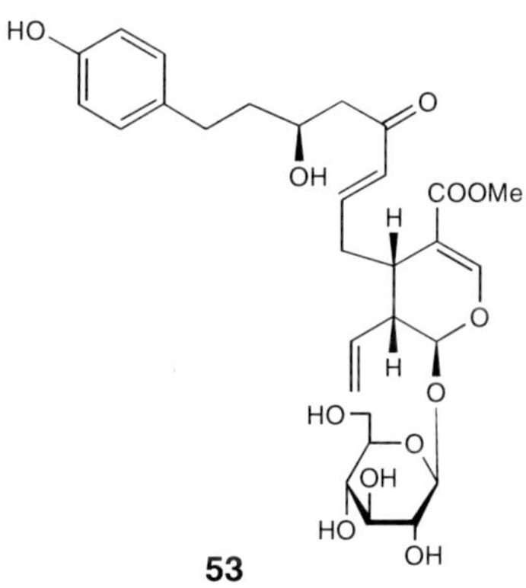

**53**

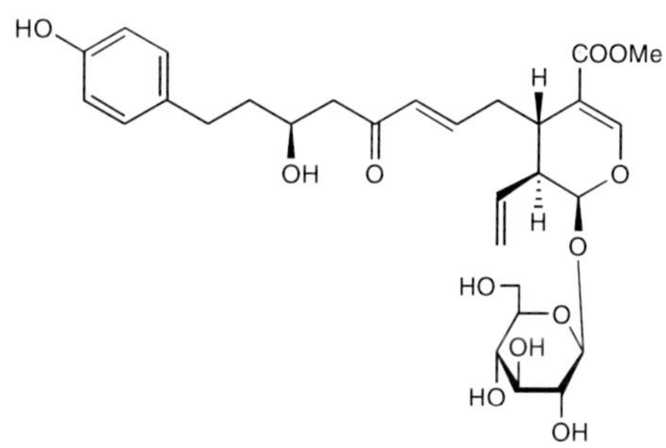

**54**

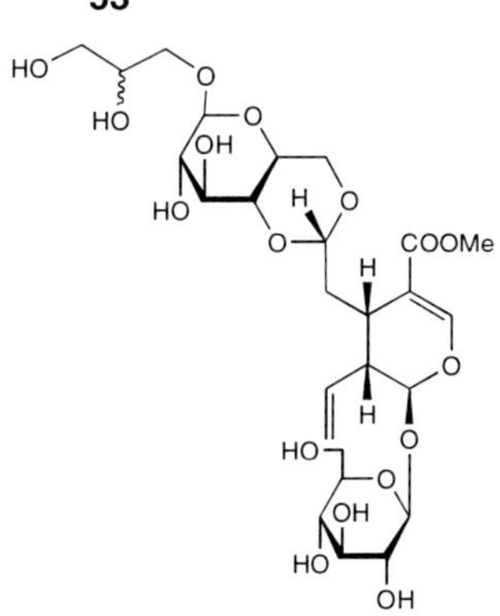

**55**

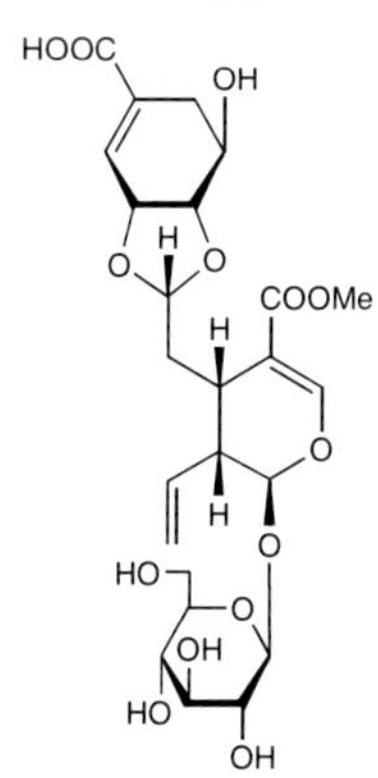

**56**

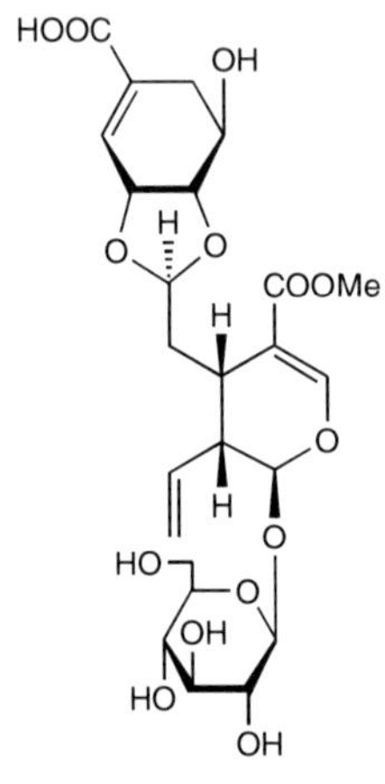

**57**

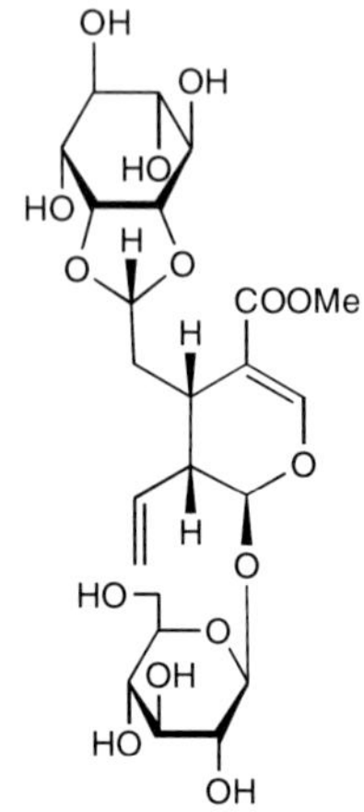

**58**

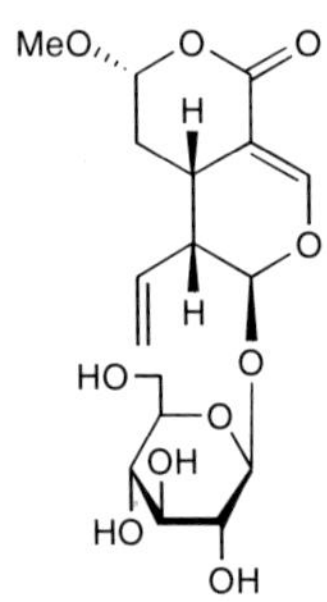

**59**

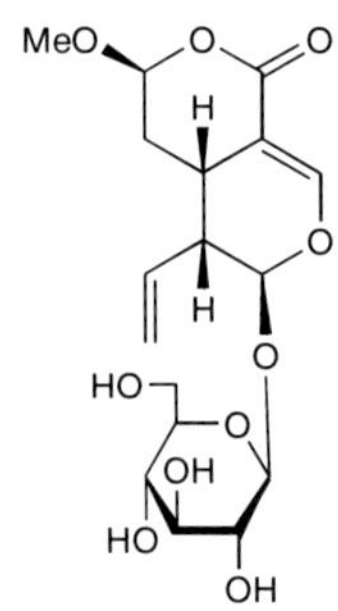

**60**

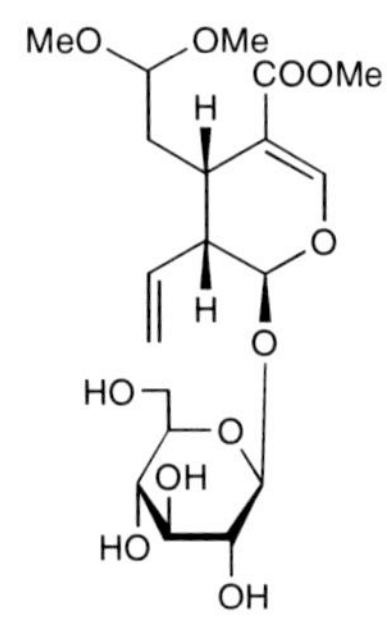

**61**

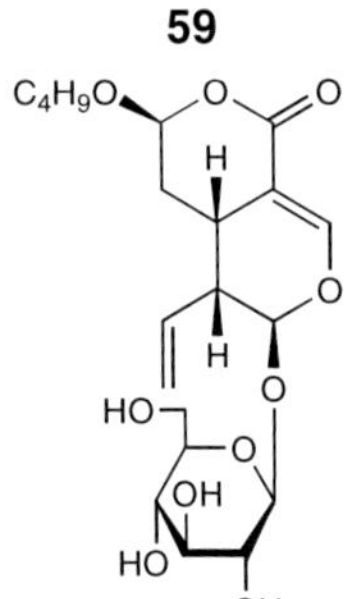

**62**

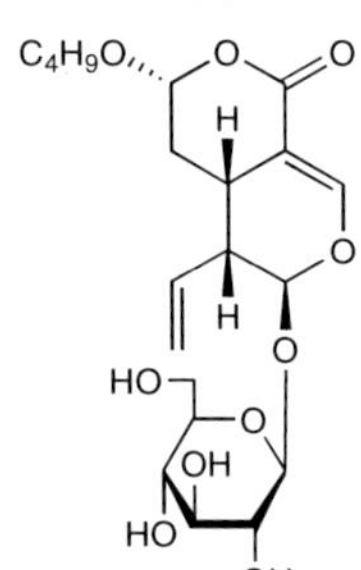

**63**

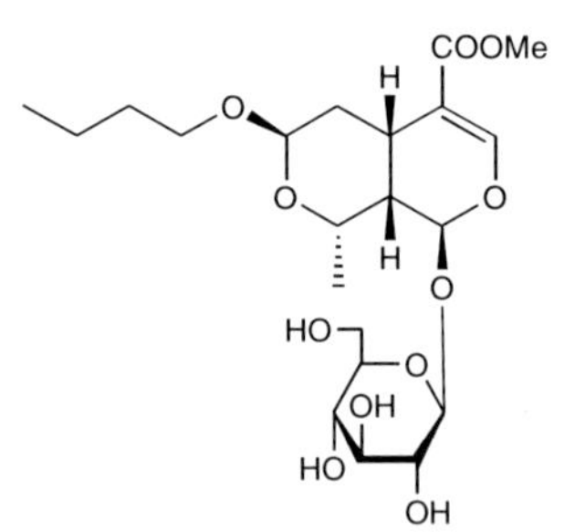

**64**

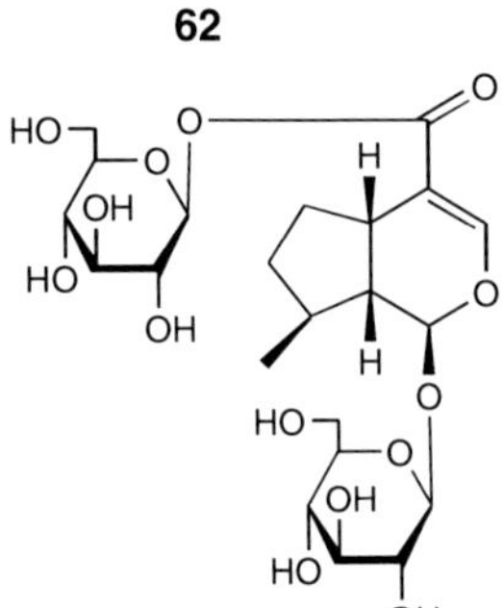

**65**

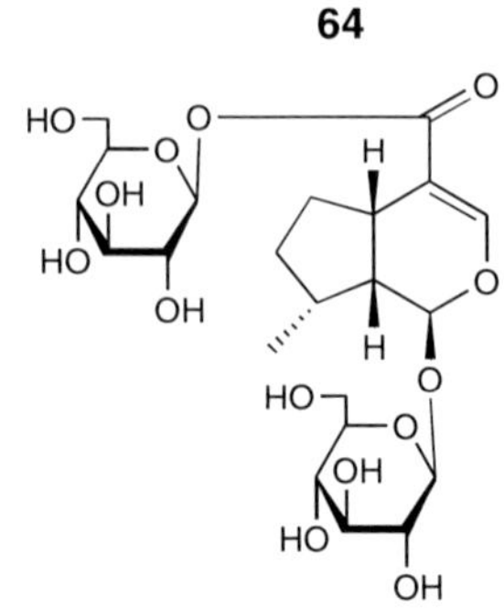

**66**

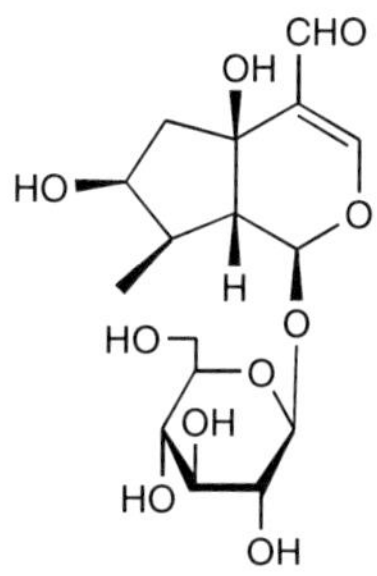

**67**

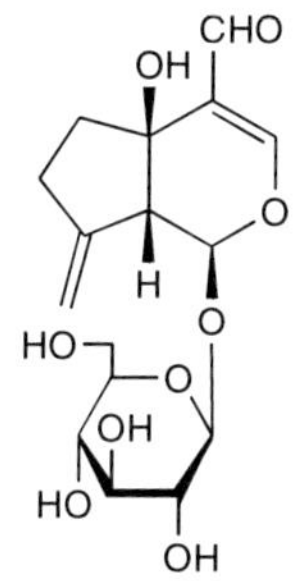

**68**

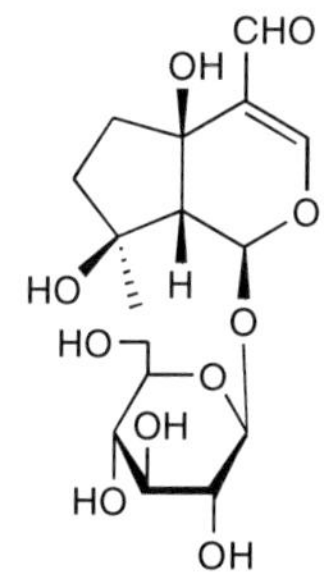

**69**

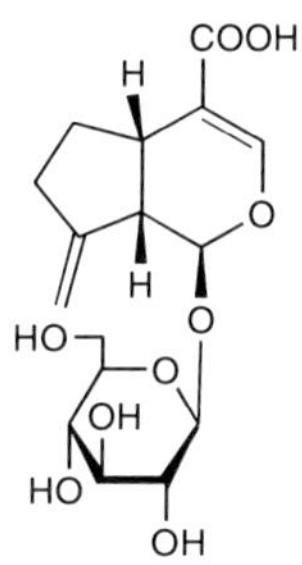

**70**

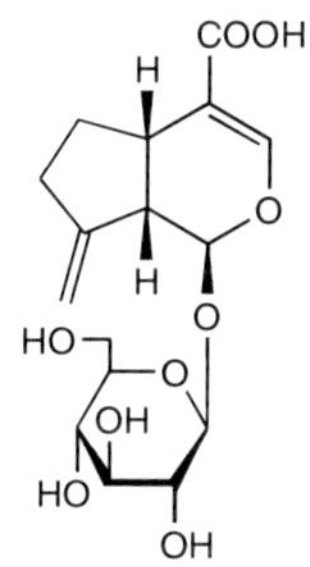

**71**

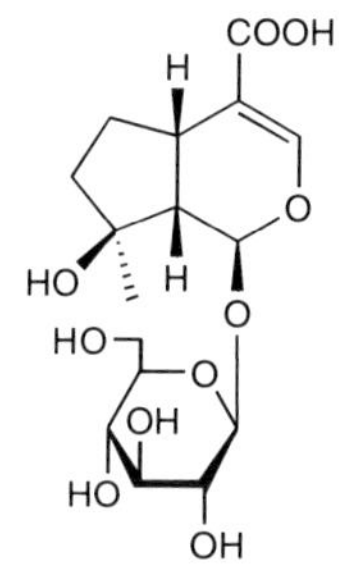

**72**

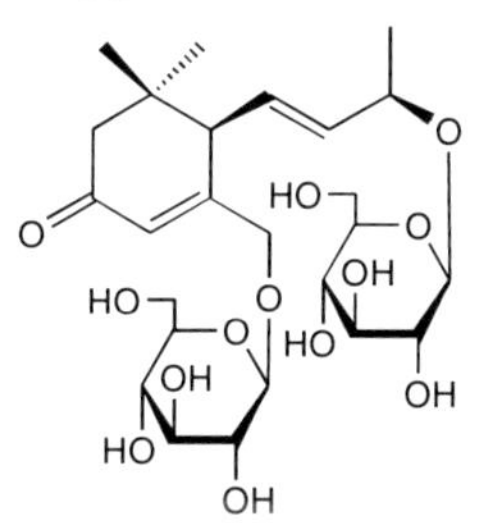

**73**

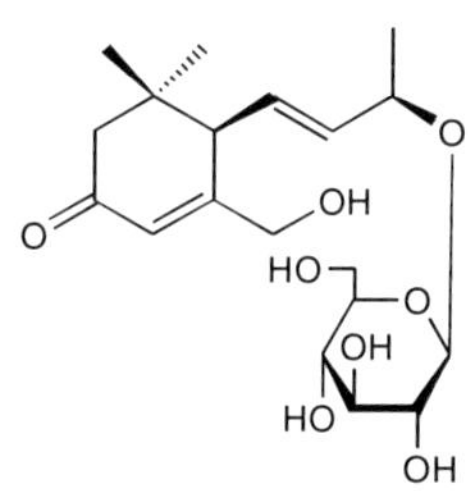

**74**

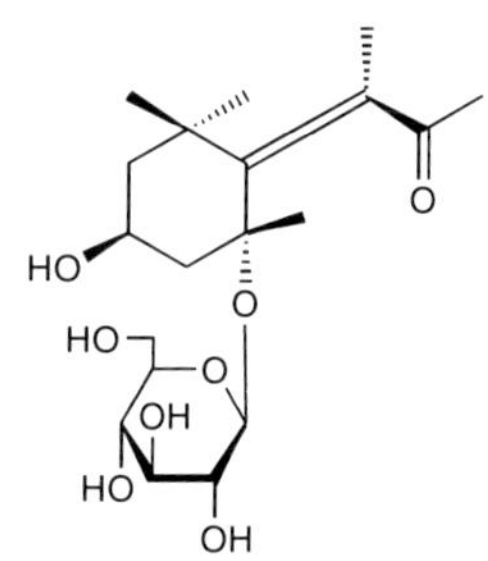

**75**

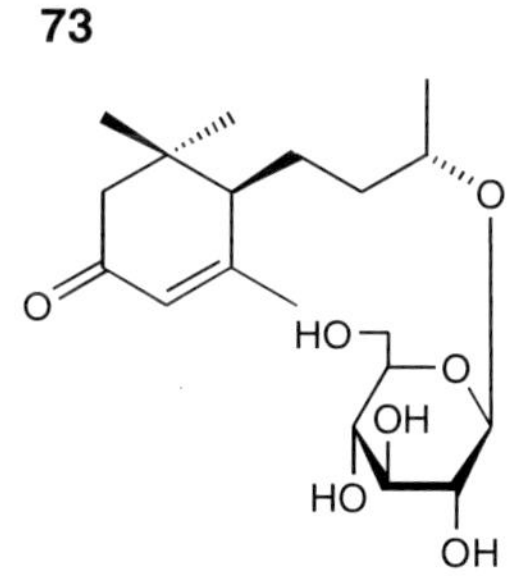

**76**

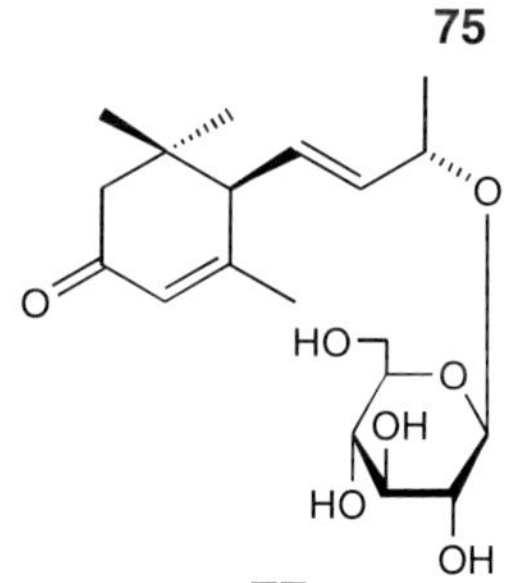

**77**

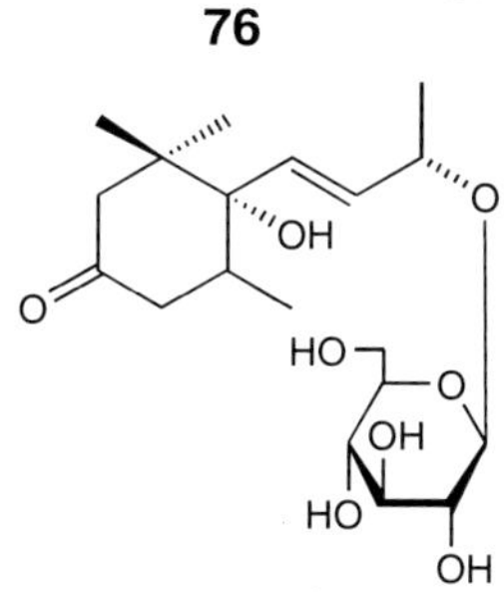

**78**

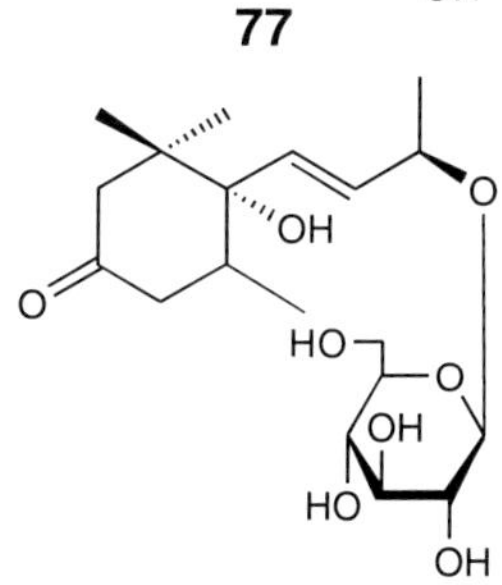

**79**

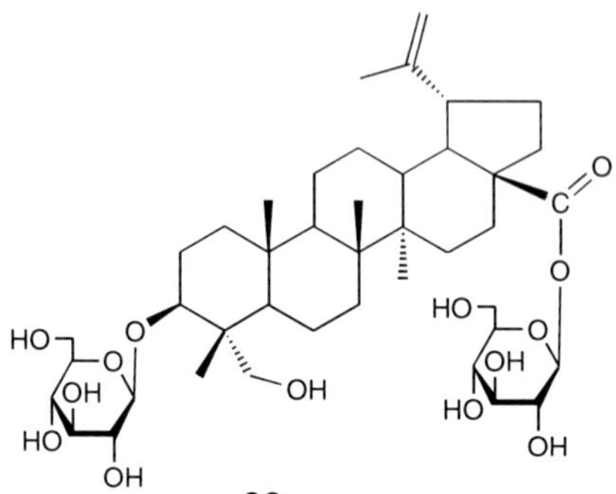

80

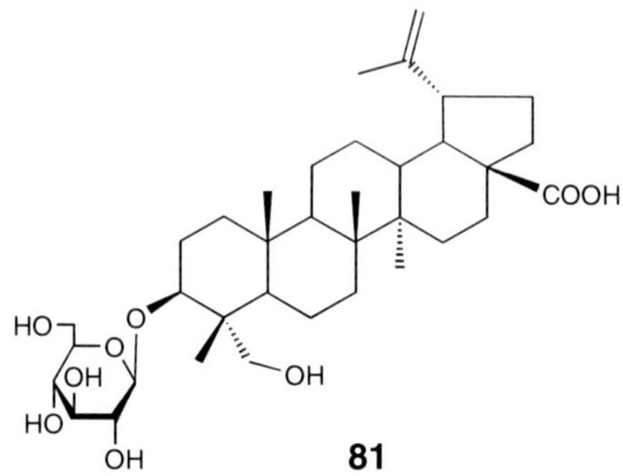

81

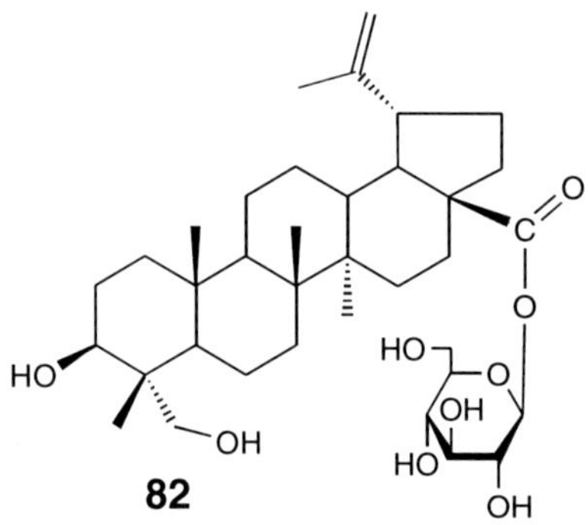

82

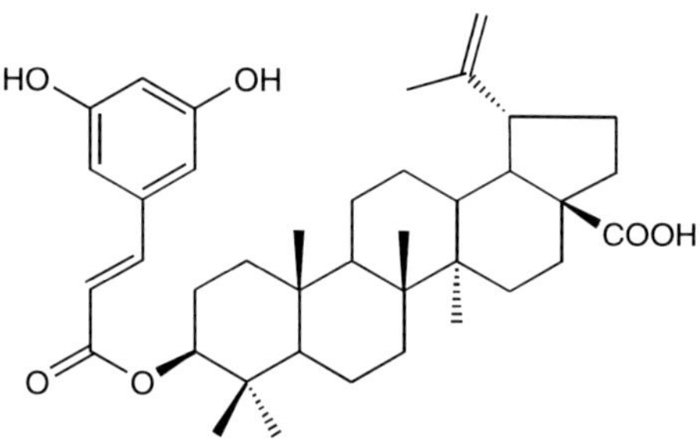

83

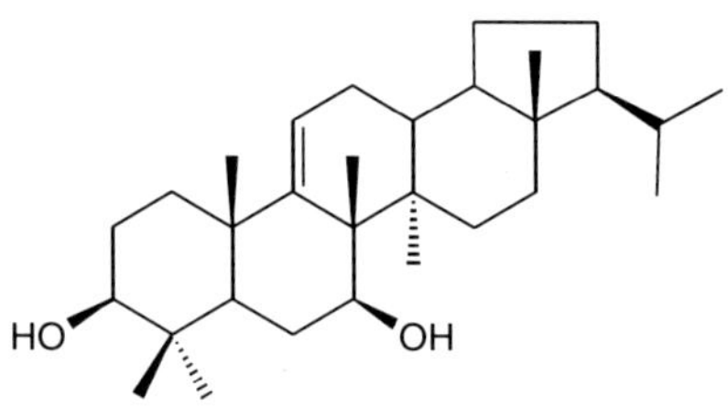

84

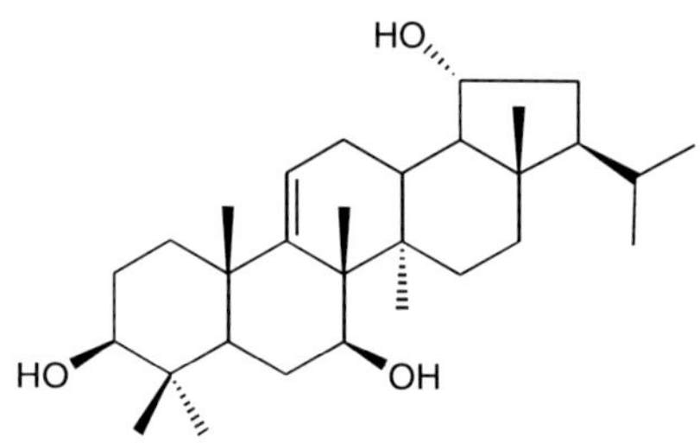

85

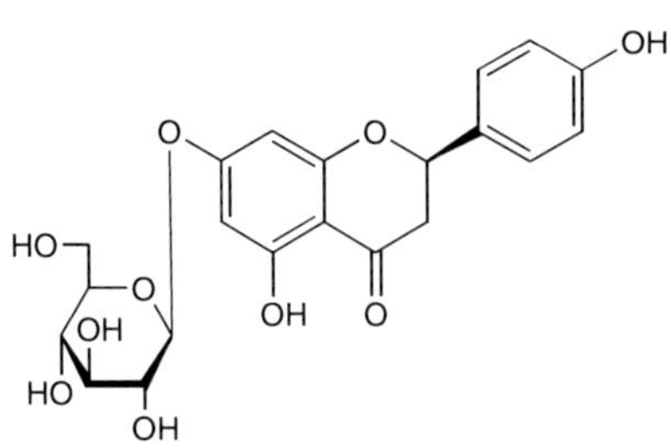

86

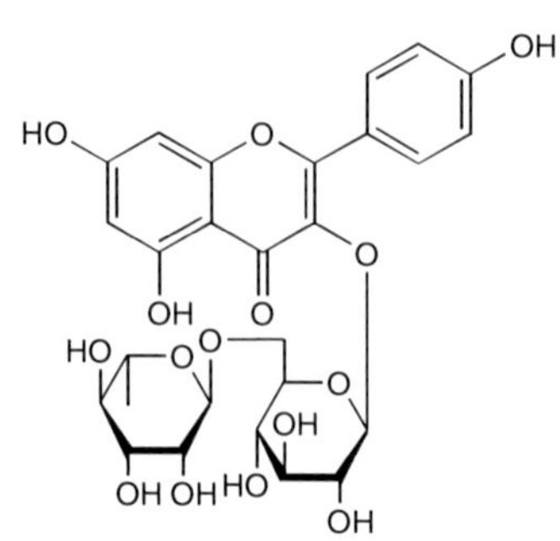

87

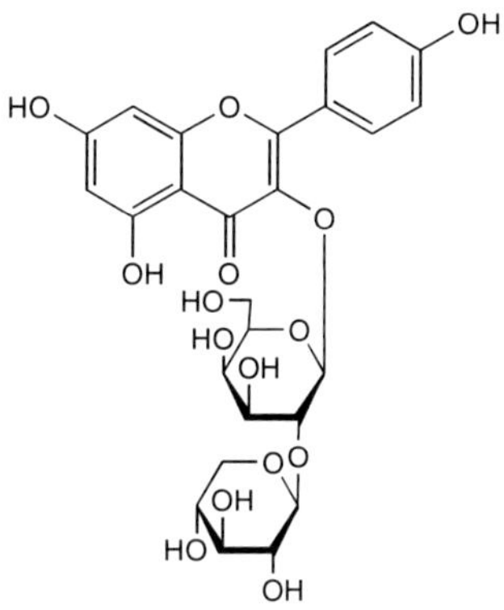

**88**

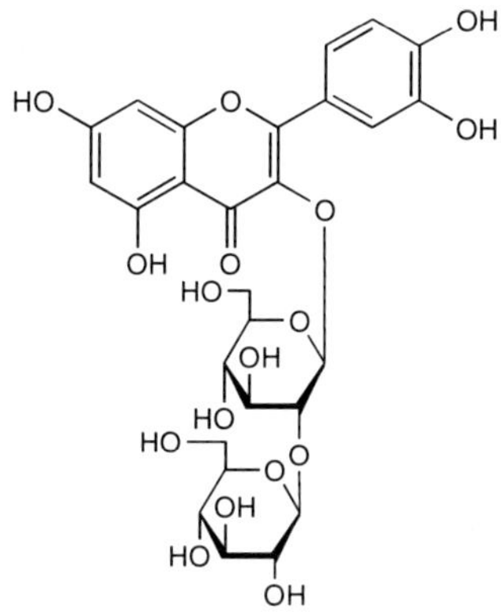

**89**

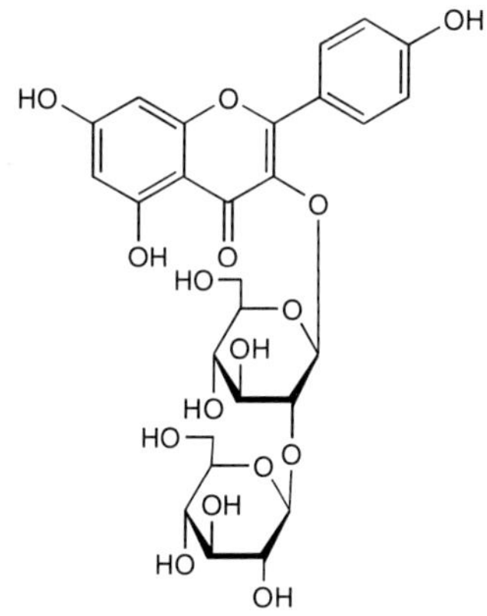

**90**

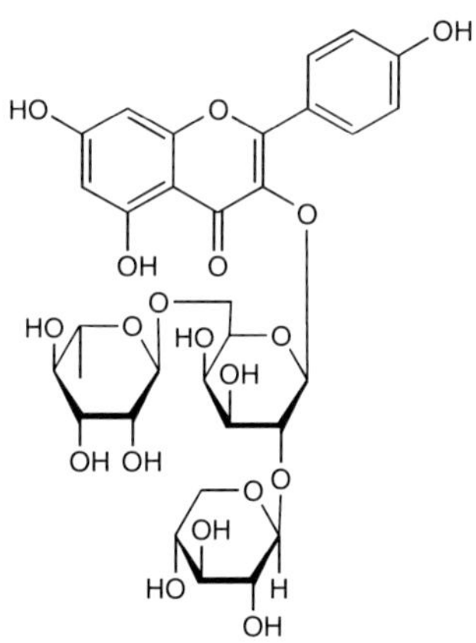

**91**

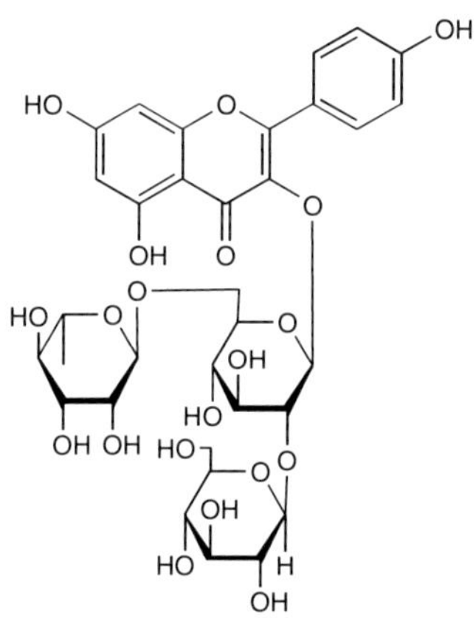

**92**

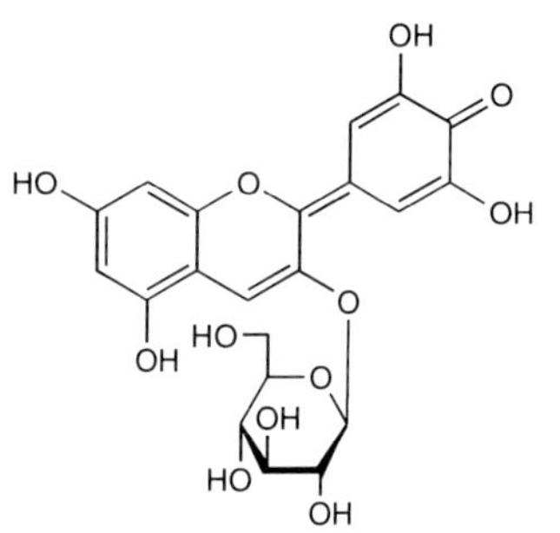

**93**

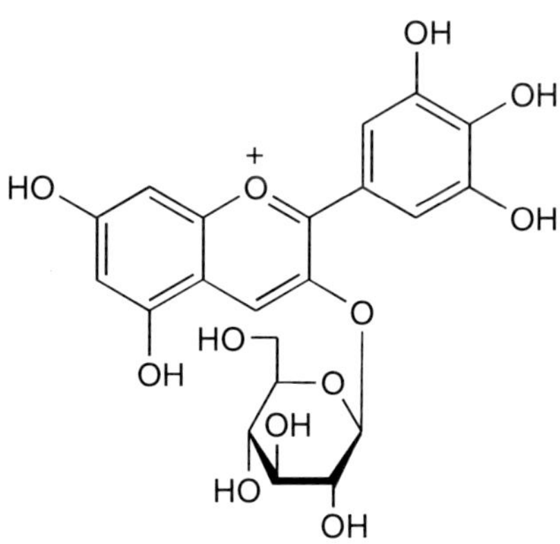

**94**

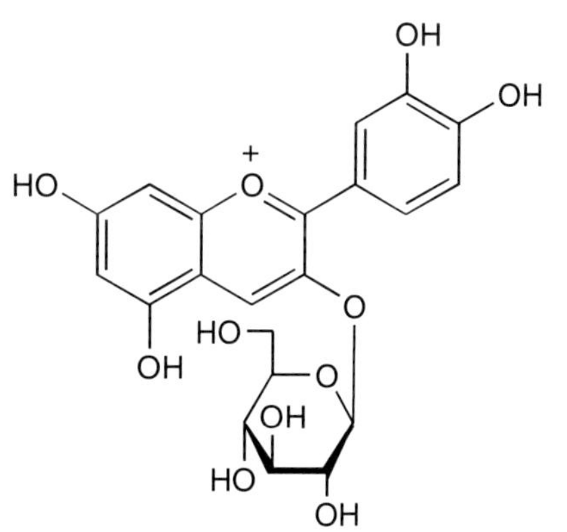

**95**

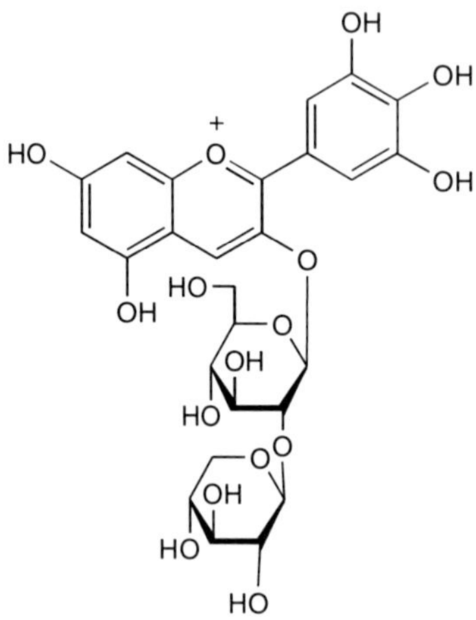

96

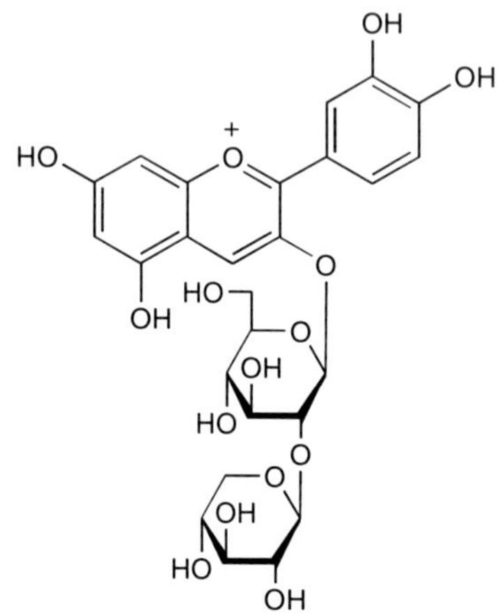

97

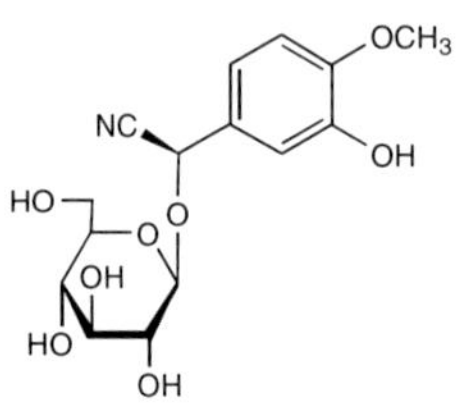

98

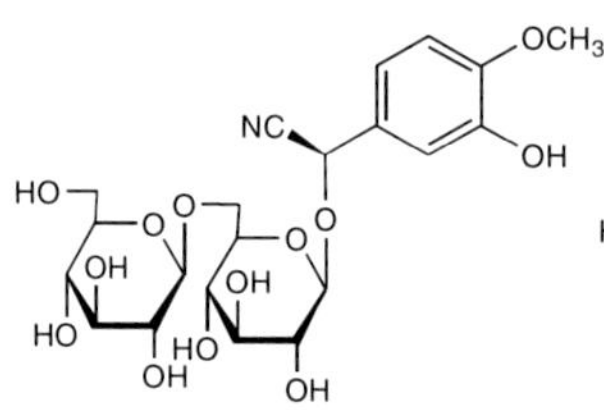

99

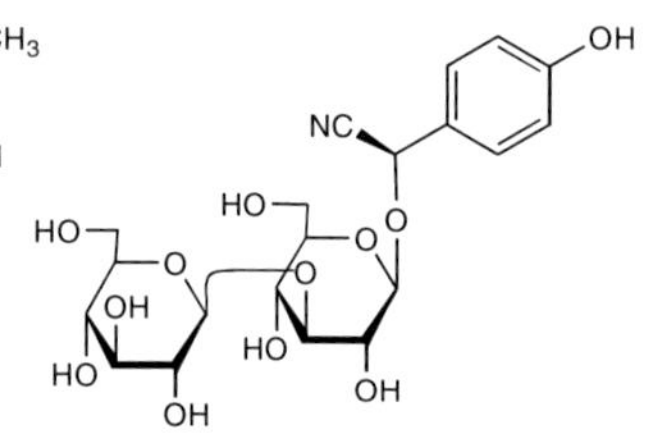

100

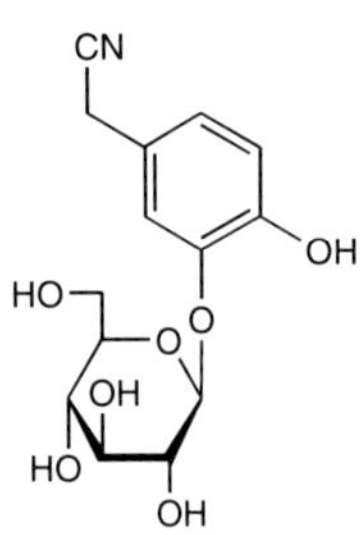

101

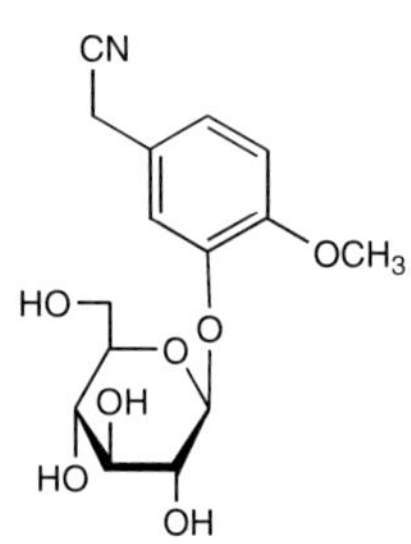

102

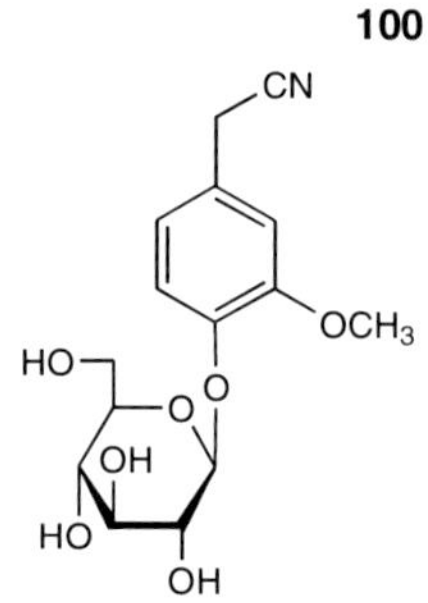

103

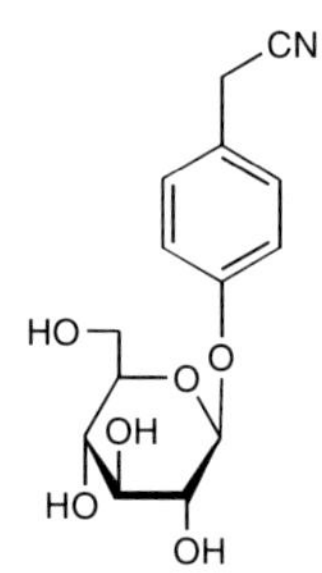

104

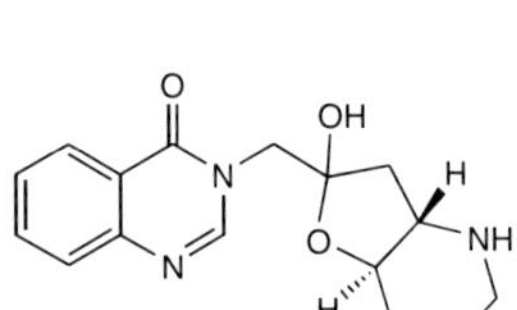

105

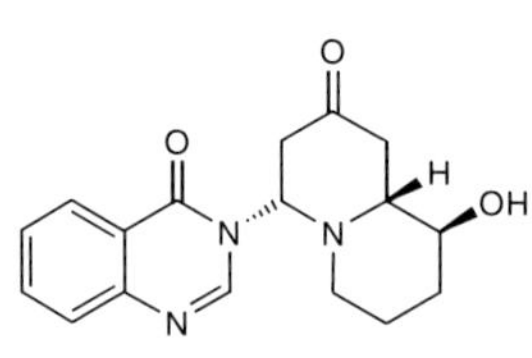

106

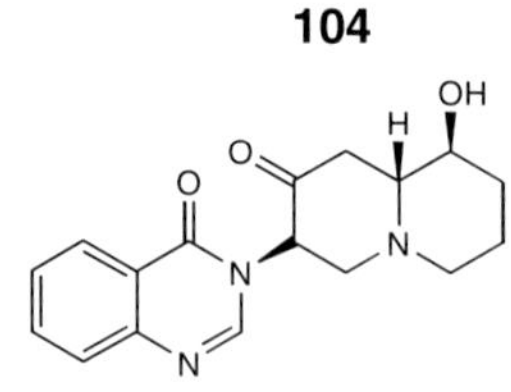

107

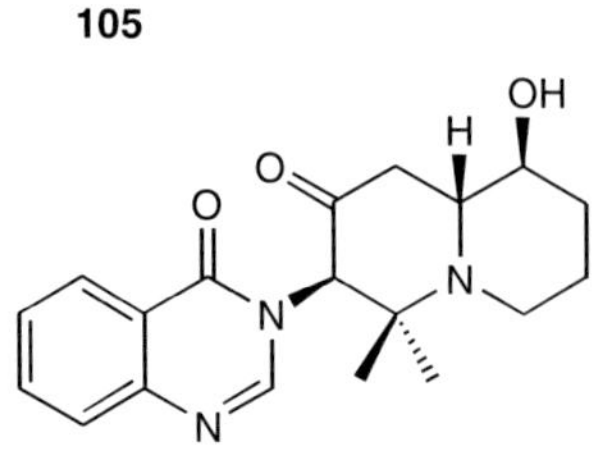

108

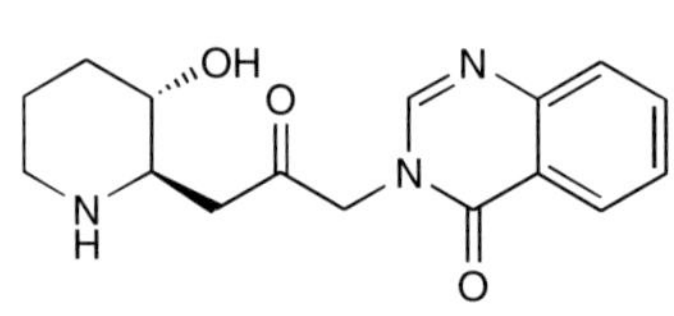

109

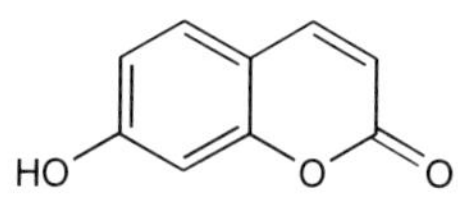

**110**

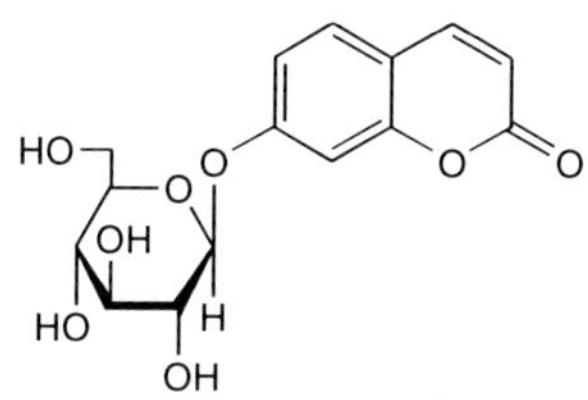

**111**

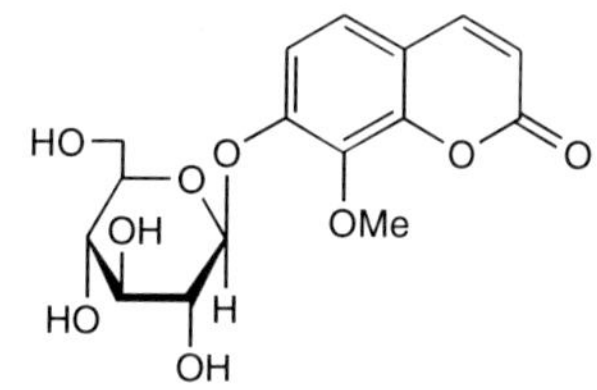

**112**

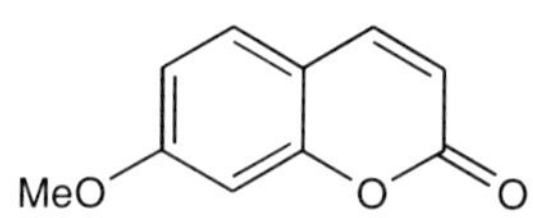

**113**

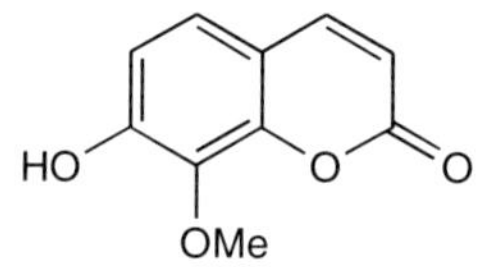

**114**

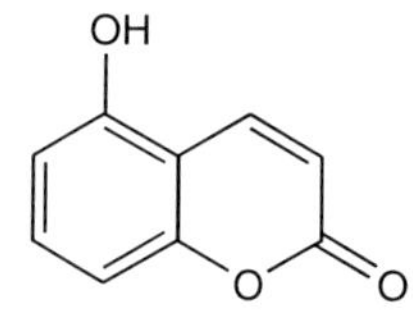

**115**

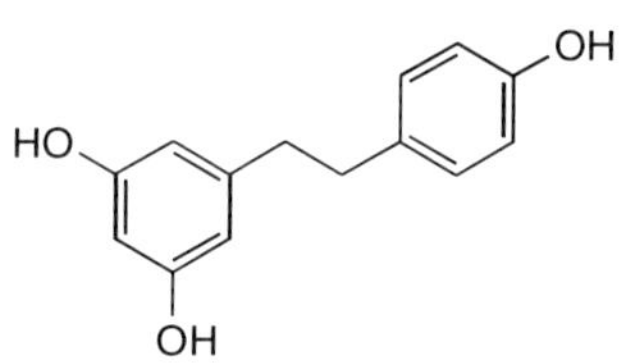

**116**

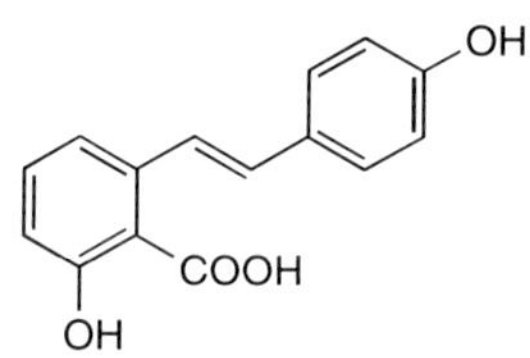

**117**

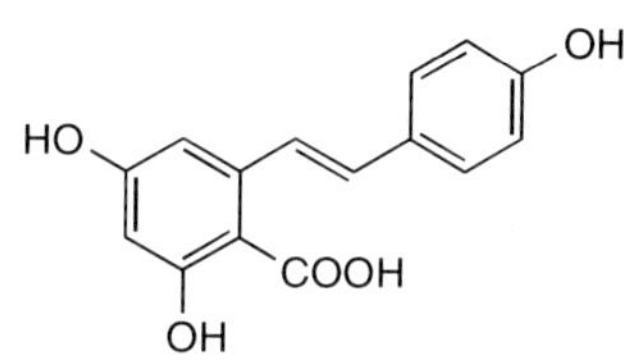

**118**

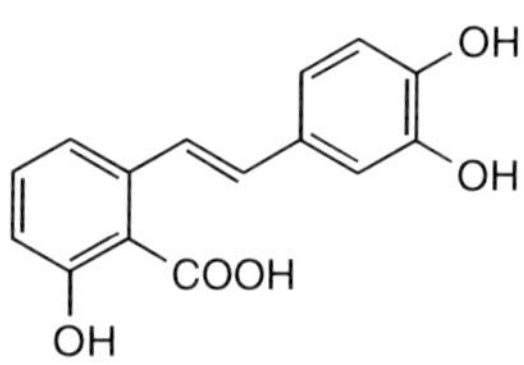

**119**

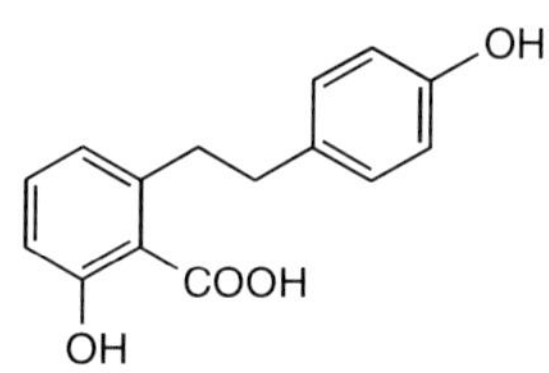

**120**

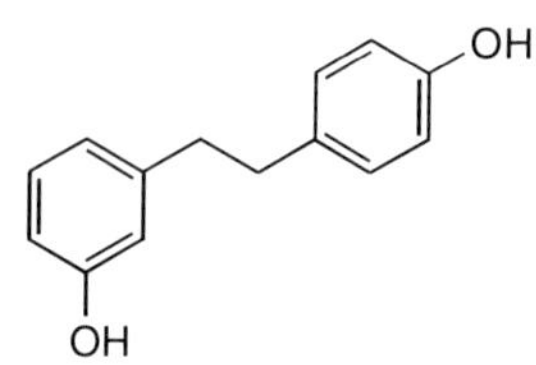

**121**

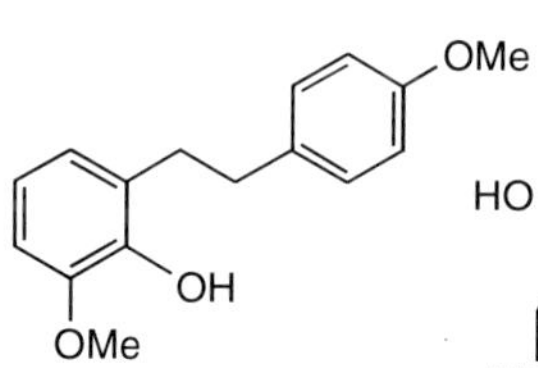

**122**

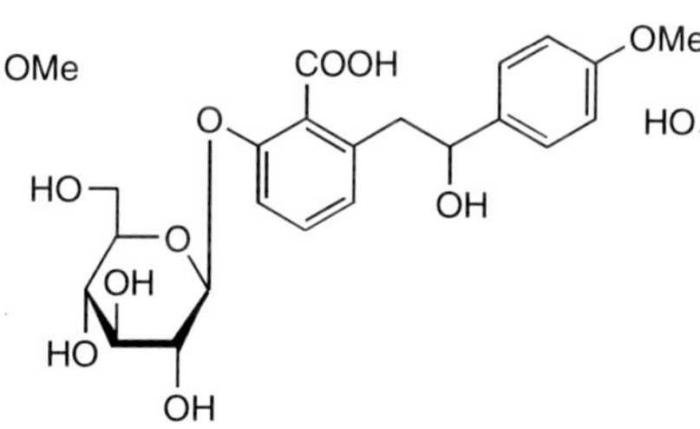

**123**

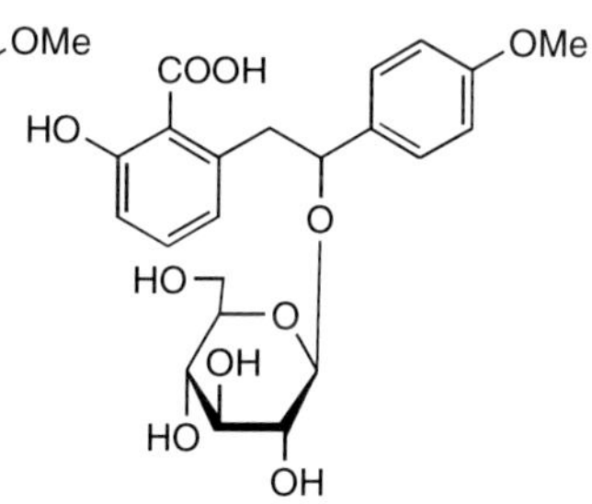

**124**

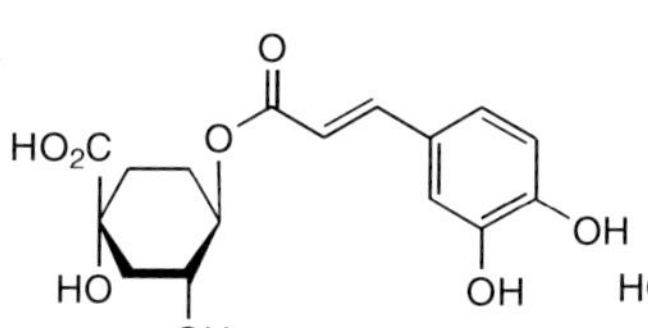

**125**

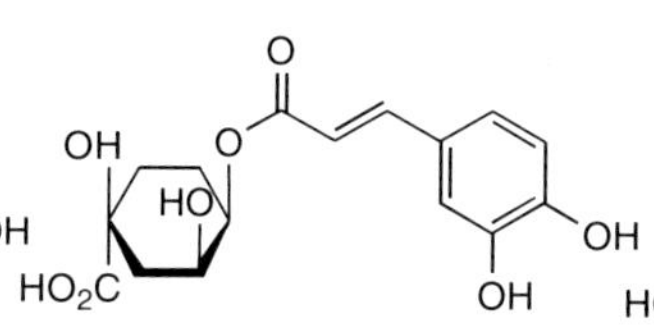

**126**

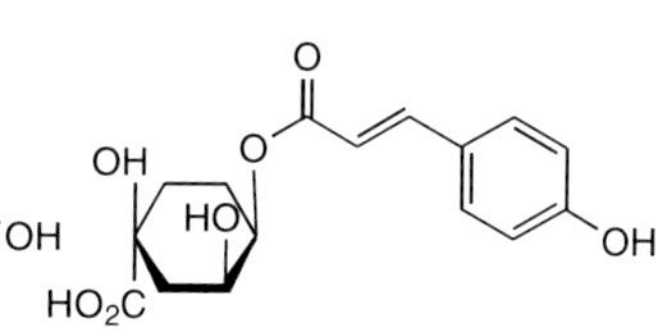

**127**

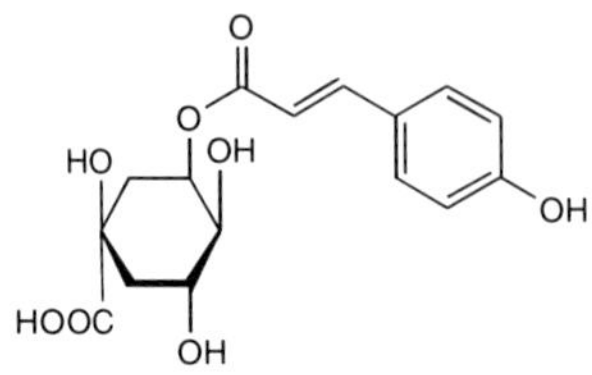

**128**

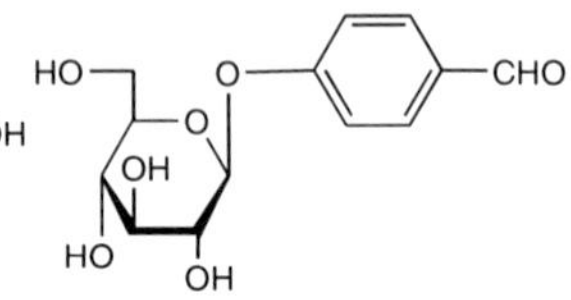

**129**

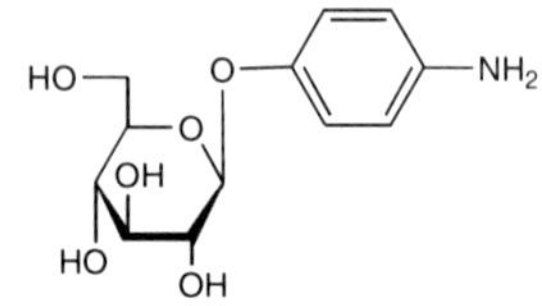

**130**

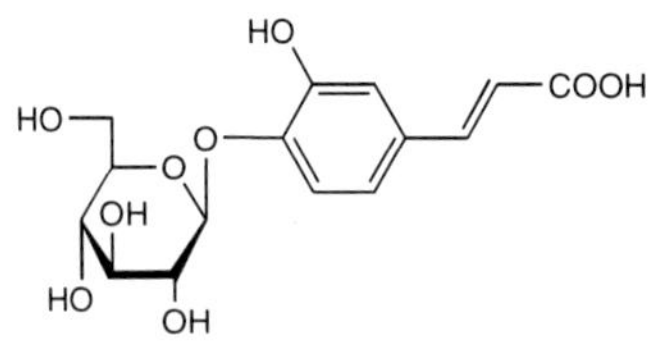

**131**

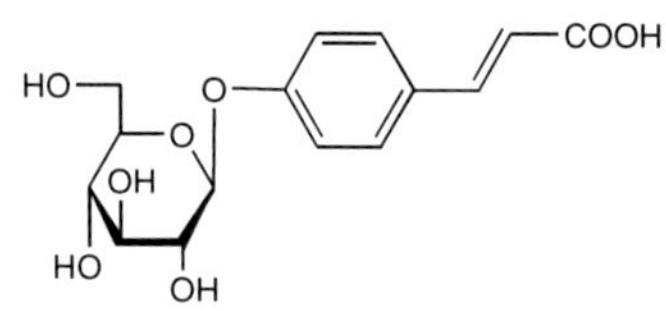

**132**

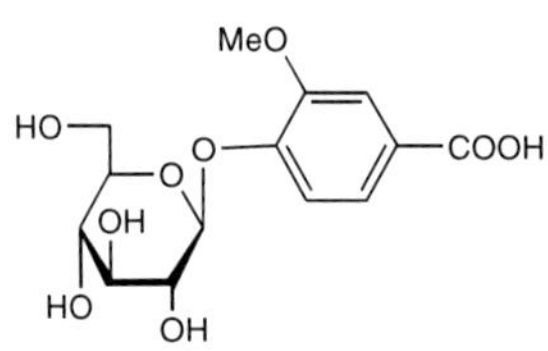

**133**

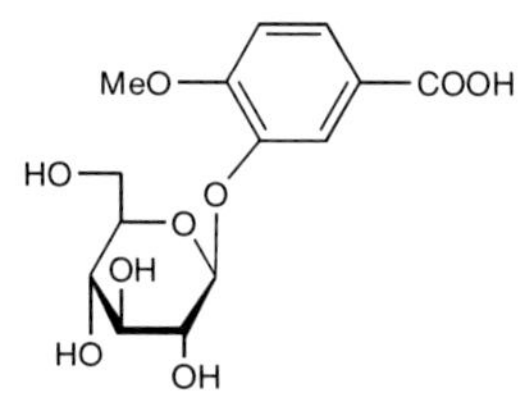

**134**

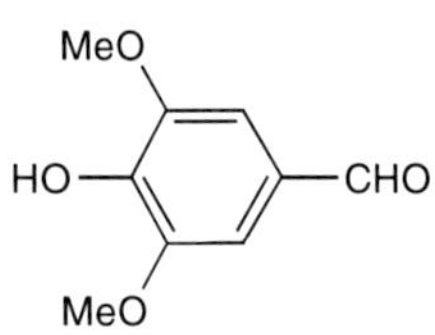

**135**

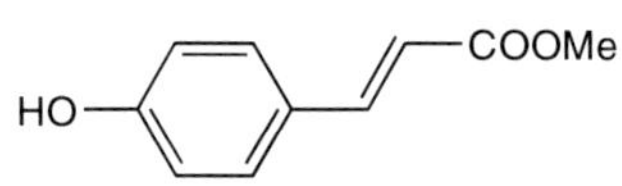

**136**

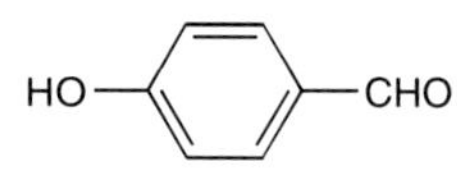

**137**

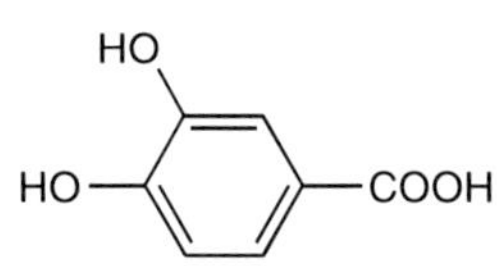

**138**

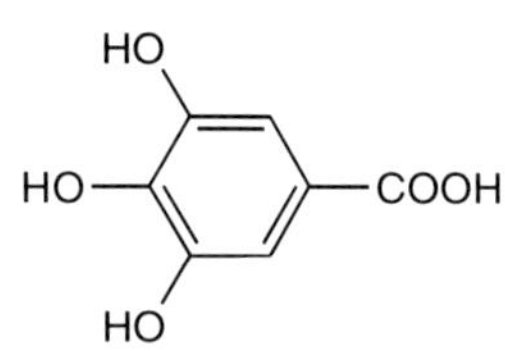

**139**

# 3. Biologocal Activities

## 3.1. Antiallergic Activity

Yoshikawa et al investigated the antiallergic activities of isocoumarins from *H. macrophylla* on the antigen-induced contraction of tracheal chain isolated from sensitized guinea pig (Schults-Dale model) and histamine-induced contraction *in vitro* [1, 2, 4, 60]. The methanolic extract from HDF showed antiallergic activity on Schultz-Dale reactions with the results that the contractile responses of methanolic extract was significantly different from the control and the positive control. Compound 1, 2, 3, 5, 6, 7, 8, 9, 11, 22, 23, 33, 34 and 36 exhibited inhibitory activities antiallergic on the Schulta-Dale reaction than two commercial antiallergic agents, tranilast and disodium cromoglycate (DSCG). Among these compounds, 1, 2 and 23 showed more potent inhibitory activities on the Schulta-Dale reaction than the other pinciples. 3, 5, 6, 7, 8, 9 exhibited more potent inhibitory activities on the Schulta-Dale reaction than two commercial antiallergic agents, tranilast and disodium cromoglycate (DSCG), while 1, 2, 6, 7, 8, 9, 23 and 34 showed little inhibitory for histamine induced contraction. Passive cutaneous anaphylaxis (PCA) reaction was used as the antiallergic activity model for the test of anti-tpye I allergic activity *in vivo* [62]. The methanolic extract from HDF showed little inhibitory for the PCA reaction. However, compound 1, 11, 23, 30, 31 and 36 were found to significantly inhibit the PCA reaction, and 1 showed the most potent inhibitory effect. Compound 11 was also considered to be the principal antiallergic component in the processed leaves after taking into account their contents.

### 3.2. Immunomodulatory Activity

Immunomodulatory activities of compounds 1, 2, 3, 6, 7, 8, 9, 11, (±) phyllodulcin, 23 and 119 on lymphocyte proliferation were investigated using splenocytes and lymph node cells [3]. 1 significantly suppressed B lymphocyte proliferation stimulated by lipopolysaccharide (LPS) at $10^{-5}$ M. However, 1 at a lower concentration ($10^{-6}$ M) and the other compounds, except 11 and 119, tended to potentiate B lymphocytes proliferation ($10^{-5}$ M). In addition, 1 significantly suppressed T lymphocyte proliferation stimulated by concanavalin A (Con A), but did not suppressed phytohemagglutinin (PHA). These results suggested that 1 acted on both T and B lymphocytes and might be have a suppressive mechanism different from the known immunosuppresants. The cytotoxicity of 1 for splenocytes seemed weaker than known immunosuppresants resulting from viability tests using MTT assay and the measurement of lactate dehydrogenase (LDH) released in the medium. Moreover, 1 suppressed antigen-specific T lymphocytes proliferation in mice lymph node cells immunized with keyhole limpet hemocyanin (KLH). These findings showed that 1 has a suppressive effect on lymphocyte activation. In addition, the immunomodulatory effects of antiallergic constituents from HDF on splenocyte proliferation in mice were investigated [72]. The results suggested that 1 and 11 significantly suppressed T lymphocyte proliferation induced by Con A. 1 also suppressed B lymphocyte proliferation induced by lipopolysaccharide, but other constituents induced significant increases. These inhibitory effects of 1 on splenocyte proliferation seemed to contribute to the suppressive effect on type IV allergy.

### 3.3. Inhibitory Effect on Histamine Release

Compounds 42 and 43 exhibited an inhibitory effect on histamine release from rat mast cells induced by an antigen-antibody reaction [16]. They inhibit the histamine release from rat peritoneal exudate cells induced by an antigen-antibody reaction in a concentration dependent manner ($10^{-5}$-$3\times10^{-4}$ M). 1, 24 and 25 were found to inhibit the histamine release from rat peritoneal exudate cells induced by antigen-antibody reaction in a concentration dependent manner ($10^{-6}$-$10^{-4}$ M) [48]. Their inhibitory activities corresponded to that of the commercial antiallergic agent amlexanox, while another antiallergic agent, tranilast lack the activity.

The structure-activity relationships of isocoumarins, phthalides and stilbenes isolated from HDF and related compounds for the inhibition of histamine release in rat peritoneal mast cells were investigated. The activities of isocoumarins such as 1 and 2 were more potent than those of dihydroisocoumarins such as 11 and 9. The double bond at the 3-position seemed to be essential to potentiate the activity. The hydroxyl groups at the 8-, 3'- and 4'-positions of isocoumarin were essential for the activity, while the hydroxyl group at the 6-position was scarcely needed. Since the activities of benzylidenephthalides such as 23 were more potent than those of 24 and 25, the presence of a double bond at the 3-position was needed to increase the activity. Moreover, the hydroxyl group at the 8-position was essential for the activity. 1, 2 and 23 completely inhibited histamine release by pretreatment at 100 mM for 1 to 15 min, whereas DSCG inhibited histamine release only following 1-min pretreatment at 1000 mM. These results suggested that the mechanisms of the inhibitory effect of thunberginols are different from that of DSCG.

### 3.4. Inhibitory Effects on Degranulation and Release of Cells

Basophils as well as mast cells play important roles in both immediate and late-phase reactions of type I allergy. Aggregation of FcεRI by antigens results in tyrosine phosphorylations, $Ca^{2+}$release from intracellular $Ca^{2+}$ stores and influx *via* $Ca^{2+}$ release-activated $Ca^{2+}$ channels (CRAC). Elevation of intracellular free $Ca^{2+}$ level ($[Ca^{2+}]_i$) plays an essential role in the degranulation process. Furthermore, mast cells and basophils concomitantly synthesize and release a variety of cytokines including interleukin (IL)-3, IL-4, granulocyte-macrophage colony-stimulating factor (GMCSF), tumor necrosis factor $\alpha$ (TNF-$\alpha$), and these cytokines induce the late-phase reactions, production of IgE, *etc.*

The effects of the isolated compounds 1, 2, 6, 8, 9, 11, (±) phyllodulcin, 23 and 119 on the degranulations and/or releases of TNF-$\alpha$ and IL-4 *via* FcεRI signaling in rat basophilic leukemia (RBL-2H3) cells were investigated [65]. The results demonstrated that the 3-phenylisocoumarins 1 and 2 and a benzylidenephthalide 23 substantially inhibited the degranulation by antigen and calcium ionophore A23187, and the releases of TNF-$\alpha$ and IL-4 by antigen in RBL-2H3 cells. With regard to structural requirements of the 3-phenylisocoumarins for the activity, the 3, 4-double bond was essential for the strong activity and the 6-hydroxyl group and lactone ring enhanced the activity. The active compounds 1, 2, and 23 inhibited increase in ($[Ca^{2+}]_i$) in RBL-2H3 cells induced by antigen, but not by calcium ionophore A23187.

The synthesized 3-substituted isocoumarins and the principle compounds thunberginols were evaluated for their anti-allergic activities, in which compound 2 exhibited the highest inhibitory potency on the degranulation of RBL-2H3 cells induced by antigen [66]. The structure-activity relationship studies were carried out to determine the necessary substituents on the 3-phenylisocoumarin skeleton for inhibitory activity.

The role of 1 on the signal transmission pathway for rat mast cell degranulation was assessed [67]. Through investigating the celluar distribution of 1 using fluorescent microscopy, indicated that 1 is immediately incorporated into cells and distributes in cytosol around the nucleus. Through investigating the effect of 1 on mast cell protein tyrosine phosphorylation, which is part of the signal transduction cascade for degranulation. Compound 1 non-specifically inhitited the tyrosine phosphorylation induced by compound 48/80 (Co. 48/80), at 10 to 100 μm, and orthovanadate/hydrogen peroxide at more than 50 μm in a dose-dependent manner. 1 suppressed the rise in $Ca^{2+}$ induced by antigens, ionomycin and thapsigargin, while 1 did not suppress the rise induced by Co. 48/80. This result suggests that 1 mainly inhibits extracellular $Ca^{2+}$ influx, but 1 does not act on the intracellular $Ca^{2+}$ mobilization from the endoplasmic reticulum. The investigation of the influence of 1 on the cytoskeleton and memebrane changes using mast cells and erythrocytes indicated that 1 inhibited the cytoskeletal assembly formation in dicyanovinyl julolidin-loaded mast cells induced by Co.48/80. Moreover, 1 suppressed the hypotonic hemolysis of erythrocytes, from 3 to 1000 μm, in a dose-dependent manner. All results suggest that inhibition of protein tyrosine phosphorylation, extracellular $Ca^{2+}$ influx and cytoskeletal assembly formation, and membrane stabilization are involved in the inhibitory effect of 1 in mast cell degranulation.

### 3.5. Antimicrobial and Antifungal Activities

Compounds from the genus of *Hydrangeae* also exhibited antimicrobial activity against oral bacteria [1, 2, 4, 10]. Compound 1, 2, 6, 7, 8, 9, 11, (±) phyllodulcin, 23 and 34 showed antimicrobial activities against two oral bacteria [Bacteroides melaninogenicus; minimum inhibitory concentration (MIC): 1(5 ppm), 2(10 ppm), 6(10 ppm), 7(10 ppm), 9(20 ppm), 11(10 ppm), 23(10 ppm), 8(50 ppm), (±) phyllodulcin (100 ppm), 34(100 ppm); Fusobacterium nucleatum; MIC: 1(10 ppm), 2(10 ppm), 6(10 ppm), 7(10 ppm), 23(10 ppm), 9(20 ppm), 11(5 ppm), 8(30 ppm), 23(100 ppm), 34(100 ppm)]. Flavonols (kaempferol and quercetin), as the references also showed antimicrobial activities, but their activities were weaker than that of 11. However, compound 31, 33, 38, 89, 90, 92 were found to lack antimicrobial activities (MIC$>$128 ppm). These results indicated that the antimicrobial ativities of isocoumarin and flavonol were decreased remarkably by glycosidation and methylation of the hydroxy group. Results showed that the main principle compound 11 and (±) phyllodulcin exhibited antimicrobial activity for various periodontopathic bacteria (MIC: 5-25 ppm) and (±) phyllodulcin shoghteen other bacteria and fungi, including cariogenic bacteria and pathogenic fungi (MIC$\geq$100). (±) phyllodulcin possessed an extremely weak antibacterual activity, inhibiting the growth of Escherichia coli and Staphylococcus aureus at concentrations over 5000 μg/ml.

Compound 11 and (±) phyllodulcin possessed antifungal activity, inhibiting the growth of the molds completely at concentrations from 25 to 200 μg/ml. 20 or 21 was also found to cause a lowering of the antifungal activity. All modifications of (±) phyllodulcin, that is, acetylation of the two hydroxyl groups (17, 18) resulted in a decrease of its antifungal activity. Similarly, structural modification of 11, i.e., methylation of the hydroxyl group at the 4'-position or the replacement of the ring oxygen by an imino group, was also found to cause a lowering of the antifungal activity. Optically active isomers of (±) phyllodulcin and 17 showed stronger activity than the racemate.

### 3.6. Anti-Diabetic Activity

Diabetes mellitus is a chronic and often undiagnosed or inadequately treated disease. The prevalence of diabetes is growing at an alarming rate. In the course of searching for anti-diabetic constituents from natural medicines, the ethyl acetate fraction of the processed leaves of HDF used as sweetening agent for diabetic patients, promoted the accumulation of triglyceride in 3T3-L1 cells [increment: 23.4 ± 4.1% ( $p < 0.01$) at 10 μg/mL]. Through examining for the effects on adipogenesis in 3T3-L1 cells, found 11, (±) phyllodulcin, 1, and 114 promoted the accumulation of TG in the cells in a concentration-dependent manner, with 11 and 114 exhibiting stronger activity. Compounds 1, 2, 23 reduced TG levels at a high concentration (100 μM), suggesting that the cytotoxic effects at high concentrations [63].

With regard to the structural requirements for the activity, a hydroxyl group at the 6-position and the 8-*O*-glucopyranosyl moiety reduced the activity [inhibition at 30 μM: 11 (29.9 %) > 9 (10.5 %) and 6 (-17.6 %); (±) phyllodulcin (23.8 %) > 36 (-3.0 %) and 8 (3.4 %); 1 (15.7 %) > 2 (7.2 %)]. The effect of 11 on the release of adiponectin into the medium, the gene expression of adiponectin, PPARγ2, interleukin 6 (IL-6), and glucose transporter 4 (GLUT4), and the uptake of 2-deoxyglu-cose were examined. Results showed 1 caused a concentration-dependent increase in the release of adiponectin in the medium and expression of adipo-nectin, PPARγ2, and GLUT4 mRNA, while 11 decreased the expression of IL-6 mRNA which was associated with insulin resistance. On the other hand, the reference compound, troglitazone, did not induce the expression of PPARγ2 mRNA and its inhibitory activity against the expression of IL-6 mRNA was weak. All results suggested that the mechanism of action by 11 was different from that of troglitazone, although the mechanism involved should be studied further. Furthermore, 1 (3-100 μM) concentration-dependently enhanced the uptake of 2-deoxyglucose in 3T3-L1 cells. Through examining the effects of 11 on blood glucose, TG, and free fatty acid (FFA) levels in KK-$A^y$ mice, as a model of type II diabetes, 11 significantly lowered blood glucose levels 2 weeks after its administration at 200 mg/kg/d. In addition, 11 at a dose of 200 mg/kg/d decreased FFA levels, which was believed to be one of the factors inducing insulin resistance, similar to troglitazone. So 11, (±) phyllodulcin, 1, and 114 from the processed HDF promoted adipogenesis in 3T3-L1 cells. Compound 11, a principal constituent, significantly increased the amount of adiponectin released into the medium and levels of adiponectin, PPARγ2, and GLUT4 mRNA, while 11 decreased the expression of IL-6 mRNA. Furthermore, 11 significantly lowered blood glucose and FFA levels 2 weeks after its administration at a dose of 200 mg/kg/d. This is the first report that 3-phenyldihydroisocoumarins have an anti-diabetic effect, and 11 is a promising seed compound with antidiabetic activity.

The improvement of diabetic complications such as lipid lowering and anti-oxidative potential of HDF was studied in streptozotocin-induced diabetic rats [64]. Male Sprague-Dawley rats were divided into 3 groups after induction of streptozotocin (STZ)-diabetes: normal control; diabetic control; dia- betic-HDF supplement (hot water extract 40 g/kg diet); and fed experimental diets for 3weeks. Serum glucose and insulin concentrations, serum lipid profile, intraperitoneal glucose tolerance test, and liver cytosolic antioxi-dant enzyme levels were measured. The HDF supplement significantly decreased serum glucose concentration, increased insulin level, and improved glucose homeostasis in diabetic control rats. The total cholesterol and triglyceride concentrations in the serum and liver were markedly reduced by HDF treatment in STZ-diabetic rats.

Moreover, low density lipoprotein (LDL)-, VLDL-, and high density lipoprotein (HDL)-cholesterol levels were ameliorated in HDF supplemented diabetic rats. Decreased fecal excretions of cholesterol, triglyceride, and bile acid in diabetic rats were significantly increased by HDF consumption. HDF supplement reversed the effects of the oxidative stress system of liver in diabetic rats. Lipid peroxidation of diabetic rats, assessed by thiobarbi-turic acid reactive substance (TBARS), as well as superoxide dismutase (SOD) activities, were significantly increased, and glutathione contents were decreased in diabetic rats. HDF supplement reverted these parameters to near normal value. The data suggested that HDF supplement could be used to improve the glucose and lipid me- tabolism as well as to reduce the imbalance between the generation of reactive oxygen species (ROS) and thescavenging enzyme activity in preventing diabetic complications.

## 3.7. Hepatoprotective and Cholagoic Activity

There was a simple and efficient process of obtaining an extract of HDF that had strong efficacy on D-galactosamine (GalN)-induced liver injury *in vivo* and *in vitro* [69]. The water extract (WE), ethanol extract (EE), and water-ethanol extract (WEE) were prepared twice with water, 60 % ethanol or water- 60 % ethanol for 24 hours, respectively.

The results demonstrated that some fractions of WEE of HDF inhibited GalN-induced hepatocyte death in vitro. Not the WE, but the WEE of HDF inhibited liver Injury in rats.

The WEE of HDF suppressed the GalN-induced increase of serum ALT activity and the extent of increase in the serum enzyme activity was known to be parallel to that of liver injury, so the results indicate that WEE of HDF might prevent GalN-induced liver injury in vivo. Since the mechanism for GalN-induced liver injury was known to share some similarities with that induced by a virus or alcohol, it could be presumed that HDF would protect against virus- or alcohol-induced liver injury.

Meanwhile the WEE of HDF inhibited ethanol and lipopolysaccharide-induced liver injury in rats. Inhibition of tumor necrosis factor-α was speculated to be involved in the action of WEE. It has revealed in this work that not the WE and the WEE of HDF inhibited liver injury and this suggested that the active compounds were likely to be lipophilic. The active compounds were thought to be more concentrated in WEE than in EE, because, in the former, water-soluble compounds that would show no effect have been removed. The reason why the hepatoprotective activity of HDF has not previously been identified was probably that only the WE of HDF been used.

In the liver injury model we used here, although the milk thistle extract and turmeric powder weakly suppressed the increase of serum ALT activities, the activity of WEE of HDF was much more potent at the same dosage.

The compounds 11, (±) phyllodulcin along with the MeOH extract HDF were investigated the cholagoic activity on the basis of the data of bile secretion and the inhibitory effects on hypercholesterolemia induced by high cholesterol diet in rats [60]. It was found that the active constituents were contained in the lipophilic portion of the methanolic extract. The effect of methanolic extract of HDF on the cholagoic activity was significantly different from the control and was uniform with the positive control based on the data. However, the lipophilic compounds 11 and (±) phyllodulcin showed the little cholagoic activities with the effect on the bile secretion in rats.

### 3.8. Anti-Malarial Activities

The prevalence of malaria in tropical zones worldwide, together with the lack of vaccine and the appearance of a strain of malaria parasite resistant to commercially available anti-malarial drugs, makes the search for new anti-malarials a global demand. This circumstance prompted us to explore new anti-malarial candidates with high selectivity index between the parasite and mammalian host cells from natural sources.

The MeOH extract and several compounds isolated from HDF were shown to demonstrate not only in vitro anti-malarial activity against Plasmodium falciparum, but also little cytotoxic property against human epidermoid carcinoma KB 3-1 cells [25]. Three flavonol glycoside1 87, 88 and 91 from the genus of *Hydrangeae* with several naturally occurring flavonol glycosides were assessed as the active constituents for anti-malarial potency.

The anti-malarial potency against P. falciparum and the cytotoxic activity against KB 3-1 cells of the three flavonol glycosides 87, 88 and 91 were investigated. Each of them exhibited characteristic anti-malarial activity: in particular, approximately 60 % of proliferation of the parasite was inhibited even at the concentration of 0.5 ng/mL. On the other hand, these flavonol glycosides have little influence on the growth of KB 3-1 representing the host cell.

In order to examine the structure-activity relationship, several flavonol glycosides, rutin, isorhamnetin 3-$\beta$-*O*-rutinoside, multinoside A, multinoside A acetate and isoquercitrin as well as flavonols, quercetin and kaempferol, were assessed for anti-malarial activity. Despite a slight reduction of activity at the concentration of 0.1 and 0.05 ng/mL of isorhamnetin 3-$\beta$-*O*-rutinoside, each flavonol diglycoside showed a similar biological score. In the case of flavonol monoglycosides (quercitrin and isoquercitrin), the anti-malarial activities particularly decreased at 5 and 0.5ng/mL concentrations in comparison with the other flavonol diglycosides. On the contrary, the two flavonols, quercetin and kaempferol showed no anti-malarial potency at the concentration of 5 ng/mL. It is curious that most of the flavonol glycosides displayed no definite growth inhibitory activity against P. falciparum in a concentration-dependent manner. As a result of preliminary examination, exposure of 87 and 4 at higher concentration (0.5 mg/mL) rather resulted in decline of activity. Additionally, little growth inhibition was observed at fairly low concentration of 87 (0.5 pg/mL). In conclusion, three flavonol glycosides, 87, 88 and 91 were assessed as the anti-malarial principles. Although these flavonol glycosides could not accomplish complete inhibition of proliferation of P. falciparum, the significant low concentration with no cytotoxic effect gave rise to more than 50 % growth inhibition against the malaria parasite.

Moreover, the anti-malarials activities of the water extract and the total base extract from the leaves of *Hydrangeae* macrophylla (Thb) DC. forma hortensiawils were investigated [68]. The results demonstrated that both of the extracts were possessed the insignificant anti-malarials activities.

*H. chinensis* was used in treatment of malaria and cardiovascular diseases. Alkaloids compounds 105, 108 and 109 isolated from *H. chinensis* and the roots of *H. arten* were detected to be the active principles against malaria [47].

### 3.9. Differentiation Inducing Activities

Difrerentiation inducers are of potential interest for thr treatment of human cancers promoting the terminal differentiation of certain human tumor cells. In the course of searching for differentiation inducers against leukemic cells,the methanolic extract of HDF was recognized to have differentiation inducing activities. Activity guided separation of the extract was carried out using M1cells. The compounds 1, 2, 6, 7, 9, 11, 12, 23, 34, 36 and 38 isolated from HDF were investigated the differentiation inducing activity against M1cells [12]. Among these compounds, dihydroisocoumarins and isocoumarins showed the activities at a concentration of greater than 200 μM against M1cells. The compound 7 and 9, having an *O*-diphenol group, showed the higher activities than that of 6 and 11. Compared with 11 and 9, 6 and 7, having a 6-hydroxy group in their structures, induced a larger amount of M1cells into phagocytic cells. It seemed that highly oxygenated isocoumarins have higher activities. Compound 11 has differentiation inducing activity against M1cells and has less cytotoxicity. As it was one of the main compounds of HDF and presented in a high proportion, the crude drug could be valuable as a differentiation inducer source.

The dihydroisocoumarins glycosides 34, 36 and 38 were all inactive in the induction of the differentiation of M1cells. These compounds showed hardly any antiproliferative activities, and a more than 70 % of growth ratio was found in 300 μM of these compound treated groups. Glycosides were good proinducers.

### 3.10. Antiulcer Activities

Gastric erosions were induced by oral administration of HCl/EtOH solution and the erosion index and the inhibition used as the evaluation index. The MeOH extract of HDF and the leaves of H. macrophylla along with the two principle constituents 11 and (±) phyllodulcin were assessed the inhibitions on the antiylcer activities [60]. The results suggested that both of the extracts possessed high antiulcer activities, and were significantly different from the control. However, 11 and (±) phyllodulcin were found to show little antiulcer activities.

## Conclusion

Pharmacological investigations have revealed that the extracts of the genus of *Hydrangeae* are highly bioactive. Further scientific research of their components is needed to verify the bioactivities of this genus. Since the processing Hydrangea Dulcis Folium even is listed in the Japanese Pharmacopoeia VIII in Japan, the research mainly focus on the components and bioactivities of the plants of HDF. Phytochemical investigation of

*Hydrangeae* has shown that isocoumarins and secoididoid glucosides are the main constituents, and the cyanogenic glycosides compounds are the specified type constituents. Except for the HDF, there are still several plants of *Hydrangeae* that need to be studied phytochemically and biologically with the aim to elucidate the components and verify their bioactivities.

# References


[1] Yoshikawa M, Uchida E, Chatani N, Kobayashi H, Naitoh Y, Okuno Y, Matsuda H, Yamahara J, Murakami N, *Chem. Pharm. Bull.* 40, 3352, (1992).

[2] Yoshikawa M, Uchida E, Chatani N, Murakami N, Yamahar J, *Chem. Pharm. Bull.* 40, 3121, (1992).

[3] Shimoda H, Matsuda H, Yamahara J, Yoshikawa M, *Biol. Pharm. Bull.* 21, 809, (1998).

[4] Yoshikawa M, Matsuda H, Shimoda H, Shimada H, Harada E, Naitoh Y, Miki A, Yamahara J, Murakami N, *Chem. Pharm. Bull.* 44, 1440, (1996).

[5] Nozawa K, Yamada M, Tsuda Y, Kawai K.I. Nakajima S, *Chem. Pharm. Bull.* 29, 2689, (1981).

[6] Asahina Y, Asano J, *Yakugaku Zasshi.* 49, 117, (1929).

[7] Yoshikawa M, Ueda T, Shimoda H, Murakami T, Yamahara J, Matsuda H, *Chem. Pharm. Bull.* 47, 383, (1999).

[8] Yoshikawa M, Murakani T, Ueda T, Shimoda H, Yamahara J, Matsuda H, *Heterocycles.* 50, 411, (1999).

[9] Asahina Y, Ueno S, *Yakugaku Zasshi.* 408, 146, (1916).

[10] Yoshikawa M, Harada E, Naitoh Y, Inoue K, Matsuda H, Shimoda H, Yamahara J, Murakami N, *Chem. Pharm. Bull.* 42, 2225, (1994).

[11] Günes M, Speicher A, *Tetrahedron.* 59, 8799, (2003).

[12] Umehara K, Matsumoto M, Nakamura M, Miyase T, Kuroyanagi M, Nouchi H, *Chem. Pharm. Bull.* 48, 566, (2000).

[13] Suzuki H, Ikeda T, Matsumoto T, Noguchi M. *Agric. Biol. Chem.* 41, 1815, (1977).

[14] Ueno Y. *Yakugaku Zasshi.* 57, 602, (1937).

[15] Kikuchi M, Kakuda R, Kikuchi M, Yaoita Y, *Chem. Pharm. Bull.* 56, 610, (2008).

[16] Matsuda H, Shimoda H, Uemura T, Ueda T, Yamahara J, Yoshikawa M, *Chem. Pharm. Bull.* 47, 1753, (1999).

[17] Yoshikawa M, Ueda T, Matsuda H, Yamahara J, Murakami N, *Chem. Pharm. Bull.* 42, 1691, (1994).

[18] Inouye H, Takeda Y, Uesato S, Uobe K, Hashimoto T, *Tetrahedron Lett.* 21, 1059, (1980).

[19] Sakai H, Kakuda R, Yaoita Y, Kikuchi M, *J. Nat. Med.* 61, 226, (2007).

[20] Song F. R, Lee Y. H, Yang Y. L, Hsieh P. W, Khalil A. T, Chen C. Y, Wu Y. C, *J. Nat. Prod.* 66, 1245, (2003).

[21] Yamaguchi K, Ishii H, *Chem. Pharm. Bull.* 29, 3421, (1981).

[22] Uesato S, Hashimoto T, Takeda Y, Inouye H, *Chem. Pharm. Bull.* 30, 4222, (1982).

[23] Kikuchi M, Kakuda R, Yaoita Y, *Heterocycles.* 76, 313, (2008).

[24] Kondo T, Toyama-Kato Y, Yoshida K, *Tetrahedron Lett.* 46, 6645, (2005).

[25] Murakami N, Mostaqul H. M, Tamura S, Itagaki S, Horii T, Kobayashi M, *Bioorg. Med. Chem. Lett.* 11, 2445, (2001).
[26] Nakamura S, Wang Z. B, Xu F. M, Matsuda H, Wu L. J, Yoshikawa M, *Tetrahedron Lett.* 50, 4639, (2009).
[27] Wang Z. B, Gao H.Y, Yang C.J, Sun Z, Wu L.J, *Helv. Chim. Acta.* 94, 847, (2011).
[28] Eckermann C, Schrőder G, Eckermann S, Strack D, Schmidt J, Schinerder B, Schrőder J, *Phytochemistry.* 62, 271, (2003).
[29] Patnam R, Chang F. R, Chen C. Y, Kuo R. Y, Lee Y. H, Wu Y. C, *J. Nat. Prod.* 64, 948, (2001).
[30] Wang Q, Matsuda H, Matsuhira K, Nakamura S, Yuan D, Yoshikawa M, *Biol. Pharm. Bull.* 30, 388, (2007).
[31] Ito D, Shinkai Y, Kato Y, Kondo T, Yoshida K, *Biosci. Biotechnol. Biochem.* 73, 1054, (2009).
[32] Yoshida K, Ito D, Shinkai Y, Kondo T, *Phytochemistry.* 69, 3159, (2008).
[33] Shinichi U, Toshihiro H, Kenichi U, Yoshio T, Hiroyuki I, Heihachiro T, Tohru E, Tennen Yuki Kagobutsu Toronkai KOen Yoshishu, 24th (Osaka Japan). (1981), pp9-16.
[34] Pryce R.J, *Phytochemistry.* 10, 2679(1971).
[35] Takeda K, Kariuda M, Itoi H, *Phytochemistry.* 24, 2251, (1985).
[36] Takeda K, Yamashita T, Takahashi A, Timberlake C. F, *Phytochemistry.* 29, 1089, (1990).
[37] Takeda K, Kubota R, Yagioka C, *Phytochemistry.* 24, 1207, (1985).
[38] Uesato S, Takeda Y, Hashimoto T, Uobe K, Inouye H, *Helv. Chim. Acta.* 67, 2111, (1984).
[39] Suzuki H, Matsumoto T, Noguchi M, *Phytochemistry.* 15, 555, (1976).
[40] Gorham J, *Phytochemistry.* 16, 249, 1977.
[41] Towers G. H. N, Chen D, *Phytochemistry.* 3, 469, (1964).
[42] Austin D. J,Meyers M. B, *Phytochemistry.* 4, 245, (1965).
[43] Hashimoto T, Tori M, Asakawa Y, *Phytochemistry.* 26, 3323, (1987).
[44] Tachibana Y, Hashimoto Y, Hagiwara Y, Konishi T, Kurokawa N, *Yakugaku Zasshi.* 94, 1167, (1974).
[45] Yoshikawa M, Chatani N, Harada E, Nishino Y, Yamahara J, Murakami N, *Yakugaku Zasshi.* 114, 176, (1994).
[46] Akada Y, Kawano S, Gotoh M, Tanase Y, *Yakugaku Zasshi.* 99, 683, (1979).
[47] Akira I, Toshio M, Tohru S, Francis W. M, Mamoru T, *J. Nat. Med.* 61, 213, (2007).
[48] Masayuki Y, Hiromi S, Nobuhiro Y, Nobutoshi M, Hiroshi S, Johji Y, Hisashi M, *Chem. Pharm. Bull.* 44, 1890, (1996).
[49] Masayuki Y, Emiko H, Nobuhiro Y, Yasuhiro O, Osamu M, Hiroshi A, Nobutoshi M, *Chem. Pharm. Bull.* 42, 721, (1994).
[50] Masahiro K, Miyosi I, Hidenobu I, Eiji O, Koichi N, Shinichi U, Hiroyuki I, Tetsuro F, *Shoyakugaku Zasshi.* 44, 288, (1990).
[51] Takeuchi N, Ochi K, Murase M, Tobinaga S, *Chem. Pharm. Bull.* 31, 4360, (1983).
[52] Takeuchi N, Murase M, Ochi K, Tobinaga S, *Chem. Pharm. Bull.* 28, 3013, (1980).
[53] Masashi T, Noriko S, Yoko S, *Chem. Pharm. Bull.* 25, 968, (1977).
[54] Ibrahim R, *Phytochemistry.* 9, 1211, (1970).
[55] Austin D, Meyers M.B, *Phytochemistry*.4, 245, (1965).

[56] Khalifa I. I, EI-Ginidi O. D, Ammar H. A, Shafie A. A, *J. Pharm. Sci.* 28, 221, (2001).
[57] Yagi A, Mashida Y, Takata N, Nishioka I, *Chem. Pharm. Bull.* 20, 1755, (1972).
[58] Masatoshi Y, Koichi S, Kuniko H, Miyoko O, Takaji K, *Chem. Pharm. Bull.* 22, 475, (1974).
[59] Yoshikawa M, Shimada H, Yagi N, Murakami N, Shimoda H, Yamahara J, Matsuda H, *Chem. Pharm. Bull.* 44, 1890, (1996).
[60] Yamahara J, Matsuda H, Shimoda H, Ishikawa H, Kawamori S, Wariishi N, Harada E, Murakami N, Yoshikawa M, *Yakugaku Zasshi*. 114, 401, (1994).
[61] Matsuda H, Shimoda H, Yoshikawa M, *Bioorg. Med. Chem.* 7, 1445, (1999).
[62] Matsuda H, Shimoda H, Yamahara J, Yoshikawa M, *Biol. Pharm. Bull.* 22, 870, (1999).
[63] Zhang H. L, Matsuda H, Kuamahara A, Ito Y, Nakamura S, Yoshikawa M, *Bioog. Med. Chem. Lell.* 17, 4972, (2007).
[64] Kim H. K, Kim M. J, Lyu E. S, Shin D. H, *Biol. Pharm. Bull.* 32, 153, (2009).
[65] Wang Q. L, Matsuda H, Matsuhira K, Nakamura S, Yuan D, Yoshikawa M, *Biol. Pharm. Bull.* 30, 388, (2007).
[66] Kurume A, Kamata Y, Yamashita M, Wang Q. L, Matsuda H, Yoshikawa M, Kawasaki I, Ohta S, *Chem. Pharm. Bull.* 56, 1264, (2008).
[67] Matsuda H, Shimoda H, Kageura T, Yoshikawa M, *Biol. Pharm. Bull.* 22, 925, (1999).
[68] Huang S. Y, Yu G. T, *Guangxi J. Trad. Chin. Med.* 3, 44, (1988).
[69] Nakagiri R, Hashizume E, Kayahashi S, Sakai Y, Kamiya T, *Biosci. Biotechnol. Biochem.* 67, 2641, (2003).
[70] Yamamoto H, Yamamoto M, Yamamoto K, Chatani N, Yoshikawa M, *Biol. Pharm. Bull.* 21, 1003, (1998).

In: Phytochemicals
Editor: Hai-Xue Kuang

ISBN: 978-1-62417-354-7


*Chapter XVII*

# Phytochemicals in Cancer Research: A Review of Their Biological Activities and Anticancer Potentials in Preclinical and Clinical Studies

***Daniel L. Pouliquen*** **and Marc Grégoire***
Centre de Recherche en Cancérologie Nantes-Angers,
Institut de Recherche Thérapeutique de l'Université de Nantes
Inserm, Nantes, France
Université de Nantes, Nantes, France
CNRS, UM, Nantes, France

## Abstract

Since the 1960s, the observation of geographical differences in the incidence of cancers has led epidemiologists to analyze their potential relationship with fruit and vegetable intakes. The capacity of a number of natural substances present in the diet to inhibit the neoplastic effects of chemical carcinogens has been demonstrated in laboratory animals. Many phytochemicals have been isolated from plants, their structures identified and their biological activities analyzed. In recent decades, some chemical classes have been the subject of extensive studies, *in vitro* and *in vivo*, and their mechanisms of action, now well documented, open up interesting prospects for cancer prevention and treatment in patients, either alone or as adjuncts to current cancer therapies. These compounds include isothiocyanates and indoles, organosulfur compounds and curcuminoids. Other potentially anticarcinogenic molecules naturally present in food may significantly reduce the incidence of specific cancers. These consist of secondary metabolites classified according to their chemical structure: alkaloids, acetogenins, stilbenoids, simple phenols, phenolic acids, polyphenols (flavonoids, lignans, proanthocyanidins, tannins, quinones), carotenoids, terpenoids, saponins, and

* Email address: Daniel.Pouliquen@univ-nantes.fr.

others. For some of these, chemopreventive effects during the initiation, promotion and/or progression steps of carcinogenesis have already been described. The diversity of traditional plant pharmacopeias throughout the world has also provided an invaluable source of ethnopharmacological information in the search for new molecules to be tested against experimental cancer models. Finally, as some of these phytochemicals are able to target multiple deregulated cellular signaling pathways, without toxic effects in normal cells, an interesting challenge for the future is to determine the optimal mode of combining their specific potentials.

## Introduction

Phytochemicals have made major contributions to medical practice over thousands of years. The oldest records of medicinal plants are from China (Chinese pharmacopeia about 5,000 B.C.), the ancient Sumerians recorded around a thousand useful plants 4,000 years ago, and an important list of plants still in use today had already been mentioned in the Ebers Papyrus in Egypt around 1,800 B.C. The emergence of the first chemotherapeutics after the end of World War II [1] was followed by a waning interest in plant-derived pharmaceuticals, as rational drug design techniques became possible. However, this period coincided with the discovery of the biological activities of podophyllotoxin from *Podophyllum peltatum* [2], vinca alkaloids from *Catharanthus roseus* [3], camptothecin from *Camptotheca acuminata* [4], and taxol from *Taxus brevifolia* [5]. Fortunately, since the mid-1980s, a renewed interest in the use of phytochemicals for cancer prevention and treatment has been observed, which has benefited from three important events.

Firstly, the observation of geographical differences in the incidence of a number of cancers, and the changes occurring in the same area over a period of time, led some investigators of the interrelations between diet and cancer, in the mid-1960s, to suggest the implication of factors protecting the host against the effects of the carcinogenic agent [6]. The capability of natural substances to inhibit the neoplastic effects of chemical carcinogens was then demonstrated in rats [7]. With time, this finding was enriched with epidemiological data and the idea, common in Asia, that foods and medicines are of equal importance in preventing and treating disease [8] was considered. By 1981, it was estimated that a 35% reduction in the burden of fatal cancers in the US could be achieved by dietary modification [9]. In 1985, a first list of compounds which may prevent the occurrence of cancer was provided, most of these being naturally occurring food constituents belonging to more than twenty different classes of phytochemicals [10].

Secondly, exploration of the scientific literature in the last half-century reveals that the combined effort of chemists and botanists has led to the isolation and characterization of a wealth of interesting molecules from plants. Starting with the large US program of phytochemical screening conducted at the end of the 1950s at the Eastern Regional Research Laboratory in Philadelphia [11], this field of research has been helped considerably by the development of analytical techniques, such as nuclear magnetic resonance spectroscopy, during the 1980s [12]. Finally, the extension of cell-culture techniques during the same period [13] has led to a limiting of the use of laboratory rodents and to the identification, among the libraries of compounds being accumulated over the three last decades in journals relevant to phytochemistry and natural products, of a number of molecules with interesting anticancer properties.

Thirdly, the consciousness that the diversity of plant types, particularly in environments such as tropical rainforests, is huge, and the fact that only a tiny proportion has already been investigated for potential drug development, has accelerated efforts for the preservation of this biodiversity. The 1990s, in particular, were characterized by the wish to draw perspectives on issues regarding intellectual property rights with regard to drug development from natural sources, the conservation of natural habitats, and international conventions on bioprospecting [14]. In 1985, it was pointed out that 74% of all plant-derived drugs in clinical use worldwide had been discovered through follow-up investigation of folk or traditional medicinal uses (ethnomedical information). In an attempt to promote both conservation of these resources, and their use in ways that are non-destructive and sustainable, Soejarto, in 1996, highlighted the difficulties encountered in this field with the example of new cytotoxic labdane diterpenoids isolated from a rare plant, *Neouvaria acuminatissima*, which was initially found in western Luzon in the Philippines, but was finally lost [15].

Among the various classes of phytochemicals which exhibit interesting properties, three first categories of molecules have been studied extensively, opening up not only considerable prospects for cancer prevention, but also for future therapies.

# 1) Glucosinolates and Derived Compounds

## Indoles

The first demonstration of the implication of naturally occurring compounds from Brassicaceae in the prevention of carcinogenesis was made by Lee W. Wattenberg in 1975. Wattenberg studied the hydroxylation of aromatic polycyclic hydrocarbons by the microsomal mixed-function oxidase system, and after testing various dietary constituents present in a rat chow, focused on the role of indoles from cabbage and cauliflower [16]. Among three indoles studied for their inhibitory effects on the carcinogenic process, indole-3-carbinol (I3C) (Figure 1) was found to be the most potent against mammary tumor formation, when added to the diet of female rats induced with DMBA [17]. Subsequently, the ability of I3C to inhibit the hepatocarcinogenesis induced by aflatoxin B1 (AFB1), when given prior to and during carcinogen exposure, was demonstrated in the rainbow trout [18]. However, this phytochemical also promoted aflatoxin-induced hepatocarcinogenesis in the same species, when given continuously after AFB1 initiation [19]. As a consequence, the anticarcinogenic mechanisms of this "ambivalent modulator" were investigated in detail in this animal model during subsequent years [20-22].

Depending on timing and administered dose, it was suggested that the estrogenic activities of I3C acid derivatives could be involved in the observed promotion of AFB1 carcinogenesis in the trout [23]. This relatively inexpensive species, its sensitivity to important known or probable human carcinogens and low spontaneous tumor incidence, compared with rodents, allowed the confirmation that, after initiation of embryos with AFB1, I3C played the role of a tumor promoter in a wide range of doses when given in the diet after hatching [24].

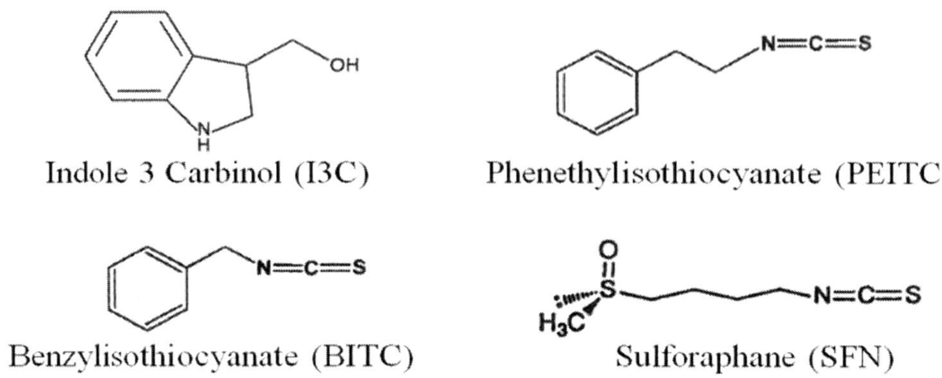


Figure 1. Chemical structures of indoles and isothiocyanates.

Conversely, when I3C was given in the diet before or at the time of initiation with carcinogen, inhibition of tumorigenesis occurred in rat models of cancers in various target organs including the liver [25], endometrium [26], tongue [27], mammary gland [28] and colon [29].

In the case of colon carcinoma induced by 2-amino-3-methyl-imidazo[4,5-*f*]quinoline (IQ), the mechanism of protection produced by a dietary equivalent of I3C in the range of 100-1000 p.p.m., involved an induction of cytochromes P450 1A1 and 1A2, and a dose-related increase of ethoxyresorufin *O*-deethylase (EROD) and methoxyresorufin *O*-demethylase (MROD) activities in the liver and colonic mucosa [30].

In the mouse, I3C also inhibited forestomach [17] and lung tumorigenesis [31]. More recently, I3C was shown to provide chemoprotection for the fetus against the lung tumorigenesis produced by the transplacental carcinogen, dibenzo[*a*,*l*]pyrene, in the mouse. The suggested mechanism of action of I3C and its acid condensation products was that they can act as blocking agents via aryl hydrocarbon receptor modulation of phase I and II enzymes and UDP-glucuronosyltransferases [32]. However, besides a strong hepatoprotective activity against various carcinogens, a detailed review of the effects of I3C *in vitro* on different tumor cell types, revealed that numerous other molecular targets were affected, including downregulation of antiapoptotic gene products, upregulation of proapoptotic proteins and inhibition of the activation of transcription factors [33].

In human clinical trials, I3C was first evaluated as a potential chemopreventive agent against breast and ovarian cancers [34]. It was later shown that, in order to prevent the development of cervical intraepithelial neoplasia (CIN), the oral administration of 200-400 mg/day I3C to patients with biopsy-proven CIN II-III, for 12 weeks, produced a statistically significant regression of CIN relative to the placebo group [35]. Following a study of the biodistribution and pharmacokinetics of I3C after oral administration in the mouse [36], phase I clinical studies were conducted in women [37].

The mechanisms of action of I3C, and its potential use for the chemoprevention of colorectal cancer, have also been reviewed recently [38]. To date, several clinical trials with I3C have been completed or are in progress, applied to the treatment of patients with prostate, breast or uterine cancers [39].

## Isothiocyanates

In preliminary observations, the finding that the stimulation of intestinal drug metabolism in rats pretreated with several indoles present in Brussels sprouts, was less than in those given Brussels sprouts or dried cabbage suggested that other dietary constituents were involved in the increased metabolism of drugs [40]. Subsequently, an experiment conducted using the same rat cancer model, confirmed the implication of benzyl isothiocyanate and closely related compounds in the reduction of the percentage of rats with tumors [41].

The sources of indoles and isothiocyanates in cruciferous vegetables are glucosinolates (GLs), which are easily converted when processed and heated. GLs, which are widely distributed in plants and especially prevalent in the order Capparales, have been isolated and characterized since the 1960s [42]. Upon autolysis, GLs are hydrolyzed to release a number of products, mainly isothiocyanates (ITCs) and nitriles [43]. GLs consist of *S*-glucopyranosyl thiohydroximates and, to date, approximately 120 different structures have been reported [44]. For the biology and biochemistry of GLs, the reader is referred to other excellent reviews [45]. The inventory of genes involved in GL biosynthesis has also recently improved our understanding of the link with primary sulfur metabolism, opening up considerable prospects in biotechnologies [46].

### *Phenethylisothiocyanate (PEITC)*

Among the studies aimed at analyzing the protection provided by isothiocyanates against the lung tumorigenesis produced by 4-(methylnitrosamino)-1-(3-pyridyl)-1-butanone (NNK) in the mouse, Chung et al. showed in 1993 that, on a molar basis, phenethyl isothiocyanate (PEITC) (Figure 1) was considerably more effective than I3C [47]. PEITC, which is, together with benzyl isothiocyanate (BITC), one of the most widely occurring ITCs, is particularly abundant in watercress and has received much attention as a possible chemopreventive agent.

Early reports demonstrated that this phytochemical inhibited lung tumorigenesis in the rat [48], and produced a favorable modification of biomarkers [49]. In this species, PEITC had inhibitory effects on N-nitrosobenzylmethylamine carcinogenesis in the esophagus [50]. These findings were completed by the demonstration of the potential of this molecule in the chemoprevention of diethylnitrosamine-induced hepatocarcinoma in the mouse [51]. The chemopreventive effects of PEITC previously observed in rats and mice were also confirmed in hamsters on lung and pancreatic tumorigenesis [52]. This molecule also acts at least in part by inhibiting the generation of reactive carcinogen metabolites which can interact with DNA, reducing tumor formation by 94% as shown in a model of *N*-nitrosomethylbenzylamine (NMBA) oral cancer in the hamster [53].

PEITC has the ability to suppress the growth of human cancer cells in culture as well as *in vivo* in xenograft assays by causing cell cycle arrest and induction of apoptosis [54]. Subsequent studies showed the role of PEITC in the induction of proapoptotic Bax and Bak proteins, *in vivo* [55], leading to disruption of the mitochondrial membrane potential and consequently to a reduced growth of prostate cancer xenografts in nude mice [56]. A recent extension of these findings has shown that the mechanism of PEITC-induced apoptosis in breast cancer cells involves the contribution of B-cell lymphoma 2-interacting mediator of cell death (Bim) and p53-upregulated modulator of apoptosis (PUMA) [57]. Other reports have revealed that the induction of apoptosis and inhibition of cell migration in prostate cancer cells is mediated by downregulation of X-linked IAP (XIAP) and / or survivin [58].

Finally, these investigations have provided a novel insight into the critical role of the growth factor adapter protein, $p66^{Shc}$, in the PEITC-induced apoptosis of human prostate cancer cells [59].

PEITC can also inhibit angiogenesis. PEITC treatment produced a decrease in survival of human umbilical vein endothelial cells in a concentration- and time- dependent manner, and was associated with the suppression of vascular endothelial growth factor (VEGF) secretion, down-regulation of VEFG receptor 2 protein levels and inactivation of prosurvival serine-threonine kinase, Akt, leading to reduced migration of human prostate cancer cells [60]. PEITC treatment also significantly inhibited tumor platelet/endothelial cell adhesion molecule expression, a marker of angiogenesis [61].

In the search for new chemoprevention strategies, as many chemical carcinogens require metabolic activation by enzymes from the cytochrome P450 superfamily, it was also important to investigate which enzymes were affected by PEITC treatment. Von Weymarn et al. showed that, among enzymes from the P450 2A subfamily which are involved in the metabolic activation of a number of nitrosamines, P450s 2A6 and 2A13, which are thought to be the enzymes responsible for the *in vivo* metabolism of the tobacco-specific carcinogen NNK, were inactivated in an NADPH-dependent manner by PEITC [62].

The fact that PEITC induced p53 transactivation *in vitro* in a dose- and time-dependent manner with accompanying apoptosis [54] suggested that ITCs may exhibit chemopreventive potential when administered during the postinitiation phase of tumorigenesis. In order to test this hypothesis, *N*-acetylcysteine (NAC) conjugate (PEITC-*NAC*) was administered by gavage after azoxymethane initiation of colon carcinoma in rats. Indeed, the conjugate inhibited aberrant crypt foci formation [63] and a second study was conducted, this time in the mouse, to investigate the mechanisms involved in this chemoprevention, revealing that the activator protein 1 (AP-1) and MAP kinase (MAPK) pathways were activated [64]. Subsequently, PEITC was shown to potently induce AP-1 activity, causing a significant elevation in the phosphorylation of extracellular signal-regulated kinase (ERK) and c-Jun N-terminal kinase (JNK) signaling pathways in human prostate cancer cells [65]. Among the other cell signaling events associated with the effects of PEITC on cancer cells, a caspase-independent downregulation of Mcl-1 and inactivation of Akt were observed in human leukemia cells, as well as in primary human acute myeloid leukemia [66]. Other responses also involved the regulation of translation initiation, in particular an increase in the expression of the eukaryotic translation initiation factor 4E-binding protein 1 (4E-BP1), and inhibition of 4E-BP1 phosphorylation, which represent new targets and a highly promising strategy for the development of anticancer agents [67].

### *Benzylisothiocyanate (BITC)*

Other ITCs, such as BITC (Figure 1), have been investigated as interesting chemopreventive agents in the field of cancer research. Findings complementary to the mechanisms of action detailed above, using this molecule, included inhibition of one member of the signal transducer and activator of transcription 3 family, STAT3 [68, 69], inhibition of histone deacetylase [70] and involvement of the PI3K/AKT/FOXO pathway [71]. As previously mentioned for PEITC, BITC also inhibits the growth of leukemia cells *in vivo* [72], and demonstrates efficacy against mammary cancer in animal models [73-75].

*Sulforaphane (SFN)*

Another ITC, sulforaphane (Figure 1), which is currently, equivalently with PEITC, considered to be the most potent chemopreventive molecule of this family, has been isolated from broccoli and its structure elucidated in 1992 [76]. SFN was initially identified as a potent inducer of phase II detoxification enzymes, and its anticarcinogenic activity against mammary tumors induced in the rat was first described in 1994 [77]. Subsequently, SFN was shown also to inhibit the phase I enzyme cytochrome P450 isoenzyme 2E1, which is responsible for the activation of nitrosamines [78]. However, more recent studies have demonstrated that SFN offers protection against tumor development during the "post-initiation" phase, and that the molecular targets vary depending upon cancer stage and target tissue. In addition to the induction of phase 2 genes and modulation of cytochrome P-450 enzymes, post-initiation, SFN can suppress cancer development through various molecular targets. The proposed mechanisms were excellently reviewed in 2007 [79] and 2008 [80], showing that SFN activates the nuclear factor erythroid 2-related factor 2 (Nrf2) and Nrf2-targeted genes, inhibits angiogenesis and metastasis, contributes to decreased cell viability via activation of the MAPK pathways ERK, JNK, and p38, downregulates NFκB activity and generates reactive oxygen species (ROS). Another, more recent review revealed that the mechanism of SFN-induced nuclear translocation of Nrf2 appeared to be different from that of PEITC, as SFN disrupts the Nrf2-Keap1 complex through Keap1 modifications [81]. SFN and other related ITCs also exert their anticancer effects via epigenetic changes, in particular inhibition of histone deacetylase (HDAC) activity. Initial demonstrations of HDAC inhibition by SFN in human colon and prostate cancer cells *in vitro* were confirmed by experiments in vivo showing similar effects in preclinical models of gastrointestinal and prostate cancer [82]. Interestingly, the conditions in which the development of spontaneous intestinal polyps were inhibited corresponded to a SFN tissue concentration in the range 3-30 μM, which is comparable to doses that inhibited HDAC activity in human colon cancer cells [83].

Most of the clinical studies conducted on ITCs have been carried out with SFN. The first randomized, double-blind, placebo-controlled phase I clinical trial of healthy volunteers was performed in 2007. Comprising three study groups given 25-100 μM GL or ITC, no significant toxicities were reported associated with ingestion of the extracts [84]. Another study conducted on residents of Qidong, China, who were at high risk for the development of hepatocarcinoma due to the consumption of food contaminated with carcinogens, showed an inverse association between urinary levels of dithiocarbamates (SFN metabolites), and aflatoxin-DNA adducts in individuals receiving glucoraphanin, the principal GL contained in broccoli sprouts [85]. A number of clinical trials are currently in progress with ITCs, among which six with SFN or broccoli sprouts including one trial treating patients with recurrent prostate cancer, and three with PEITC including one trial focusing on the prevention of lung cancer in smokers.

## 2) Organosulfur Compounds from Liliaceae

The second category is represented by organosulfur compounds from liliaceae. Although these compounds have been widely used since historical times, their potential for the chemoprevention of cancer was not documented until the mid-1980s [86]. In 1983, Belman reported that the topical application of garlic or onion oil inhibited tumor formation in mice treated with tetradecanoylphorbolacetate [87]. Five years later, Wargovitch and coauthors demonstrated that diallylsulfide (DAS), a natural organosulfur product from garlic, when given orally, inhibited the carcinogenic effects of N-nitrosomethylbenzylamine (NMBA) in rat esophagus [88]. Since then, the anticarcinogenic properties of organosulfur compounds from garlic have been confirmed *in vivo* in various carcinoma models in rats and mice, including renal [89], esophageal [90], liver [89] colon [89, 91], forestomach [92], mammary [93, 94], lung [95], and glioma [96]. Evidence for a preventive effect of garlic consumption against a number of cancers has already been documented in the epidemiologic literature, particularly for stomach and colorectal cancers [97, 98]. Even more interestingly, the use of organosulfur compounds from garlic has been suggested for the treatment of leukemias [99, 100]. The mechanisms explaining the chemopreventive effects of garlic-derived products have been reviewed over the last few years [101, 102]. The prospective clinical anticancer role of organosulfur compounds from garlic has also been examined recently [103].

Depending on the conditions of its cultivation, garlic may contain at least 33 different organosulfur compounds. The protective efficiency against DNA damage of garlic powders obtained from bulbs grown in soils with different levels of sulfur fertilization can differ [104]. Also of great interest is the finding that selenium-enriched garlic or organoselenium compounds provide more potent protection against mammary carcinogenesis in rats and greater inhibition of breast cancer cells in culture than natural garlic or the respective organosulfur analogues [105]. The allyl sulfur constituents of garlic, which comprise approximately 1% of its dry weight, are unstable and break down rapidly to produce odorous lipophilic diallyl sulfide (DAS), diallyl disulfide (DADS), diallyl trisulfide (DATS) and ajoene [106]. In addition, other hydrophilic molecules such as S-allylcysteine (SAC) and S-allylmercaptocysteine, are formed by the process of natural aging bioconversion.

## Diallyl Disulfide (DADS)

In the mid-1990s, Reddy demonstrated that, among eight molecules tested for their capacities to prevent colon carcinogenesis, DADS (Figure 2) was the most potent [107]. In athymic nude mice bearing human colon tumor cell xenografts, Sundaram and Milner confirmed this finding by showing that i.p. injection of DADS reduced tumor volume by 69%, and revealed that DADS could be combined with 5-FU to produce an even more important inhibition [108]. Subsequently, Knowles and Milner demonstrated that DADS was more effective than dipropyl disulfide or SAC in inhibiting the growth of human tumor cells [109]. In addition, a correlation was observed between the ability of DADS to suppress proliferation and a decrease in cellular thiols, an increase in intracellular calcium, $[Ca^{2+}]_i$ and the induction of apoptosis [110].

Knowles and Milner revealed in 1998 that DADS suppresses the activity of the $p34^{cdc2}$ kinase complex, which modulates the progression of cells from the G2 into the M phase of the cell cycle by promoting chromosome condensation [111]. Two years later, an investigation of the mechanisms involved in this process showed that this suppression does not result from direct interactions with the protein, but through changes in parameters conditioning the formation and conversion of the enzyme to its active form.

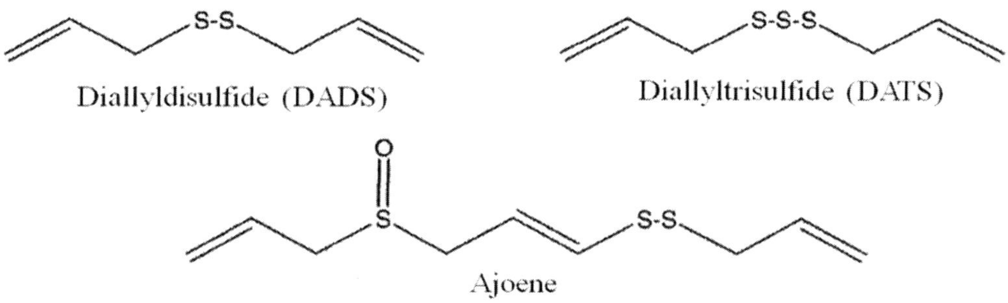


Figure 2. Chemical structures of organosulfur compounds from garlic.

In particular, DADS treatment decreased the amount of protein associated with the cyclin B1 regulatory unit, and increased the hyperphosphorylation of $p34^{cdc2}$ [112]. A number of reports have confirmed that DADS induced both cell cycle arrest and apoptosis in various cancer cells, resulting in accumulation of cells in the G2/M phase. However, p53 is not involved in the regulatory mechanisms of this process [113]. A recent study has provided further insights, revealing the effects of DADS on the cell division cycle 25 (Cdc25) family of dual-specific phosphatases. In particular, Cdc25C, one of the central targets and regulators of the G2/M checkpoint, which triggers the activation of the Cdk1/cyclin B kinase complex and leads to the progression of cells from G2 into the M phase of the cell cycle [101], is down-regulated during DADS treatment of human gastric cancer cells. This process is mediated by the checkpoint kinase, Chk1, which is accumulated in its inactive phosphorylated form [114].

Singh also demonstrated that DADS inhibits the farnesylation of $p21^{H\text{-}ras}$, which is critical for the plasma membrane association and transforming properties of the H-*ras* oncogene, through inhibition of the enzyme, HMG-CoA reductase and thus depletion of the farnesyl pyrophosphate pool [115, 116].

The DADS-mediated G2/M cell cycle arrest also correlated with activation of extracellular signal-regulated kinases (ERKs), while the activities of protein kinase C (PKC) and calcium/calmodulin-dependent protein kinase II (CaMK-II) were not affected [117].

The mechanisms of apoptosis induced by organosulfur compounds have been reviewed previously [101]. In the first studies conducted with DADS on cancer cell lines, an increase in intracellular ratio of proapoptotic versus antiapoptotic proteins was observed [118, 119]. In contrast, with cells treated with DAS and garlic extract, induction of apoptosis in DADS-treated cells was associated with an increase in the level of p53, which is involved in the activation of Bax transcription and the repression of Bcl-2 expression [118]. There is also evidence for the involvement of reactive oxygen species (ROS) in apoptosis induction by DADS, via activation of the c-Jun NH2-terminal kinase/c-Jun pathway [120]. In agreement with previous findings documenting that DADS-induced apoptosis was associated with upregulation of the $Ca^{2+}$-dependent protease, calpain and an increase in the Bax:Bcl-2 ratio and cytosolic Smac/Diablo, Karmakar et al. reported a down regulation of the inhibitors of apoptosis BIRC and of NFκB [121]. These findings confirmed in human malignant neuroblastoma cells an observation previously made on human leukemia cells, that DADS promoted caspase-3 activation in apoptosis [122].

Over the last decade, a growing interest has emerged in a new field of research, since the finding that DADS can increase acetylation of H3 and H4 histones in mouse and human leukemia cells [123], and in rat Morris hepatoma cells [124]. Subsequently, the ability of DADS to inhibit histone desacetylase (HDAC) activity was associated with the previously described cell cycle arrest in the G2/M phase in human colon tumor cell lines [125]. These demonstrations of the DADS effect on the acetylation state of histones on various cancer cells *in vitro* were then confirmed *in vivo* on rat colonocytes treated with 200 mg/kg DADS *via* intracecal perfusion or oral feeding by gavage. The authors concluded that the data generated using cDNA expression arrays suggested that the observed hyperacetylation of histones H3 and H4 by DADS could modulate the expression of a subset of genes [126, 127].

## Diallyltrisulfide (DATS)

In initial investigations, DATS (Figure 2), like DADS, was found to produce an increase in $[Ca^{2+}]_i$, leading to a suppression of the proliferation of human lung tumor cells *in vitro* [128]. Some years later, the observed G2/M phase arrest and marked decrease in Cdc25C level was explained by the promotion of proteasome-mediated degradation of this phosphatase, induced by DATS [129]. The importance of the oligosulfide chain length on the antiproliferative activity, first suggested in 2003 [130], was then confirmed in two separate studies: firstly, in 2004, with the demonstration that two human prostate cancer cell lines were relatively more sensitive to apoptosis induction by DATS than DAS or DADS [131]; secondly, in 2011, with data supporting a mechanism of mitotic arrest of cancer cells due to alteration of the microtubule network, leading the authors to conclude that the high reactivity of sulfur atoms against the thiol groups of different cellular macromolecules could control crucial regulatory functions [132].

The DATS-induced inhibition of growth of human prostate cancer xenografts observed *in vivo* in athymic mice was then associated with the induction of the proapoptotic proteins, Bax and Bak [133]. In 2008, Powolny and Singh proposed that the mechanisms of inhibition of cell cycle progression, induced by DATS in these cancer cells, would involve ROS generation through the degradation of ferritin via the JNK pathway, and Chk1-mediated inactivation of the anaphase-promoting complex/cyclosome (APC/C) [134]. This suppression of cancer cell growth correlated with apoptosis induction, in particular a) the phosphorylation of Bcl-2, leading to a reduced interaction between Bax and Bcl-2, and finally initiation of mitochondria-mediated caspase activation; and b) inactivation of Akt resulting in the translocation of the other proapoptotic protein, Bad [134]. This observation that Bax and Bak proteins are critical targets of DATS-induced apoptosis was also confirmed in human lung cancer cells, while a normal human bronchial epithelial cell line was found to be significantly more resistant to growth inhibition than cancer cells [135]. In order to investigate the structure-function relationships, a recent comparative study has shown that a series of derivatives of DATS in which the tri-thio-ether structure was kept unchanged, presented the same potential of apoptosis induction [136].

## Ajoene and Aged Garlic Extract

Ajoene (Figure 2) is another garlic compound which, besides its antithrombosis, anti-microbial and cholesterol lowering properties, has been shown to be an inhibitor of proliferation and an inducter of apoptosis in a number of different human cancer cell lines. The first investigations with this molecule indicated its cytostatic/cytotoxic properties on a B-cell lymphoma-derived cell line [137], and induction of apoptosis in human promyeloleukemic cells [138] and skin tumors [139, 140]. Subsequently, the biological actions of ajoene on several human leukemia cells were confirmed [141, 142], including its ability to enhance the apoptotic effect of chemotherapeutic drugs against resistant myeloid leukemia cells [143]. These reports led to the proposal, in 2004, of ajoene as a new antileukemia agent for acute myeloid leukemia therapy [144]. Other recent investigations with ajoene and analogues of this molecule have revealed that the *Z*-isomer is moderately more active than the E-isomer at inhibiting tumor growth, suggesting that specific protein interactions may be important [145]. In the same way, structure-activity studies of a library of twelve ajoene analogs has identified a derivative containing *p*-methoxybenzyl-substituted end groups that is twelve times more active than *Z*-ajoene, with an $IC_{50}$ of 2.1 μM on oesophageal cancer cells [146].

Ajoene is not found in freshly crushed garlic but rather in aged preparations, with the *Z*-isomer predominating in heated extracts in oil. Manufacturing processes can significantly affect the complex composition of garlic preparations. Some authors have shown that a unique preparation, called aged garlic extract (AGE) was less irritating after chronic administration to rodents while inhibiting the development of preneoplastic lesions in rat hepatocarcinogenesis [147]. Aged garlic also had a chemopreventive effect on colon carcinogenesis in the same species [148]. To confirm these effects in humans, a preliminary clinical trial using high-dose AGE as an active treatment and low-dose AGE as a control was performed on patients with colorectal adenomas precancerous lesions of the large bowel. The results, published in 2006, demonstrated that AGE significantly suppressed both the size and number of colon adenomas in patients after twelve months of high-dose treatment [149].

Among the other findings associated with the use of AGE for cancer chemoprevention or treatment, one of the most interesting concerns its impact on the immune system in the context of cancer. In 2006, Ishikawa and colleagues observed that AGE administration to patients with inoperable colorectal, liver, or pancreatic cancer produced a significant increase in the number of NK cells and the NK cell activity, compared with a placebo group [150]. It is interesting to compare this report with the results of a study published in 2001, which revealed many similarities between the immune-enhancing effects produced by AGE and those of bacillus Calmette-Guérin (BCG) immunotherapy [151]. These observations open up considerable prospects for the search of new, innovative strategies of treatment in the context of immunosuppression generally observed in patients with cancer.

## 3) Curcuminoids from Zingiberaceae

The third category corresponds to curcuminoids, which are commonly found in zingiberaceae. Among this family, the genus *Curcuma* has been used for centuries in the diet in South India and other Asian countries to combat a variety of ailments. The use of *Curcuma zedoaria*, a Chinese traditional medicine, has been reported since the 1970s for the treatment of cervical cancer [152-154]. *Curcuma longa* L, also known as turmeric, has been described as a cancer remedy in Indian natural medical literature [155].

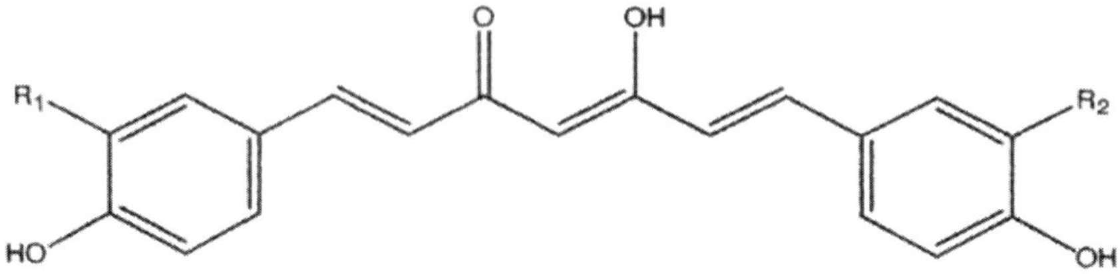


Curcumin: R1, R2 = $OCH_3$
Demethoxycurcumin: R1 = H, R2 = $OCH_3$
Bisdemethoxycurcumin: R1, R2 = H

Figure 3. Chemical structures of curcuminoids.

The first studies of the mechanism of action of curcumin (Figure 3) as a cytotoxic agent were conducted in the 1950s, showing bactericidal properties at concentrations of 1 ppm [156]. Two decades later, the addition of turmeric to mammalian cells *in vitro* was shown to arrest mitosis, alter chromosome morphology and interfere with nucleic acid synthesis [157]. However, investigations on the potential anticancer activity of turmeric really began in 1985, when Kuttan et al. demonstrated that turmeric extract inhibited the cell growth of Dalton's lymphoma cells [158].

The pharmacokinetics of curcumin was first studied in normal rats in 1978 in two independent studies. Walström and Blennow analyzed the uptake, distribution and excretion of curcumin after oral administration, and showed that if curcumin was poorly absorbed from the gut, no toxic effects were observed after doses of up to 5g/kg [159]. Holder and coauthors completed this approach by using deuterium- and tritium-labeled curcumin, demonstrating that oral and intraperitoneal doses of $^3$H-curcumin led to the fecal excretion of most of the radioactivity while intravenous and intraperitoneal doses were excreted in the bile after metabolism [160].

In 1987, Kuttan and coworkers observed that an ethanol extract of turmeric as well as an ointment of curcumin produced remarkable symptomatic relief in patients with external cancerous lesions [161]. Subsequently, Huang et al. demonstrated that curcumin was a potent inhibitor of tumor promotion by 12-*O*-tetradecanoylphorbol-13-acetate (TPA) in the two-stage initiation-promotion model in mouse skin [162]. Two years later, on the basis of their *in vitro* experiments, they proposed that c-Jun/AP-1, a key component of the cellular transcriptional machinery, could be the target of this blocking process [163]. This was the beginning of a huge number of investigations aimed at deciphering the molecular mechanisms involved in the effects of this molecule against various cancers. In parallel, preclinical studies on the use of oral curcumin for cancer chemoprevention facilitated the determination of the metabolic pathways in the rat [164]. In 2007, a large review of the subject estimated that over 1,700 papers had been published on curcumin over the last 50 years, including an extensive literature suggesting that this molecule had an important potential in the prevention and treatment of a variety of cancers [165]. At that time, the numerous targets on which the effect of curcumin was already documented included transcription and apoptosis factors, inflammatory enzymes, protein kinases, adhesion molecules, and several biochemical pathways involved in cell cycle deregulation [165]. In recent years, some important additional information has added to this considerable background.

Curcumin suppresses the activation of several transcription factors implicated in carcinogenesis, such as NF-κB [166], and AP-1 [163]. Three recent studies have confirmed the effect of curcumin on the expression of these major transcription factors *in vivo* in the mouse on xenografts of human head and neck squamous cell carcinoma [167], colorectal cancer [168] and bladder cancer [169]. To date, given that curcumin abolishes the constitutive activation of NF-κB in the tumor tissue, its mechanism of action has been clarified. Under resting conditions, NF-κB is retained in the cytoplasm in an inactive form by an inhibitor, IκB. Upon activation, IκB is phosphorylated at its α subunit by a kinase, IκK, which leads to ubiquitination and degradation of IκBα and then to NF-κB release. NF-κB translocates into the nucleus where it induces the expression of more than 200 genes, suppressing apoptosis and inducing cellular transformation, proliferation, and invasion. Curcumin blocks the IκK-mediated phosphorylation and degradation of IκBα which, thus, remains in its inactivated form in the cytoplasm and is not able to enter the nucleus to activate transcription [170].

Suppression of NF-κB by curcumin leads to suppression of the expression of cyclin D1 [171], a component subunit of cyclin-dependent kinases which are rate limiting in the progression of cells through the cell cycle, and dowregulation of enzymes involved in inflammatory (COX-2) [172] and invasion (MMPs) processes [173]. Curcumin was previously shown to upregulate the expression of some cyclin kinase inhibitors, in particular members of the Cip/Kip family, thus blocking the association of cyclin D1 with Cdk4/Cdk6 [174]. Recently, Mukhopadhyay and coworkers demonstrated that this molecule also decreased the phosphorylation of Rb and suppressed the transcription of E2F-regulated genes [175]. The regulation of the eicosanoid pathway involved in the production of inflammatory cytokines by activated monocytes and macrophages via the enzymes cyclooxygenase-2 (COX-2) and lipoxygenase (LOX) was reviewed a few years ago [176]. A recent study has documented the inhibition of the synthesis of prostaglandin E2 in a number of colorectal cancer cell lines [177]. Insights into the inhibition of the invasion process by curcumin have shown that the metastatic properties of breast cancer cell lines are stopped [178], while the transcription of metalloproteinases such as MMP-1 is significantly affected [179].

In initial studies on apoptosis, curcumin was shown to induce the apoptosis of B-cell lymphomas, a process mediated by the downregulation of several survival genes including *egr*-1, c-*myc*, *bcl*-$X_L$ and *NF-κB* as well as the tumor suppressor gene, *p53* [180]. More recently, Liu and coworkers reported that the cell-cycle arrest induced by curcumin in human glioma cells was, rather, associated to an upregulation of p53 expression. Their observations suggested the implication of ING4, a member of a new family of tumor suppression genes, *ING*, which, like p21, was also upregulated along with cell-cycle arrest [181]. However, the mechanisms by which curcumin triggers apoptosis in cancer cells appear to be more complex than initially expected. Indeed, Watson et al., when assessing the cytotoxic activity of this phytochemical against a panel of ovarian carcinoma cell lines with differing p53 status, observed that p53-independent apoptosis also occurred, involving the activation of p38 mitogen-activated protein kinase (MAPK), thus confirming the fact that curcumin activated both the extrinsic and intrinsic pathways [182]. Regarding the implication of curcumin in the extrinsic pathway, Lu and colleagues have also shown that this molecule increases the level of "death activators", such as Fas Ligand and its corresponding receptor-attached FADD adapter molecule, thus initiating the caspase cascade [183]. Other investigations have also confirmed the role of autophagy, the type II programmed cell death, in the biological effects induced by curcumin in cancer cells. The first evidence for involvement of autophagy in the suppression of the growth of malignant gliomas *in vitro* and *in vivo* was provided by Aoki and colleagues in 2007 [184]. Subsequently, Jia and coworkers demonstrated that the autophagy inhibitor, bafilomycin A1, suppressed curcumin-induced cell death in the chronic myeloid leukemia cell line, K562 [185]. In addition, O'Sullivan-Coyne et al. reported that the addition of a caspase inhibitor, zVAD, had a marginal or no effect on the viability of esophageal cancer cells treated with curcumin, indicating the predominance of a non-apoptotic form of cell death [186].

Curcumin was shown to possess anti-initiating and antipromoting activity in various experimental systems. However, some aspects of its mechanism of action remained to be elucidated *in vivo*. Garg and colleagues provided evidence that dietary curcumin modulated transcriptional regulators of phase I and phase II enzymes in benzo[*a*]pyrene-treated mice [187]. Another new and promising field of research with curcumin concerns its development as an epigenetic agent, which was recently reviewed by Fu and Kurzrock [188]. Five years ago, Shishodia and colleagues also reported some interesting observations related to chemosensitizing, radiosensitizing and radioprotective effects produced by this phytochemical [165]. Goel and Aggarwal have more recently reviewed the preclinical studies conducted in this domain, which are expected to lead to clinical trials to prove the potential of this age-old golden spice for treating cancer patients [189]. Finally, apart from the sixty-two clinical trials with curcumin to date, one of the most exciting fields of investigation for the future concerns the implication of its degradation products in its pharmacological effects [190].

# 4) Other Potentially Anticarcinogenic Molecules

## a) Alkaloids

Berberine (Figure 4), an isoquinoline alkaloid commonly used in China and other countries, is a major active compound of several species of the genus, *Coptis*. Although its main pharmacodynamic properties have long been recognized in the Chinese pharmacopeia, its antineoplastic effects have received considerable attention during the last decade, and the long story of its traditional use together with biomedical investigations have been the subject of a very interesting review [191]. The growing interest in this molecule and other related structures present in *Coptis* spp. coincided with the search, at the end of the 1990s, for new chemotherapeutic drugs dealing with the problem of multidrug resistance frequently observed in patients treated for cancer. Initially, berberine was shown to modulate the expression and function of the multidrug-resistant transporter, pgp-170, in several digestive-tract-cancer cell lines [192]. A recent report highlighting the effects of berberine on various cancer cell lines, and its mechanisms of action, has emphasized its potential to enhance the efficacy of chemotherapy/radiotherapy [193].

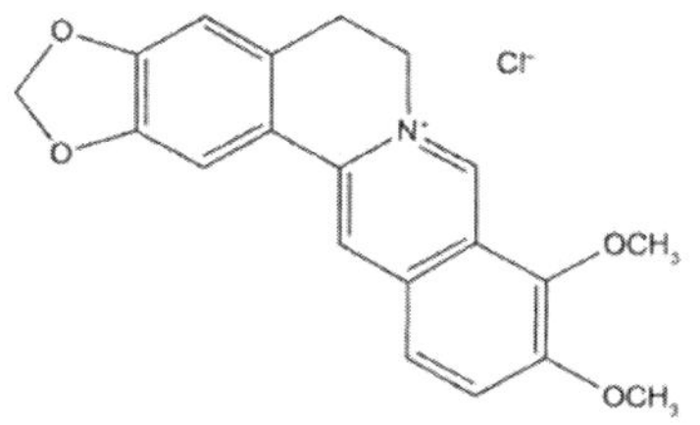

Berberine

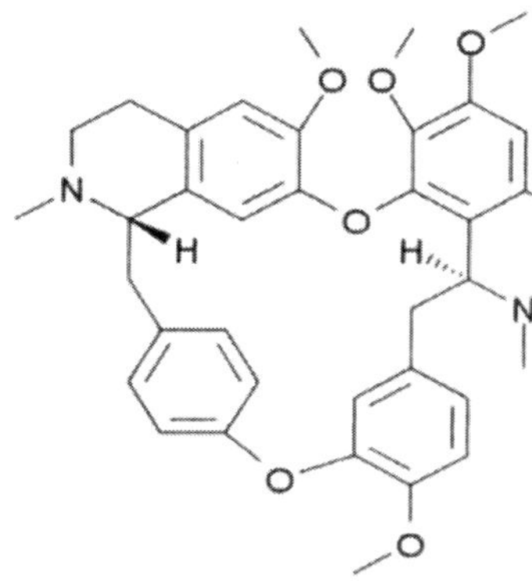

Tetrandrine

Figure 4. Chemical structures of alkaloids.

Another important molecule relevant to this chemical class is tetrandrine (Figure 4), a bis-benzylisoquinoline alkaloid also used in traditional Chinese medicine, isolated from the root of *Stephania tetrandra*, and shown to inhibit the proliferation of human leukemia [194] and hepatoblastoma cell lines [195].

The prevention of doxorubicin-induced drug resistance by tetrandrine has also been recently demonstrated in leukemia cells [196]. Finally, another possibility for overcoming the problem of multidrug resistance has been provided by the discovery of the biological properties of Pervilleines, a group of tropane alkaloid esters obtained from extracts of roots or stems of the genus, *Erythroxylum* [197]. Among this family of phytochemicals, Pervillein F provided the best potential for the induction of dose-dependent, G2/M-phase arrest in KB-V1 cells implanted in nude mice when combined with vinblastine [198].

## b) Acetogenins

Acetogenins are an important class of natural compounds found in the Annonaceae, consisting of two long hydrocarbon chains, a terminal 2,4-disubstituted γ-lactone, and a variable number of tetrahydrofuran rings. They are presented as the most potent *in vitro* inhibitors of the mitochondrial respiratory chain complex I. Among them, the adjacent bis-tetrahydrofuranic acetogenins, with a threo/cis/threo/cis/erythro relative configuration, have been described as the most potent group [199]. The first study reporting antitumor effects of a member of this family, uvaricin, isolated from *Uvaria accuminata*, was published in 1982 [200]. One decade later, one of the most potent antitumor agents belonging to this chemical class, bullatacin (Figure 5), was described [201]. Its mechanisms of action on a human hepatocarcinoma cell line, mainly based on the induction of apoptosis, have been particularly well studied, based on the determination of morphological changes using inverted phase-contrast microscopy and electron microscopy [202]. By 2003, about 350 acetogenins had been isolated and Chiu et al. reported that the concentration-dependent decrease of both intracellular cyclic AMP and cyclic GMP induced by bullatacin was parallel with the magnitude of apoptosis observed in human hepatoma cells [203].

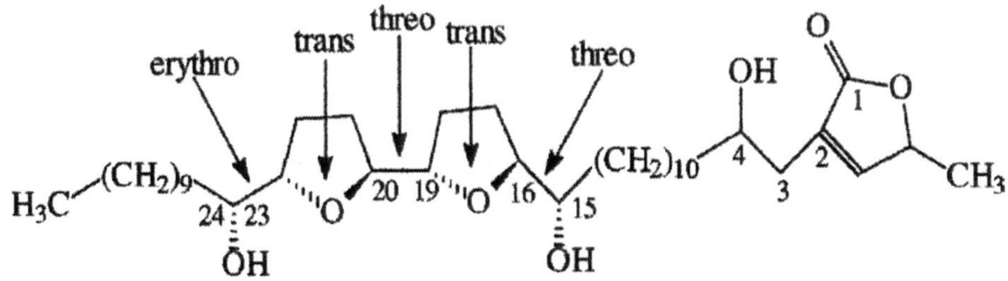


Figure 5. Chemical structure of bullatacin.

Subsequently, Liang and coworkers demonstrated that bullatacin caused an upregulation of ROS and a downregulation of the mitochondrial transmembrane potential in a concentration-dependent manner, both in a drug-sensitive and an MDR variant of a human epidermoid carcinoma cell line, while cleavage of caspase-9 and caspase-3 was observed following the release of cytochrome c from mitochondria [204]. Recent studies focusing on other sources of acetogenins have also shown that plants of this family, such as *Annona squamosa* [205] and *Asimina triloba* [206], can provide interesting alternative molecules for the induction of apoptosis in various cancer cells.

### c) Stilbenoids

Stilbenoids are hydroxylated derivatives of 1,2-diphenylethene (stilbene) which are produced in response to fungal infection of plants. The most famous molecule of this chemical class is resveratrol (Figure 6), first isolated from the roots of *Veratrum grandiflorum* in 1940, then in 1963 from the roots of *Polygonum cuspidatum*, a plant used in traditional Chinese and Japanese medicine, and finally in 1976 from *Vitis vinifera*. The discovery of resveratrol in wine later generated a growing interest, initially on the basis of its ability to inhibit the activity of cyclooxygenase I (COX-1), and then as an inhibitor of carcinogenesis at multiple stages when the topical application of resveratrol produced a 98% reduction in tumor numbers in a mouse model of skin cancer [207]. Preclinical studies of the potential of this molecule as an antitumor agent were reviewed by Baur and Sinclair in 2006, confirming that resveratrol slows tumor development through multiple complementary mechanisms, including inhibition of enzymes such as COX1, COX2, ornithine decarboxylase (ODC), cytochrome P450s and flavin monooxygenases, inhibition of angiogenesis, and induction of cell cycle arrest and apoptosis [208]. A recent report has also mentioned that, in addition to being a peroxisome proliferator-activated receptor (PPAR) activator, resveratrol is an inducer of endothelial nitric oxide synthase (eNOS) [209].

In the search for other stilbenoids of interest, additional molecules of this chemical class have been explored. For that purpose, other species of the genus have recently been analyzed, leading to the isolation of six oligostilbenoids from *Vitis amurensis*, a wild-growing grape widely distributed in Korea, China and Japan [210]. Non-toxic resveratrol oligomers include molecules such as α-viniferin, recently isolated in the genus *Carex*, which arrests the cell cycle in the S-phase in human colon cancer cells but not in normal colon cells [211].

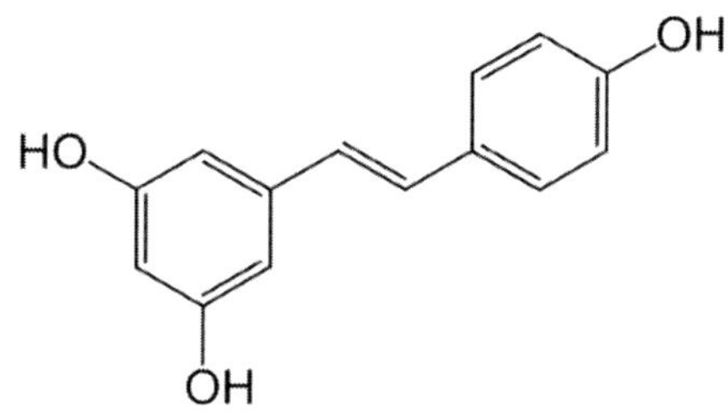


Figure 6. Chemical structure of resveratrol.

Finally, the most marked event in this field was the finding that combretastatins, a family of cis-stilbenoids isolated from *Combretum caffrum*, a species indigenous to South Africa, exhibited antineoplastic properties [212]. Ten years ago, combretastatins were shown to be potent inhibitors of microtubule assembly, causing hemorrhagic necrosis in murine subcutaneous tumors [213]. The mechanism of action of combretastatin A4 (CA4) involves its binding to tubulin, leading to cytoskeletal and then morphological changes in endothelial cells. These changes increase vascular permeability and disrupt tumor blood flow, resulting in extensive ischemic necrosis in areas that are often resistant to conventional anticancer treatments [214]. Today, among anticancer drugs that target tumor vasculature, CA4 appears to be the lead compound [215].

## d) Simple Phenols

It has been demonstrated that populations who consume virgin olive oil have a low incidence of several cancers, in particular breast cancer. Although this effect was first attributed to the high content of monounsaturated fatty acids, the importance of the minor fraction was recently emphasized, in particular the antioxidant activity of phenols. Bioavailability studies demonstrated that these minor compounds are dose-dependently absorbed in animals and humans after ingestion, accumulate in the body and, finally, modulate oxidative/antioxidative status in humans [216]. Extensive studies then focused on hydroxytyrosol (Figure 7) and, following the observation that this molecule can arrest the cell cycle in cancer cell lines, its anticancer properties have recently been reviewed [217].

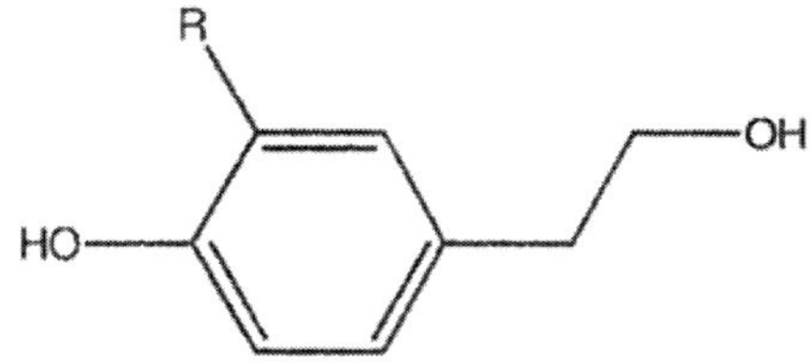


R = H: tyrosol
R = OH: hydroxytyrosol

Figure 7. Chemical structures of simple phenols from virgin olive oil.

## e) Phenolic Acids

In this class, three main molecules have been the subject of investigations as potential anticancer agents. The first one, caffeic acid, is a widespread phenolic acid present in a number of fruits, vegetables, wine, olive oil and coffee, which was demonstrated to be one of the chemical constituents of coffee that suppressed the invasion of the hepatoma cancer cell line, AH109A [218]. However, most of the interest has rather focused on its phenyl ester (Figure 8), which was extracted from honey bee propolis in 1988 [219], exhibiting inhibition of metalloproteinases 2 and 9 in human hepatocellular carcinoma cells [220]. The potential of this molecule and other phenolics as anti-angiogenic agents has recently been reviewed [221].

Another strong metalloproteinase-9 inhibitor is chlorogenic acid (Figure 8), a molecule present in coffee beans, blueberries, grapes, pears, apples and many other plants [222]. This molecule has been known to inhibit colon carcinogenesis in hamsters since 1986 [223], and coffee bean extracts rich in chlorogenic acid were reported to suppress the chemical-induced development of colonic aberrant crypt foci in rats [224]. Subsequently, this molecule was shown to inhibit 8-hydroxydeoxyguanosine formation in cellular DNA in a rat model of tongue carcinogenesis [225]. The strong, concentration-dependent inhibitory activity of chlorogenic acid, isolated from the Korean plant *Euonymus alatus*, on MMP-9 has also been confirmed by gelatin zymography [226].

Besides gallic acid (Figure 8), for which the potential has also been reviewed [221], another major molecule is ellagic acid (Figure 8), which is thought to modulate a number of subcellular signaling pathways. In initial studies, this phytochemical, isolated from *Terminalia arjuna* in India, was explored for its antimutagenic potential on strains of *Salmonella typhimurium* [227]. During the last decade, the growing interest in this molecule has been due to the finding that, among fruits, those which present the highest antioxidant activities are pomegranate (Punica granatum), blueberry and cranberry, which contain high amounts of ellagic acid.

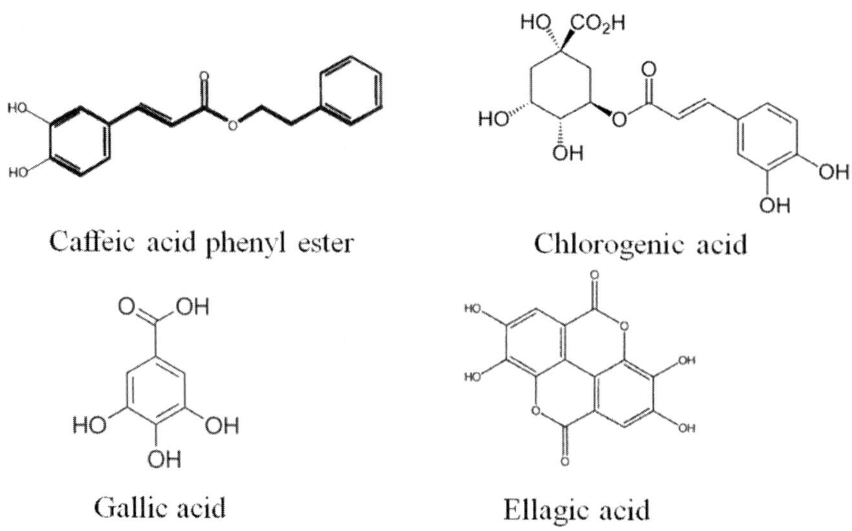


Figure 8. Chemical structures of phenolic acids.

Indeed, Albrecht and coworkers demonstrated in 2004 that the proliferation and invasion of several prostate cancer cell lines were significantly reduced in the presence of polyphenols extracted from pomegranate fruit [228]. Insights into the mechanisms of action of this molecule were obtained the following year when Malik and colleagues showed that ellagic acid induced both cell growth inhibition and apoptosis in prostate cancer cells through an action on the cyclin kinase inhibitor-cyclin-cdk machinery [229].

Since then, ellagic acid has been found to prevent the activation of the phosphoinositide-3-kinase (PI3K)/Akt pathway, resulting in modulation of its downstream Bcl-2 family proteins, elevation of cytochrome c levels and finally death of colon cancer cells in rats [230]. In a recent review, Bell and Hawthorne described the potential of pomegranate extracts and ellagic acid in the context of prostate cancer [231].

## f) Quinones

Some quinones have shown pharmacological potential as antitumor agents since the 1960s [233]. Eight major plants have recently been listed as important sources of anticancer phytochemicals of this chemical class [232]. Among these, emodin is an active component from the roots of *Polygonum cuspidatum*, which was reported to inhibit the activation of NFκB and IκB degradation in human umbilical vein endothelial cells [234]. In fact, recent investigations have shown that this molecule, which inhibits the proliferation of many cancer cells but not normal liver cells, induces caspase-dependent apoptosis via the transactivation of Specificity Protein 1 (Sp1) and inhibition of survivin expression [235]. Another interesting member of this family is tanshinone, a derivative of phenanthrene-quinone isolated from *Salvia miltiorrhiza*, which showed growth inhibitory effects on five tumor cell lines in 1997 [236].

Subsequently, it was found that tanshinone IIA, the most abundant diterpene quinone found in the root extract, induced apoptosis in human leukemia cell lines [237]. However, new developments in the chemistry and biology of the bioactive constituents of this plant have revealed that other molecules such like neo-tanshinlactone, have potent selective anti-breast-cancer activity [238]. Finally, tanshinone I (Figure 9) was found to be the most potent of these molecules in inhibiting the proliferation of lung cancer cell lines [239]. Embelin (Figure 9) is a benzoquinone isolated from *Embelia ribes*, which was first reported to show antiproliferative effects against methylcholanthrene-induced fibrosarcoma in rats [240].

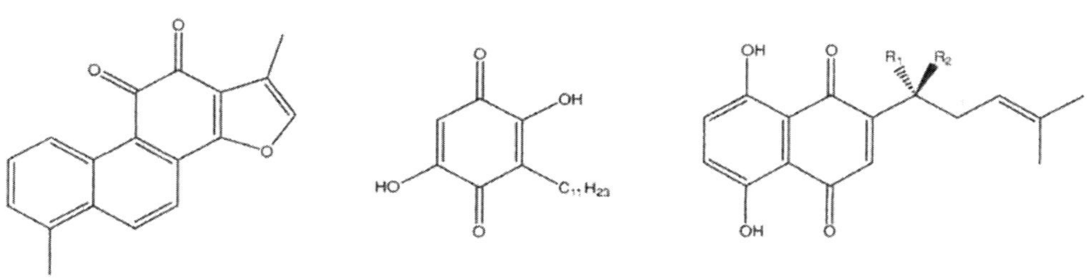

Tanshinone I Embelin Shikonin

Figure 9. Chemical structures of quinones.

The potential of embelin in chemoprevention was then confirmed against DENA/PB-induced hepatocarcinogenesis in the same species [241]. Investigations of the mechanisms of action involved demonstrated that embelin abrogated both inducible (by diverse stimuli) and constitutive NFκB activation, and down-regulated gene products involved in cell survival, proliferation, invasion, and metastasis of cancer cells [242]. This benzoquinone also suppressed the osteoclastogenesis induced by multiple myeloma and by breast cancer cells through an inhibition of the receptor activator of the NFκB ligand (RANKL), a process which opens up interesting prospects for the treatment of cancer-linked bone loss [243]. Another plant which has long been used in traditional Asian medicine, *Lithospermum erythrorhizon*, has shown chemopreventive activity via targeting of NFκB and AP-1 [244]. The interest in shikonin (Figure 9), the active quinone isolated from this plant, is due to the recent finding that this phytochemical decreases the mRNA and the enzymatic activity of steroid sulfatase, an enzyme which regulates the biosynthesis of estrogen within breast tumors, in breast cancer cells [245]. To give just one more example of the potential and diversity of quinones in cancer research, denbinobin, a phenanthrenequinone isolated from an orchidae, *Dendrobium nobile*, has already been shown to be a potent inhibitor of TNFα and NFκB activation. Denbinobin also blocks TAK1 activity upstream of IKK activation and sustains activation of the mitogen-activated kinases (MAPKs) ERK1 and 2, p38 and JNK1 and 2 [246].

### g) Polyphenols

Flavonoids are a group of more than 4,000 molecules having in common a basic skeleton of phenylchromone. According to their detailed structure, they can be classified into flavones, flavanones, isoflavones and flavans (Figure 10) [232]. Due to the widespread occurrence of flavonoids in plants, a number of epidemiological and animal studies have focused, primarily, on associations between vegetable and fruit consumption and the incidence of some cancers, finally indicating that flavonoids are structurally similar to steroid hormones. As a consequence, an original association between flavonoid intake and steroid hormone-dependent cancers (breast, prostate and colon) was made in 2002 [247]. Subsequently, epidemiological evidence, such as the much lower incidence of these cancers in Asians, who consume more vegetables, fruits and tea, than in populations in the Western hemisphere has tended to confirm this postulate. A first review of the different mechanisms underlying the potential anticancer effects of flavonoids emphasized their ability to target cell-surface signal-transduction enzymes and angiogenesis factors [248]. More recently, emerging potential mechanisms for flavonoids have included alteration of the Wnt/β-catenin and EGF/AKT pathways [249]. With this background, the last few years have been characterized by an explosion of studies relevant to the phytochemistry and metabolic engineering of flavonoids for the chemoprevention of some cancers [250].

In the search for phytochemicals which could affect the deleterious effects of autocrine/paracrine growth factors which facilitate the progression of prostate cancer, a good example is provided by the biological properties of quercetin, the main flavonoid of onions.

Quercetin presents a wide spectrum of pharmacological properties, including the downregulation of serine proteases involved in prostate cancer progression, of growth factors and their receptors, and of proliferative signaling molecules [251].

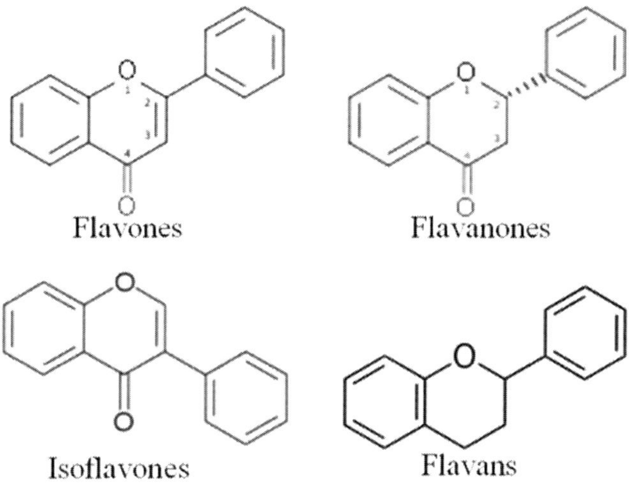


Figure 10. Chemical structures of flavonoids.

The ability of flavonoids to overcome the metastatic cascade, including cell-cell attachment, tissue-barrier degradation, migration, invasion, cell-matrix adhesion and angiogenesis, has also been reviewed recently [252]. Epidemiological studies have suggested that a combination of dietary flavonoids with diverse molecular mechanisms could have enhanced anticancer efficacy at physiological concentrations. Applied to the case of phytoestrogens and prostate cancer, a recent investigation combining genistein, quercetin and biochanin A has confirmed this assumption [253].

Tannins are water-soluble polyphenolic compounds which can be divided into two subgroups according to whether they are hydrolysable (gallo- and ellagi-tannins) or condensed (proanthocyanidins) [232]. These phytochemicals, particularly catechins, have been thought to play an important role in the chemoprevention of cancers since the discovery of the antimutagenic properties of (-)-epigallocatechin gallate (EGCG) (Figure 11), a tannin belonging to first subgroup, which is present in high amounts in green tea [254]. In initial investigations on animal models of cancer, EGCG inhibited tumor promotion in mouse skin and duodenal carcinogenesis [255]. The many mechanisms responsible for the activity of EGCG *in vivo* have already been well documented [256]. However, as for a number of other anticancer phytochemicals, successive investigations of the various biochemical pathways involved in its action have revealed a much more complex pattern than initially expected. For example, epigenetic studies have documented the fact that EGCG inhibits DNA methylation, suggesting its potential use for the reversal of related gene-silencing in the prevention of carcinogenesis [257]. Subsequently, the modifications of epigenetic events were studied in detail on a human epidermoid carcinoma cell line, showing that EGCG treatment decreased the levels of 5-methylcytosine, mRNA and protein levels of DNA methyltransferase 1, 3a and 3b [258]. A recent review of the various mechanisms of action of EGCG on cancer cells has reported that blocking of carcinogenesis occurs via effects on a wide array of signal transduction pathways including JAK/STAT, MAPK, PI3K/AKT, Wnt and Notch [259].

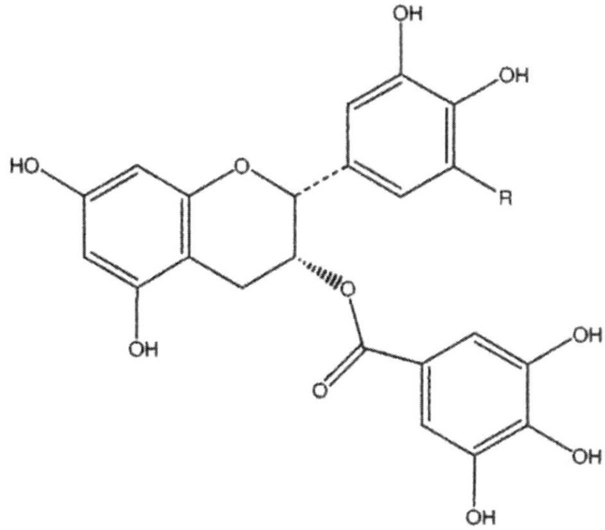


Figure 11. Chemical structure of (-)-epigallocatechin gallate (EGCG).

Besides green tea, other important sources of tannins are grape and berry extracts. The ability of these extracts to inhibit growth and stimulate apoptosis of human oral, breast, colon, and prostate tumor cells has been documented both *in vitro* and *in vivo* [260].

Blueberry phytochemicals, in particular, have been the subject of important recent studies, producing a wide combination of biological effects on triple-negative breast cancer cell lines *in vitro* and *in vivo*, including a decrease in cell proliferation, cell motility, MMP-9 activity and urokinase-type plasminogen activator secretion [261]. Lignans are a large group of fiber-associated phenolic compounds widely distributed in edible plants, which result from tail-tail linkage of two coniferyl or sinapyl alcohol units [232]. The interest in these structures initially came from the observation that non human primates appear to be remarkably resistant to the carcinogenic effect of estrogens, thus some researchers investigated the possible occurrence of lignans in the urine of chimpanzees consuming their regular diet. They found that the lignan, enterolactone, was excreted in very large amounts, also identified another lignan, enterodiol, and suggested that these molecules could play a role in maintaining resistance against the carcinogenic effects of estrogens [262]. Subsequently, evidence from epidemiological studies has tended to confirm that dietary fibers, as well as fat, have the potential to affect breast cancer risk [263]. A first review of the implications of lignans in cancer prevention, by Thompson in 1998, reported that mammalian lignans can be produced from plant precursors such as secoisolariciresinol diglycoside (SDG) and matairesinol (Figure 12) via the action of bacteria in the human or animal colon. In particular, experimental studies in animals and humans have demonstrated the anticancer effects of flax seed and its SDG [264]. Due to their extremely low concentrations in plants, alternative production systems, based on biotechnology, have been developed for anticancer lignans such as aryltetralin and arylnaphthalene lignans [265]. Among enterolignans which are converted by intestinal microbiota from ingested plant lignans, the major biologically active molecules are thought to be enterodiol (END) and enterolactone (ENL), which are suggested to be associated with low breast cancer risk [266].

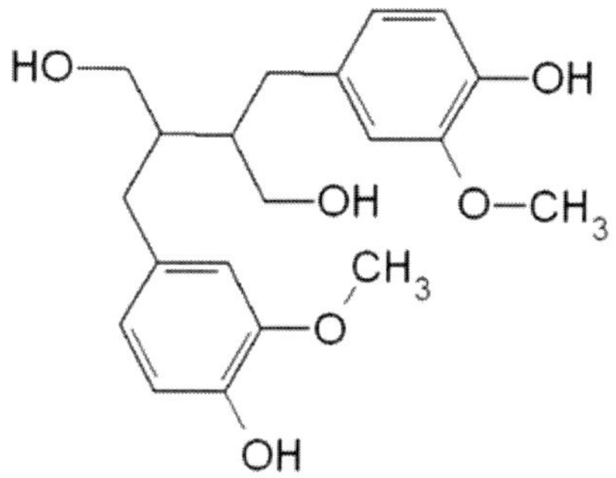

Secoisolariciresinol

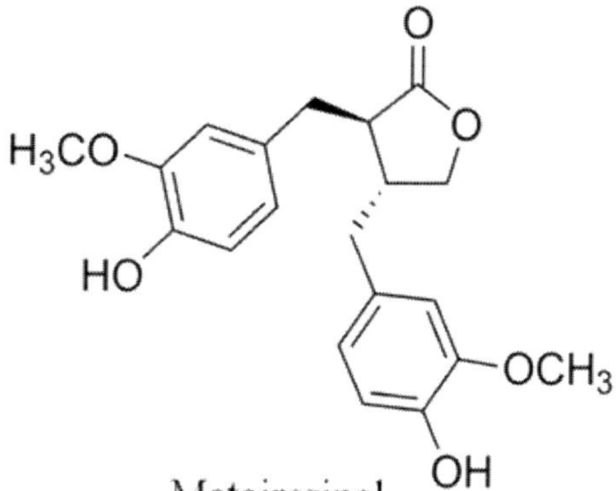

Matairesinol

Figure 12. Chemical structures of lignans.

## h) Carotenoids

The interest in carotenoids for the prevention of cancers began in the 1980s after the demonstration that mice and rats fed a diet supplemented with an "unlimited amount of red carrots" developed DMBA-induced skin or subcutaneous tumors at a lower rate than did the animals receiving a nonsupplemented diet [267]. This result was subsequently confirmed one decade later by Mathews-Roth, reporting that carotenoids with vitamin A activity, such as *β*-carotene (Figure 13) protected mice against DMBA-induced skin tumors, while phytoene and canthaxanthin treatment had no effect [268]. The enthusiasm was then tempered when the conclusion of a working group of experts negatively evaluated the cancer-preventive potential of carotenoid supplementation [269]. In 2008, a systematic review of the association between carotenoids and lung cancer concluded that *β*-carotene supplementation was not associated with a decrease in the risk of developing lung cancer [270]. In these studies however, care must be taken about a possible confusion between supplementation and intake of carotenoids from natural sources.

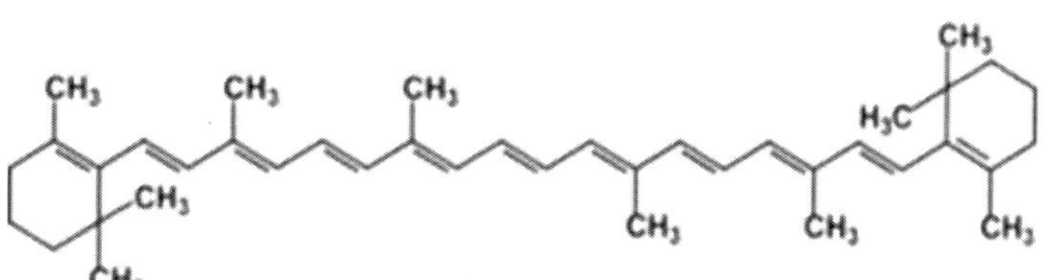

*β*-carotene

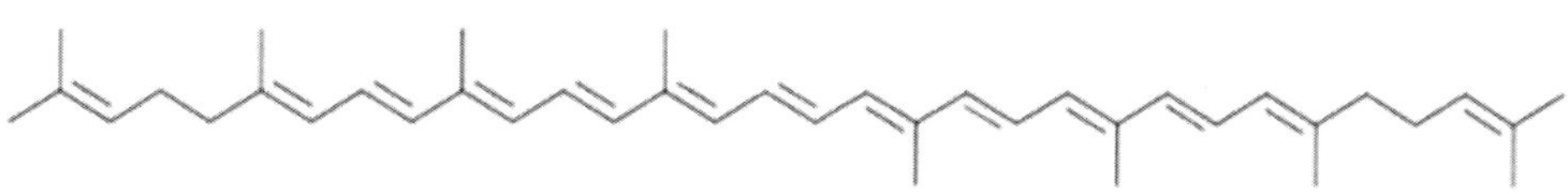

Lycopene

Figure 13. Chemical structure of carotenoids.

Indeed, sources of carotenoids in nature are generally complex, consisting of mixtures of various phytochemicals. Many natural carotenoids, apart from $\beta$-carotene have been shown to have anticarcinogenic activity, and some have shown more potent activities than $\beta$-carotene, including $\alpha$-carotene, lutein, zeaxanthin, lycopene, $\beta$-cryptoxanthin, fucoxanthin, astaxanthin, capsanthin, crocetin and phytoene [271].

Saffron, for example, has also shown interesting chemopreventive properties together with cytotoxic activity on various tumor cells lines and primary cells from surgically excised tumor samples [272]. Among the different molecules present in saffron, dimethyl-crocetin was reported to be the most active [273].

Lycopene (Figure 13), an important carotenoid from tomato, with no provitamin A activity, was early-on studied for its potential protective effects against some types of cancer, in particular prostate cancer [274].

A first review of the epidemiological literature, in 1999, reported that, among 72 studies identified, 57 had concluded an inverse association between tomato intake or blood lycopene level and the risk of cancer at a defined anatomic site [275]. Lycopene is a promising compound for cancer chemoprevention, which is now being tested in clinical studies for its considerable ROS scavenging activity, but its complex mechanisms of action also include an increase in gap junctional communication, which is suppressed during carcinogenesis [276].

### i) Terpenoids

The idea that terpenoids, like other dietary nutrients, could influence the activities of hepatic drug-metabolizing enzymes was early-on investigated in laboratory animals, with important consequences in the following decades for cancer chemoprevention in humans [277]. Among monoterpenoids, the particular case of *d*-limonene (Figure 14), which comprises more than 90% of citrus fruit peel oil, requires some attention. In 1988, Elson and colleagues demonstrated that this phytochemical was effective in reducing the average number of mammary carcinomas that developed in DMBA-treated rats when the terpene was fed during the initiation or the promotion/progression stage of carcinogenesis [278]. Two years later, this team provided additional evidence that monoterpenoids could contribute to the chemoprevention and chemotherapy of mammary tumors [279]. By 1994, the complete regression of mammary carcinomas by *d*-limonene appeared to involve tissue redifferentiation, and the chemopreventive activity of this molecule during promotion/progression was attributed in part to inhibition of the post-translational isoprenylation of growth-controlling small G proteins, such as p21ras [280]. In her review of 1997, Pamela L. Crowell documented the mechanisms of action of dietary monoterpenes, confirming the role of these phytochemicals as direct inhibitors of prenyl-protein transferase enzymes [281].

However, monoterpenes also appear to act through additional mechanisms such as effects on mevalonate metabolism and tumor remodeling or redifferentiation through an increased expression of the mannose-6-phosphate/insulin-like growth factor II receptor and transforming growth factor $\beta$1 [282].

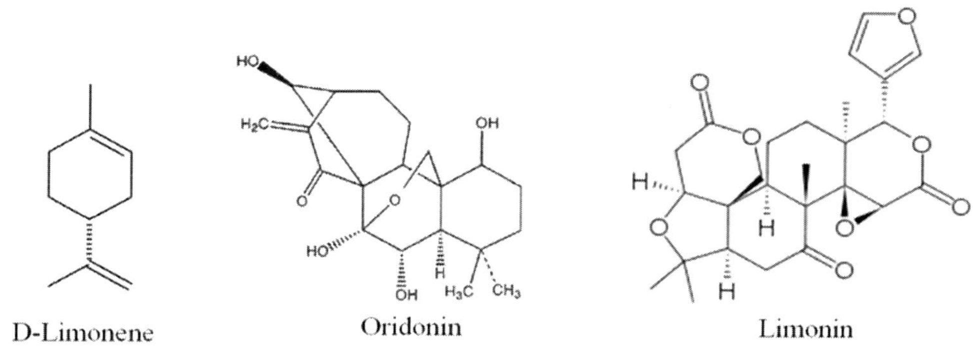


Figure 14. Chemical structures of terpenoids.

Among the other monoterpenoids of interest, compounds which have been identified in the oil extracted from lemon grass (*Cymbopogon citratus*, a Thai medicinal plant), such as citral, geraniol, methyl heptenone and $\beta$-myrcene, inhibited the formation of azoxymethane-induced DNA adducts and aberrant crypt foci in the rat colon [283].

Another subclass with potential active compounds is diterpenoids, which are represented, for example, by oridonin (Figure 14), a molecule extracted and purified from a Chinese medicinal plant, *Rabdosia rubescens*. Laboratory and clinical data based on the use of this phytochemical suggest that oridonin induces growth inhibition and apoptosis in various human cancer cells, and particularly through a disruption of the mitochondrial membrane potential, and activation of caspase-3 in human leukemia cells [284].

Tripernoids, another large subclass of molecules synthesized in many plants by the cyclization of squalene, present considerable structural diversity [285]. The particular case of saponins will be treated in a separate paragraph below. Among the numerous families comprising this subclass, one which was investigated early-on for potential inhibition of carcinogenesis is the limonoid family, a group of oxidized triterpenes responsible for the bitter taste in citrus products. As early as 1989, limonin and nomilin (Figure 14) were demonstrated to be effective inhibitors of benzo[*a*]pyrene-induced forestomach tumors in mice [286], while other members of this family, isolated from Meliaceae, exhibited cytotoxic activities against leukemia cells [287]. Additional structures of interest, which have previously been explored for their anticancer properties include glycyrrhetinic acid [288], lantadenes [289], betulinic acid derivatives [290], ursolic and oleanolic acids [291, 292] and many others [293].

## j) Saponins

In plants, a number of triterpenoids occur in the form of glucosides. Among the various structures which have been studied so far for their interesting anticancer properties, tubeimoside III, isolated from the traditional Chinese medicinal plant *Bolbostemma paniculatum*, demonstrated potent anti-tumorigenic activity *in vivo* in the mouse, the hypothesized mechanism of action including a modulation of cellular phospholipid metabolism [294]. Another group of compounds is represented by cumingianosides, obtained from *Dysoxylum cumingianum*, which exhibited significant cytotoxicities against leukemia and melanoma cell lines [295]. The potential cancer-preventive properties of ginsenosides from *Panax ginseng*, both in experimental models and epidemiological studies, have also attracted much attention over several decades, in particular the ability to inhibit tumor-associated angiogenesis [296]. Additional triterpenoid glycosides of interest include wistariasaponins (from the knots of *Wistaria brachybotrys*), gleditsiasaponin C and gymnocladussaponin G (from the fruits of *Gleditsia japonica* and *Gymnocladus chinensis*, respectively), and damaranesaponins from other species of the genus *Panax* [297]. Some authors have also emphasized the fact that the carbohydrate composition of some saponins influences their cytotoxic and antineoplastic actions [298]. These investigations have led to the observation that virgaureasaponin E, isolated from *Solidago virgaurea*, induced cytotoxic macrophages, leading to *in vivo* antitumoral effects of this molecule in the allogeneic sarcoma 180 tumor model and the syngeneic DBA/2MC.SC-1 fibrosarcoma tumor model in the mouse [299].

### k) Others

This section will include carbohydrates and proteins. Lectins are proteins or glycoproteins which bind to cell membranes in a specific and reversible fashion, without displaying enzymatic activity. In recent decades, most attention has focused on preparations of *Viscum album* (mistletoe), a toxic plant which was considered to be sacred by the Celts in ancient times. Since the pioneering work of Evans and Preece in 1973 [300], a growing interest in this plant has emerged, and an important review of its biological properties by Rentea et al. in 1981 confirmed the previous finding that repeated injections of an extract, Iscador, produced a proliferation of cortical thymocytes in rats [301]. However, at that time, the nature of the active components of Iscador was not clearly determined. Five years later, the immunomodulatory effects of Iscador were investigated, showing that after a single intravenous infusion to patients with breast cancer, several immunological parameters were changed, including a significant enhancement of the phagocytic activity of granulocytes and an increase in natural killer (NK) and antibody-dependent, cell-mediated cytotoxicity activities [302]. A major improvement in the understanding of the mechanisms involved was obtained in 1989, when Hajto and coworkers demonstrated that small and nontoxic doses of the mistletoe extract, Iscador, and in particular its carbohydrate-binding B chain, were able to activate tumoricidal effector mechanisms by increasing the activity and recruitment of NK cells in peripheral blood of rabbits [303]. In a second study, these investigators also provided evidence that human mononuclear cells exposed to this lectin increased their secretion of tumor necrosis factor $\alpha$, and interleukins 1 and 6 [304]. In addition, they showed that this enhancement of secretion was caused by selective protein-carbohydrate interaction, and that cytokine levels in the sera of patients were increased after injection of optimal lectin doses [304]. Subsequently, another complementary study conducted in patients with breast cancer reported increased levels of plasma $\beta$-endorphin after administration of this lectin, an observation which correlated with an improved quality of life of this group of patients [305]. Bantel and colleagues also observed that mistletoe lectin I induced caspase-dependent apoptosis in two human leukemic T- and B cell lines, and potentiated the cytotoxic effect of chemotherapeutic drugs such as etoposide [306]. Finally, a systematic literature review of publications on cancer patients treated with Iscador revealed that, although a publication bias limited the evidence, studies with positive effects of *Viscum album* extracts on cancer-patient survival were accumulating [307].

Among other plants screened for their ability to mediate potent antitumor immunity, *Aeginetia indica*, a plant parasitic on the roots of Japanese pampa grasses, has provided promising biological properties. Interestingly, like *Viscum album*, extracts from the seeds of this plant present a subtle combination of mitogenic (on thymocytes) and antitumor activity *in vivo* which appears to involve carbohydrate and proteins, respectively [308]. During the past two decades, a number of additional plants have been explored as sources of original polysaccharides involved as biomodulators in the immune response against tumors. Among them, a leguminous tree from the Brazilian flora, *Anadenanthera colubrina*, contains an acidic heteropolysaccharide which dramatically induced tumoricidal activities in peritoneal macrophages, when injected intraperitoneally to mice, producing a decrease in sarcoma 180 growth [309].

## 5) Future Prospects: From Traditional Pharmacopeiae Screening to the Search for Synergistic Effects

The above review gives a notion of the considerable resource we have to hand to improve the chemoprevention and treatment of cancers. In addition to continuing the characterization of the biological effects of new, potential phytochemicals of interest, we must now define how to use correctly this potential for optimal results.

Although libraries of phytochemicals are being accumulated, random selection screening of plants for potential pharmacological activity is an enormous task. On the other hand, the diversity of traditional plant pharmacopeias all over the world has provided an invaluable source of ethnopharmacological information in the search for additional new molecules to be tested against experimental models of cancers. This interesting approach was first emphasized thirty years ago by Farnsworth and Kass [310], and was considerably extended two decades later [311]. Since then, these efforts are ongoing in many countries, for example in Argentina [312], Bangladesh [313], South Africa [314], Brazil [315] and China [316].

Finally, as some phytochemicals are able to target different, deregulated cellular signaling pathways, without toxic effects in normal cells, an interesting challenge for the future is to determine the optimal way to combine their specific potentials. Some authors have already observed that the combination of two potent phytochemicals belonging to different chemical classes, such as PEITC and curcumin, significantly reduced the growth of human prostate cancer cell xenografts in athymic mice, while PEITC or curcumin alone had little effect [317]. Other reports have documented how an experimental diet containing carotenoids, isothiocyanates, organosulfur compounds, curcuminoids and others, significantly decreased the incidence of experimental lymphoma in mice [318] and both the incidence and aggressiveness of experimental glioma in rats [319], compared with animals given diets without phytochemicals. These findings have important implications for the management of metabolic, systemic effects associated with the growth of aggressive tumors [320]. As a number of these phytochemicals have also been shown recently to be potentially useful as epigenetic drugs, another interesting prospect concerns their use, alone or in combination, to produce cell death in cancer cells, for the induction of tumor-specific immune responses against some types of cancers for which current treatments are of limited efficacy [321].

## References


[1] L. S. Goodman, M. M. Wintrobe, W. Dameshek, M. J. Goodman, A. Gilman, M. T. McLennan, *J. Am. Med. Assoc.,* 132, 126 (1946).

[2] M. G. Kelly, J.L. Hartwell, *J. Nat. Cancer Inst.*, 14, 967 (1954).

[3] I. S. Johnson, J. G. Armstrong, M. Gorman, J. P. Jr. Burnett, *Cancer Res.,* 23, 1390 (1963).

[4] M. C. Wani, M. E. Wall, *J. Org. Chem.,* 34, 1364 (1969).

[5] M. C. Wani, H. L. Taylor, M. E. Wall, P. Coggon, A. T. McPhail, *J. Am. Chem. Soc.,* 93, 2325 (1971).

[6] L. W. Wattenberg, *Cancer Res.* 26, 1520 (1966).

[7] L. W. Wattenberg, M. A. Page, J. L. Leong, *Cancer Res.* 28, 934 (1968).

[8] T. K. Yun, Ann. N Y Acad. Sci., 889, 157 (1999).

[9] R. Doll, R. Peto, *J. Natl. Cancer Inst.*, 66, 1191 (1981).

[10] L. W. Wattenberg, *Cancer Res.* 45, 1 (1985).

[11] M. E. Wall, *Med. Res. Rev.,* 18, 299 (1998).

[12] Atta-ur-Rahman, *Nuclear Magnetic Resonance Basic Principles*, Springer-Verlag, New York, 1986.

[13] R. I. Freshney, *Animal cell culture a practical approach*, IRL Press, Oxford (1986).

[14] C. Gyllenhaal, J.D. McChesney, *J. Nat. Prod.,* 59, 334 (1996).

[15] D. D. Soejarto, *J. Ethnopharmacol.,* 51, 1 (1996).

[16] L. W. Wattenberg, *Cancer Res.* 35, 3326 (1975).

[17] L. W. Wattenberg, W. D. Loub, *Cancer Res.* 38, 1410 (1978).

[18] J. E. Nixon, J. D. Hendricks, N. E. Pawlowski, C. Pereira, R. O. Sinnhuber, G. S. Bailey, *Carcinogenesis*, 5, 615 (1984).

[19] G. S. Bailey, J. D. Hendricks, D. W. Shelton, J. E. Nixon, N. E. Pawlowski, *J. Natl. Cancer Inst.* 78, 931 (1987).

[20] R. H. Dashwood, D. N. Arbogast, A. T. Fong, J. D. Hendricks, G. S. Bailey, Carcinogenesis, 9, 427 (1988).

[21] R. H. Dashwood, A. T. Fong, D. E. Williams, J. D. Hendricks, G. S. Bailey, *Cancer Res.* 51, 2362 (1991).

[22] R. H. Dashwood, A. T. Fong, D. N. Arbogast, L. F. Bjeldanes, J. D. Hendricks, G. S. Bailey, *Cancer Res.* 54, 3617 (1994).

[23] A. Oganesian, J. D. Hendricks, C. B. Pereira, G. A. Orner, G. S. Bailey, D. E. Williams, *Carcinogenesis*, 20, 453 (1999).

[24] D. E. Williams, G. S. Bailey, A. Reddy, J. D. hendricks, A. Oganesian, G. A. Orner, C. B. Pereira, J. A. Swenberg, *Toxicol. Pathol.* 31, 58 (2003).

[25] T. Tanaka, Y. Mori, Y. Morishita, A. Hara, T. Ohno, T. Kojima, H. Mori, *Carcinogenesis*, 11, 1403 (1990).

[26] T. Kojima, T. Tanaka, H. Mori, *Cancer Res.*, 54, 1446 (1994).

[27] T. Tanaka, T. Kojima, Y. Morishita, H. Mori, *Jpn. J. Cancer Res.*, 83, 835 (1992).

[28] C. J. Grubbs, V. E. Steele, T. Casebolt, M. M. Juliana, I. Eto, L. M. Whitaker, K. H. Dragnev, G. J. Kelloff, R. L. Lubet, *Anticancer Res.*, 15, 709 (1995).

[29] D. Guo, H. A. J. Schut, C. D. Davis, E. G. Snyderwine, G. S. Bailey, R. H. Dashwood, *Carcinogenesis*, 16, 2931 (1995).

[30] M. Xu, H. A. J. Schut, L. F. Bjeldanes, D. E. Williams, G. S. Bailey, R. H. Dashwood, *Carcinogenesis*, 18, 2149 (1997).

[31] M. A. Morse, S. D. LaGreca, S. G. Amin, F-L. Chung, *Cancer Res.* 50, 2613 (1990).

[32] Z. Yu, B. Mahadevan, C. V. Löhr, K. A. Fischer, M. A. Louderback, S. K. Krueger, C. B. Pereira, D. J. Albershardt, W. M. Baird, G. S. Bailey, D. E. Williams, *Carcinogenesis*, 27, 2116 (2006).

[33] B. B. Aggarwal, H. Ichikawa, *Cell Cycle*, 4, 1201 (2005).

[34] H. L. Bradlow, D. W. Sepkovic, N. T. Telang, M. P. Osborne, *Ann. NY Acad. Sci.,* 768, 180 (1995).

[35] M. C. Bell, P. Crowley-Nowick, H. L. Bradlow, D. W. Sepkovic, D. Schmidt-Grimminger, P. Howell, E. J. Mayeaux, A. Tucker, E. A. Turbat-Herrera, J. M. Mathis, *Gynecol.Oncol.*, 78, 123 (2000).

[36] M. J. Anderton, M. M. Manson, R. D Verschoyle, A. Gescher, J. H. Lamb, P. B. Farmer, W. P. Steward, M. L. Williams, *Clin. Cancer Res.*, 10, 5233 (2004).

[37] G. A. Reed, K. S. Peterson, H. J. Smith, J. C. Gray, D. K. Sullivan, M.S. Mayo, J. A. Crowell, A. Hurwitz, *Cancer Epidemiol. Biomarkers Prev.*, 14, 1953 (2005).

[38] D. R. McGrath, A. D. Spigelman, *Expert. Opin. Ther. Targets*, 12, 729 (2008).

[39] http://clinicaltrials.gov/ct2/results?term=indole+3+carbinol.

[40] E. J. Pantuck, K. C. Hsiao, W. D. Loub, L. W. Wattenberg, R. Kuntzman, A. H. Conney, *J. Pharmacol. Exp. Ther.,* 198, 278 (1976).

[41] L. W. Wattenberg, *Cancer Res.* 41, 2991 (1981).

[42] A. Kjaer, *Fortschr. Chem. Org. Naturst.* 18, 122 (1960).

[43] G. R. Fenwick, R. K. Heaney, W. J. Mullin, *CRC Crit. Rev. Food Sci. Nutr.*, 18, 123 (1983).

[44] J. W. Fahey, A. T. Zalcmann, P. Talalay, *Phytochemistry*, 56, 5 (2001).

[45] B. A. Halkier, J. Gershenzon, *Annu. Rev. Plant. Biol.,* 57, 303 (2006).

[46] I. E. Sønderby, F. Geu-Flores, B. A. Halkier, *Trends Plant Sci.,* 15, 283 (2010).

[47] F. L. Chung, M. A. Morse, K. I. Eklind, Y. Xu, *Ann. NY Acad. Sci*, 686, 186 (1993).

[48] M. A. Morse, C. X. Wang, G. D. Stoner, S. Mandal, P. B. Conran, S. G. Amin, S. S. Hecht, F. L. Chung, *Cancer Res.*, 49, 549 (1989).

[49] S. S. Hecht, N. Trushin, J. Rigotty, S. G. Carmella, A. Borukhova, S. Akerkar, A. Rivenson, *Cancer Epidemiol. Biomarkers Prev.,* 5, 645 (1996).

[50] G. D. Stoner, D. T. Morrissey, Y. H. Heur, E. M. Daniel, A. J. Galati, S. A. Wagner, *Cancer Res.,* 51, 2063 (1991).

[51] M. A. Pereira, *Anticancer Res.,* 15, 1953 (1995).

[52] A. Nishikawa, F. Furukawa, C. Uneyama, S. Ikezaki, Z-Y. Tanakamuru, F-L. Chung, M. Takahashi, Y. Hayashi, *Carcinogenesis*, 17, 1381 (1996).

[53] D. B. Solt, K-w. Chang, I. helenowski, A. W. Rademaker, *Cancer Lett.,* 202, 147 (2003).
[54] C. Huang, W. Y. Ma, J. Li, S. S. Hecht, Z. Dong, *Cancer Res.,* 58, 4102 (1998).
[55] D. Xiao, S. V. Singh, *Cancer Res.,* 62, 3615 (2002).
[56] D. Xiao, Y. Zeng, S. Choi, K. L. Lew, J. B. Nelson, S. V. Singh, *Clin. Cancer Res.,* 11, 2670 (2005).
[57] E-R. Hahm, S. V. Singh, *Mol. Carcinogenesis*, 1 (2011).
[58] K. Sakao, S. Desineni, E-R. Hahm, S. V. Singh, *Prostate*, (2011).
[59] D. Xiao, S. V. Singh, *Cancer Res.*, 70, 3150 (2010).
[60] D. Xiao, S. V. Singh, *Cancer Res.*, 67, 2239 (2007).
[61] T. S. Hudson, S. N. Perkins, S. D. Hursting, H. A. Young, Y. S. Kim, T-C. Wang, T. T. Y. Wang, *Int. J. Oncol.*, 40, 1113 (2012).
[62] L. B. von Weymarn, J. A. Chun, P. F. Hollenberg, *Carcinogenesis*, 27, 782 (2006).
[63] F. L. Chung, C. C. Conaway, C. V. Rao, B. S. Reddy, *Carcinogenesis*, 21, 2287 (2000).
[64] Y-M. Yang, C. C. Conaway, J. W. Chiao, C-X. Wang, S. Amin, J. Whysner, W.Dai, J. Reinhardt, F-L. Chung, *Cancer Res.,* 62, 2 (2002).
[65] C. Xu, G. Shen, X. Yuan, J-h. Kim, A. Gopalkrishnan, Y-S. Keum, S. Nair, A-N. Kong, *Carcinogenesis*, 27, 437 (2006).
[66] N. Gao, A. Budhraja, S. Cheng, E-H. Liu, J. Chen, Z. Yang, D. Chen, Z. Zhang, X. Shi, *Cell Death and Disease*, 2, e140 (2011).
[67] J. Hu, J. Straub, D. Xiao, S. V. Singh, H-S. Yang, N. Sonenberg, J. Vatsyayan, *Cancer Res.,* 67, 3569 (2007).
[68] R. P. Sahu, S. K. Srivastava, *J. Natl. Cancer Inst.*, 101, 176 (2009).
[69] S. R. Boreddy, R. P. Sahu, S. K. Srivastava, *PLoS ONE*, 6, e25799 (2011).
[70] S. Batra, R. P. Sahu, P. K. Kandala, S. R. Boreddy, S. K. Srivastava, *Mol. Cancer Ther.*, 9, 1596 (2010).
[71] S. R. Boreddy, K. C. Pramanik, S. K. Srivastava, *Clin Cancer Res.,* 17, 1784 (2011).
[72] M. F. Tsou, C-T. Peng, M-C. Shih, J-S. Yang, C-C. Lu, J-H. Chiang, C-L. Wu, J-P. Lin, C. Lo, M-J. Fan, J-G. Chung, *Leukemia Res.,* 33, 1505 (2009).
[73] R. Warin, W. H. Chambers, D. M. Potter, S. V. Singh, *Cancer Res.,* 69, 9473 (2009).
[74] R. Warin, D. Xiao, J. A. Arlotti, A. Bommareddy, S. V. Singh, *Mol. Carcinog.,* 49, 500 (2010).
[75] A. Sehrawat, S. V. Singh, *Cancer Prev. Res.*, 4, 1107 (2011).
[76] Y. Zhang, P. Talalay, C-G. Cho, G. H. Posner, *Proc. Natl. Acad. Sci. USA*, 89, 2399 (1992).
[77] Y. Zhang, T. W. Kensler, C-G. Cho, G. H. Posner, P. Talalay, *Proc. Natl. Acad. Sci. USA* 91, 3147 (1994).
[78] S. Barcelo, J. M. Gardiner, A. Gescher, J. K. Chipman, *Carcinogenesis*, 17, 277 (1996).
[79] Y. Zhang, L. Tang, *Acta Pharmacol Sin.*, 28, 1343 (2007).
[80] J. D. Clarke, R. H. Dashwood, E. Ho, *Cancer Lett.,* 269, 291 (2008).
[81] K. L. Cheung, A-N. Kong, *AAPS J.*, 12 (2009).
[82] R. H. Dashwood, E. Ho, *Nutr. Rev.,* 66, S36 (2008).
[83] H. Nian, B. Delage, E. Ho, R. H. Dashwood, *Environ. Mol. Mutagen*, 50, 213 (2009).
[84] T. A. Shapiro, J. W. Fahey, A. T. Dinkova-Kostova, W. D. Holtzclaw, K. K. Stephenson, K. L. Wade, L. Ye, P. Talalay, *Nutr. Cancer*, 55, 53 (2006).

[85] T. W. Kensler, J-G. Chen, P. A. Egner, J. W. Fahey, L. P. Jacobson, K. K. Stephenson, L. Ye, J. L. Coady, J-B. Wang, Y. Wu, Y. Sun, Q-N. Zhang, B-C Zhang, Y-R. Zhu, G-S. Qian, S. G. Carmella, S. S. Hecht, L. Benning, S. J. Gange, J. D. Groopman, P. Talalay, *Cancer Epidemiol. Biomarkers Prev.,* 14, 2605 (2005).

[86] C. K. Pushpendran, T. P. Devasagayam, G. J. Chintalwar, A. Banerji, J. Eapen, *Experientia*, 36, 1000 (1980).

[87] S. Belman, *Carcinogenesis*, 4, 1063 (1983).

[88] M. J. Wargovich, C. Woods, V. W. S. Eng, L. C. Stephens, K. Gray, *Cancer Res.,* 48, 6872 (1988).

[89] S. Takahashi, K. Hakoi, H. Yada, M. Hirose, N. Ito, S. Fukushima, *Carcinogenesis*, 13, 1513 (1992).

[90] M. J. Wargovich, O. Imada, L. C. Stephens, *Cancer Lett.*, 64, 39 (1992).

[91] B. S. Reddy, C. V. Rao, A. Rivenson, G. Kelloff, *Cancer Res*. 53, 3493 (1993).

[92] V. L. Sparnins, G. Barany, L. W. Wattenberg, *Carcinogenesis*, 9, 131 (1988).

[93] C. Ip, D. J. Lisk, H. J. Thompson, *Carcinogenesis*, 17, 1979 (1996).

[94] H. Mori, S. Sugie, W. Rahman, N. Suzui, *Cancer Lett.* 143, 195 (1999).

[95] J. Y. Hong, Z. Y. Wang, T. J. Smith, S. Zhou, S. Shi, J. Pan, C. S. Yanq, *Carcinogenesis*, 13, 901 (1992).

[96] E. Perkins, J. Calvert, J. A. Lancon, A. D. Parent, J. Zhang, *Mol. Brain Res.,* 111, 42 (2003).

[97] A. T. Fleischauer, L. Arab, *J. Nutr.*, 131, 1032S (2001).

[98] S. N. T. Ngo, D. B. Williams, L. Cobiac, R. J. Head, *J. Nutr.,* 137, 2264 (2007).

[99] J-S. Yang, L-F. Kok, Y-H. Lin, T-C. Kuo, J-L. Yang, C-C. Lin, G-W. Chen, W-W. Huang, H-C. Ho, J-G. Chung, *Anticancer Res.*, 26, 219 (2006).

[100] G. Hodge, S. Davis, M. Rice, H. Tapp, B. Saxon, T. Revesz, *Biologics*, 2, 143 (2008).

[101] A. Herman-Antosiewicz, S. V. Singh, *Mutat. Res*. 555, 121 (2004).

[102] Y. Shukla, N. Kalra, *Cancer Lett.*, 247, 167 (2007).

[103] H. T. Hassan, *Anticancer Agents Med. Chem.,* 11, 247 (2011).

[104] V. Singh, C. Belloir, M. H. Siess, A. M. Le Bon, *Nutr. Cancer*, 55, 178 (2006).

[105] A. Tsubura, Y. C. Lai, M. Kuwata, N. Uehara, K. Yoshizawa, *Anticancer Agents Med.*, 11, 249 (2011).

[106] E. Block, *Sci. Am.*, 252, 114 (1985).

[107] B. S. Reddy, *Prev. Med.,* 25, 48 (1996).

[108] S. G. Sundaram, J. A. Milner, *J. Nutr.*, 126, 1355 (1996).

[109] S. G. Sundaram, J. A. Milner, *Biochim. Biophys. Acta*, 1315, 15 (1996).

[110] S. G. Sundaram, J. A. Milner, *Carcinogenesis*, 17, 669 (1996).

[111] L. M. Knowles, J. A. Milner, *Nutr. Cancer*, 30, 169 (1998).

[112] L. M. Knowles, J. A. Milner, *Carcinogenesis*, 21, 1129 (2000).

[113] H. J. Jo, J. D. Song, K. M. Kim, Y. H. Cho, K. H. Kim, Y. C. Park, *Oncol. Rep.*, 19, 275 (2008).

[114] H. Ling, L. Wen, X. X. Ji, Y. L. Tang, J. He, H. Tan, H. Xia, J. G. Zhou, Q. Su, *Braz. J. Med. Biol. Res.*, 43, 271 (2010).

[115] S. V. Singh, R. R. Mohan, R. Agarwal, P. J. Benson, X. Hu, M. A. Rudy, H. Xia, A. Katoh, S. K. Srivastava, H. Mukhtar, V. Gupta, H. A. Zaren, *Biochem. Biophys. Res. Commun.*, 225, 660 (1996).

[116] S. V. Singh, *J. Nutr.*, 131, 10465 (2001).

[117] L. M. Knowles, J. A. Milner, *J. Nutr.*, 133, 2901 (2003).
[118] Y.S. Hong, Y. A. Ham, J. H. Choi, J. Kim, *Exp. Mol. Med.*, 32, 127 (2000).
[119] H. Nakagawa, K. Tsuta, K. Kiuchi, H. Senzaki, K. Tanaka, K. Hioki, A. Tsubura, *Carcinogenesis*, 22, 891 (2001).
[120] G. Filomeni, K. Aquilano, G. Rotilio, M. R. Ciriolo, *Cancer Res.*, 63, 5940 (2003).
[121] S. Karmakar, N. L. Banik, S. J. Patel, S. K. Ray, *Apoptosis*, 12, 671 (2007).
[122] K. B. Kwon, S. J. Yoo, D. G. Ryu, J. Y. Yang, H. W. Rho, J. S. Kim, J. W. Park, H. R. Kim, B. H. Park, *Biochem. Pharmacol.*, 63, 41 (2002).
[123] M. A. Lea, V. M. Randolph, M. Patel, *Int. J. Oncol.*, 15, 347 (1999).
[124] M. A. Lea, V. M. Randolph, *Anticancer Res.,* 21, 2841 (2001).
[125] N. Druesne, A. Pagniez, C. Mayeur, M. Thomas, C. Cherbuy, P-H. Duée, P. Martel, C. Chaumontet, *Carcinogenesis*, 25, 1227 (2004).
[126] N. Druesne-Pecollo, C. Chaumontet, A. Pagniez, P. Vaugelade, A. Bruneau,M. Thomas, C. Cherbuy, P-H. Duée, P. Martel, *Biochem. Biophys.Res.Commun.,* 354, 140 (2007).
[127] N. Druesne-Pecollo, P. Latino-Martel, *Anticancer Agents Med. Chem.*, 11, 254 (2011).
[128] K. Sakamoto, L. D. Lawson, J. A. Milner, *Nutr. Cancer*, 29, 152 (1997).
[129] D. Xiao, S. V. Singh, *Proc. Am. Assoc. Cancer Res.*, 44, 425 (2003).
[130] K. Lew, D. Xiao, Y. Zheng, S. V. Singh, *Proc. Am. Assoc. Cancer Res.*, 44, 197 (2003).
[131] D. Xiao, S. A. Choi, D. E. Johnson, V. Vogel, C. S. Johnson, D. L. Trump, Y. Lee, S. V. Singh, *Oncogene*, 23, 5594 (2004).
[132] C. Cerella, M. Dicato, C. Jacob, M. Diederich, *Anticancer Agents Med. Chem.*, 11, 267 (2011).
[133] D. Xiao, K. L. Lew, Y-A. Kim, Y. Zeng, E-R. Hahm, R. Dhir, S. V. Singh, *Clin. Cancer Res.*, 12, 6836 (2006).
[134] A. A. Powolny, S. V. Singh, *Cancer Lett.*, 269, 305 (2008).
[135] D. Xiao, Y. Zeng, E-R. Hahm, Y-A. Kim, S. Ramalingam, S. V. Singh, *Environ. Mol. Mutagen.*, 50, 201 (2009).
[136] M. Chen, B. Li, X. Zhao, H. Zuo, X. He, Z. Li, X. Liu, L. Chen, *Mol. Cell Biochem.*, 363, 75 (2012).
[137] K. Scharfenberg, T. Ryll, R. Wagner, K. G. Wagner, *J. Cell Physiol.*, 158, 55 (1994).
[138] V. M. Dirsch, A. L. Gerbes, A. M. Volmar, *Mol. Pharmacol.*, 53, 402 (1998).
[139] T. Nishikawa, N. Yamada, A. Hattori, H. Fukuda, T. Fujino, *Biosci. Biotechnol. Biochem.,* 66, 2221 (2002).
[140] C. M. Tilli, A. J. Stavast-Kooy, J. D. Vuerstaek, M. R. Thissen, G. A. Krekels, F. C. Ramaekers, H. A. Neumann, *Arch. Dermatol. Res.,* 295, 117 (2003).
[141] V. M. Dirsch, D. S. Antlsperger, H. Hentze, A. M. Volmar, *Leukemia*, 16, 74 (2002).
[142] M. Li, J. M. Min, J. R. Cui, L. H. Zhang, K. Wang, A. Valette, C. Davrinche, M. Wright, J. Leunq-Tack, *Nutr. Cancer*, 42, 241 (2002).
[143] N. Ahmed, L. Laverick, J. Sammons, H. Zhang, D. Maslin, H. T. Hassan, *Anticancer Res.*, 21, 3519 (2001).
[144] H. T. Hassan, *Leukemia Res.*, 28, 667 (2004).
[145] C. H. Kaschula, R. Hunter, M. I. Parker, *Biofactors*, 36, 78 (2010).
[146] C. H. Kaschula, R. Hunter, N. Stellenboom, M. R. Caira, S. Winks, T. Ogunleye, P. Richards, J. Cotton, K. Zilbeyaz, Y. Wang, V. Siyo, E. Ngarande, M. I. Parker, *Eur. J. Med. Chem.*, doi:10.1016/j.ejmech. 2012.01.058 (2012).

[147] N. Uda, N. Kashimoto, I. Sumioka, E. Kyo, S-i. Sumi, S. Fukushima, *J. Nutr.,* 136, 855S (2006).
[148] T. Katsuki, K. Hirata, H. Ishikawa, N. Matsuura, S-i. Sumi, K. Itoh, *J. Nutr.,* 136, 847S (2006).
[149] S. Tanaka, K. Haruma, M. Yoshihara, G. Kajiyama, K. Kira, H. Amagase, K. Chayama, *J. Nutr.,* 136, 821S (2006).
[150] H. Ishikawa, T. Saeki, T. Otani, T. Suzuki, K. Shimozuma, H. Nishino, S. Fukuda, K. Morimoto, *J. Nutr.*, 136, 816S (2006).
[151] D. L. Lamm, D. R. Riggs, *J. Nutr.*, 131, 1067S (2001).
[152] D. Lu, *Zhong Cao Yao Tong Xun (Chin. Tradition Herb. Drug)* 2, 69 (1972).
[153] Chen Guan Hospital, *Xin Yi Yao Xue Za Zhi (J. Tradition Chin. Med.)* 3, 116 (1976).
[154] D. Lu, *Xin Yi Yao Xue Za Zhi (J. Tradition Chin. Med.)* 3, 109 (1977).
[155] K. M. Nadkarni, *Indian Materia Medica* (1976). Curcuma longa, Popular Prakashan, Bombay, pp414-417.
[156] C. Ramaprasad, M. Sirsi, *J. Sci. Ind. Res.,* 156, 239 (1956).
[157] R. C. Srimal, B. N. Dhawan, *J. Pharm. Pharmacol.* 25, 447 (1973).
[158] R. Kuttan, P. Bhanumathy, K. Nirmala, M. C. George, *Cancer Lett.* 29, 197 (1985).
[159] B. Wahlström, G. Blennow, *Acta Pharmacol. Toxicol.* 43, 86 (1978).
[160] G. M. Holder, J. L. Plummer, A.J. Ryan, *Xenobiotica*, 8, 761 (1978).
[161] R. Kuttan, P. C. Sudheeran, C. D. Josph, *Tumori*, 73, 29 (1987).
[162] M-T. Huang, R. C. Smart, C-Q. Wong, A. H. Conney, *Cancer Res.* 48, 5941 (1988).
[163] T-S. Huang, S-C. Lee, J-K. Lin, *Proc. Natl. Acad. Sci. USA*, 88, 5292 (1991).
[164] C. Ireson, S. Orr, D. J. Jones, R. Verschoyle, C. K. Lim, J. L. Luo, L. Howells, S. Plummer, R. Jukes, M. Williams, W.P. Steward, A. Gescher, *Cancer Res.*, 61, 1058 (2001).
[165] S. Shishodia, M. M. Chaturvedi, B. B. Aggarwal, *Curr. Probl. Cancer*, 31, 243 (2007).
[166] S. Singh, B. B. Aggarwal, *J. Biol. Chem.*, 270, 24995 (1995).
[167] D. Wang, M. S. Veena, K. Stevenson, C. Tang, B. Ho, J. D. Suh, V. M. Duarte, K. F. Faull, K. Mehta, E. S. Srivatsan, M. B. Wang, *Clin. Cancer Res.*, 14, 6228 (2008).
[168] A. B. Kunnumakkara, P. Diagaradjane, P. Anand, K. B. Harikumar, A. Deorukhkar, J. Gelovani, S. Guha, S. Krishnan, B. B. Aggarwal, *Int. J. Cancer*, 125, 2187 (2009).
[169] S. T. Tharakan, T. Inamoto, B. Sung, B. B. Aggarwal, A. M. Kamat, *Biochem. Pharmacol.*, 79, 218 (2010).
[170] R. Wilken, M. S. Veena, M. B. Wang, E. S. Srivatsan, *Mol. Cancer*, 10, 12 (2011).
[171] A. Mukhopadhyay, S. Banerjee, L. J. Stafford, C. Xia, M. Liu, B. B. Aggarwal, *Oncogene*, 21, 8852 (2002).
[172] S. M. Plummer, K. A. Holloway, M. M. Manson, R. J. Munks, A. Kaptein, S. Farrow, *Oncogene*, 18, 6013 (1999).
[173] B. Gupta, B. Ghosh, *Int. J. Immunopharmacol.,* 21, 745 (1999).
[174] M. J. Park, E. H. Kim, I. C. Park, H. C. Lee, S. H. Woo, J. Y. Lee, Y. J. Hong, C. H. Rhee, S. H. Choi, B. S. Shim, S. H. Lee, S. I. Hong, *Int. J. Oncol.,* 21, 379 (2002).
[175] A. Mukhopadhyay, S. Banerjee, L. J. Stafford, C. Xia, M. Liu, B. B. Aggarwal, *Cell Cycle*, 6, 2953 (2007).
[176] C. V. Rao, *Adv. Exp. Med. Biol.*, 595, 213 (2007).
[177] A. Shehzad, S. Khan, O. Shehzad, Y. S. Lee, *Drugs Today*, 46, 523 (2010).

[178] S. Yodkeeree, C. Ampasavate, B. Sung, B. B. Aggarwal, P. Limtrakul, *Eur. J. Pharmacol.*, 627, 8 (2010).
[179] Q. Liu, W. T. Loo, S. C. Sze, Y. Tong, *Phytomedicine*, 16, 916 (2009).
[180] S. S. Han, S. T. Chung, D. A. Robertson, D. Ranjan, S. Bondala, *Clin. Immunol.*, 93, 152 (1999).
[181] E. Liu, J. Wu, W. Cao, J. Zhang, W. Liu, X. Jiang, X. Zhang, *J. Neurooncol.,* 85, 263 (2007).
[182] J. L. Watson, A. Greenshields, R. Hill, A. Hilchie, P. W. Lee, C. A. Giacomantonio, D. W. Hoskin, *Mol. Carcinogenesis*, 49, 13 (2010).
[183] H. F. Lu, K. C. Lai, S. C. Hsu, H. J. Lin, M. D. Yang, Y. L. Chen, M. J. Fan, J. S. Yang, P. Y. Cheng, C. L. Kuo, J. G. Chung, *Oncol. Rep.*, 22, 97 (2009).
[184] H. Aoki, Y. Takada, S. Kondo, R. Sawaya, B. B. Aggarwal, Y. Kondo, *Mol. Pharmacol.,* 72, 29 (2007).
[185] Y. L. Jia, J. Li, Z. H. liang, *J. Asian Nat. Prod. Res.,* 11, 918 (2009).
[186] G. O'Sullivan-Coyne, G. C. O'Sullivan, T. R. O'Donovan, K. Piwocka, S.L. McKenna, *Br. J. Cancer*, 101, 1585 (2009).
[187] R. Garg, S. Gupta, G. B. Maru, *Carcinogenesis*, 29, 1022 (2008).
[188] S. Fu, R. Kurzrock, *Cancer*, 116, 4670 (2010).
[189] A. Goel, B. B. Aggarwal, *Nutr. Cancer*, 62, 919 (2010).
[190] L. Shen, H-F. Ji, *Trends Mol. Med.,* 18, 138 (2012).
[191] J. Tang, Y. Feng, S. Tsao, N. Wang, R. Curtain, Y. Wang, *J. Ethnopharmacol.,* 126, 5 (2009).
[192] H-L. Lin, T-Y. Liu, C-W. Wu, C-W. Chi, *Brit. J. Cancer*, 81, 416 (1999).
[193] W. Tan, Y. Li, M. Chen, Y. Wang, *Int. J. Nanomedicine*, 6, 1773 (2011).
[194] Y. L. Lai, Y. J. Chen, T. Y. Wu, S. Y. Wang, K. H. Chang, C. H. Chung, M. L. Chen, *Anticancer Drugs*, 9, 77 (1998).
[195] P-L. Kuo, C-C. Lin, *Life Sci.*, 73, 243 (2003).
[196] H. Shen, W. Xu, Q. Chen, Z. Wu, H. Tang, F. Wang, *J. Cancer Res. Clin. Oncol.,* 136, 659 (2010).
[197] Q. Mi, B. Cui, G.L. Silva, D. Lantvit, E. Lim, H. Chai, M. G. Hollingshead, J. G. Mayo, A. D. Kinghorn, J. M. Pezzuto, *Cancer Lett.,* 184, 13 (2002).
[198] Q. Mi, B. Cui, D. Lantvit, E. Reyes-Lim, H. Chai, J. M. Pezzuto, A. D. Kinghorn, S. M. Swanson, *Anticancer Res.*, 23, 3607 (2003).
[199] I. Rovo, N. DePedro, E. Estornell, D. Cortes, F. Peláez, J. R. Tormo, *Oncol. Res.,* 13, 521 (2003).
[200] S. D. Jolad, J. J. Hoffmann, K. H. Schram, J. R. Cole, *J. Org. Chem.*, 47, 3151 (1982).
[201] K. I. Ahammadsahib, R. M. Hollingworth, J. P. McGovren, Y. H. Hui, J. L. McLaughlin, *Life Sci.,* 53, 1113 (1993).
[202] H-W. Chih, H-F. Chiu, K-S. Tang, F-R. Chang, Y-C. Wu., *Life Sci.,* 69, 1321 (2001).
[203] H-F. Chiu, T-T. Chih, Y-M. Hsian, C-H. Tseng, M-J. Wu, Y-C. Wu, *Biochem. Pharmacol.*, 65, 319 (2003).
[204] Y-J. Liang, X. Zhang, C-L. Dai, J-Y. Zhang, Y-Y. Yan, M-S. Zeng, L-M. Chen, L-W. Fu, *J. Biomed. Biotechnol.*, 867123. Epub 2009 Jul 21.
[205] Y. Chen, J. W. Chen, X. Li, *J. Nat. Prod.,* 74, 2477 (2011).
[206] J. L. McLaughlin, *J. Nat. Prod.,* 71, 1311 (2008).

[207] M. Jang, L. Cai, G. O. Udeani, K. V. Slowing, C. F. Thomas, C. W. Beecher, H. H. Fong, N. R. Farnsworth, A. D. Kinghorn, R. G. Mehta, R. C. Moon, J. M. Pezzuto, *Science*, 275, 218 (1997).

[208] J. A. Baur, D. A. Sinclair, *Nature Rev. Drug Discov.*, 5, 493 (2006).

[209] R. Nakata, S. Takahashi, H. Inoue, *Biol. Pharm. Bull.*, 35, 273 (2012).

[210] D. T. Ha, Q. C. Chen, T. M. Hung, U. J. Youn, T. M. Ngoc, P. T. Thuong, H. J. Kim, Y. H. Seong, B. S. Min, K. Bae, *Arch. Pharm. Res.*, 32, 177 (2009).

[211] A. Gonzalez-Sarrias, S. Gromek, D. Niesen, N. P. Seeram, G. E. Henry, *J. Agric. Food Chem.*, 59, 8632 (2011).

[212] G. R. Pettit, S. B. Singh, M. L. Niven, E. Hamel, J. M. Schmidt, *J. Nat. Prod.*, 50, 119 (1987).

[213] S. E. Holwell, P. A. Cooper, M. J. Thompson, G. R. Pettit, L. W. 3rd Lippert, S. W. Martin, M. C. Bibby, *Anticancer Res.*, 22, 3933 (2002).

[214] C. M. West, P. Price, *Anticancer Drugs*, 15, 179 (2004).

[215] M. Marrelli, F. Conforti, G. A. Statti, X. Cachet, S. Michel, F. Tillequin, F. Menichini, *Curr. Med. Chem.*, 18, 3085 (2011).

[216] T. Weinbrenner, M. Fito, R. de la Torre, G. T. Saez, P. Rijken, C. Tormos, S. Coolen, M. F. Albaladejo, S. Abadenes, H. Schroder, J. Marrugat, M. I. Covas, *J. Nutr.* 134, 2314 (2004).

[217] S. Granados-Principal, J. L. Quiles, C. L. Ramirez-Tortosa, P. Sanchez-Rovira, M. C. Ramirez-Tortosa, *Nutr. Reviews*, 68, 191 (2010).

[218] K. Yagasaki, Y. Miura, R. Okauchi, T. Furuse, *Cytotechnology*, 33, 229 (2000).

[219] D. Grunberger, R. Banerjee, K. Eisinger, E. M. Oltz, L. Efros, M. Caldwell, V. Estevez, K. Nakanishi, *Experientia*, 44, 230 (1988).

[220] K. W. Lee, N. J. Kang, J. H. Kim, K. M. Lee, D. E. Lee, H. J. Hur, H. J. Lee, *Genes Nutr.*, 2, 319 (2008).

[221] C-J. Weng, G-C. Yen, *Cancer Treat. Rev.*, doi:10.1016/j.ctrv. 2011.03.001 (2011).

[222] E. Sondenheimer, *Bot. Rev.*, 30, 677 (1964).

[223] H. Mori, T. Tanaka, H. Shima, T. Kuniyasu, M. Takahashi, *Cancer Lett.*, 30, 49 (1986).

[224] H. Mori, S. Sugie, T. Tanaka, H. Makita, N. Yoshimi, *Environ. Mut. Res. Commun.*, 18, 73 (1996).

[225] H. Kasai, S. Fukada, Z. Yamaizumi, S. Sugie, H. Mori, *Food Chem. Toxicol.*, 38, 467 (2000).

[226] U-H. Jin, J-Y. Lee, S-K. Kang, J-K. Kim, W-H. Park, J-G. Kim, S-K. Moon, C-H. Kim, *Life Sci.*, 77, 2760 (2005).

[227] S. Kaur, I. S. Grover, S. Kumar, *Indian J. Exp. Biol.*, 35, 478 (1997).

[228] M. Albrecht, W. Jiang, J. Kumi-Diaka, E. P. Lansky, L. M. Gommersall, A. Patel, R. E. Mansel, I. Neeman, A. A. Geldof, *J. Med. Food*, 7, 274 (2004).

[229] A. Malik, F. Afaq, S. Safaraz, V. M. Adhami, D. N. Syed, H. Mukhtar, *Proc. Natl. Acad. Sci. USA*, 102, 14813 (2005).

[230] S. Umesalma, G. Sudhandiran, *Eur. J. Pharmacol.*, doi:10.1016/j.ejphar. 2011.03.036 (2011).

[231] C. Bell, S. Hawthorne, *J. Pharm. Pharmacol.*, 60, 139 (2008).

[232] W-Y. Huang, Y-Z. Cai, Y. Zhang, *Nutr. Cancer*, 62, 1 (2010).

[233] K. Rao, T. J. McBride, J. J. Oleson, *Cancer Res.*, 28, 1952 (1968).

[234] A. Kumar, S. Dhawan, B. B. Aggarwal, *Oncogene*, 17, 913 (1998).

[235] J-A. Shin, J-H. Shim, J-G. Jeon, K-H. Choi, E-S. Choi, N-P. Cho, S-D. Cho, *Oral Diseases*, 17, 162 (2011).
[236] S. Y. Ryu, C. O. Lee, S. U. Choi, *Planta Med.,* 63, 339 (1997).
[237] H. J. Sung, S. M. Choi, Y. Yoon, K. S. An, *Exp. Mol. Med.*, 31, 174 (1999).
[238] X. Wang, S. L. Morris-Natschke, K-H. Lee, *Med. Res. Rev.*, 27, 133 (2007).
[239] Y. Li, Y. Gong, L. Li, H. M. Abdolmaleky, J-R. Zhou, *Mol. Carcinogenesis*, doi: 10.1002/mc.21888. (2012).
[240] M. Chitra, E. Sukumar, V. Suja, C. S. Devi, *Chemotherapy*, 40, 109 (1994).
[241] M. Sreepriya, G. Bali, *Fitoterapia*, 76, 549 (2005).
[242] K. S. Ahn, G. Sethi, B. B. Aggarwal, *Mol. Pharmacol.*, 71, 209 (2007).
[243] S. Reuter, S. Prasad, K. Phromnoi, R. Kannappan, V. R. Yadav, B. B. Aggarwal, *Mol. Cancer Res.*, 8, 1425 (2010).
[244] K. Y. Han, T. H. Kwon, T. H. Lee, S-J. Lee, S-H. Kim, J. Kim, *BMB reports*, 41, 328 (2008).
[245] Y. Zhang, R-Q. Qian, P-P. Li, *Cancer Lett.*, 284, 47 (2009).
[246] G. Sanchez-Duffhues, M. A. Calzado, A. Garcia de Vinuesa, G. Appendino, B. L. Fiebich, U. Loock, A. Lefarth-Risse, K. Krohn, E. Muñoz, *Biochem. Pharmacol.*, 77, 1401 (2009).
[247] R. S. Rosenberg Zand, D. J. A. Jenkins, E. P. Diamandis, *J. Chromatogr.,* 777, 219 (2002).
[248] C. Kanadaswami, L-T. Lee, P-P.H. Lee, J-J. Hwang, F-C. Ke, Y-T. Huang, M-T. Lee, *In vivo*, 19, 895 (2005).
[249] N. G. Amado, B. F. Fonseca, D. M. Cerqueira, V. M. Neto, J. G. Abreu, *Life Sci.*, 89, 545 (2011).
[250] A-M. Boudet, *Phytochemistry*, 68, 2722 (2007).
[251] K. Senthilkumar, R. Arunkumar, P. Elumalai, G. Sharmila, D. N. Gunadharini, S. Banudevi, G. Krishnamoorthy, C. S. Benson, J. Arunakaran, *Cell Biochem. Funct.,* 29, 87 (2011).
[252] C-J. Weng, G-C. Yen, *Cancer Metastasis Rev.*, doi: 10.1007/s10555-012-9347-y (2012).
[253] R. Kumar, V. Verma, A. Jain, R. K. Jain, J. P. Maikhuri, G. Gupta, *J. Nutr. Biochem.*, 22, 723 (2011).
[254] T. Kada, K. Kaneko, S. Matsuzaki, T. Matsuzaki, Y. Hara, *Mutat. Res.*, 150, 127 (1985).
[255] Y. Fujita, T. Yamane, M. Tanaka, K. Kuwata, J. Okuzumi, T. Takahashi, H. Fujiki, T. Okuda, *Jpn. J. Cancer Res.,* 80, 503 (1989).
[256] C. S. Yang, P. Maliakal, X. Meng, *Rev. Pharmacol. Toxicol.,* 42, 25 (2002).
[257] M. Z. Fang, Y. Wang, N. Ai, Z. Hou, Y. Sun, H. Lu, W. Welsh, C. S. Yang, *Cancer Res.,* 63, 7563 (2003).
[258] V. Nandakumar, M. Vaid, S. K. Katiyar, *Carcinogenesis*, 32, 537 (2011).
[259] B. N. Singh, S. Shankar, R. K. Srivastava, *Biochem. Pharmacol.*, 82, 1807 (2011).
[260] V. Nandakumar, T. Singh, S. K. Katiyar, *Cancer Lett.,* 269, 378 (2008).
[261] L. S. Adams, S. Phung, N. Yee, N. P. Seeram, L. Li, S. Chen, *Cancer Res.*, 70, 3594 (2010).
[262] H. Adlercreutz, P. I. Musey, T. Fotsis, C. Bannwart, K. Wähälä, T. Mäkelä, G. Brunow, T. Hase, *Clin. Chim. Acta*, 158, 147 (1986).

[263] D. P. Rose, *Nutr. Cancer*, 13, 1 (1990).
[264] L. U. Thompson, *Bailleries Clin. Endocrinol. Metab.*, 12, 691 (1998).
[265] I. Ionkova, *Mini Rev. Med. Chem.*, 11, 843 (2011).
[266] N. M. Saarinen, A. Wärri, M. Airio, A. Smeds, S. Mäkelä, *Mol. Nutr. Food Res.*, 51, 857 (2007).
[267] A. C. Dorogokupla, E. G. Troitzkaia, L. K. Adilgireieva, S. F. Postolnikov, Z. P. Chekrygina, *Zdravookhr. Kazakh.*, 10, 32 (1973).
[268] M. M. Mathews-Roth, *Oncology*, 39, 33 (1982).
[269] H. Vainio, M. Rautalahti, *Cancer Epidemiol. Biomarkers Prev.*, 7, 725 (1998).
[270] L. Gallicchio, K. Boyd, G. Matanovski, X. Tao, L. Chen, T. K. Lam, M. Shiels, E. Hammond, K. A. Robinson, L. E. Caulfield, J. G. Herman, E. Guallar, A. J. Alberg, *Am. J. Clin. Nutr.*, 88, 372 (2008).
[271] H. Nishino, M. Murakoshi, T. Ii, M. Takemura, M. Kuchide, M. Kanazawa, X. Y. Mou, S. Wada, M. Masuda, Y. Ohsaka, S. Yogosawa, Y. Satomi, K. Jinno, *Cancer Metast. Rev.*, 21, 257 (2002).
[272] S. C. Nair, B. Panikkar, K. R. Panikkar, *Cancer Lett.*, 57, 109 (1991).
[273] S. C. Nair, S. K. Kurumboor, J. H. Hasegawa, *Cancer Biotherapy*, 10, 257 (1995).
[274] W. Stahl, H. Sies, *Arch. Biochem. Biophys.*, 336, 1 (1996).
[275] E. Giovannucci, *J. Natl. Cancer Inst.*, 91, 317 (1999).
[276] M. Kelkel, M. Schumacher, M. Dicato, M. Diederich, *Free Radic. Res.*, 45, 925 (2011).
[277] D. V. Parke, H. Rahman, *Proc. Biochem. Soc.*, (1969).
[278] C. E. Elson, T. H. Maltzman, J. L. Boston, M. A. Tanner, M. N. Gould*Carcinogenesis*, 9, 331 (1988).
[279] M. N. Gould, W. D. Wacker, T. H. Maltzman, *Prog. Clin. Biol. Res.*, 47, 255 (1990).
[280] P. L. Crowell, M. N. Gould, *Crit. Rev. Oncog.*, 5, 1 (1994).
[281] P. L. Crowell, *Breast Cancer Res. Treat.*, 46, 191 (1997).
[282] P. L. Crowell, *J. Nutr.*, 129, 775S (1999).
[283] R. Suaeyun, T. Kinouchi, H. Arimochi, U. Vinitketkummuen, Y. Ohnishi, *Carcinogenesis*, 18, 949 (1997).
[284] J-J. Liu, R-W. Huang, D-J. Lin, X-Y. Wu, J. Peng, X-L. Pan, Q. Lin, M. Hou, M-H. Zhang, F. Chen, *Am. J. Hematol.*, 81, 86 (2006).
[285] J. D. Connolly, R.A. Hill, *Nat. Prod. Rep.*, 27, 79 (2010).
[286] L. K. T. Lam, S. Hasegawa, *Nutr. Cancer*, 12, 43 (1989).
[287] H. Itokawa, Z-S. Qiao, C. Hirobe, K. Takeya, *Chem. Pharm. Bull.*, 43, 1171 (1995).
[288] W. Logemann, F. Lauria, G. Cudkowicz, J. Franceschini, *Nature*, 187, 607 (1960).
[289] A. Inada, T. Nakanishi, H. Tokuda, H. Nishino, O.P. Sharma, *Planta Med.*, 63, 272 (1997).
[290] R. Mukherjee, V. Kumar, S. K. Srivastava, S. K. Agarwal, A. C. Burman, *Anticancer Agents Med. Chem.*, 6, 271 (2006).
[291] Z. Ovesná, K. Kozics, D. Slameňová, *Mutat. Res.*, 600, 131 (2006).
[292] R. E. De Angel, S. M. Smith, R. D. Glickman, S. N. Perkins, S. D. Hursting, *Nutr. Cancer*, 62, 1074 (2010).
[293] W. N. Setzer, M. C. Setzer, *Mini Rev. Med. Chem.*, 3, 540 (2003).
[294] L. Yu, T. Yu, R. Ma, *Carcinogenesis*, 16, 3045 (1995).
[295] T. Fujioka, A. Sakurai, K. Mihashi, Y. Kashiwada, I-S. Chen, K-H. Lee, *Chem. Pharm. Bull.*, 45, 68 (1997).

[296] K. Sato, M. Mochizuki, I. Saiki, Y.C. Yoo, K. Samukawa, I. Azuma, *Biol. Pharm. Bull.*, 17, 635 (1994).

[297] Waller and Yamasaki, *Saponins used in traditional and modern medicine* (1996). T. Konoshima, Anti-tumor-promoting activities of triterpenoid glycosides; cancer chemoprevention by saponins, Plenum Press, New York, pp87-100.

[298] G. Bader, B. Plohmann, K. Hiller, G. Franz, *Pharmazie*, 51, 414 (1996).

[299] B. Plohmann, G. Bader, K. Hiller, G. Franz, *Pharmazie*, 52, 953 (1997).

[300] M. R. Evans, A. W. Preece, *Bristol Med. Chir. J.*, 88, 17 (1973).

[301] R. Rentea, E. Lyon, R. Hunter, *Lab. Invest.,* 44, 43 (1981).

[302] T. Hajto, *Oncology*, 43, 51 (1986).

[303] T. Hajto, K. Hostanska, H-J. Gabius, *Cancer Res.*, 49, 4803 (1989).

[304] T. Hajto, K. Hostanska, K. Frei, C. Rordorf, H-J. Gabius, *Cancer Res.*, 50, 3322 (1990).

[305] B-M. Heiny, J. Beuth, *Anticancer Res.*, 14, 1339 (1994).

[306] H. Bantel, I.H. Engels, W. Voelter, K. Schulze-Osthoff, S. Wesselborg, *Cancer Res.*, 59, 2083 (1999).

[307] T. Ostermann, C. Raak, A. Büssing, *BMC Cancer*, 9, 451 (2009).

[308] J-G. Chai, M. Okamoto, T. Bando, H. Nagasawa, H. Hisaeda, T. Sakai, K. Himeno, M. Sato, S. Ohkubo, *Immunopharmacology*, 30, 209 (1995).

[309] M. P. Moretão, A. R. Zampronio, P. A. J. Gorin, M. Iacomini, M. B. M. Oliveira, *Immunol. Lett.*, 93, 189 (2004).

[310] N. R. Farnsworth, C. J. Kaas, *J. Ethnopharmacol.*, 3, 85 (1981).

[311] J. G. Graham, M. L. Quinn, D. S. Fabricant, N. R. Farnsworth, *J. Ethnopharmacol.*, 73, 347 (2000).

[312] [M. J. Ruffa, G. Ferraro, M. L. Wagner, M. L. Calcagno, R. H. Campos, L. Cavallaro, *J. Ethnopharmacol.*, 79, 335 (2002).

[313] L. V. Costa-Lotufo, M. T. H. Khan, A. Ather, D. V. Wilke, P. C. Jimenez, C. Pessoa, *J. Ethnopharmacol.*, 99, 21 (2005).

[314] G. Fouche, G. M. Cragg, P. Pillay, N. Kolesnikova, V. J. Maharaj, J. Senabe, *J. Ethnopharmacol.*, 119, 455 (2008).

[315] M. L. de Mesquita, J. E. de Paula, C. Pessoa, M. O. de Moraes, L. V. Costa-Lotufo, R. Grougnet, S. Michel, *J. Ethnopharmacol.*, 123, 439 (2009).

[316] S. Wang, X. Wu, M. Tan, J. Gong, W. Tan, B. Bian, M. Chen, Y. Wang, *J. Ethnopharmacol.*, 140, 33 (2012).

[317] T. O. Khor, Y-S. Keum, W. Lin, J-H. Kim, R. Hu, G. Shen, C. Xu, A. Gopalakrishnan, B. Reddy, X. Zheng, A. H. Conney, A-N. T. Kong, *Cancer Res.,* 66, 613 (2006).

[318] D. Pouliquen, F. Foussard, G. Tanguy, J. Roux, Y. Malthièry, *Cell. Mol. Biol.,* 47, 947 (2001).

[319] D. Pouliquen, C. Olivier, E. Hervouet, F. Pedelaborde, E. Debien, M-T. Le Cabellec, C. Gratas, T. Homma, K. Meflah, F.M. Vallette, J. Menanteau, *Int. J. Cancer*, 123, 288 (2008).

[320] D. L. Pouliquen, *Curr. Opin. Clin. Nutr. Metab. Care*, 10, 475 (2007).

[321] F. Guillot, B. Boutin, C. Blanquart, J-F. Fonteneau, M. Robard, M. Grégoire, D. Pouliquen, *Vaccine*, 29, 5534 (2011).

# Index


**A**

abstraction, 91, 168
abuse, 16
acarbose, 4
access, 4, 132
accessions, 219, 220, 221, 222, 223
accounting, 50
acetaminophen, 101, 107, 111
acetic acid, 28, 30, 94, 159, 228, 234
acetogenins, 357, 371, 372
acetone, 17, 149, 161, 255, 256, 328, 334
acetonitrile, 159, 221
acetophenone, 255, 256
acetylation, 31, 179, 349, 365
acetylcholine, 234
acetylcholinesterase, 273
acidic, 298, 383
activation state, 181
active compound, 9, 69, 121, 123, 148, 348, 351, 370, 381
acute lung injury, 180
acute myeloid leukemia, 362, 367
ADA, 274
adalimumab, 27
adaptation, 206
adaptogen, 208, 307, 312
additives, 199
adenine, 192, 193, 309
adenocarcinoma, 267
adenosine, 192
adenosine triphosphate, 192
adhesion, 176, 191, 232, 362, 368, 377
adipocyte, 8
adiponectin, 8, 16, 81, 350
adipose tissue, 8, 177, 310
adjunctive therapy, 4
ADP, 100, 274
adrenal glands, 311
adrenaline, 235
adsorption, 298
adults, 2, 33
advanced glycation end products (AGEs), 4
adverse effects, 5, 16, 70, 179, 296
adverse event, 273
AFB1, 359
afferent nerve, 31
aflatoxin, 359, 363
Africa, 5, 220
agar, 140, 274
age, 2, 192, 193, 194, 197, 370
agencies, 219
aggregation, 182, 235, 238, 274
aggressiveness, 383
aging process, 191, 194
agonist, 9
Agricultural Research Service (ARS), 220
agriculture, vii, 29, 223, 244
air quality, 196
alanine, 9, 310
alanine aminotransferase, 310
albumin, 176, 275
alcoholism, 272, 311, 312
alcohols, 95
aldehydes, 13, 93, 197
algae, 5, 88
algorithm, 134, 135, 138, 145, 146
ALI, 180
alkaloids, 1, 6, 9, 11, 13, 17, 18, 19, 20, 202, 203, 251, 284, 293, 317, 318, 324, 357, 358, 371
allergens, 176
allergic inflammation, 251

allergic reaction, 239
allergic rhinitis, 225
allergy, 175, 347, 348
aloe, 11
alpha-tocopherol, 207
ALT, 272, 310, 351
alternative medicine, 276, 312
alveolar macrophage, 183
amino, 4, 197, 273, 360
amino acid(s), 6, 9, 14, 18, 19, 48, 125, 232, 240, 241
amplitude, 308
amputation, 2
amylase, 10, 183
analgesic, 125, 182, 184, 234, 250, 268, 296, 308, 309
anaphylaxis, 275, 347
anatomic site, 380
androgen, 237
androgens, 236
anemia, 225
anger, 209
angiogenesis, 7, 191, 202, 362, 363, 372, 376, 377, 382
angiotensin II, 273
antagonism, 93, 308
anthocyanin, 8, 74, 160, 161, 167, 318
anti-angiogenic agents, 374
antibiotic, 138, 144, 334
antibody, 347, 382
anticancer activity, 139, 368
anticancer drug, 6, 109, 115, 239, 373
anticoagulant, 235
antidepressant, 121, 123, 125, 139, 144
antigen, 176, 195, 239, 346, 347, 348
antihypertensive drugs, 4
anti-inflammatories, 204
anti-inflammatory agents, 10, 178
anti-inflammatory drugs, 177
antimony, 253
antioxidative activity, 202, 270
antipyretic, 235, 250
antitrombotic, 70
antitumor, 87, 178, 184, 237, 269, 295, 307, 371, 372, 375, 383
antitumor agent, 371, 372, 375
antitumor immunity, 383
anus, 26
anxiety, 209
AOC, 75, 77
APC, 366
apoptosis, 3, 32, 77, 92, 101, 197, 238, 269, 275, 307, 312, 361, 362, 364, 365, 366, 367, 368, 369, 371, 372, 375, 378, 381, 382
apples, 31, 374
aqueous solutions, 125, 184
arabinogalactan, 232, 271
arabinoside, 163, 233
Argentina, 25, 383
arginine, 192
aromatic compounds, 153
aromatic rings, 64, 203
arrest(s), 77, 361, 365, 366, 368, 369, 371, 372, 373
arrhythmia, 273
ARS, 219, 220, 223
arsenic, 253
arteries, 3
artery, 3, 81, 236, 273
arthritis, 178, 179, 181, 183, 184, 225, 235, 251, 296
aryl hydrocarbon receptor, 360
ascorbic acid, 15, 19, 59, 60, 97, 100, 149, 151, 192
ASI, 268
Asia, 26, 201, 208, 220, 224, 247, 248, 258, 276, 295, 296, 317, 358
Asian countries, 152, 367
ASL, 244
aspartic acid, 9
assessment, 69
asthma, 175, 176
astringent, 14
astrocytes, 179, 180
atherosclerosis, 3, 4, 56, 109, 175, 176
atoms, 107, 366
atopic dermatitis, 205
atopic eczema, 195
ATP, 192, 195, 236
atrophy, 193
attachment, 377
autoimmune disease(s), 87, 175
autolysis, 361
autooxidation, 108
avoidance, 3
azadirachtin, 11, 88

**B**

B1 (AFB1), 359

bacillus, 238, 367
Bacillus subtilis, 60, 138, 140, 141, 142, 143
back pain, 296
bacteria, 6, 28, 56, 57, 65, 66, 88, 90, 138, 139, 140, 141, 142, 143, 146, 175, 207, 274, 334, 349, 378
bacterial strains, 139, 141
bacteriostatic, 56, 65
Bangladesh, 383
base, vii, 82, 138, 192, 193, 308, 334, 335, 352
basophils, 348
behaviors, 168, 287
Beijing, 241, 242, 276, 277, 314
beneficial effect, 30, 31, 32, 40, 50, 76, 79, 81, 175, 181, 185, 202, 239, 270, 273
benefits, 27, 41, 42, 44, 45, 47, 48, 49, 56, 63, 70, 79, 132, 133, 147, 148, 151, 153, 168, 176, 199, 201
benign, 193, 237, 240
benign prostatic hyperplasia, 237
benign prostatic hypertrophy, 240
benzene, 109
beta-carotene, 17, 203
beverages, 74, 148
BHC, 101, 107
Bhutan, 248
bias, 383
bile, 182, 351, 368
bile acids, 44
binding globulin, 237
bioactive agents, 17, 81
bioassay, 17
bioavailability, 31, 33, 70, 76, 77, 81, 199, 334
biochemistry, 82, 361
bioconversion, 364
biodiversity, 5, 69, 359
bioenergy, 191, 192
biological activities, 10, 29, 97, 179, 204, 234, 284, 292, 357, 358
biological activity, 1, 70, 81, 88, 197
biological systems, 62, 63
biologically active compounds, 42, 70, 125
biomarkers, 27, 77, 79, 81, 176, 177, 199, 361
biomolecules, 63
biopolymers, 271
biopsy, 360
biosynthesis, 5, 20, 89, 90, 178, 331, 335, 361, 376
biotechnology, 378
black tea, 31
bladder cancer, 369
bleaching, 62, 168
bleeding, 30, 181, 235
bleeding time, 235
blindness, 2
blood, 2, 3, 7, 9, 11, 13, 14, 15, 16, 17, 18, 19, 20, 27, 76, 79, 80, 100, 108, 176, 177, 178, 191, 235, 236, 238, 239, 269, 276, 309, 350, 373, 380
blood circulation, 235
blood clot, 177
blood flow, 7, 176, 373
blood plasma, 310
blood pressure, 3, 79, 80, 236, 238
blood urea nitrogen, 239, 269
blood vessels, 3, 7, 176
BMI, 80, 81
body fluid, 78
body weight, 15, 29, 31, 80, 81, 180
bone, 115, 238, 239, 251, 267, 273, 277, 376
bone form, 239, 273
bone growth, 251, 277
bone marrow, 115, 267
bones, 247, 250
bowel, 25, 26, 28, 45, 183, 367
brain, 179, 180, 194, 238, 247, 250, 272, 273
Brazil, 39, 201, 383
breast cancer, 77, 361, 364, 369, 373, 376, 378, 382
breathing, 196
bronchitis, 57, 181, 247, 250
Bulgaria, 123
bursitis, 176

## C

$Ca^{2+}$, 273, 348, 349, 364, 365, 366
cabbage, 152, 157, 359, 361
cacao, 31
caffeine, 308
calcium, 10, 69, 70, 239, 348, 365
calibration, 220, 290, 306
canals, 123
cancer cells, 77, 195, 203, 269, 274, 307, 361, 362, 365, 366, 367, 369, 372, 375, 376, 377, 381, 384
cancer stem cells, 220
candidates, 33, 208, 268, 352
capillary, 132, 221, 235, 309
capsule, 234, 240

carbohydrate(s), 6, 7, 10, 12, 13, 16, 17, 18, 19, 39, 46, 240, 382, 383
carbohydrate metabolism, 2, 14
carbon, 17, 44, 71, 91, 101, 107, 122, 128, 129, 134, 144, 145, 146, 149, 201, 238, 272, 310, 330, 332
carbon dioxide, 122, 128, 129, 134, 144, 145, 146, 149, 201
carbon tetrachloride, 101, 107, 238, 272, 310
carbonyl groups, 95
carboxyl, 298, 331, 332
carboxylic acid(s), 263, 271, 329, 335
carcinogen, 359, 360, 361, 362
carcinogenesis, 32, 358, 359, 361, 364, 369, 372, 374, 377, 380, 381
carcinogenicity, 148
carcinoma, 352, 360, 362, 364, 369, 372, 377
cardiotonic, 159
cardiovascular disease, 29, 39, 56, 69, 81, 92, 93, 108, 147, 148, 168, 353
cardiovascular risk, 79, 80, 81
carotene, 44, 69, 70, 109, 114, 115, 116, 149, 153, 201, 240, 252, 379, 380
carotenoids, 6, 40, 41, 44, 49, 88, 115, 116, 125, 199, 201, 357, 379, 380, 383
cartilage, 296, 312
cascades, 181, 195
caspases, 274
catabolism, 195
cataract, 13
Catharanthus roseus, 358
cation, 91
C-C, 386, 387, 390
CD8+, 139
cDNA, 366
cell culture, 8, 77, 199, 271, 384
cell cycle, 77, 269, 361, 364, 365, 366, 369, 372, 373
cell death, 92, 197, 220, 238, 361, 370, 384
cell differentiation, 198, 237, 239, 273
cell division, 365
cell invasion, 268
cell line(s), 7, 77, 97, 109, 181, 183, 273, 307, 365, 366, 367, 369, 370, 371, 372, 373, 374, 375, 377, 378, 382
cell membranes, 101, 207, 382
cell metabolism, 208
cell signaling, 197, 362
cellular energy, 192
cellular homeostasis, 198
cellular immunity, 268
cellular signaling pathway, 358, 383
cellulose, 42, 133
central nervous system, 180, 275, 308
cerebral edema, 182
cervical cancer, 367
cervical intraepithelial neoplasia, 360
cervix, 77
challenges, 50
chaos, 150, 169
chemical properties, 149
chemical reactivity, 198
chemical structures, 6, 25, 93, 95, 97, 107, 115, 261, 271
chemicals, 18, 28, 57, 74, 176, 209, 285
chemiluminescence, 269
chemokines, 175
chemoprevention, 29, 184, 207, 360, 361, 362, 363, 367, 368, 376, 377, 380, 383, 394
chemopreventive agents, 362
chemotaxis, 181
chemotherapeutic agent, 109
chemotherapy, 370, 380
children, 27, 195, 236
China, 18, 147, 150, 152, 155, 200, 201, 202, 205, 220, 225, 227, 228, 240, 241, 242, 243, 245, 247, 248, 251, 252, 253, 254, 257, 276, 283, 284, 285, 295, 296, 315, 317, 318, 329, 333, 334, 358, 363, 370, 372, 383
Chinese medicine, 239, 242, 243, 371
chirality, 332
chitin, 232
chloral, 308
chloroform, 184
chlorophyll, 240
cholangitis, 183
cholesterol, 3, 4, 6, 7, 11, 15, 44, 45, 78, 79, 80, 197, 271, 309, 351, 366
cholesteryl esters, 4
choline, 14, 125
chondrocyte, 296
chromatograms, 287, 291
chromatography, 219
chromium, 11
chromosome, 76, 192, 238, 364, 368
chronic diseases, 8, 39, 44, 69, 81, 147, 148, 168, 176
chronic granulomatous disease, 235
chronic illness, 27, 175, 176
cigarette smoke, 195

circulation, 16, 77, 247, 250, 308
cirrhosis, 182
classes, 50, 69, 70, 74, 107, 149, 160, 259, 324, 357, 358, 359, 383
classification, 228
cleavage, 372
clinical application, 225, 247, 251, 275, 312
clinical symptoms, 27
clinical trials, 14, 32, 33, 70, 79, 178, 184, 208, 360, 363, 370
cluster of differentiation, 26
clusters, 31
CNS, 182, 275
CO2, 125, 128, 129, 132, 134, 138, 140, 144, 145, 146, 149, 262
coding, 192
coffee, 102, 374
colitis, 26, 28, 29, 30, 31, 32, 267
collagen, 191, 193, 194, 195, 201, 296
colon, 26, 29, 30, 31, 44, 45, 50, 77, 267, 360, 362, 363, 364, 366, 367, 372, 374, 375, 376, 378, 381
colon cancer, 29, 45, 363, 372, 375
colon carcinogenesis, 29, 364, 367, 374
color, 12, 40, 47, 48, 62, 63, 116, 125, 162, 201, 318, 323, 335
colorectal cancer, 30, 360, 364, 369
commercial, 56, 74, 144, 146, 151, 162, 347, 348
common symptoms, 183
communication, 88, 380
community, 5
comparative analysis, 133
competition, 6
compilation, vii
complement, 199
complexity, 20, 190
complications, 1, 2, 3, 4, 5, 7, 13, 19, 20, 27, 178, 239, 350, 351
composites, 5
composition, 13, 27, 41, 46, 50, 51, 70, 93, 124, 128, 139, 142, 149, 159, 160, 162, 205, 233, 241, 243, 298, 367, 382
condensation, 4, 90, 332, 360, 364
conditioning, 364
conduction, 7
configuration, 30, 48, 139, 371
conifer, 100
conjugated dienes, 203
conjugation, 69, 70, 91
consciousness, 359
conservation, 136, 189, 359
constipation, 19
construction, 194
consumers, 4, 40
consumption, 39, 40, 44, 46, 47, 56, 69, 74, 77, 78, 79, 80, 81, 129, 132, 138, 148, 168, 179, 309, 351, 363, 364, 376
contact dermatitis, 209
contour, 205
control group, 79, 81, 235, 310
controversial, 203, 207
conversion rate, 149
COOH, 259, 260, 261, 262
cooking, 46, 47, 48, 50, 70, 151, 154
cooling, 60, 220
copper, 10, 11
coronary heart disease, 2, 40, 308
correlation(s), 56, 61, 63, 65, 76, 77, 81, 90, 107, 134, 150, 151, 160, 238, 364
correlation coefficient, 76
cortical neurons, 101
corticosteroids, 33, 176
cortisol, 13
cosmetic(s), 189, 199, 200, 201, 204, 205, 207, 208, 209, 225, 240, 247
cost, 5, 28, 134, 149, 178
cotton, 308
cough, 235
coumarins, 15, 248, 251, 317, 324, 334
counterbalance, 78
country of origin, 5
creatine phosphokinase, 308
creatinine, 239
Croatia, 123
crop(s), 220, 223
CRP, 176, 177, 183
crystalline, 133
crystals, 239
cultivars, 42, 43, 74, 160, 161, 162, 219, 221, 223
cultivation, 179, 364
cultivation conditions, 318
cultural conditions, 161
culture, 9, 29, 60, 76, 94, 96, 97, 98, 99, 110, 114, 116, 199, 205, 358, 361, 364
curcumin, 7, 12, 33, 168, 182, 183, 184, 199, 202, 368, 369, 370, 383
cure, 6, 27, 108, 178, 199, 236, 284
cyclooxygenase, 26, 29, 177, 179, 182, 196, 273, 311, 369, 372
cyclophosphamide, 109

cyclosporine, 27
cysteine, 192, 196, 197, 232
cytochrome(s), 101, 193, 360, 362, 363, 372, 375
cytokines, 32, 77, 175, 176, 177, 179, 180, 182, 183, 184, 185, 195, 198, 270, 272, 348, 369
cytokinesis, 76
cytomegalovirus, 239
cytoplasm, 208, 369
cytosine, 194
cytoskeleton, 349
cytotoxicity, 139, 237, 274, 312, 347, 353, 382
Czech Republic, 145

## D

damages, 31, 181, 196, 198
danger, 197
decomposition, 62, 122, 132, 204
decortication, 48
defence, 148, 189
defense mechanisms, 1, 6, 88, 197
deficiency, 2, 29, 56
deficit, 3
deforestation, 251
degradation, 74, 122, 128, 149, 160, 179, 193, 195, 209, 235, 296, 334, 366, 369, 370, 375, 377
degradation process, 149
dehydration, 90, 191
demonstrations, 363, 366
denaturation, 197
dendritic cell, 267
deoxyribose, 160
Department of Agriculture, 51
depolymerization, 48
deposits, 239
depression, 9, 272, 273
deprivation, 273
depth, 190
deregulation, 369
derivatives, 12, 14, 15, 30, 42, 44, 71, 95, 121, 123, 125, 130, 152, 160, 161, 162, 167, 181, 196, 197, 198, 202, 204, 207, 208, 233, 234, 299, 330, 331, 359, 366, 372, 381
dermatitis, 176
dermatology, 57
dermis, 190, 191, 193, 194, 195, 196
desiccation, 133
desorption, 298
destruction, 2, 296
detection, 161, 293, 306
detection system, 75
detoxification, 182, 189, 193, 198, 202, 363
developing countries, 3, 5, 26
diabetes, 1, 2, 3, 4, 5, 7, 9, 12, 13, 18, 19, 20, 29, 40, 56, 80, 92, 111, 116, 147, 148, 168, 175, 176, 179, 182, 225, 247, 250, 271, 309, 310, 314, 350
Diabetes mellitus (DM), 2
diabetic nephropathy, 275
diabetic patients, 3, 4, 5, 9, 350
diarrhea, 26, 27, 29, 183
diastolic blood pressure, 80
diet, 1, 3, 4, 5, 15, 40, 41, 46, 47, 74, 116, 153, 175, 177, 178, 350, 351, 357, 358, 359, 360, 367, 378, 379, 383
dietary fiber, 44, 378
dietary habits, 177
dietary intake, 70, 77
diffusion, 7, 129, 132, 133, 274
digestibility, 48
digestion, 20, 44, 45, 271
dilation, 176
direct action, 76, 78
discomfort, 179, 183
disease activity, 29, 181
disease progression, 27
diseases, vii, 1, 6, 25, 26, 29, 40, 55, 56, 81, 87, 88, 92, 108, 123, 139, 147, 148, 150, 152, 176, 178, 179, 182, 183, 270, 284, 296, 309
disgust, 333
disorder, 5, 25, 284
dissociation, 198, 307
distillation, 57, 123, 125, 128
distilled water, 58, 59
distortions, 197
distribution, 27, 57, 75, 76, 199, 225, 227, 238, 252, 253, 254, 255, 256, 257, 258, 334, 348, 368
diterpenoids, 247, 248, 251, 267, 359, 381
diuretic, 14, 57
diversity, 6, 91, 125, 150, 169, 358, 359, 376, 381, 383
diverticulitis, 176
DNA, 63, 76, 78, 83, 92, 93, 94, 101, 148, 189, 192, 193, 194, 197, 202, 203, 204, 205, 206, 207, 211, 212, 235, 238, 269, 273, 361, 363, 364, 374, 377, 381
DNA damage, 78, 101, 194, 195, 203, 204, 364
DNA repair, 205

docosahexaenoic acid, 197
DOI, 21, 119
donors, 192
dopamine, 125, 272
dopaminergic, 101
dosage, 16, 94, 310, 351
double bonds, 44, 114, 204, 207
dough, 46
down-regulation, 77, 180, 183, 268, 274, 296, 362
drinking water, 180
drug design, 358
drug metabolism, 361
drug resistance, 56, 371
drug treatment, 17
drugs, 4, 5, 7, 17, 25, 27, 30, 56, 60, 176, 177, 178, 181, 184, 187, 225, 233, 234, 276, 334, 352, 359, 361, 367, 370, 383, 384
dry matter, 43
drying, 47, 133, 148
dyslipidemia, 79, 80
dysplasia, 193

## E

East Asia, 2, 296
Eastern Europe, 208
ECM, 191, 193
ecology, 29, 247, 251
Ecuador, 151
eczema, 123, 225
edema, 30, 109, 308
education, 3, 5
egg, 93, 308
Egypt, 358
elastin, 191, 193, 195, 197
electrical resistance, 311
electron, 3, 62, 63, 76, 88, 97, 139, 149, 150, 161, 168, 192, 198, 209
electron microscopy, 371
electrons, 62, 149, 192, 193, 199
elucidation, 70, 81, 307, 332
emission, 219, 220
employment, 2, 56
emulsions, 75
endangered, 251
endocrine, 2, 14
endoscopy, 27
endosperm, 40, 41, 44, 47
endothelial cells, 3, 205, 362, 373, 375
energy, 39, 41, 149, 192, 196, 198, 310
engineering, 243, 376
enuresis, 236
environment, 6, 191, 197, 198, 199, 220, 221, 226
environmental effects, 223
environmental factors, 26, 28, 70, 71
environmental stress, 193, 206, 270
environments, 233, 359
enzymatic activity, 376, 382
enzyme(s), 7, 8, 9, 11, 14, 17, 29, 44, 74, 77, 90, 91, 94, 95, 100, 101, 107, 108, 109, 116, 139, 148, 149, 168, 177, 182, 185, 192, 197, 199, 202, 205, 206, 208, 270, 271, 275, 310, 350, 360, 362, 363, 364, 365, 368, 369, 370, 372, 376, 380
epidemiologic, 26, 32, 364
epidemiologic studies, 32
epidermis, 183, 190, 191, 193, 194, 195, 196, 207
epigenetic modification, 208
epinephrine, 14, 274
epithelial cells, 30, 267
epithelium, 30, 312
equilibrium, 134, 136, 137, 138
equipment, 134
erosion, 191, 353
erythrocyte sedimentation rate, 176
erythrocytes, 93, 349
erythropoietin, 273
ESI, 160, 285, 286, 307
esophageal cancer, 370
esophagus, 361, 364
ESR, 176, 183
essential fatty acids, 4, 223
ester, 19, 29, 47, 71, 95, 205, 252, 253, 254, 258, 262, 263, 271, 309, 327, 329, 332, 333, 335, 374
ester bonds, 48
estrogen, 193, 200, 236, 237, 376
ethanol, 13, 31, 43, 49, 109, 128, 134, 145, 149, 153, 233, 235, 237, 271, 272, 275, 287, 290, 298, 351, 368
ethyl acetate, 236, 237, 239, 268, 350
ethylene, 203
etiology, 25, 26, 28, 181
eukaryotic, 362
Europe, 26, 122, 123, 144, 187, 205, 240
European Union, 45
everyday life, 19
evidence, 4, 5, 28, 31, 32, 33, 148, 178, 223, 267, 332, 365, 370, 376, 378, 380, 382
evil, 123

evolution, 40
excision, 192, 203
excitability, 308
excitation, 16
excitotoxicity, 111
excretion, 275, 368
exercise, 78, 269
expectorant, 275
experimental condition, 43, 111
exploitation, 39, 241, 251
exposure, 3, 4, 27, 32, 191, 193, 194, 195, 272, 352, 359
external environment, 202
extracellular matrix, 191, 202
extraction, vii, 40, 47, 60, 74, 75, 121, 122, 123, 125, 128, 129, 130, 131, 132, 133, 134, 136, 137, 138, 140, 143, 144, 145, 146, 147, 148, 149, 153, 168, 184, 232, 233, 240, 290, 298, 311, 312
extrusion, 48
exudate, 347

**F**

facial muscles, 195
factories, 192
families, 5, 72, 73, 91, 97, 153, 203, 205, 381
farmers, 223
fasting, 2, 17, 19
fasting glucose, 15, 80, 81
fat, 2, 3, 10, 12, 41, 45, 78, 178, 191, 219, 222, 223, 240, 269, 272, 378
fatty acids, 9, 13, 16, 44, 125, 177, 192, 193, 219, 220, 222, 223, 241, 251, 255, 256, 257, 259
fear, 228
female rat, 183, 359
fermentation, 44, 46, 50
ferric ion, 76, 108
ferritin, 196, 366
ferrous ion, 63
fertilization, 364
fetus, 360
fever, 178, 309
fiber(s), 1, 5, 19, 39, 40, 41, 45, 193, 226, 233, 240, 378
fibroblasts, 100, 161, 191, 192, 195, 196, 199, 202, 205, 206
fibromyalgia, 184
fibrosarcoma, 375, 382
fibrosis, 272
filters, 199
filtration, 58, 269
financial support, 81
fish, 177
fission, 307
flame, 233
flavonoids, 1, 6, 8, 11, 13, 15, 18, 20, 29, 30, 39, 46, 56, 64, 70, 71, 74, 75, 76, 124, 125, 129, 149, 151, 152, 153, 160, 161, 162, 166, 167, 169, 179, 199, 203, 204, 207, 219, 220, 221, 222, 223, 228, 229, 234, 247, 251, 276, 357, 376, 377
flavonol, 71, 73, 125, 160, 161, 167, 317, 333, 349, 352
flavor, 40, 46, 47, 48, 247, 248
flora, 28, 383
flour, 39, 40, 43, 46, 47, 53
flowers, 71, 123, 124, 125, 128, 180, 259, 267, 317, 318
fluid, 88, 122, 128, 131, 134, 144, 145, 148, 149, 168, 181
fluid extract, 122, 128, 134, 144, 145, 148, 149, 168
folate, 10
folic acid, 69, 70
food additives, 151
Food and Drug Administration (FDA), 41, 45, 51
food industry, 56, 74, 122, 128
food poisoning, 333
food products, 41, 50
force, 65, 234, 238, 308
formation, 3, 4, 39, 46, 59, 63, 65, 89, 90, 91, 93, 101, 111, 115, 161, 177, 192, 194, 195, 196, 197, 200, 201, 202, 272, 308, 309, 331, 349, 359, 361, 362, 364, 374, 381
formula, 46, 59, 60, 89, 206
fragments, 193
France, 48, 123, 189, 357
free radicals, 3, 44, 56, 61, 62, 63, 64, 101, 108, 116, 148, 149, 151, 191, 196, 197, 198, 199, 208, 209, 273, 308, 311
freedom, 204
frequency of diabetes, 2
freshwater, 204
friction, 47
fructose, 9
fruits, vii, 9, 18, 30, 32, 41, 42, 45, 46, 50, 57, 69, 70, 71, 74, 75, 76, 77, 78, 79, 80, 81, 108, 109, 111, 116, 128, 147, 148, 150, 151, 152, 153,

154, 159, 162, 163, 168, 177, 200, 204, 228, 251, 267, 271, 274, 374, 376, 382
fucoxanthin, 380
fulminant hepatitis, 272
functional food, 40, 70, 184, 223, 271, 276
fungal infection, 372
fungus(i), 5, 6, 88, 90, 101, 108, 138, 139, 209, 349
fusion, 194

## G

GABA, 273
gallbladder disease, 182
ganglion, 116
gastric mucosa, 109
gastric ulcer, 123, 247, 251, 311
gastritis, 28, 176
gastrointestinal tract, 26, 31, 33, 77, 78, 94, 181
gel, 11, 202, 233, 269
gelatin zymography, 374
gene expression, 8, 32, 100, 148, 178, 183, 197, 205, 267, 268, 350
gene regulation, 197
genes, 8, 28, 32, 178, 195, 196, 197, 205, 361, 363, 366, 369
genetic background, 193
genome, 270
genomic instability, 194
genotype, 221, 223
genus, 12, 56, 72, 99, 122, 123, 125, 144, 204, 225, 228, 241, 247, 248, 250, 251, 252, 259, 261, 263, 265, 267, 275, 295, 296, 299, 306, 312, 317, 318, 323, 324, 330, 332, 333, 334, 349, 352, 353, 367, 370, 371, 372, 382
Germany, 123, 220, 235, 285
germination, 50
gill, 299
ginger, 153, 175, 178, 179, 185
gingivitis, 176, 296
ginseng, 16, 251, 293, 307, 382
gland, 360
glioma, 364, 369, 383
global demand, 352
glomerulus, 3
glucagon, 4
gluconeogenesis, 17
glucose, 2, 3, 4, 5, 7, 8, 9, 11, 13, 14, 15, 16, 17, 18, 19, 20, 41, 71, 79, 81, 116, 192, 194, 204, 232, 271, 299, 309, 332, 334, 350, 351
glucose oxidase, 116
glucose tolerance, 12, 310, 350
glucose tolerance test, 310, 350
glucoside, 9, 10, 15, 109, 111, 113, 160, 161, 162, 163, 164, 165, 166, 180, 200, 229, 233, 252, 253, 268, 317, 324, 325, 326, 327, 328, 329, 330, 332, 333, 334, 335
glucosinolates, 161, 361
GLUT, 8
GLUT4, 7, 350
glutamate, 94, 111
glutamic acid, 9
glutathione, 32, 48, 78, 94, 101, 107, 109, 148, 183, 184, 192, 208, 270, 275, 351
glycated LDL, 4
glycogen, 17, 269, 310
glycogenesis, 17
glycopeptides, 1, 6
glycoproteins, 382
glycosaminoglycans, 191, 195
glycoside, 11, 15, 30, 71, 76, 109, 111, 165, 167, 205, 207, 264, 311, 318
glycosylation, 69, 70, 167
grass, 39, 158, 227, 228, 381
grasses, 383
green alga, 90
growth, 6, 15, 27, 57, 60, 61, 77, 88, 139, 141, 191, 193, 195, 197, 220, 236, 237, 239, 268, 269, 273, 274, 307, 309, 349, 352, 353, 361, 362, 364, 366, 368, 370, 375, 376, 378, 380, 381, 383
growth arrest, 197
growth factor, 191, 193, 195, 237, 362, 376, 380
guanine, 193
guidelines, 44, 60

## H

habitats, 251
hair, 192, 228
hairless, 123
half-life, 193
harvesting, 123, 125
HDAC, 363, 366
healing, 121, 123, 202, 204
health, 2, 5, 6, 40, 41, 44, 45, 47, 48, 49, 51, 56, 63, 69, 70, 79, 81, 82, 87, 122, 147, 148, 152, 168, 175, 178, 179, 182, 219, 223, 240, 275, 276
health care, 5, 240

health condition, 175
health effects, 6, 69, 70, 122
health promotion, 69, 81, 82
health risks, 148
heart attack, 2, 87
heart disease, 3, 45, 51, 79, 175, 179, 247, 250
heat shock protein, 180, 196
Helicobacter pylori, 139
helium, 221
hematochezia, 26
hematocrit, 27
heme, 29, 31, 109, 192
heme degradation, 29
heme oxygenase, 31, 109, 192
hemicellulose, 44
hemisphere, 56, 376
hemodialysis, 78
hemoglobin, 309
hemorrhagic stroke, 177
hepatic fibrosis, 310
hepatic injury, 238
hepatitis, 176, 182, 310
hepatitis a, 251
hepatitis b, 272
hepatocarcinogenesis, 359, 367, 376
hepatocellular carcinoma, 374
hepatocytes, 102, 178, 182
hepatoma, 268, 366, 371, 374
hepatotoxicity, 111, 272, 310
heptane, 221
herbal medicine, 181, 184, 241, 242, 243, 292, 315
heterogeneity, 5, 6
hexane, 17
high density lipoprotein, 17, 351
high fat, 78, 271
high serum glucose, 2
hippocampus, 273
histamine, 234, 239, 275, 309, 311, 317, 324, 346, 347, 348
histology, 183, 228
histone, 31, 366
histone deacetylase, 179, 362, 363
histones, 192, 365
history, 40, 225, 240
HIV-1, 239
HIV-2, 239
HLA, 28
HO-1, 26, 29, 109, 192, 196
homeostasis, 2, 14, 198, 350
hormone(s), 2, 88, 191, 193, 236, 237, 376
hormones, 88, 191, 236, 376
host, 27, 352, 358
HPLC-UV, 159, 161
human body, 6, 70, 77, 81, 150, 199
human health, 6, 29, 31, 32, 150, 169, 220, 223
human leukemia cells, 362, 365, 366, 367, 381
human skin, 100, 202, 205, 208
hyaline, 47
hybrid, 237
hydrocarbons, 44, 93, 94, 95, 125, 259, 359
hydrogen, 4, 62, 76, 91, 100, 101, 109, 116, 149, 167, 193, 197, 204, 209, 220, 330, 349
hydrogen abstraction, 62
hydrogen peroxide, 4, 100, 101, 116, 167, 193, 197, 220, 349
hydrolysis, 128
hydroperoxides, 204
hydrophilicity, 124, 149
hydrophobicity, 65
hydroxyl, 4, 61, 150, 160, 161, 167, 197, 230, 257, 332, 334, 350
hydroxyl groups, 64, 98, 109, 114, 133, 203, 204, 348, 349
hyperactivity, 275
hypercholesterolemia, 80, 81, 351
hyperglycaemia, 4, 57
hyperglycemia, 2, 3, 4, 5, 14, 15, 79, 271, 309
hyperinsulinemia, 2
hyperlipidemia, 19
hyperplasia, 236, 309
hypertension, 178, 247, 250
hypertrophy, 240
hypodermis, 190, 191
hypotensive, 236
hypothermia, 296
hypothesis, 92, 197, 362
hypoxia, 107, 196, 272, 273, 308
hypoxia-inducible factor, 107


## I


IBD, 25, 26, 27, 28, 29, 30, 31, 32, 33, 181, 183
ibuprofen, 177
ICAM, 81
icon, 123
ideals, 5, 310
identification, vii, 40, 147, 148, 159, 160, 169, 228, 293, 358
IFN, 31, 237, 270, 272

IFNγ, 26
IL-8, 30, 32, 179, 195, 267
imbalances, 3
immortality, 108
immune function, 8, 237
immune response, 175, 190, 195, 198, 269, 383, 384
immune system, 28, 44, 175, 181, 191, 197, 239, 269, 367
immunity, 159, 181, 191
immunoglobulin(s), 176, 270
immunomodulatory, 247, 250, 270, 271, 317, 324, 347, 382
immunostimulatory, 276
immunosuppresants, 347
immunosuppression, 181, 195, 203, 367
immunotherapy, 367
improvements, 79
in vivo, 19, 29, 30, 32, 33, 45, 76, 77, 78, 93, 94, 100, 102, 107, 109, 115, 116, 139, 180, 199, 201, 202, 203, 204, 208, 239, 268, 272, 307, 309, 347, 351, 357, 361, 362, 363, 364, 366, 369, 370, 377, 378, 381, 383
incidence, 25, 26, 40, 357, 358, 359, 373, 376, 383
income, 4
India, 151, 152, 201, 220, 248, 367, 374
indirect effect, 77
individuals, 8, 363
inducer, 206, 353, 363, 372
induction, 78, 181, 182, 183, 184, 195, 205, 237, 270, 350, 353, 360, 361, 363, 364, 365, 366, 367, 371, 372, 384
inductor, 101
industrialization, 40
industries, 40, 88
industry, 88, 122, 225, 245, 247
infection, 176
inflammation, 3, 8, 26, 27, 28, 29, 30, 32, 79, 108, 175, 176, 177, 178, 179, 180, 181, 182, 183, 184, 185, 193, 194, 198, 199, 206, 207, 235, 267, 268, 273, 309, 310
inflammatory bowel disease, 25, 28, 33, 175, 176, 179, 181, 183
inflammatory disease, 32, 175, 176, 177, 179, 181, 200
inflammatory mediators, 180, 183
inflammatory responses, 27, 175, 177, 180, 197
infliximab, 27
ingestion, 29, 363, 373
ingredients, 19, 41, 47, 151, 152, 193, 198, 199, 01, 228, 232, 236, 247, 251, 267, 277
inhibitor, 125, 195, 200, 272, 366, 368, 369, 370, 372, 374, 375, 376
initiation, 62, 358, 359, 360, 362, 363, 366, 368, 380
injections, 221, 235, 382
injury(ies), 15, 18, 29, 31, 32, 101, 109, 175, 176, 202, 238, 270, 273, 284, 308, 310, 351
inositol, 48, 332
insects, 88
insertion, 228
insulin, 1, 2, 3, 4, 5, 7, 8, 9, 12, 13, 14, 15, 16, 17, 18, 19, 20, 44, 195, 271, 350, 380
insulin dependent diabetes mellitus (IDDM), 2
insulin resistance, 2, 3, 271, 350
insulin sensitivity, 271
insulin signaling, 8
integrity, 65, 191
intellectual property rights, 359
intercellular adhesion molecule, 205
interface, 65, 191
interferon, 31, 270
interferon gamma, 26
interferons, 175
interleukin-8, 267
interleukine, 203
interrelations, 358
intervention, 77, 78, 79, 80, 81, 180
intestine, 77, 94, 308
intracellular calcium, 364
ionization, 91, 161, 198
ionizing radiation, 148
ions, 6, 59, 62, 63, 64, 90, 202, 204, 307
IRC, 196
iron, 10, 63, 65, 202, 204
irradiation, 49
irritable bowel syndrome, 183
ischemia, 273, 308, 311
Islam, 35
isoflavonoids, 30
isolation, 159, 160, 169, 276, 358, 372
isomerization, 90, 91
isomers, 160, 161, 205, 349
isoprene, 87, 88, 97, 107, 114
issues, 359
Italy, 55, 57, 58, 60, 61, 83, 123, 153, 243

## J

Japan, 200, 247, 248, 251, 253, 257, 258, 277, 286, 296, 314, 318, 333, 353, 355, 372
jaundice, 182, 310
Java, 205
joint pain, 235
joint swelling, 184

## K

$K^+$, 236
kaempferol, 18, 71, 125, 160, 161, 162, 163, 165, 219, 220, 221, 222, 252, 254, 327, 328, 333, 349, 352
keratin, 197
keratinocytes, 190, 191, 192, 193, 195, 199, 202, 206, 207
keratosis, 193
ketones, 93
kidney, 57, 247, 248, 250, 275, 307, 309
kidney stones, 239
kinetics, 137, 138, 198
knots, 382
Korea, 247, 248, 251, 253, 257, 277, 296, 309, 310, 372

## L

lactate dehydrogenase, 269, 347
lactic acid, 269
lactoferrin, 27
lamella, 133
Langerhans cells, 190, 191, 193, 196
large intestine, 30
L-arginine, 95
laryngitis, 176
Latin America, 5
LC-MS, 129
LC-MS/MS, 129
LDL, 4, 15, 44, 75, 79, 80, 93, 101, 149, 167, 168, 238, 271, 351
lead, 16, 20, 27, 50, 64, 91, 92, 114, 128, 137, 138, 181, 194, 196, 197, 308, 370, 373
learning, 273
lesions, 107, 367, 368
leucine, 9
leucocyte, 184
leukemia, 225, 274, 348, 362, 366, 367, 370, 371, 375, 381, 382
leukocytes, 176, 180
levator, 311
liberation, 274
life expectancy, 2
lifestyle changes, 3
lifetime, 41
ligand, 376
light, 122, 123, 125, 194, 200
light scattering, 283, 284, 293
lignans, 41, 44, 56, 228, 230, 237, 247, 248, 251, 276, 357, 378, 379
lignin, 44
linear polymers, 41
linoleic acid, 7, 13, 18, 93, 100, 153, 160, 161, 240, 252, 258
lipid metabolism, 4, 8, 272
lipid oxidation, 63, 65, 75, 311
lipid peroxidation, 3, 30, 31, 56, 60, 62, 63, 64, 65, 78, 92, 93, 94, 95, 100, 101, 107, 109, 111, 160, 161, 182, 183, 184, 195, 196, 197, 200, 270, 308, 309, 310
lipids, 3, 6, 15, 40, 78, 80, 148, 182, 189, 191, 192, 194, 197, 201, 209
lipoproteins, 4, 148, 181
liposomes, 60, 75, 94, 116, 202
liquid chromatography, 50, 161, 219, 220, 293
Listeria monocytogenes, 140, 141, 142, 143
liver, 2, 15, 94, 101, 107, 176, 228, 238, 247, 248, 250, 270, 271, 308, 309, 350, 351, 360, 364, 367
liver cancer, 182
liver cells, 182, 310, 375
liver damage, 272
liver disease, 182
liver spots, 191
localization, 31
locus, 184
longevity, 190, 198, 207, 208, 209
low density lipoproteins (LDL), 4
low risk, 148, 168
low-density lipoprotein, 79, 93, 149
low-grade inflammation, 183
LTB4, 26, 30
lumen, 31
lung cancer, 268, 307, 363, 366, 375, 379
lupus, 176, 309
lutein, 8, 17, 44, 69, 70, 114, 115, 116, 149, 380
lycopene, 114, 115, 116, 203, 380
lymph node, 347
lymphocytes, 31, 76, 78, 83, 270, 307, 347

lymphoma, 27, 361, 367, 368, 383
lysine, 31, 192, 194, 241

## M

machinery, 194, 368, 375
macromolecules, 148, 366
macronutrients, 10, 41
macrophages, 3, 4, 178, 180, 183, 185, 234, 267, 268, 269, 271, 307, 369, 382, 383
magnesium, 10, 39, 90
magnitude, 371
Maillard reaction, 4, 47
majority, 28, 123, 200
malaria, 88, 352, 353
Malaysia, 248
malignant cells, 76
malnutrition, 5
mammal, 196
mammalian cells, 139, 368
mammals, 17, 90
man, 235, 274
management, 1, 3, 4, 5, 6, 12, 19, 20, 27, 383
manganese, 10, 11, 201
manipulation, 7
mannitol, 257
manufacturing, 162
manure, 223
marketing, 201
marrow, 157
mass, 27, 124, 129, 131, 132, 134, 136, 137, 138, 161, 259
mass spectrometry, 161
mast cells, 267, 275, 347, 348, 349
materials, vii, 45, 58, 122, 128, 132, 134, 148, 150, 153, 169, 200
matrix, 47, 77, 133, 191, 193, 194, 195, 203, 268, 377
matrix metalloproteinase, 77, 193, 268
matter, 12, 42, 128, 145, 192, 196
MCP, 81, 182
MCP-1, 81
measurement, 63, 76, 108, 149, 153, 169, 287, 290, 347
measurements, 167
medical, vii, 5, 122, 178, 358, 367
medicine, vii, 1, 5, 29, 55, 56, 57, 97, 108, 116, 121, 123, 151, 159, 162, 177, 182, 225, 228, 240, 243, 245, 251, 276, 284, 292, 295, 296, 307, 309, 310, 312, 313, 314, 317, 318, 367, 372, 376, 394
Mediterranean, 57, 177, 182
MEK, 181
melanin, 194, 205, 208
melanoma, 77, 382
melatonin, 48
mellitus, 1, 2, 80, 310, 350
melon, 9, 156
membranes, 132, 139, 182, 183, 192, 200, 207
memory, 180, 273
menstruation, 284
mercury, 253
meridian, 247, 248
mesenchyme, 191
messages, 192
messengers, 196
metabolic disorders, 2
metabolic disturbances, 8
metabolic pathways, 368
metabolic syndrome, vii, 78, 79, 80, 81, 84
metabolism, 1, 6, 7, 8, 28, 33, 70, 76, 77, 81, 111, 148, 182, 185, 191, 193, 199, 202, 236, 273, 275, 361, 362, 368, 380, 381
metabolites, 1, 6, 7, 9, 15, 28, 56, 65, 77, 88, 102, 181, 204, 207, 357, 361, 363
metabolized, 77, 101
metabolizing, 7, 380
metal ions, 56, 63
metalloproteinase, 374
metals, 149, 202
metastasis, 77, 269, 363, 376
meter, 220
metformin, 1, 4, 5, 6
methanol, 55, 57, 58, 59, 62, 63, 64, 65, 111, 133, 134, 143, 146, 149, 159, 161, 162, 220, 236, 268, 270, 273, 285, 286, 290, 298
methodology, 42
methyl group, 125, 197, 207
methyl groups, 207
methylation, 167, 197, 349, 377
Mexico, 49
$Mg^{2+}$, 90
MHC, 237
mice, 8, 9, 16, 28, 29, 30, 32, 94, 102, 107, 108, 111, 180, 182, 234, 235, 237, 238, 269, 270, 271, 272, 274, 307, 308, 309, 311, 347, 350, 361, 364, 366, 370, 371, 379, 381, 383
microbiota, 27, 44, 378
microcirculation, 201, 205

micronucleus, 76
micronutrients, 10, 29, 39, 175
microorganism(s), 27, 56, 57, 139, 140, 142, 203, 274
microscopy, 349, 371
microsomes, 100, 309
microstructure, 228
middle lamella, 132, 133
migrant populations, 26
migration, 179, 182, 184, 290, 291, 309, 361, 362, 377
Ministry of Education, 81, 225, 283, 295
Missouri, 312
mitochondria, 3, 100, 110, 189, 191, 192, 207, 308, 366, 372
mitochondrial DNA, 192
mitogen, 8, 26, 178, 180, 181, 182, 267, 271, 369, 376
mitogens, 270
mitosis, 368
mixing, 47, 59
MMP, 195, 196, 205, 268, 296, 369, 374, 378
MMP-9, 374, 378
MMPs, 193, 195, 196, 198, 205, 369
model system, 93, 167, 201
modelling, 134, 136, 145, 146
models, 7, 16, 27, 28, 30, 31, 32, 33, 75, 97, 100, 109, 134, 179, 180, 183, 199, 311, 358, 360, 362, 363, 364, 377, 382, 383
moderate activity, 274
modifications, 27, 91, 116, 133, 191, 192, 198, 202, 208, 349, 363, 377
moisture, 12, 48
moisture content, 48
molds, 349
molecular biology, 276
molecular oxygen, 3, 194
molecular structure, 203
molecular weight, 9, 31, 48, 64, 160, 182, 232, 312
molecules, 1, 3, 5, 6, 20, 25, 28, 30, 32, 41, 59, 90, 97, 176, 184, 190, 192, 194, 196, 198, 199, 200, 203, 204, 207, 208, 244, 312, 357, 358, 359, 364, 368, 372, 374, 375, 376, 378, 380, 381, 383
Mongolia, 240, 248
monoclonal antibody, 27
monocyte chemoattractant protein, 182
monomers, 17, 161
monosaccharide, 287, 311
monoterpenoids, 380, 381
monounsaturated fatty acids, 177, 373
morbidity, 2, 26, 27
morning stiffness, 184
morphology, 228, 368
mortality, 2, 26, 31, 267, 308
mosaic, 197
motif, 87, 89, 197
mRNA, 8, 31, 268, 307, 350, 376, 377
mtDNA, 192, 194, 203
mucosa, 29, 30, 31, 45, 183, 209, 271, 360
multi-factorial approach, 3
multiple myeloma, 376
multiples, 90, 298
mutagen, 238
mutagenesis, 92, 275
mutant, 237, 307, 311
mutation, 238
myocardial infarction, 80, 81, 259, 308
myocardial ischemia, 273, 308, 311

## N

$Na^+$, 236
NADH, 100, 192, 193
natural compound, 115, 208, 371
natural habitats, 359
natural killer cell, 268
necrosis, 25, 373
negative effects, 78
nematode, 198, 208, 269
Nepal, 248
nephritis, 176, 309
nephropathy, 2, 3
nerve, 7
nerve growth factor, 238
nervous system, 247, 250, 295, 312
nervousness, 16
Netherlands, 172
neuralgia, 296
neuroblastoma, 101, 275, 365
neurodegenerative diseases, 4, 32, 87, 92, 147
neurological disease, 182
neurons, 116
neuropathy, 2, 3
neurotransmitter, 125, 273
neurotransmitters, 193
neutrophils, 3, 181
NH2, 365
nicotinamide, 125, 192, 309

nitrates, 30
nitric oxide, 3, 26, 30, 81, 101, 179, 181, 196, 197, 198, 200, 267, 372
nitric oxide synthase, 26, 30, 196, 200, 372
nitrogen, 94, 197, 269
nitrogen dioxide, 116
nitrosamines, 362, 363
NK cells, 269, 367, 382
NMR, 161, 162, 259, 285, 286, 333
nonane, 124
non-enzymatic antioxidants, 192
non-insulin dependent diabetes, 2, 80, 81
non-polar, 149
non-steroidal anti-inflammatory drugs (NSAIDs), 177, 234
norepinephrine, 273
North America, 293, 295, 296
Nrf2, 95, 100, 109, 202, 206, 270, 363
nuclear magnetic resonance, 358, 384
nucleic acid, 310
nucleic acid synthesis, 368
nucleophiles, 56
nucleus, 207, 208, 229, 233, 299, 334, 349, 369
null, 49
nutrient, 39, 45, 51, 152, 219
nutrients, 6, 8, 10, 17, 45, 50, 74, 78, 191, 219, 334, 380
nutrition, 27, 29, 42, 51, 191, 223, 240
nutritional status, 27

## O

obesity, 9, 40, 57, 79, 175, 176, 177, 184, 271, 272
observed behavior, 6
occlusion, 273
Oceania, 295, 296
octane, 124
ocular diseases, 116
ODS, 283, 286, 287, 290
oedema, 180
oesophageal, 367
oil, 8, 11, 12, 13, 40, 57, 93, 123, 124, 125, 128, 134, 137, 138, 139, 145, 146, 154, 159, 162, 179, 182, 204, 259, 364, 367, 380, 381
oleic acid, 18, 101, 219, 223, 240, 257, 258
oligomers, 48, 200, 372
olive oil, 177, 373, 374
operating costs, 74
opportunities, 2
optical density, 59
optimization, 79, 122, 134, 145, 146, 312
oral hypoglycemic agents, 5
organ, 94, 198
organelle(s), 191, 192
organic solvents, 128, 149
organism, 6, 147, 237, 310
organs, 88, 189
oriental medicine, 251
ornithine, 372
osmosis, 132, 133
osmotic stress, 196
osteoarthritis, 175, 179, 184, 296
osteoclastogenesis, 376
osteoporosis, 273
ototoxicity, 274
ovarian cancer, 360
overharvesting, 251
overproduction, 87
overtraining, 310
overweight, 8
ox, 44, 79, 93, 95, 161, 259, 353
oxalate, 239
oxidation, 9, 56, 78, 80, 91, 93, 100, 148, 149, 150, 152, 160, 161, 192, 193, 196, 197, 202
oxidative damage, 3, 56, 63, 65, 78, 116, 148, 150, 197, 308, 310
oxidative reaction, 74, 204
oxidative stress, 3, 15, 56, 63, 78, 79, 87, 88, 92, 94, 95, 100, 101, 102, 107, 109, 111, 115, 116, 147, 148, 150, 152, 189, 192, 194, 197, 199, 208, 270, 272, 351
oxygen, 3, 30, 61, 62, 71, 76, 108, 115, 116, 147, 148, 149, 150, 169, 192, 194, 195, 196, 197, 203, 204, 206, 238, 268, 273, 308, 311, 349
oxygen consumption, 3, 308
oxygen consumption rate, 308
ozonation, 196
ozone, 194, 195, 196

## P

p53, 197, 269, 307, 361, 362, 365, 369
Pacific, 2, 224
pain, 26, 27, 94, 176, 177, 179, 183, 184, 189, 228, 234, 235, 308, 309, 310
palliative, 25
pancreas, 14, 180, 183, 310
pancreatic cancer, 220, 223, 367
pancreatitis, 180, 183, 184

parallel, 351, 368, 371
parasite, 88, 352
parasites, 6, 90
parents, 5
partial differential equations, 134
partition, 239
pasta, 40, 41, 47, 48
pasteurization, 74
patents, 240
pathogenesis, 3, 4, 28, 29, 31, 92, 111, 176, 177
pathology, 3, 27
pathophysiological, 32, 87
pathophysiology, 28, 148
pathways, 3, 8, 13, 90, 109, 176, 180, 185, 192, 198, 202, 267, 295, 362, 363, 369, 376, 377
PCA, 347
PCR, 307
penicillin, 139, 273
peptidase, 4
peptide(s), 4, 16, 19, 275, 309
peptides, 16
perfusion, 366
periodontal disease, 296
peripheral blood, 270, 382
peripheral vascular disease, 2
peritonitis, 184
permeability, 14, 176, 184, 206, 309, 373
peroxidation, 56, 63, 64, 65, 93, 94, 100, 101, 102, 167, 168, 183, 351
peroxynitrite, 150
pests, 6
petroleum, 237, 239
pH, 59, 61, 159
phagocyte, 3
phagocytic cells, 353
phagocytosis, 269
pharmaceutical, 5, 20, 88, 121, 122, 123, 209, 225, 241, 247
pharmaceuticals, 5, 128, 358
pharmacokinetics, 360, 368
pharmacological research, 274
pharmacology, 247, 251, 312, 315
pharmacotherapy, 3
phenol, 32, 82, 94, 97, 107, 108, 230
phenolic compounds, 18, 39, 40, 42, 46, 48, 49, 50, 56, 62, 64, 65, 70, 74, 75, 76, 78, 81, 133, 149, 151, 152, 153, 159, 160, 161, 162, 166, 169, 180, 378
phenylalanine, 9, 167
Philippines, 248, 359
phosphate(s), 9, 17, 48, 59, 60, 89, 90, 91, 192, 193, 207, 239, 309, 380
phosphorus, 10
phosphorylation, 30, 31, 139, 178, 179, 180, 192, 268, 349, 362, 366, 369
photosensitizers, 194
phototoxicity, 139
physical activity, 3, 177
physical characteristics, 149
physical phenomena, 134
physicochemical characteristics, 199
physiological, 52, 198
physiological mechanisms, 44
physiology, 196
phytosterols, 11, 69, 70, 88
phytotherapy, 1, 243
PI3K, 8, 31, 109, 178, 180, 197, 362, 375, 377
PI3K/AKT, 362, 377
pigmentation, 32, 190
pilot study, 183
pistil, 123
placebo, 7, 8, 79, 80, 183, 235, 360, 363, 367
plant sterols, 44
plant type, 331, 359
plaque, 3
plasma levels, 29, 30, 79, 80
plasma membrane, 65, 365
plasminogen, 378
platelet aggregation, 7, 81, 235, 274
platelet count, 27
platelets, 182
pleurisy, 184
pneumonia, 57
Poland, 48, 123
polar, 149, 181, 298
polarity, 133, 236
pollinators, 6, 88
pollution, 151, 194
polymerase chain reaction, 307
polymerization, 48, 71, 202
polymers, 7, 31, 71, 88
polypeptide, 9, 232
polyphenols, 7, 8, 15, 19, 29, 31, 32, 33, 45, 56, 57, 74, 97, 148, 160, 162, 168, 180, 199, 201, 202, 203, 204, 357, 375
polyps, 363
polysaccharide(s), 1, 6, 7, 16, 18, 28, 45, 65, 133, 181, 228, 232, 234, 236, 237, 247, 248, 251, 254, 269, 271, 272, 383
polystyrene, 60

population, 5, 40, 81, 202
population group, 78
Portugal, 162
positive correlation, 64, 79, 148, 168
potassium, 10, 59, 238
potato, 152, 200
poultry, 177, 245
predators, 6
preparation, 49, 74, 140, 183, 185, 234, 237, 240, 367
preservation, 50, 74, 359
prevention, 4, 5, 6, 20, 32, 62, 69, 77, 81, 88, 93, 111, 147, 150, 178, 271, 272, 273, 310, 357, 358, 359, 363, 368, 371, 377, 378, 379
primary cells, 380
primary tumor, 77
principles, 6, 14, 16, 181, 185, 352, 353
probands, 78
probiotics, 209
progenitor cells, 194
pro-inflammatory, 3, 29, 32, 177, 179, 180, 181, 184, 185, 196, 197, 198, 205, 272
project, 81
proliferation, 77, 193, 196, 197, 236, 237, 239, 269, 296, 307, 308, 310, 324, 347, 352, 364, 366, 367, 369, 371, 375, 376, 378, 382
proline, 13, 192
promoter, 198, 359
propagation, 195
prophylactic, 28, 251, 272
prostaglandins, 177, 182, 311
prostate cancer, 77, 361, 362, 363, 366, 375, 376, 377, 380, 383
proteasome, 194, 366
protection, 29, 32, 56, 63, 65, 101, 109, 111, 168, 179, 189, 200, 201, 203, 207, 209, 272, 296, 310, 360, 361, 363, 364
protective mechanisms, 109
protective role, 109, 115, 116, 310
protein kinase C, 139, 183, 196, 365
protein kinases, 178, 180, 267, 271, 368
protein oxidation, 92
protein synthesis, 310
proteins, 4, 6, 9, 17, 29, 40, 48, 56, 133, 148, 175, 176, 179, 182, 189, 191, 194, 195, 196, 197, 201, 208, 209, 232, 269, 360, 361, 365, 366, 375, 380, 382, 383
proteoglycans, 191
Pseudomonas aeruginosa, 60, 90, 140, 141, 142, 143, 144
psoriasis, 175, 225
public health, 2, 25
public health problem, 2
public interest, 5
pulp, 10, 155, 156, 159, 160
PUMA, 361
purification, vii, 298
purity, 122, 128, 233, 299
pyoderma gangrenosum, 183
pyrimidine, 194, 203, 205
pyrophosphate, 88, 90, 91, 92, 365


## Q


quality control, 283, 306
quality of life, 2, 382
quantification, vii, 42, 169, 283
quercetin, 7, 10, 11, 18, 30, 71, 125, 129, 145, 152, 160, 161, 163, 164, 165, 167, 219, 220, 221, 222, 223, 252, 253, 328, 349, 352, 376, 377
quinone(s), 357, 375, 376


## R


race, 220
racemization, 167
radiation, 194, 196, 200, 203
radicals, 4, 30, 43, 61, 63, 76, 78, 93, 94, 95, 101, 108, 149, 152, 167, 192, 194, 196, 197, 198, 200, 238, 273
radiotherapy, 370
Ramadan, 187
raw materials, 298
reactant, 176
reaction mechanism, 91
reaction medium, 150
reaction rate, 204
reaction rate constants, 204
reactions, 46, 50, 63, 74, 90, 139, 150, 181, 196, 197, 209, 239, 346, 348
reactive groups, 109
reactive oxygen, 3, 32, 56, 92, 94, 109, 147, 191, 208, 268, 270, 275, 351, 363, 365
reactivity, 63, 366
receptors, 7, 19, 78, 193, 195, 237, 271, 376
recognition, 4
recombination, 192
recovery, 208, 283, 291, 292, 306
rectum, 26

red blood cells, 232
red wine, 31
reducing sugars, 4, 194
regeneration, 7, 14, 194, 202
regions of the world, 12, 26, 153
regression, 290, 360, 380
regression equation, 287
regulations, 41, 45, 296
rehabilitation, 243
rehydration, 133
relevance, 88, 199
relief, 177, 179, 368
remission, 27, 183
renal failure, 2
repair, 14, 192, 194, 199, 203, 205, 209
repetitions, 192
replication, 192, 194
repression, 365
reproduction, 6
requirements, 6, 41, 348, 350
researchers, vii, 1, 45, 88, 97, 150, 180, 228, 234, 237, 238, 241, 251, 276, 378
residues, 151, 192, 196, 331
resins, 12, 57, 88, 182
resistance, 6, 206, 207, 208, 239, 269, 271, 350, 370, 371, 378
resolution, 287, 290
resources, 147, 160, 225, 241, 248, 276, 312, 359
respiration, 308
respiratory distress syndrome, 175
respiratory problems, 181
respiratory syncytial virus, 239
response, 7, 44, 81, 130, 133, 176, 178, 180, 182, 195, 196, 206, 234, 270, 284, 372
restoration, 197
resveratrol, 7, 8, 32, 33, 199, 207, 372, 373
reticulum, 349
retina, 3, 116
retinoblastoma, 269
retinopathy, 2, 3, 4, 13
rheumatoid arthritis, 175, 176, 179, 183, 184, 235, 275, 284, 309
rhizome, 201
riboflavin, 194
ribose, 309
rings, 114, 162, 207, 371
risk(s), 2, 3, 5, 27, 44, 45, 78, 79, 81, 132, 147, 152, 175, 177, 202, 363, 378, 379, 380
risk factors, 2, 3, 79, 81
rodents, 358, 359, 367
Romania, 123
room temperature, 58, 59, 60, 238
root(s), 16, 97, 100, 101, 107, 108, 111, 123, 134, 145, 157, 158, 175, 180, 200, 227, 228, 230, 232, 233, 234, 236, 237, 238, 239, 240, 247, 248, 251, 259, 267, 268, 271, 275, 276, 283, 284, 285, 287, 293, 292, 293, 296, 300, 301, 302, 303, 304, 305, 306, 309, 310, 311, 329, 371, 372, 375, 383
rosacea, 206
routes, 89, 91
rubber, 88
rural areas, 196
Russia, 247, 248, 251

## S

safety, 5, 28
*Salmonella*, 90, 140, 141, 142, 143, 274, 374
saponin, 6, 9, 111, 271, 283, 293, 298, 307, 310
sarcoidosis, 175
SAS, 221
saturated fat, 45, 177, 223, 269
scavenger receptor pathway, 4
scavengers, 56, 95, 97, 167, 200, 203
science, 29, 34, 51, 150, 242, 243, 245
scientific investigations, 20
seafood, 177
sebum, 191, 197
Second World, 29
secrete, 275
secretion, 2, 6, 8, 13, 16, 20, 175, 179, 180, 185, 195, 196, 235, 238, 267, 351, 362, 378, 382
sedative(s), 123, 159, 247, 250, 251
sedimentation, 27, 60, 61, 240
seed, 10, 11, 19, 50, 71, 76, 83, 88, 108, 128, 151, 159, 201, 204, 207, 208, 220, 221, 223, 350, 378
seedling(s), 9, 61, 220
selectivity, 77, 128, 352
selenium, 10, 69, 70, 253, 364
seminal vesicle, 311
senescence, 191, 197, 204, 318, 323, 335
sensations, 177
senses, 239
sensitivity, 139, 140, 197, 292, 359
sepal, 123, 318, 323, 328, 329, 333, 335
septum, 184
Serbia, 69, 81, 121, 123, 144
serine, 9, 362, 376

serotonin, 125
serum, 2, 4, 13, 14, 15, 16, 17, 18, 44, 78, 81, 101, 179, 183, 237, 269, 270, 271, 272, 309, 310, 311, 350, 351
serum albumin, 309
severe stress, 199
sex hormones, 237
shape, 48, 228
shear, 48
shock, 179, 196, 267, 274
shoot(s), 207, 295, 296
shortage, 5, 235
showing, 19, 40, 45, 46, 49, 74, 75, 79, 80, 83, 93, 100, 283, 363, 364, 368, 377, 382
shrubs, 56, 226, 296
Siberia, 207, 253, 257
side chain, 4, 44, 167, 207, 332, 334, 335
side effects, 5, 25, 177, 178, 181, 209, 270
signal transduction, 349, 377
signaling pathway, 8, 25, 77, 195, 196, 197, 362, 374
signals, 176, 191, 197
significant morbidity, 2, 27
signs, 28, 29, 190, 205, 209
simulation, 134, 136
skeleton, 71, 168, 203, 259, 332, 348, 376
skin, 123, 182, 183, 189, 190, 191, 192, 193, 194, 195, 196, 197, 198, 199, 200, 201, 203, 204, 205, 206, 207, 208, 209, 228, 367, 368, 377, 379
skin cancer, 202, 220, 223, 372
skin diseases, 66
sleep deprivation, 273
small intestine, 7, 28, 44, 271
smog, 196
smoking, 3, 79, 148
smooth muscle cells, 109, 183
social costs, 25
social image, 209
society, 2
sodium, 10, 25, 28, 29, 109, 207, 221, 275, 308
software, 286
solid phase, 134, 137, 138
solubility, 125, 129, 136, 139, 149
solution, 58, 59, 60, 61, 63, 228, 287, 295, 353
solvation, 132
solvents, 125, 143, 149, 285, 290
somatic cell, 192
South Africa, 155, 156, 373, 383
Soviet Union, 220, 223
Spain, 87, 175
specialists, vii
spectrophotometry, 161
spectroscopy, 161, 162, 219, 220, 259, 358
spleen, 270, 276
sponge, 311
Sprague-Dawley rats, 350
sprouting, 50
squamous cell carcinoma, 369
stability, 40, 56, 192, 206
stabilization, 349
stable radicals, 199
standard deviation, 58, 59, 60, 130, 133, 290
standardization, 150, 169
starch, 40, 41, 44, 46
stasis, 247, 250
state(s), 29, 78, 80, 139, 151, 176, 197, 366
statin, 80
stem cells, 194
sterile, 221
steroids, 6, 9, 10, 181, 236
sterols, 6, 15, 41, 44, 201, 228
stimulation, 12, 44, 202, 203, 237, 361
stimulus, 40
stock, 287
stomach, 44, 77, 364
storage, 5, 40, 69, 70, 74, 75, 220
stress, 3, 93, 100, 107, 109, 111, 147, 148, 152, 189, 191, 192, 193, 195, 196, 197, 198, 199, 202, 207, 208, 209, 251, 269, 270, 272
stress response, 202, 272
stressors, 199, 208
striatum, 273
strictures, 26
stroke, vii, 2, 79
stroma, 236
structural characteristics, 97
structural modifications, 209
style, 177
subacute, 182
subgroups, 377
substitution(s), 64, 167, 168, 330
substrate(s), 8, 17, 28, 56, 93, 150
sugarcane, 234
sulfate, 25, 28, 29
sulfur, 180, 361, 364, 366
sunspots, 209
supplementation, 31, 32, 79, 80, 81, 178, 203, 238, 273, 379
suppository, 240

suppression, 30, 31, 178, 180, 362, 364, 366, 369
surface area, 189
surface layer, 191, 196
survival, 269, 362, 369, 376, 383
susceptibility, 146
sustainable development, 201
Sweden, 41, 45
swelling, 132, 133, 176, 179, 308
symmetry, 290
symptomology, 183
symptoms, 1, 5, 26, 27, 179, 181, 206, 228, 310, 333
synaptic plasticity, 273
syndrome, 16, 79, 80, 209
synergistic effect, 41, 308, 309
synthesis, 2, 7, 8, 9, 12, 29, 89, 90, 100, 182, 189, 192, 195, 201, 205, 208, 269, 271, 275, 310, 311, 369
systolic blood pressure, 80

## T

T cell, 26, 31, 139, 195, 237, 270
T lymphocytes, 190, 237, 347
Taiwan, 227, 318, 334
tannins, 10, 11, 13, 15, 18, 48, 56, 64, 69, 70, 71, 125, 130, 160, 168, 199, 201, 204, 252, 357, 377, 378
target, 2, 8, 29, 33, 176, 178, 189, 192, 194, 195, 198, 199, 273, 290, 358, 360, 363, 368, 373, 376, 383
target organs, 360
TCC, 60, 140, 141, 142, 143, 144
TCR, 26, 31
techniques, 71, 128, 133, 149, 159, 169, 307, 358
technology(ies), 48, 50, 51, 74, 122, 128, 232, 242, 243, 245, 276, 298
teeth, 226
telomere, 194
temperature, 48, 74, 125, 129, 132, 134, 136, 137, 138, 149, 189, 240, 268, 286, 290, 298
template molecules, 2
tension, 115, 202
terpenes, 1, 11, 13, 87, 88, 139
testing, 27, 33, 60, 150, 169, 199, 359
testosterone, 193
tetrahydrofuran, 230, 371
texture, 46, 47, 48, 209
TGF, 269
Thailand, 155, 248, 258
therapeutic agents, vii, 28
therapeutic approaches, 88
therapeutic benefits, 177
therapeutic effects, 7, 235, 296
therapeutic use, 179
therapy, 1, 3, 5, 17, 25, 27, 28, 80, 176, 178, 181, 182, 183, 271, 272, 367
thermal degradation, 128
thermoregulation, 189
thiazolidinediones, 4
threats, 191
threonine, 9, 192, 362
thrombosis, 274
thromboxanes, 177, 182
thrombus, 235
thymine, 194
thyroid, 15
Tibet, 241, 245
TIMP, 195
TIMP-1, 195
tissue, 3, 9, 31, 32, 76, 115, 123, 132, 133, 138, 177, 190, 191, 194, 198, 220, 236, 239, 269, 272, 308, 309, 310, 311, 363, 369, 377, 380
TNF, 25, 29, 30, 376
TNF-alpha (TNF-α), 16, 27, 31, 179, 180, 181, 182, 183, 184, 195, 198, 235, 238, 267, 268, 272, 274, 275, 307, 348
tobacco, 362
tobacco smoke, 195
tocopherols, 44, 48, 192, 201, 207
tonic, 14, 251, 296
tonsillitis, 235
torsion, 234, 308
total cholesterol, 7, 15, 79, 80, 238, 351
toxic effect, 358, 368, 383
toxic substances, 94, 101, 272
toxicity, 27, 111, 148, 238, 268, 308
TPA, 268, 368
trace elements, 4, 11, 234, 241
trade, 205
training, 308, 310
traits, 228
transcription, 8, 29, 32, 107, 109, 181, 208, 235, 268, 362, 365, 368
transcription factors, 183, 195, 196, 198, 360, 369
transcripts, 30
transducer, 362
transduction, 192, 376
transformation(s), 77, 194, 236, 268, 369
transforming growth factor, 202, 380

transition metal, 62, 63, 65, 76
translation, 362
translocation, 139, 179, 180, 208, 363, 366
transmission, 348
transport, 7, 8, 16, 19, 139, 191, 192, 271
trauma, 176
trial, 7, 79, 80, 183, 363, 367
triggers, 27, 365, 369
triglycerides, 15, 17, 79, 80, 146, 271
trypsin, 183, 267
tryptophan, 49, 165, 241, 269, 273
tuberculosis, 57, 90
tumor, 16, 77, 115, 239, 247, 250, 267, 295, 312, 348, 351, 359, 360, 361, 362, 363, 367, 368, 369, 372, 373, 375, 377, 382, 384, 394
tumor cells, 237, 269, 307, 353, 364, 366, 378, 380
tumor development, 363, 372
tumor growth, 367
tumor metastasis, 269
tumor necrosis factor, 16, 267, 348, 351, 382
tumor progression, 77
tumorigenesis, 360, 361, 362
tumors, 88, 179, 307, 361, 363, 367, 373, 376, 379, 380, 381, 383, 384
Turkey, 55, 57, 66
turnover, 14, 197
type 2 diabetes, 2, 39, 270, 272, 310
tyrosine, 272, 348, 349
tyrosine hydroxylase, 273

## U

ulcerative colitis, 25, 26, 32, 33, 176, 183
ultrasound, 132, 133, 142, 143, 145, 149
UN, 239, 269
underlying mechanisms, 3
uniform, 161, 351
United Kingdom (UK), 20, 41, 57, 66
United States (USA), 41, 51, 67, 76, 82, 123, 144, 146, 155, 210, 211, 216, 220, 223, 285, 386, 389, 391
urban, 196
urea, 269
uric acid, 239
urinary retention, 236
urine, 236, 378
urokinase, 378
USDA, 51, 71, 82, 219, 220, 223
uterine cancer, 360
UV, 58, 59, 161, 162, 190, 191, 192, 193, 195, 199, 203, 204, 206, 209
UV radiation, 194
UVA irradiation, 205

## V

vaccine, 352
validation, 283
valine, 9
vanadium, 11
variables, 290
variations, 30, 189
varieties, 42, 57, 69, 153, 225, 248, 295, 318
vascular endothelial growth factor (VEGF), 196, 362
vascular system, 191
vasculature, 373
vasodilator, 238
VCAM, 81
vegetables, vii, 32, 41, 45, 116, 147, 148, 150, 151, 152, 153, 154, 159, 161, 162, 163, 168, 177, 204, 240, 241, 296, 361, 374, 376
vehicles, 202
vein, 3, 362, 375
very low density lipoprotein, 17
vessels, 4, 238
vibration, 133
Vietnam, 248, 258
viral infection, 139
viruses, 175
vitamin A, 69, 70, 201, 203, 234, 379
vitamin C, 10, 17, 18, 45, 74, 75, 125, 196, 234
vitamin C, 74, 240
vitamin D, 189, 193
vitamin E, 7, 39, 40, 41, 44, 182, 196, 207
vitamin K, 10
vitamins, 4, 6, 10, 11, 12, 29, 39, 41, 48, 69, 70, 148, 168, 177, 199, 201, 334
vitiligo, 202
VLDL, 351
vomiting, 183

## W

walking, 184
Washington, 67, 187, 277
waste, 6, 267
wastewater, 49

water, 46, 47, 49, 61, 65, 125, 133, 139, 143, 148, 149, 153, 159, 161, 168, 172, 191, 200, 207, 220, 234, 235, 236, 237, 238, 239, 240, 269, 270, 272, 285, 286, 287, 290, 298, 299, 350, 351, 352, 377
water absorption, 46
wavelengths, 194
wealth, vii, 358
weight control, 3
weight gain, 8, 271
weight loss, 26, 29, 183, 184
well-being, 27, 209
wells, 61
Western countries, 26
wheat germ, 237
white blood cell count, 176
white blood cells, 176, 181, 235
WHO, 2, 5, 169
windows, 123
withdrawal, 197
wood, 57, 111
workers, 5, 131, 134
World War I, 358
worldwide, 1, 2, 26, 41, 56, 152, 153, 178, 225, 247, 352, 359
wound healing, 32, 139, 159

## X

xanthones, 15, 124, 125
xenografts, 307, 361, 364, 366, 369, 383

## Y

yang, 313, 315
yeast, 61, 139, 140
yield, 121, 123, 124, 125, 128, 129, 130, 132, 133, 134, 136, 137, 138, 149, 330
yolk, 93

## Z

zinc, 10, 11, 39, 90, 221